MODERN APPROACHES TO QUALITY CONTROL

Edited by **Ahmed Badr Eldin**

Modern Approaches To Quality Control
Edited by Ahmed Badr Eldin

Published by InTech
Janeza Trdine 9, 51000 Rijeka, Croatia

Publishing Process Manager Niksa Mandic

Technical Editor Teodora Smiljanic

Cover Designer InTech Design Team

Additional hard copies can be obtained from orders@intechopen.com

Modern Approaches To Quality Control, Edited by Ahmed Badr Eldin
p. cm.
ISBN 978-953-307-971-4

For EU product safety concerns:
IN TECH d.o.o., Prolaz Marije Krucifikse Kozulić 3, 51000 Rijeka, Croatia,
info@intechopen.com or visit our website at intechopen.com.

Meet the editor

Dr. Ahmed Badr Eldin is currently employed at Sigma Pharmaceutical Corp., Egypt. As a Methodology Manager he performs analytical methods development, stability, validation and auditing functions.
The position is supported with his academic study through Masters degree in analytical chemistry, Zagazig University, Certified Quality Manager from ASQ - USA, Total Quality Management diploma from the American University in Cairo and Certified auditor for ISO 9001 QMS by IRCA.
His academic and work experience has taught him how to analyze and evaluate problems, communicate effectively with a wide range of people, make decisions and use organizational administrative skills.

Contents

Preface

Our pharmaceutical and medical organizations are living organisms need feeding, growth and development to be competitive and can strive to survive otherwise it will degenerate by the tough competition from the surroundings. Quality is a major tool for agile and competitive organization. Quality need to be monitored and to be built into the product so it needs quality control which need highly accurate, precise, sensitive and valid analytical procedures to judge the quality throughout the manufacturing process.

Rapid advance have been made in the last decade in the quality control procedures and techniques, most of the existing books try to cover specific techniques with all of their details. The aim of this book is to demonstrate varieties of quality control processes in pharmaceutical and medical fields. We have covered wide range of techniques and procedures.

We have started from general definitions and basic information about quality, good laboratory practices, decision in Q.C, evaluation of the Q.C data, planning for Q.C. then we have gone through a varieties of quality control issues which are helpful for different workers in the pharmaceutical and medical field.

Dr. Ahmed Badr Eldin,
Sigma Pharmaceutical Corp.,
Egypt

Part 1

Quality, General Definitions

1

IA-Quality - General Concepts and Definitions

Ahmed Badr Eldin
Sigma Pharmaceutical Corp.,
Egypt

1. Introduction

The Meanings of "Quality." Of the many meanings of the word "quality," two are of critical importance to managing for quality:

- "Quality" means those *features of products* which meet customer needs and thereby provide customer satisfaction. In this sense, the meaning of quality is oriented to income. The purpose of such higher quality is to provide greater customer satisfaction and, one hopes, to increase income.
  However, providing more and/or better quality features usually requires an investment and hence usually involves increases in costs. Higher quality in this sense usually "costs more."
- "Quality" means *freedom from deficiencies*—freedom from errors that require doing work over again (rework) or that result in field failures, customer dissatisfaction, customer claims, and so on. In this sense, the meaning of quality is oriented to costs, and higher quality usually "costs less."

Satisfaction and Dissatisfaction Are Not Opposites. Customer *satisfaction* comes from those features which induce customers to buy the product. *Dissatisfaction* has its origin in deficiencies and is why customers complain. Some products give little or no dissatisfaction; they do what the producer said they would do. Yet they are not salable because some competing product has features that provide greater customer satisfaction. The early automated telephone exchanges employed electromagnetic analog switching methods. Recently, there was a shift to digital switching methods, owing to their superior product features. As a result, analog switching systems, even if absolutely free from product deficiencies, were no longer salable.

Thus Quality can evolve several definitions such as:

- customer satisfaction and loyalty ;
- Do right things right.
- providing a product which is 'fit for the purpose';
- providing an acceptable product at an acceptable cost;
- a standard which can be accepted by both the supplier and the customer.
- the totality of features or characteristics of a product that bear on its ability to satisfy a given need.
- Fitness for use.

Big Q And Little Q. Definitions of words do not remain static. Sometimes they undergo extensive change. Such a change emerged during the 1980s. It originated in the growing quality crisis and is called the concept of "Big Q."

Table 1 shows how the quality "umbrella" has been broadening dramatically. In turn, this broadening has changed the meanings of some key words. Adoption of Big Q grew during the 1980s, and the trend is probably irreversible. Those most willing to accept the concept of Big Q have been the quality managers and the upper managers. Those most reluctant have been managers in the technological areas and in certain staff functions.

Topic	Content of little Q	Content of big Q
Products	Manufactured goods	All products, goods, and services, whether for sale or not
Processes	Processes directly related to manufacture of goods	All process manufacturing support; business, etc.
Industries	Manufacturing	All industries, manufacturing, service, government, etc., whether for profit or not
Quality is viewed as:	A technological problem	A business problem
Customer	Clients who buy the products	All who are affected, external and internal
How to think about quality	Based on culture of functional departments	Based on the universal trilogy
Quality goals are included:	Among factory goals	In company business plan
Cost of poor quality	Costs associated with deficient manufactured goods	All costs that would disappear if everything were perfect
Evaluation of quality is based mainly on:	Conformance to factory specifications, procedures, standards	Responsiveness to customer needs
Improvement is directed at:	Departmental performance	Company performance
Training in managing for quality is:	Concentrated in the quality department	Companywide
Coordination is by:	The quality manager	A quality council of upper managers

Source: *Planning for Quality*, 2d ed. (1990). Juran Institute, Inc., Wilton, CT, pp. 1–12.

Table 1. Contrast, Big Q and Little Q.

2. Quality: the financial effects

The Effect on Income. Income may consist of sales of an industrial company, taxes collected by a government body, appropriations received by a government agency, tuitions received by a school, and donations received by a charity. Whatever the source, the amount of the income relates in varying degrees to the features of the product produced by the recipient. In many markets, products with superior features are able to secure superior income, whether through higher share of market or through premium prices. Products that are not competitive in features often must be sold at below-market prices.

Product deficiencies also can have an effect on income. The customer who encounters a deficiency may take action of a cost-related nature: file a complaint, return the product, make a claim, or file a lawsuit. The customer also may elect instead (or in addition) to stop buying from the guilty producer, as well as to publicize the deficiency and its source. Such actions by multiple customers can do serious damage to a producer's income.

The Effect on Costs. The cost of poor quality consists of all costs that would disappear if there were no deficiencies—no errors, no rework, no field failures, and so on. This cost of poor quality is shockingly high. In the early 1980s, it was estimated that within the U.S. manufacturing industries, about a third of the work done consisted of redoing what had already been done. Since then, estimates from a sample of service industries suggest that a similar situation prevails in service industries generally.

Deficiencies that occur prior to sale obviously add to producers' costs. Deficiencies that occur after sale add to customers' costs as well as to producers' costs. In addition, they reduce producers' repeat sales.

3. How to manage for quality: the Juran trilogy

To attain quality, it is well to begin by establishing the "vision" for the organization, along with policies and goals. Conversion of goals into results (making quality happen) is then done through managerial processes—sequences of activities that produce the intended results. Managing for quality makes extensive use of three such managerial processes:

1. 1- Quality planning
- 2- Quality control
- 3- Quality improvement

These processes are now known as the "**Juran trilogy**."

A summery for the 3 process is illustrated in table 2.

Fig. 1. Juran Triology.

Quality planning	Quality control	Quality improvement
Establish quality goals	Evaluate actual performance	Prove the need
Identify who the customers are	Compare actual performance with quality goals	Establish the infrastructure
Determine the needs of the customers	Act on the difference	Identify the improvement projects
Develop product features that respond to customers' needs		Establish project teams
Develop processes able to produce the product features		Provide the teams with resources, training, and motivation to: Diagnose the causes Stimulate remedies
Establish process controls; transfer the plans to the operating forces		Establish controls to hold the gains

Table 2. The three universal processes of managing for quality. [*Adapted from Juran, J.M. (1989). The Quality Trilogy: A Universal Approach to Managing for Quality. Juran Institute, Inc., Wilton, CT.*]

Inspection and Inspectors. The concepts of inspection and inspectors are of ancient origin. Wall and jewelry paintings in Egyptian tombs show the inspections used during stone construction projects. The measuring instruments included the square, level, and plumb bob for alignment control. Surface flatness of stones was checked by "boning rods" and by threads stretched across the faces of the stone blocks.

Safety and Health of the Citizens. Early forms of protection of safety and health were after-the fact measures. The Code of Hammurabi (c. 2000 B.C.) prescribed the death penalty for any builder of a house that later collapsed and killed the owner. In medieval times, the same fate awaited the baker who inadvertently had mixed rat poison with the flour.

The Industrial Revolution. The Industrial Revolution began in Europe during the mid eighteenth century. Its origin was the simultaneous development of power-driven machinery and sources of mechanical power. It gave birth to factories that soon outperformed the artisans and small shops and made them largely obsolete.

The Twentieth Century and Quality. The twentieth century witnessed the emergence of some massive new forces that required responsive action. These forces included an explosive growth in science and technology, threats to human safety and health and to the environment, the rise of the consumerism movement, and intensified international competition in quality.

An Explosive Growth in Science and Technology. This growth made possible an outpouring of numerous benefits to human societies: longer life spans, superior communication and transport, reduced household drudgery, new forms of education and entertainment, and so on. Huge new industries emerged to translate the new technology into these benefits. Nations that accepted industrialization found it possible to improve their economies and the well-being of their citizenry.

The new technologies required complex designs and precise execution. The empirical methods of earlier centuries were unable to provide appropriate product and process designs, so process yields were low and field failures were high. Companies tried to deal with low yields by adding inspections to separate the good from the bad. They tried to deal with field failures through warranties and customer service. These solutions were costly, and they did not reduce customer dissatisfaction. The need was to prevent defects and field failures from happening in the first place.

Threats to Human Safety and Health and to the Environment. With benefits from technology came uninvited guests. To accept the benefits required changes in lifestyle, which, in turn, made quality of life dependent on continuity of service. However, many products were failure-prone, resulting in many service interruptions. Most of these were minor, but some were serious and even frightening—threats to human safety and health, as well as to the environment.

Thus the critical need became quality.

Expansion of Government Regulation of Quality. Government regulation of quality is of ancient origin. At the outset, it focused mainly on human safety and was conducted "after the fact"—laws provided for punishing those whose poor quality caused death or injury. Over the centuries, there emerged a trend to regulation "before the fact"—to become preventive in nature. This trend was intensified during the twentieth century. In the field of human health, laws were enacted to ensure the quality of food, pharmaceuticals, and medical devices. Licensing of practitioners was expanded. Other laws were enacted relating to product safety, highway safety, occupational safety, consumer protection, and so on.

Growth of government regulation was a response to twentieth-century forces as well as a force in its own right. The rise of technology placed complex and dangerous products in the hands of amateurs—the public. Government regulation then demanded product designs that avoided these dangers.

To the companies, this intervention then became a force to be reckoned with.

4. The rise of the consumerism movement

4.1 How to think about quality

Consumers lacked expertise in technology. Their senses were unable to judge which of the competing products to buy, and the claims of competing companies often were contradictory. When products failed in service, consumers were frustrated by vague warranties and poor service.

"The system" seemed unable to provide recourse when things failed. Individual consumers were unable to fight the system, but collectively they were numerous and hence potentially powerful, both economically and politically. During the twentieth century, a "consumerism" movement emerged to make this potential a reality and to help consumers deal more effectively with these problems. This same movement also was successful in stimulating new government legislation for consumer protection.

Intensified International Competition in Quality. Cities and countries have competed for centuries. The oldest form of such competition was probably in military weaponry. This competition then intensified during the twentieth century under the pressures of two world wars. It led to the development of new and terrible weapons of mass destruction. A further stimulus to competition came from the rise of multinational companies. Large companies had found that foreign trade barriers were obstacles to export of their products. To get around these barriers, many set up foreign subsidiaries that then became their bases for competing in foreign markets, including competition in quality. The most spectacular twentieth-century demonstration of the power of competition in quality came from the Japanese. Following World War II, Japanese companies discovered that the West was unwilling to buy their products—Japan had acquired a reputation for making and exporting shoddy goods. The inability to sell became an alarm signal and a stimulus for launching the Japanese quality revolution during the 1950s. Within a few decades, that revolution propelled Japan into a position of world leadership in quality. This quality leadership in turn enabled Japan to become an economic superpower. It was a phenomenon without precedent in industrial history.

5. Quality to center stage

The cumulative effect of these massive forces has been to "move quality to center stage." Such a massive move logically should have stimulated a corresponding response—a revolution in managing for quality. However, it was difficult for companies to recognize the need for such a revolution—they lacked the necessary alarm signals. Technological measures of quality did exist on the shop floors, but managerial measures of quality did not exist in the boardrooms. Thus, except for Japan, the needed quality revolution did not start until very late in the twentieth century. To make this revolution effective throughout the world, economies will require many decades—the entire twenty-first century. Thus, while the twentieth century has been the "century of productivity," the twenty-first century will be known as the "century of quality." The failure of the West to respond promptly to the need for a revolution in quality led to a widespread crisis. The 1980s then witnessed quality initiatives being taken by large numbers of companies.

Most of these initiatives fell far short of their goals. However, a few were stunningly successful and produced the lessons learned and role models that will serve as guides for the West in the decades ahead.

Lessons Learned. Companies that were successful in their quality initiatives made use of numerous strategies. Analysis shows that despite differences among the companies, there was much commonality—a lengthy list of strategies was common to most of the successful companies. These common strategies included:

Customer focus: Providing customer satisfaction became the chief operating goal.

Quality has top priority: This was written into corporate policies.

Strategic quality planning: The business plan was opened up to include planning for quality.

6. IB-quality control - general concept

Quality Control Defined. "Quality control" is a universal managerial process for conducting operations so as to provide stability—to prevent adverse change and to "maintain the status quo."

To maintain stability, the quality control process evaluates actual performance, compares actual performance to goals, and takes action on the difference.

Quality control is one of the three basic managerial processes through which quality can be managed. The others are quality planning and quality improvement,. The Juran trilogy diagram (Figure 2) shows the interrelation of these processes. Figure 2 is used also to describe the relationships between quality planning, quality improvement, and quality

control and the fundamental managerial processes in total quality management. What is important for this section is to concentrate on the two "zones of control." In Figure 2 we can easily see that although the process is in control in the middle of the chart, we are running the process at an unacceptable level of waste. What is necessary here is not more control but improvement—actions to change the level of performance. After the improvements have been made, a new level of performance has been achieved. Now it is important to establish new controls at this level to prevent the performance level from deteriorating to the previous level or even worse. This is indicated by the second zone of control. The term "control of quality" emerged early in the twentieth century (Radford 1917, 1922). The concept was to broaden the approach to achieving quality, from the then-prevailing after-the-fact inspection, to what we now call "defect prevention." For a few decades, the word "control" had a broad meaning which included the concept of quality planning. Then came events which narrowed the meaning of "quality control." The "statistical quality control" movement gave the impression that quality control consisted of using statistical methods. The "reliability" movement claimed that quality control applied only to quality at the time of test but not during service life. In the United States, the term "quality control" now often has the narrow meaning defined previously.
The term "total quality management" (TQM) is now used as the all-embracing term.

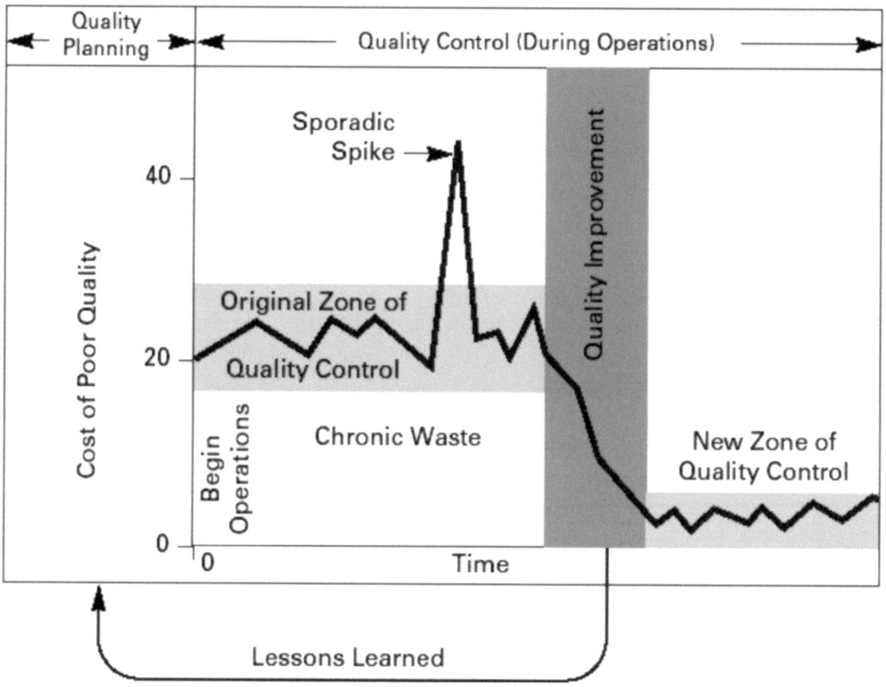

Fig. 2. The Juran trilogy diagram. (*Juran Institute, Inc., Wilton, CT.*)

In Europe, the term "quality control" is also acquiring a narrower meaning. Recently, the European umbrella quality organization changed its name from European Organization for Quality Control to European Organization for Quality. In Japan, the term "quality control" retains a broad meaning.

Their "total quality control" is roughly equivalent to our term "total quality management." In 1997 the Union of Japanese Scientists and Engineers (JUSE) adopted the term total quality management (TQM) to replace total quality control (TQC) to more closely align themselves with the more common terminology used in the rest of the world. The quality control process is one of the steps in the overall quality planning sequence,. Figure 3 shows the input-output features of this step. In Figure 3 the input is operating process features developed to produce the product features required to meet customer needs. The output consists of a system of product and process controls which can provide stability to the operating process.

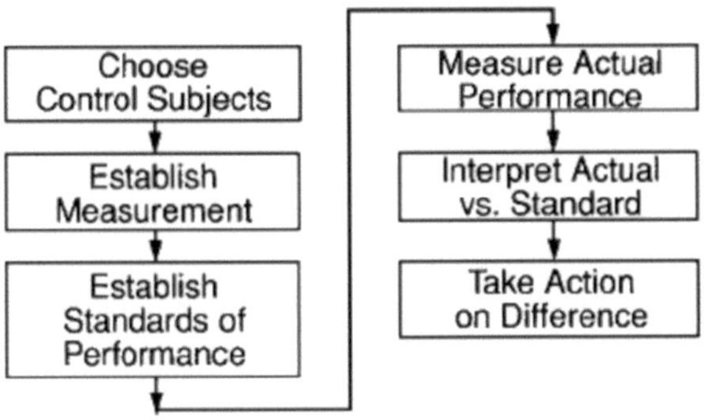

Fig. 3. The input-output diagram for the quality control process.

7. The relation to quality assurance

Quality control and quality assurance have much in common. Each evaluates performance. Each compares performance to goals. Each acts on the difference.

However they also differ from each other. Quality control has as its primary purpose to maintain control. Performance is evaluated during operations, and performance is compared to goals during operations. The resulting information is received and used by the operating forces. Quality assurance's main purpose is to verify that control is being maintained. Performance is evaluated after operations, and the resulting information is provided to both the operating forces and others who have a need to know. Others may include plant, functional, or senior management; corporate staffs; regulatory bodies; customers; and the general public.

The Feedback Loop. Quality control takes place by use of the feedback loop. A generic form of the feedback loop is shown in Figure 4. The progression of steps in Figure 4 is as follows:

1. A sensor is "plugged in" to evaluate the actual quality of the control subject—the product or process feature in question. The performance of a process may be determined directly by evaluation of the process feature, or indirectly by evaluation of the product feature—the product "tells" on the process.
2. The sensor reports the performance to an umpire.
3. The umpire also receives information on what is the quality goal or standard.
4. The umpire compares actual performance to standard. If the difference is too great, the umpire energizes an actuator.
5. The actuator stimulates the process (whether human or technological) to change the performance so as to bring quality into line with the quality goal.

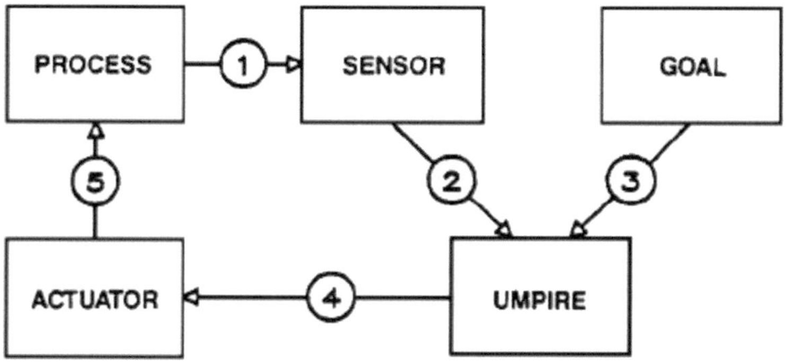

Fig. 4. The generic feedback loop. (*Making Quality Happen, Juran Institute, Inc., senior executive workshop, p. F-3, Wilton, CT.*)

8. The elements of the feedback loop

The feedback loop is a universal. It is fundamental to any problem in quality control. It applies to all types of operations, whether in service industries or manufacturing industries, whether for profit or not. It applies to all levels in the hierarchy, from the chief executive officer to the work force, inclusive. However, there is wide variation in the nature of the elements of the feedback loop. In Figure 5 a simple flowchart is shown describing the quality control process with the simple universal feedback loop imbedded.

The Process. In all of the preceding discussion we have assumed a process. This may also be human or technological or both. It is the means for producing the product features, each of which is a control subject. All work is done by a process which consists of an input, labor, technology, procedures, energy, materials, and output. For a more complete discussion of process.

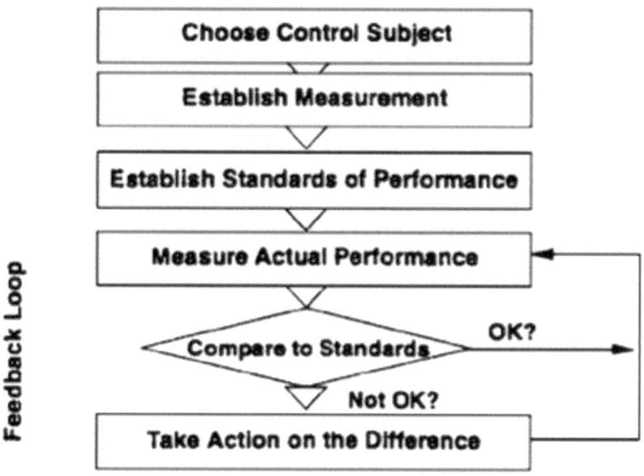

Fig. 5. The qualty control process. (*"Quality Control", Leadershitp for the Quality Century, Juran Institute, Inc., senior executive workshop, p. 2, Wilton, CT.*)

9. Deming chain reaction

The so-called Deming Chain Reaction was actually borrowed from a model that Walter Shewhart developed. He probably borrowed the idea from another thinker. Basically the idea was for management to move away from thinking about quality as a desirable outcome, to thinking about quality as a competitive strategy. Competitive strategy as a concept has been around for centuries. A person selling an item similar to that sold by another can compete on price, by selling it for less money. Perhaps the seller may try to compete by adding extras, gift-wrapping, for example. Technical companies compete by being technology leaders and being on the cutting edge of new developments. There are no end to methods to compete. But some methods are more effective in the long run than others. It is not a mistake that Deming's first published book on the subject was entitled "On Quality, Productivity and Competitive Position". In the book, he sets forth the reasons why emphasis on quality leads to productivity improvement and how that is a very effective competitive strategy in the long run. Phil Crosby in the early 80s in his book, "Quality is Free" pointed out that improving quality lowered cost. But Deming had shown this to the Japanese 30 years earlier. And, Deming pointed out the benefits of developing a competitive strategy based on quality. One of the problems in talking about quality is that many people have pre-conceived notions of quality is. For some it is meeting specifications. Joseph Juran defines it as 'meeting customer requirements'.

Fig. 6.

Zero Defects was Crosby's nostrum, but is really just another way of saying quality is meeting specfications. Deming's ideas are much broader than that and are, perhaps, best captured with the phrase 'continual improvement'. This term connotes the ongoing nature of the strategy. According to Deming, quality is not a state to be achieved in manufacturing, but is, rather, an ongoing company-wide effort at continual improvement. What Bill Conway called "the process - the way everyone thinks, talks, works and acts every day." After all the nonsense is stripped away, the fact is that Japanese automakers (Toyota, Honda and Nissan) make better cars than American automakers (GM, Ford and Chrysler) and have done now for years. Buyers are not idiots. They understand value, and realize that better quality at the same or lower cost is excellent value. End of story. But the implications of the story are not just confined to the auto industry. Cameras, computers, appliances, power tools, earthmoving equipment, and more have fallen from America's basic manufacturing industries to a legacy of plant closings, job losses and dwindling revenues and profits.

10. Quality control: what is new?

Recent decades have witnessed a growing trend to improve the effectiveness of quality control by formal adoption of modern concepts, methodologies, and tools. These have included: Systematic planning for quality control, with extensive participation by the operating personnel Formal application of the feedback loop, and establishment of clear responsibility for the associated
decisions and actions Delegation of decisions to the work force through self-control and self-inspection Wide application of statistical process control and the associated training of the operating personnel A structured information network to provide a factual basis for decision making A systematic process for corrective action in the event of sporadic adverse change Formal company manuals for quality control, with periodic audits to ensure up-to-dateness and conformance.

11. Quality control in pharmaceutical industries

11.1 Responsibilities of quality control unit

a. There shall be a quality control unit that shall have the responsibility and authority to approve or reject all components, drug product containers, closures, in-process materials, packaging material, labeling, and drug products, and the authority to review production records to assure that no errors have occurred or, if errors have occurred, that they have been fully investigated. The quality control unit shall be responsible for approving or rejecting drug products manufactured, processed, packed, or held under contract by another company.
b. Adequate laboratory facilities for the testing and approval (or rejection) of components, drug product containers, closures, packaging materials, in-process materials, and drug products shall be available to the quality control unit.
c. The quality control unit shall have the responsibility for approving or rejecting all procedures or specifications impacting on the identity, strength, quality, and purity of the drug product.
d. The responsibilities and procedures applicable to the quality control unit shall be in writing; such written procedures shall be followed.

12. References

Phillips-Donaldson, Debbie, American Society for Quality 37 (5): 25–39(2000).

Bunkley, Nick, Joseph Juran, Pioneer in Quality Control, Dies", New York Times, 37,March 3 (2008).

Dr. Joseph M. Juran JURAN'S QUALITY HANDBOOK McGraw-Hill, p. 79, (2004).

Parasuraman, A., Zeithami, Valarie A., and Berry, Leonard L, Journal of Marketing, Fall, pp. 41–50 (1985).

Koura, Kozo Societas Qualitas, Japanese Union of Scientists and Engineers, Tokyo, p. 180 - 186 (1991).

Dawes, Edgar W., and Siff, Walter, ASQC Annual Quality Congress Transactions, pp. 810–816(1993).

2

Evaluating Quality Control Decisions: A Simulation Approach

Mohamed K. Omar[1] and Sharmeeni Murugan[2]
[1]*Nottingham University Business school Malaysia*
[2]*Faculty of Engineering & Technology Multimedia University Malaysia*

1. Introduction

Quality has become one of the core factors for almost all manufacturing and service companies that aim to achieve customer satisfaction. Therefore, improving quality is considered to be one of the efforts that companies consider a must to attain customer loyalty in today's complex global competitive environment. Studies concluded that any serious endeavour to improve quality will lead to an increase of cost of the product or service. Obviously, improving quality has its own costs. As a result, measuring cost of quality is very important as it provides valuable insights into the different cost of quality components. Thus, favourable returns on investment maybe achieved. For this fact, the quality cost concept was introduced and implemented in many manufacturing and service companies. The first model of cost of quality was introduced by Feigenbaum (1956) known as the P-A-F model which consists of prevention, appraisal and failure cost. Feigenbaum (1991) categorized the model into two major areas: costs of control (costs of conformance), and costs of failure of controls (costs of non-conformance), which since then used by numerous research studies (for example, Sumanth and Arora (1992), Burgess (1996), Purgslove and Dale (1995), Gupta and Campbell (1995), Chang et al. (1996), Sorqvist (1997)).
Most of the reported literature does not provide a single universal definition for cost of quality (Dale and Plunkett (1999)). However, cost of quality is usually best understood in terms of the sum of costs of conformance and the costs of non-conformance which was first introduced by Crosby (1979). Here, cost of conformance is known as the costs associate with quality requirement for achieving specific quality standards for a product or service. On the other hand, cost of non-conformance is known as the cost of failure to deliver the required standard of quality for a product or service. From the voluminous literature, one may categorized the cost of quality models into five generic models which are: P-A-F model, Crosby model, opportunity or intangible cost model, process cost models and ABC (activity based costing) model. Traditional cost accounting approaches typically used to measure cost of quality has been reported in the literature to have serious limitations when dealing with the components of intangible costs (see Son (1991), Chiadamrong (2003) and Sharma (2007)).
As manufactures continue to improve their factories, they discover that existing cost measure systems should be updated and no matter how sophisticated and reliable these economic evaluation measure maybe, such problems still remain if unreliable cost information is not obtained as inputs for these economic evaluation, Chiadamrong (2003).

It is true to state that cost of quality modelling will provides more accurate approach to determine the cost involved in any quality control activities. However, the challenge does not end here. The cost of quality model must be used to determine the cost of improving activities associated with quality control strategies introduced to improve customer expectations. Therefore, a realistic cost of quality estimation could be determined that allows managers to show the economic benefit or otherwise of that specific quality control strategy. In other words, a quality control improvement strategy can only be justified if the increase in profitability is sufficient to cover the costs involved in the implementation. Once the cost of quality model is developed, a simulation can be used to determine the impact of any quality control strategy that a company wish to investigate. Among the strategies that manufacturing companies may consider investigating would be the allowable defect rate in some process or operations. In this case, simulation could be used to calculate the impact of defective rate of that operation on the overall profitability and productivity of the manufacturing system. Moreover, simulation could be used to study the system before and after some quality control improvement policy. Once the study is completed, a true picture about the cost of that policy could be determined as well as the impact of that policy on the overall defect rate.

This chapter is organized in the following manner; the literature review is presented in section 2 and followed by cost of quality model in section 3. Problem and solution methodology are presented in section 4. Experiment design and model verification, results and discussions are presented in sections 5 and 7 respectively. Finally, conclusions are presented in section 7.

2. Literature review

In its simplest definition of cost of quality, the American Society for Quality Control (ASQC (1971)) and the BS6143 Part 2 (1990) define cost of quality as the costs incurred in ensuring quality, together with the loss incurred when quality is not achieved. Feigenbaum (1956, 1961) introduces the so called PAF model in which cost of quality was classified into four components, prevention, appraisal and failure (internal and external) costs. In Plunkett and Dale (1987) survey, it is stated that literature suggests that most of the researchers use the PAF model for measuring cost of quality. However, Schiffauerova and Thomson (2006) reported that PAF model concept is not the only one since other models were found in the literature to be developed, discussed and used as detailed in Table 1. It is worth noting that Table 1 was originally developed by Schiffauerova and Thomson (2006) and updated by the authors.

The importance of cost of quality has been reported in many research works, Moyer and Gilmore (1979) reported that cost of quality could reach as high as 38% of sales and Albright and Roth (1992) estimated that the cost of quality may represents 30% of all manufacturing costs in the United States. Moreover, Harry and Schroeder (2000) asserted that most companies would find that cost of quality, if properly evaluated, falls somewhere between 15 and 25% of total sale-rather that 3-7% that often assumed. In other study, Giakatis et al. (2001) report that cost of quality represents considerable portion of company's total costs. More recently, Kent (2005) estimated that the turnover rates faced by companies are between 5 to 15% of the overall cost of quality. The benefits of implementing cost of quality system in any profitable organization has been reported extensively in the cost of quality literature, for example, Prickett and Rapley (2001) highlighted four common benefits that any organization is bound to gain from implementing cost of quality system :(1) it will be

Generic model	Cost/ activity categories	Examples of publications describing, analyzing or developing the model
P-A-F models	Prevention + appraisal + failure	Feigenbaum (1956), Purgslove and Dale (1995), Merino (1988), Chang *et al.*(1996), Sorqvist (1997), Plunkett and Dale (1988),Tatikonda and Tatikonda (1996), Bottorff (1997), Gupta and Campbell (1995), Burgess (1996), Dawes (1989), Sumanth and Arora (1992), Morse (1983), Weheba and Elshennawy (2004), etc.
Crosby's model	Conformance + non-conformance	Suminsky (1994) and Denton and Kowalski (1988)
Opportunity or intangible cost models	Prevention + appraisal + failure + opportunity	Sandoval-Chavez and Beruvides (1998) and Modarres and Ansari (1987)
	Conformance + non-conformance + opportunity	Carr (1992) and Malchi and McGurk (2001)
	Tangibles + intangibles	Juran *et al.*(1975)
	P-A-F (failure cost includes opportunity cost)	Heagy (1991) and Chiadamrong (2003)
Process cost models	Conformance + non-conformance	Ross (1977), Marsh (1989), Goulden and Rawlins (1995), Crossfield and Dale (1990)
ABC models	Value added + non-value added	Cooper (1988), Cooper and Kaplan (1988), Tsai (1998), Jorgenson and Enkerlin (1992), Dawes and Siff (1993) and Hester (1993)

Table 1. Generic cost models and cost categories adopted from Andrea and Thomson (2006).

able to focus upon areas of poor performance that need improvements, (2) it will have the opportunity to monitor the progress of ongoing improvement activities, (3) it will have an opportunity to plan for quality improvement and (4) it will be able to communicate better within the organization for improving the overall of the quality control.
Moreover, Schiffauerova and Thomson (2006) in their extensive literature review on cost of quality indicate that companies that use cost of quality programs have been quite successful in reducing cost of quality and improving the quality for the customer.
Although the impact of implementing cost of quality systems on increasing profit of any organization is obvious, Yang (2008) reported that the literature on cost of quality systems implementation indicates that most of companies do not know the true cost of their own quality. Despite that 82% of companies in the United Sates are involved in quality programs, only 33% actually compute the cost of quality (Harry and Schroeder, 2000), and in north-east England 66% of organization do not make use of quality costing as reported by Prickett and Rapley (2001). Some studies have highlighted reasons for the lack of implementing cost of quality systems in practice, Harry and Schroeder (2000) state that many significant quality related costs cannot be captured by most types of accounting systems. Chen and Yang (2002) related the difficulties to measure cost of quality to the fact that there is a lack of adequate methods for determining the financial consequences of poor quality. Moreover, Chiadamrong (2003) has concluded that there is a widespread belief that quality cost cannot be measured in practical terms because traditional cost accounting systems have not been adapted to quantify the value of quality. The need for quantifying cost of quality as stated by Yang (2008) and has been reported in the literature by many researchers, (Feigenbaum (1956); Juran (1952, 1989); Krishnan et al. (2000); Giakatis et al. (2001); Prickett and Rapley (2001), Chen and Yang (2002). However, there are evidences in the literature that clearly indicate that quantifying cost of quality has been neglected by most of organization (Harry and Schroeder (2000) and Omachonu et al. (2004)).
Quality performance is not something that can be readily altered in practice therefore one cannot test with the actual system. Most of the operations system are interconnected and subjected to both variability and complexity. Hence it is impossible to predict the performance of operation systems that are potentially subjected to variability, interconnectedness and complexity. Simulation has been used to overcome this problem and to investigate on the effect of quality loss financially and also examines the effect of different quality strategies on the financial aspect. Simulation provides a flexible technique to model an extensive array of issues that arise relating to quality and manufacturing. The flexibility of simulation methods permitted the invention of models with greater complexity than analytical techniques. Burgess (1996) constructed a simulation model based on system dynamics nature where the model was incorporated with the traditional P-A-F element. The model has facilitated the precise examination of the major relationships concerning conformance quality and costs at the organizational level. Gardner et al. (1995) examines the quality improvement in a manufacturing system by using simulation approach. The modelling was more complicated as it allows the defective parts to move along the assembly operations to examine the impact on the profitability and productivity.
Tannock (1995, 1997) emphasizes the significance of process capability in the selection of quality control strategy and revealed the economic advantages of control charting where special or assignable causes exist. Clark and Tannock (1999) investigate the use of simulation model to estimate the quality cost associated with multi manufacturing system setup and quality control strategies. This approach was validated and aligned with actual costs at a

case study company. Although there are voluminous literatures written on cost of quality, very few literatures were written on tracing the invisible element of cost of quality and also the method in measuring the element of cost of quality.
The literature review presented so far indicates that estimating cost of quality is not a simple and straight forward issue. Moreover, the literature shows that there are several methods that can be used to determine cost of quality. Moreover, simulation stands as a favourite approach that might be used to investigate quality improvements justification in terms of profitability for a specific cost of implantation. Obviously, cost of quality model must first be developed and then simulation is followed to investigate the impact of some quality strategies and or improvements.

3. The cost of quality model

In this section of the chapter we, intend to provide a brief description of the model developed by Son and Lie (1991) and the way it was modified to be suitable for our research idea. First, the notations of the model are presented.

Notations

t	sampling interval
n	sample size per sampling
Θ	average number of assignable cause per hour
d_1	mean time spent to identify an assignable cause
d_2	mean time spent to correct an assignable cause
g	time required to sample a product
t_o	time inspection of the acceptance sampling
c_{as}	cost of investigating acceptance sampling
Γ	probability of process in control state 2
β	Probability of process in control state 2
τ	time taken for assignable cause to occurs
T1	in control period
T2	time period until the assignable is detected for the first time
T3	time required to investigate the true alarm
T4	time to correct the assignable cause

3.1 Description of the model

Son and Lie (1991) considered a small manufacturing system which consists of a machining area (sampling inspection) and a final inspection area (100% inspection) as indicated in Figure 1. However; they assume that 100% inspection is not always possible. They explained that a complete check of a component part may require the part removal from the fixture; the removal makes it difficult to realign the part to its original position.
Therefore, they assume that sampling inspection is assumed for component parts (in-process inspection), and 100% inspection for finished parts (final inspection). Throughout the machining process, the feature of the product quality is monitored by an x-bar chart which consists of a center line, lower, limit and an upper limit. A sample size in use is taken and inspected at a specific time interval. During machine process, when the sample mean fall outside of the control limit in the x-bar chart, an investigation is made during an average time period to check if the alarm is true or false. The false alarm occurs if the process is in

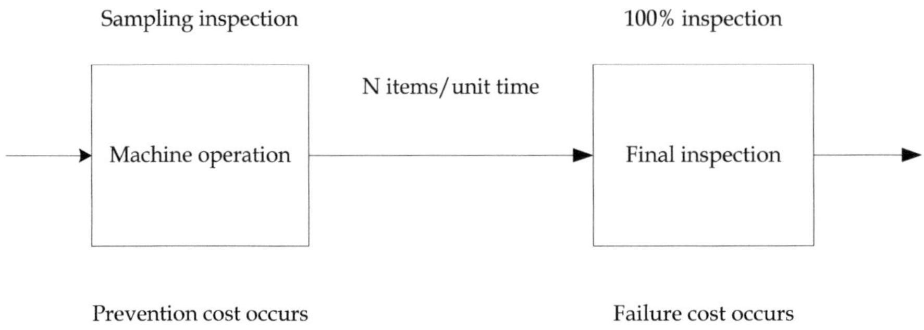

Fig. 1. Simplified manufacturing system.

control and on the other hand, the true alarm occurs if an assignable cause of specific magnitude makes the process out of control. In their modelling, first they calculate the cycle time, and then quality cost (prevention and failure) per cost was determined. In the next section, equations for all components are stated, obviously, readers interested in the detail steps of deriving these equations should referred to the original article.

3.1.1 Cycle time calculations

The assignable cause occurs according to the Poisson process with mean rate Θ. The period of the process is in control is represented by T_1. T_1 follows an exponential distribution with mean $1/\Theta$.
The expected length of a cycle;

$$E(T_c) = E(T_1) + E(T_2) + E(T_3) + E(T_4) \tag{1}$$

Which can be obtained using equation 2

$$= 1/\Theta + (t + t\Gamma/(1-\beta)-\tau) + (gn + d_1) + d_2 \tag{2}$$

3.1.2 Prevention cost per cycle

There are three types of prevention costs that occur at the machining area during a cycle of time. The first cost is associated with inspection works and denoted as (C_{p1}), the second cost is representing the cost of investigating the false alarms (C_{p2}), and the third cost is associated with adjusting the assignable cause (C_{p3}). Then, the expected prevention cost per cycle (C_p) can be determined by summing up these three components as in equation 3.

$$E(C_p) = E(C_{p1}) + E(C_{p2}) + E(C_{p3}) \tag{3}$$

3.1.3 Failure cost per cycle

The failure cost has three components; the first component is the cost of rework per cycle (C_{f1}), the second component is the scrap cost per cycle (C_{f2}) and the third component is the external failure cost during a cycle of time (C_{f3}). Then, the expected failure cost per cycle (C_F) can be determined by summing up these three components as in equation 4.

$$E\,(C_F) = E(C_{f1}) + E(C_{f2}) + E(C_{f3}) \tag{4}$$

3.1.4 Acceptance sample cost per cycle

Costs associated with sampling inspections prior entering the manufacturing system are not considered in the model developed by Son and Lie (1991). Our idea consists of the fact that companies often receive a shipment of material from a supplier and need to ascertain that the quality of the shipment. If it is impractical to inspect every item in the shipment, then a sampling plan is the (n, c) is used. In this plan, n items are chosen (without replacement) from a batch of shipped material. If c or fewer of the sample items are defective, then the batch is accepted; otherwise, the batch is rejected. Therefore, there will be a cost incurred during inspection of the sample before entering the manufacturing system. The expected number of samples entering the manufacturing system is n whereas t_o is representing the time taken to inspect the sample. Let (C_{aS}) be the cost of investigating acceptance sampling during a cycle of time. Then the expected cost associated with this activity is given by equation 5.

$$E\,(C_A) = C_{as} n t_o \qquad (5)$$

Hence, the total cost of quality can be represented by summing up all the cost components (1) through (5) and is presented by equation 6.

$$E\,(C_Q) = E(C_P) + E(C_F) + E(C_A) \qquad (6)$$

Where $E(C_P)$ is representing the prevention cost, $E(C_F)$ is representing the failure costs and $E(C_A)$ is representing the acceptance sampling costs.

4. The problem and solution methodology

The improvement made to the cost of quality model developed by Son and Lie (1991) that was described in section 3.1.4 adds acceptance sampling plans. Almost all quality control managers in the manufacturing firms develop and implement such plans. As a result an element of appraisal cost is incurred prior to the commencement of the production activities. Moreover, unlike the work reported by Son and Lie (1991) that consider a single stage manufacturing system, this research work considers a two-stage manufacturing system.

Therefore, the problem considered by this chapter can be described as: : a manufacturing system that consists of two stages, incoming raw materials inspection carried out according to some quality plan and as a result an element of appraisal cost in incurred prior to the commencement of the production activity. Once orders are realized by the company, raw materials are brought into the shop floor and at this stage; preventive and failure costs are incurred. In order to investigate the cost of quality for this manufacturing system, we have adopted the four strategies for inspection and removal of defectives across a range of detection rates reported by Gardner et.al (1995). The strategies are summarized in the following manner:

1. Inspection and removal of defectives based on acceptance sampling prior to assembly points
2. Inspection and removal of defectives at completion of finished product only,
3. Inspection and removal of defectives prior to assembly points,
4. Inspection and removal of defectives following every operations, and

The motivation for creating the above strategies (quality control decisions) and then using simulation tool is to investigate the impact of these strategies on cost of quality. One may conclude that the result of implementing any of these decisions is obvious. For example, the strategy that calls to carry out the inspection works at the final stage of the production and then removal of defectives will minimize the inspection costs, however, the increase of defectives at the final stage of manufacturing is inevitable. Moreover, inspection and removal of defectives prior to the assembly line will increase the quality prevention cost component and indirectly reduce the failure cost component. However, without the development of the cost of quality model and the simulation works, the magnitude of cost of quality could not be determined and the impact would not be known. It is not a secret that many firms sacrifices parts of the quality control steps under pressure to reduce cycle time.

Once the cost of quality model and its three components is developed, a simulation model using @Risk spreadsheet simulation software is developed for the two-stage process. The popularity using spreadsheet technology among practitioners justifies the use of the proposed @Risk simulation software.

5. Experiment design

Inspection time was considered as the main factor for calculating cost of quality. This is important as in real-life manufacturing systems; production line managers consider cycle time to be an important factor. Moreover, quite often, quality control activities or part of the activities would be sacrificed in order to attain a desired cycle time.

In order to simplify the experiment, it was decided to use the same fractions of defective rate to every operation in the manufacturing system for a given trial. The fractions of defective rate are divided into five categories: .001, 0.005, 0.010, 0.050 and 0.100. Every trial will run for about 10000 times of the simulation runs as indicated in Table 2.

Inspection and defective removal strategies	Fraction Detection per Operation				
	0.001	0.005	0.01	0.05	0.1
Acceptance sampling	10000 run	10000 run	10000 run	10000 run	10000 run
Completion of finish product	10000 run	10000 run	10000 run	10000 run	10000 run
Prior to assembly	10000 run	10000 run	10000 run	10000 run	10000 run
Following every operation	10000 run	10000 run	10000 run	10000 run	10000 run

Table 2. Experiment design for simulation run.

Therefore there will be twenty combinations of strategies and defective rates. All the combinations will run for 10000 iterations each. It is worth to note that during the design of the sampling acceptance procedure, it was assumed that the company would receives a batch of 10,000 items which require assembly activities using processes one and two respectively. Moreover, it is assumed that the quality engineers have developed a quality plan which is described as the (n, c) plan. In an (n, c) plan, n items are chosen (without replacement) from the batch of shipped material whereas c is the maximum number of defective items that a sample could have. If the number of the defective items in the sample is fewer or equal to c, then the batch is accepted otherwise the batch is rejected. For example,

in process 1, a sample of 90 items will be inspected and the entire 1000 items are accepted if 0, 1, 2, 3, 4, or 5 defective items are found in the sample otherwise the batch is rejected.
On the other hand, for process 2, a sample of 40 items will be inspected and the entire batch will be accepted if 0, 1, 2 or 3 defective items are found in the sample otherwise the batch will be rejected. The idea here is to set the Actable Quality Level (AQL) to be equal to 1% and the Lot Tolerance Percentage Defective (LTPD) to be equal to 5%. Once AQL and the LTPD are set, a simulation run is carried out to determine the cost of quality associated with the quality control strategies (decisions) detailed in section 4 in this chapter. It is worth noting and as indicated earlier, the process under consideration consists of two processes. In order to avoid computational complexities; it was decided to apply and conduct simulation runs for each management strategies to each process separately and then combined the two processes to examine the overall findings. The numerical values which were used in the simulation model are portrayed in Tables 3 and 4.

Manufacturing system		**Process 1**	**Process 2**
	Symbol	**Value**	**Value**
Average number of assignable cause per hour	Θ	0.001	0.001
Lower specification level	LSL	0.79	0.99
Upper specification level	USL	0.81	1.01
Average height of the product population	μ	0.80	1.00
Standard deviation of the product population	σ	0.007	0.015
Magnitude of an assignable cause	δ	5	7
Value to decide control limit	q	3	7
Sampling interval	t	5	20
Sample size per sampling	n	90	40
Mean time spent to identify an assignable cause	d_1	0.5	0.5
Mean time spent to correct an assignable cause	d_2	2	2.5
Time required to sample a product	g	0.05	10

Table 3. Numeric values for processes 1 and 2.

Strategy 1. Inspection and removal of defectives based on acceptance sampling prior to assembly points.
This quality control strategy calls for the removal of defectives at the sampling stage and no quality control to be conducted after that stage. The total cost of quality associated with this strategy would be determined based on the following equation:

$$TCOQ_1 = C_{as} n t_o$$

Manufacturing system		Process 1	Process 2
	Symbol	Value	Value
Number of products produced at the machining area per hour	N	40	20
Error rate of misclassifying a good part into bad	e_1	0.01	0.01
Error rate of misclassifying a bad part as good	e_2	0.005	0.005
Rate of restoring a defect to a good part	w	0.95	0.95
Fixed sampling cost per sampling interval	a_1	25	30
Variable sampling cost per unit product	a_2	8	5
Cost of investigating a false alarm	C_{fs}	80	80
Cost of correcting an assignable cause	C_{cr}	150	150
Cost of reworking a good part because of misclassification	C_g	100	100
Cost of reworking a defective part	C_b	200	200
Cost of scrapping a defective part that cannot be restored	C_s	75	75
Cost of dissatisfying a customer by selling a defective part	C_a	100	100

Table 4. Numeric values for processes 1 and 2.

Strategy 2. Inspection and removal of defectives at completion of finished product only.
This strategy calls for inspection of items among completion of the two processes. Therefore, it means that there will be no quality control during the process of making the products. Hence, the cost of quality that will be incurred in this strategy is meanly failure costs. However, it is assumed here that samples have to undergo visual inspection before the items enter into the manufacturing system. Therefore, the total costs of quality associates on this strategy are;

$$TCOQ_1 = E(C_A) + E(C_F) = E(C_A) + E(C_{f1}) + E(C_{f2}) + E(C_{f3})$$

Strategy 3. Inspection and removal of defectives prior to assembly points.
This strategy requires that items must be inspected and defectives are removed prior to assembly. In this case, since the inspection is done at the end of the process, the cost of quality is simply the prevention cost which is incurred during inspection of the material. However, parts with defects could be produced since there is no quality control during and at the end of the assembly. Therefore, there will be some failure costs which is required to convert defective items into good items. Moreover, inspection costs before the items enter the manufacturing system have to be considered. Therefore, the total costs of quality associates on this strategy are;

$$TCOQ_2 = E(C_A) + E(C_{p1}) + E(C_F) = E(C_A) + E(C_{p1}) + E(C_{f1}) + E(C_{f2}) + E(C_{f3})$$

Strategy 4. Inspection and removal of defectives following every operations.
This strategy calls for inspection and removal of defectives following every operation. Hence, the inspection will take place at the end of every operation to determine if there are defectives and perform the necessary corrections (rework). As a result, there will be an element of failure costs. Moreover, since the system includes inspection costs before items enter the manufacturing system, an additional inspection cost will be incurred. The total costs of quality associates on this strategy are;

$$TCOQ_3 = E(C_A) + E(C_{p1}) + E(C_{p2}) + E(C_F) = E(C_A) + E(C_{p1}) + E(C_{p2}) + E(C_{f1}) + E(C_{f2}) + E(C_{f3})$$

Each mathematical model that describe each strategy is developed in MS Excel spreadsheet for each process separately and then for the combinations of the two stages. The simulation software @Risk is imbedded and a simulation of 10000 iterations is carried out. This way, the impact of each of the strategy can be investigated and the impact on cost of quality can be determined.

6. Model verification, results and discussions

6.1 Model verification

A problem of model verifications arises when using simulation approach in decision evaluation. The literature referred to model verifications procedure developed by Naylor and Finger (1967) that ensures the developed model is free from any illogical error. In order to accomplish the task, the formulation and its parameters provided by Son and Lie (1991) were used to compute the prevention and failure costs reported by the authors. A simulation run was carried out and the results of the simulation model were compared with the results reported by Son and Lie (1991). The results are presented in Table 5.

Son and Lie (1991) results			The Simulation results		
Total length of a cycle, T_C (hours)	Prevention cost per cycle, C_P ($)	Failure cost per cycle, C_F ($)	Total length of a cycle, T_C (hours)	Prevention cost per cycle, C_P ($)	Failure cost per cycle, C_F ($)
$E[T_1]$= 100.00	$E(C_{P1}]$= $312.83	$E(C_{F1}]$= $3292.96	$E[T_1]$= 100	$E(C_{P1}]$= $312.83	$E(C_{F1}]$= $3292.96
$E[T_2]$= 4.28	$E[C_{P2}]$ = $2.63	$E[C_{F2}]$= $214.21	$E[T_2]$= 4.28	$E[C_{P2}]$ = $2.63	$E[C_{F2}]$= $214.21
$E[T_3]$= 0.55	$E[C_{P3}]$= $100.00	$E[C_{F3}]$= $32.29	$E[T_3]$= 0.55	$E[C_{P3}]$= $100.00	$E[C_{F3}]$= $32.29
$E[T_4]$ = 2			$E[T_4]$ = 2		
$E[T_C]$= 106.83	$E[C_P]$= $415.46	$E[C_F]$= $3539.46	$E[T_C]$= 106.83	$E[C_P]$= $415.46	$E[C_F]$= $3359.46

Table 5. Comparison results.

Table 5 indicates that the results obtained from the simulation model match exactly the results reported by Son and Lie (1991). The results from this finding indicate that the simulation model is an accurate presentation of that reported by Son and Lie (1991). Hence, it is concluded that the model is ready for investigation the evaluating the quality control decisions.

6.2 Results and discussions

6.2.1 Acceptance sampling prior to assembly points

As indicted earlier, the quality control engineers develop quality control plan.

In this inspection as mentioned in section 4, the quality control engineers have developed a quality control plan based on the concept of producer's risk which is related to acceptable quality level (AQL) and the consumer's risk which is related to tolerance percentage defective (LTPD).

The summary of the statistics that resulted from the simulations runs shows that the mean was found to be equal to 0.994 with a tile of 15% and the maximum is equal to 1 with a tile of 5%. Since it was concluded that the sample is good, since 99.94% of the sample is within 1% of defectives, then the items are considered to be acceptable. On the other hand, the remaining 0.06% of the sample contributes to produce risk. This means that the plan has a risk of losing 0.06% of good sample. It is worth noting that the skewness has a negative value that is -40.7941969 indicating that distribution of the sample exceedingly to the left.

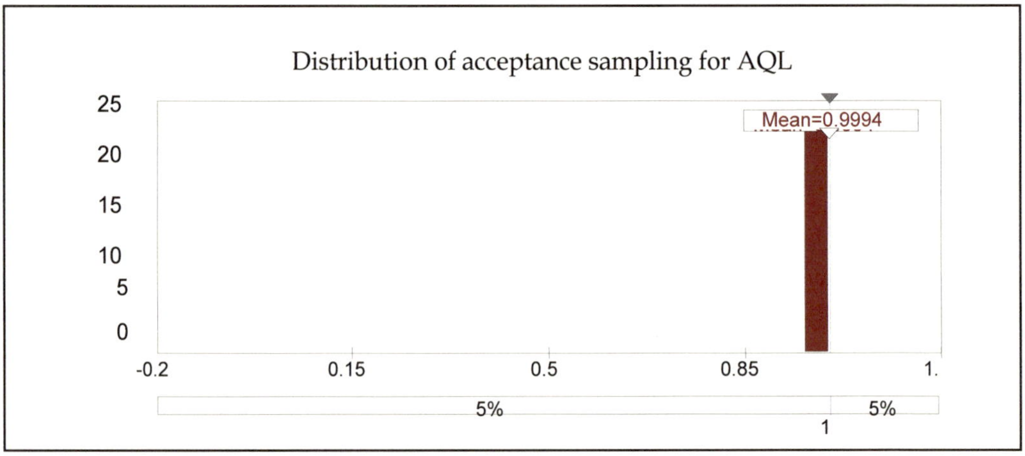

Fig. 2. Distribution of acceptance sampling for AQL.

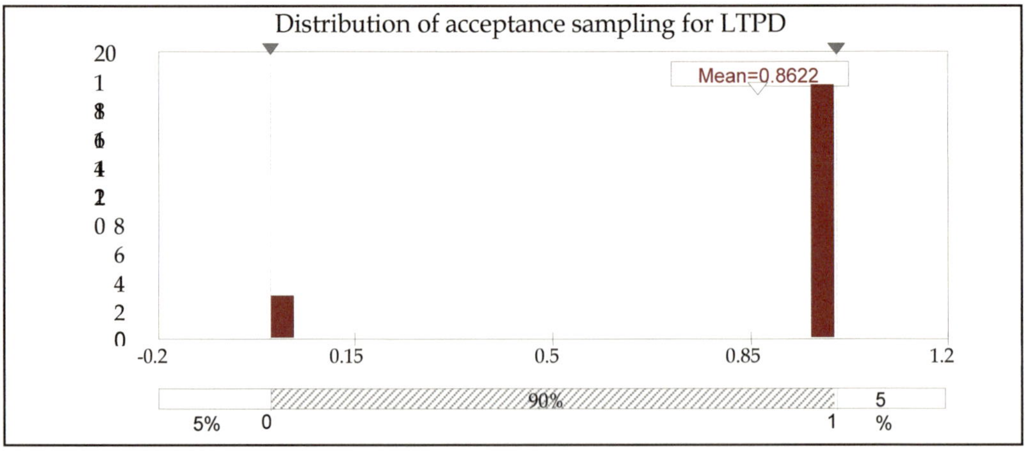

Fig. 3. Distribution of acceptance sampling for LTPD.

Figure 2 was constructed to display the summary statistics, the mean value for the acceptable quality level (AQL) over the tile percentage. From the chart it could be seen that the mean of the sample is equal to 0.9994 which very close to the value of 1 which indicate that the risk is equal to 0.00006.

As for the risk associated with low tolerance percentage defect (LTPD), the statistics summary results indicate that the mean has a value of 0.8622 which says that 86.22% of the sample size has defects and would be accepted as good parts. Therefore, such situation indicates that the risk would be about 86.22% of the sampling plan. Figure 3 summarizes the summary statistics. In Figure 3, the distribution is skewed towards the right represents the lot tolerance percentage defects with a mean of 86.22%.

6.2.2 Inspection upon completion of assembly

In this case inspection and removal will took place upon completion of assembly activities. Simulation runs resulted in obtaining the cost of quality associated with strategy of inspection upon completion of assembly activities at each level of defects rate per operation is presented in Table 6.

Cost of Quality (x 1000)					
Defective rate per operation (%)	Time interval (t, hr) and time inspection of a product (g, hr), Cost is in MU=Monetary Unit				
	t=20 g=10	t=17 g=9	t=15 g=8	t=10 g=7	t=8 g=6
0.001	3558.95	3398.49	3239.98	3075.65	2917.15
0.005	1989.88	1829.28	1670.68	1506.19	1347.65
0.01	1794.33	1633.55	1474.85	1310.16	1151.55
0.05	1642.42	1480.26	1320.76	1154.48	995.41
0.10	1628.56	1464.93	1304.54	1136.43	976.81

Table 6. Cost of quality at different values of defect rate, time interval and inspection time.

Table 6 reveals the total cost of quality is decreasing as we move towards the right side of the table. The reason for that is the fact that the more time is allocated for inspection work, the more of the defects will be found. This finding could be seen for example when considering the total cost at 0.001 defects rate at interval time of 20 hours and inspection time for 10 hours. Considering that example reveals that the cost has decrease by 27%. Moreover, Table 6 indicates that the maximum impact on total cost of quality resulted from implanting this strategy occurred at defect rate 0.001 with time equal to 20 hours and inspection time of 10 hours. On the other hand, the minimum cost of quality occurs at a defect rate of 0.10 with 8 hours interval time of inspection. The total cost of quality behaviour at various interval times for each defect rate is presented in Figure 4.

Figure 4 reveals that there is a large gap between the curve for the defects rate at 0.001 and the curve represents the defect rate at 0.005. As it could be seen, there is a drastic drop and the drop is due to the increase of the defects that will cause the time taken to inspect the defect is shorter and thus the time to correct the assignable cause shorter. Hence, it will reduce the total cost of quality in this situation compared to the lower defect rate. Moreover, the difference of the total cost of quality at defect rate of 0.05 and defect rate of 0.10 is small. Therefore, the curves represent this defect rate is overlapping with each other's and the curve that represents the defect rate of 0.10 has the smallest total cost of quality.

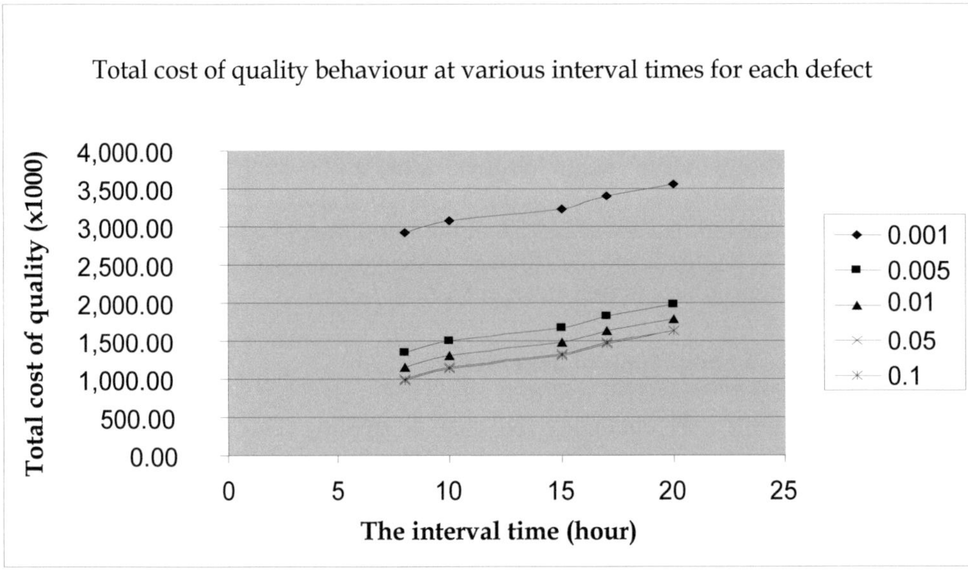

Fig. 4. Cost of quality behaviour at various interval times for each defect rate.

6.2.3 Inspection prior to assembly operations

This quality control strategy calls for inspection to be carried out at the commencement of the assembly process; the cost of quality at various defective rates and inspection interval times is presented in Table 7.

Cost of Quality (x 1000)					
Defective rate per operation (%)	Time interval (t, hr) and time inspection of a product (g, hr), Cost is in MU=Monetary Unit				
	t=20 g=10	t=17 g=9	t=15 g=8	t=10 g=7	t=8 g=6
0.001	3073.71	2937.75	2803.23	2669.58	2539.01
0.005	1715.11	1577.40	1441.37	1301.44	1166.24
0.01	1545.79	1407.72	1271.42	1130.55	994.72
0.05	1414.24	1274.82	1137.69	994.84	858.15
0.10	1402.23	1261.52	1123.60	979.09	841.87

Table 7. Cost of quality at different values of defect rate, time interval and inspection time.

Table 7 reveals that the maximum total cost of quality occurs at defect rate of 0.001 and time interval of 20 hours and inspection time of 10 hours. On the other hand, the minimum total cost of quality occurs at defective rate of .10, time interval of 8 hours and inspection time of 6 hours. As it could be noticed that as the time of inspection become shorter, total cost of quality is reduced. The relationship between total cost of quality, the time interval and defective rate is presented in Figure 5.

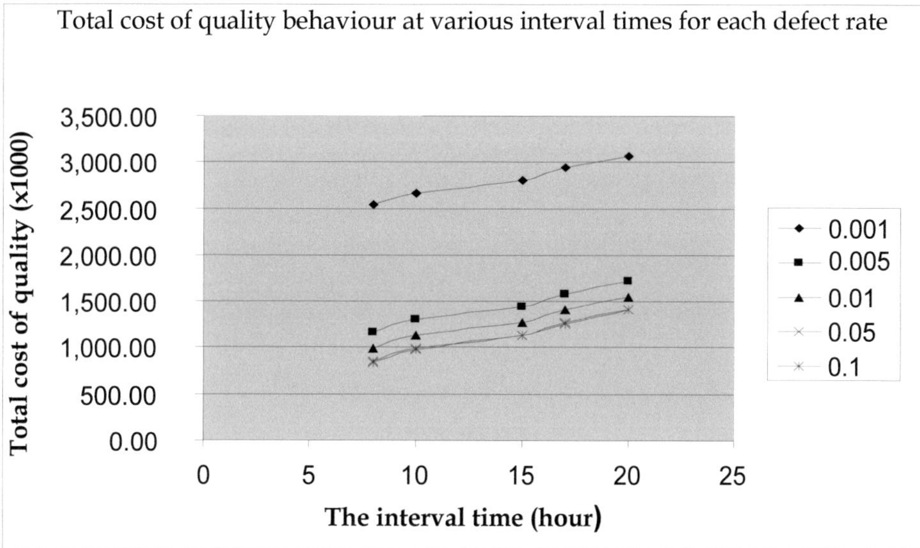

Fig. 5. Cost of quality behaviour at various interval times for each defect rate.

6.2.4 Inspection following every assembly operations

This strategy calls for inspection works to be carried out at the end of the process for every operation. The cost of quality at various defective rates and inspection interval times is presented in Table 8.

Cost of Quality (x 1000)					
Defective rate per operation (%)	Time interval (t, hr) and time inspection of a product (g, hr), Cost is in MU=Monetary Unit				
	t=20 g=10	t=17 g=9	t=15 g=8	t=10 g=7	t=8 g=6
0.001	3062.25	2924.25	2787.93	2646.61	2510.30
0.005	1712.85	1574.73	1438.34	1296.88	1160.53
0.01	1544.67	1406.40	1269.92	1128.29	991.88
0.05	1414.03	1274.57	1137.40	994.40	857.61
0.10	1402.11	1261.39	1123.45	978.88	841.60

Table 8. Cost of quality at different values of defect rate, time interval and inspection time.

Table 8 reveals that the maximum total cost of quality occurs at defective rate of 0.001, interval time of 20 hours and inspection time of 10 hours. On the other hand, the minim total cost of quality occurs at defective rate of 0.10, time interval of 8 hours and inspection time of 6 hours.. The relationship between total cost of quality, the time interval and defective rate is presented in Figure 6.

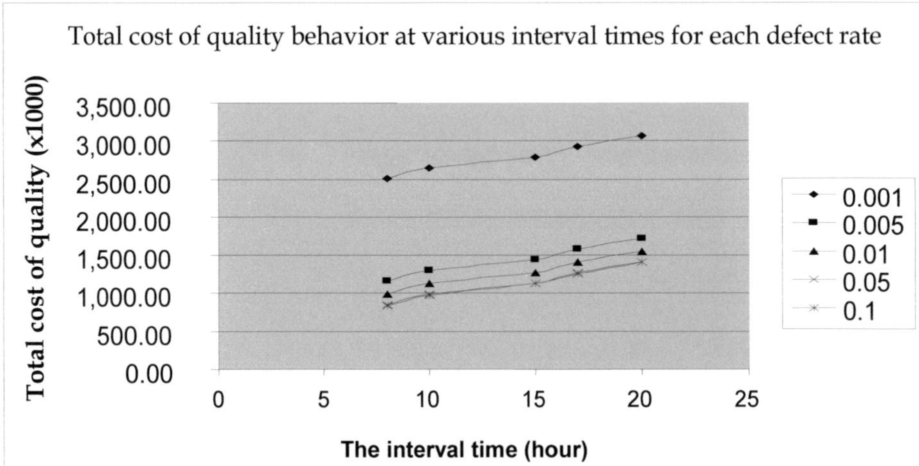

Fig. 6. Cost of quality behaviour at various interval times for each defect rate.

Figure 6 reveals that as defective rate per operation is increased, the total cost of quality decreases. This is because as the defective rate is increased, the time taken to detect the defect becomes shorter. Hence, the time taken to convert the process out of control to in control state will be shorter and finally the impact of cost of quality will be smaller.

In order to provide a good picture to the reader, Figure 7 was developed and a graphical presentation that shows the relationship between total cost of quality, and detection rate per operation.

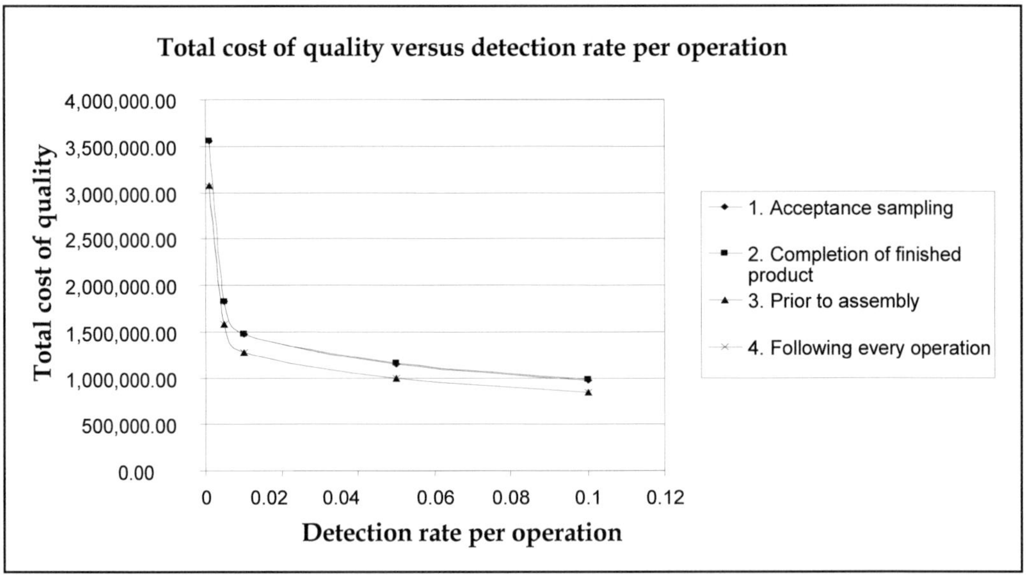

Fig. 7. Cost of quality for the four strategies.

Figure 7 indicates that total cost of quality is at its maximum value when the detection rate is zero, and then steadily decreased as the window of inspection increases. Moreover, inspection following every assembly operations has the lowest total cost of quality. This is because as the appraisal cost increases, the failure costs will simultaneously reduced since less defect rate occurs and hence less items is sent for rework.

6.2.5 Investigation of strategies combination

Since simulation modelling can provide a window of opportunity to investigate important quality control decisions and their impact on cost of quality. It is thought that the quality control management may wish to consider some combination of strategies for each process rather than having a single strategy for the whole assembly process. In this case, one has to determine the possible quality strategies combinations that could be considered for the simulation run. Let R be the possible strategies combination outcome, then we would have:

$$R = \sum S_{i,j} \tag{7}$$

Where

i = A strategy in process 1 i = 1, 2, 3, 4 and j = A strategy in process 2 j = 1, 2, 3, 4

Inspection and defective removal strategies	Fraction Detection Rate per Operation				
	0.001	0.005	0.01	0.05	0.1
$S_{1,1}$	4,845.07	2,137.34	1,651.57	1,224.13	1,023.91
$S_{1,2}$	4,854.02	2,141.85	1,655.16	1,226.89	1,026.21
$S_{1,3}$	4,368.78	1,889.97	1,451.73	1,067.25	891.27
$S_{1,4}$	4,357.32	1,887.30	1,450.23	1,066.82	891.01
$S_{2,1}$	4,848.23	2,138.03	1,651.92	1,224.21	1,023.93
$S_{2,2}$	4,857.18	2,142.53	1,655.51	1,226.97	1,026.23
$S_{2,3}$	4,371.95	1,890.65	1,452.08	1,067.32	891.30
$S_{2,4}$	4,360.48	1,887.98	1,450.58	1,066.89	891.03
$S_{3,1}$	4,622.03	2,090.06	1,630.57	1,217.32	1,023.83
$S_{3,2}$	4,630.98	2,094.56	1,634.16	1,220.08	1,026.13
$S_{3,3}$	4,145.74	1,842.68	1,430.73	1,060.44	891.20
$S_{3,4}$	4,134.28	1,840.00	1,429.23	1,060.01	890.93
$S_{4,1}$	4,472.97	2,052.62	1,605.53	1,209.65	1,016.16
$S_{4,2}$	4,481.93	2,057.12	1,609.12	1,212.41	1,018,.47
$S_{4,3}$	3,996.69	1,805.25	1,405.69	1,052.76	883.53
$S_{4,4}$	3,985.23	1,802.57	1,404.19	1,052.33	883.26

Table 9. Total cost of quality for the assembly system.

Table 9 shows all possible quality strategies combinations with their respective cost at each selected fraction detection rate for each operation. As it could be seen from Table 9, the combination of {2, 2} strategies have caused the system to incur the largest cost of quality. This is the case when the management decided to use the quality strategy in which quality control inspection work is only to be carried at the end of the production runs. On the other hand, one can see from Table 9 that that adopting combination of {4, 4} strategies have caused the system to incur the least cost of quality. As it could be easily seen that {4, 4} strategies calls for quality control works to be carried after each operation and as a result, completely eliminating the failure costs. There are many important management implications illustrated in all the examples presented in this chapter. First, quality costs are very large when quality activities time window is sacrificed to reduce the total cycle time. Secondly, failure costs are very large and managers should completely avoid these costs since there is no trade off that exists with these costs. Another important issue is the fact that using simulation to measure and understand cost of quality has provide managers with opportunity to rank their process in terms of cost of quality and the cost consequences that resulted from adopting a specific set of strategies .

7. Conclusions

In this chapter, an analytical model reported in the literature for cost of quality computations was considered and modified to include an important component of cost of quality. The model is then used to develop a simulation model for a two-stage manufacturing system. Moreover, quality control strategies common in the manufacturing community were used in the simulation works to investigate their impact on cost of quality. The results indicate that some of these practiced strategies when combined with the detection periods will significantly increase cost of quality. Furthermore, the results indicate simulation works can be used to understand and measure cost of quality. An interesting venue for further research is to use simulation to investigate a real-industrial application to investigate the cost of quality; the authors are currently considering such idea.

8. References

Albright, T.L. and Roth, H. P (1992) .The measurement of quality cost: an alternative paradigm, *Accounting Horizons,* June, 15, 1992.

ASQC . (1971) Quality costs, what and how? Milwaukee: WI: ASQC Quality Press.

Bottorff, D. L.(1997). CoQ systems: the right stuff, *Quality Progress,* March, 33.

BS 6143 Part 2. (1990) Guide to determination and use of quality-related costs. London: BSI.

Burgess, T.F. (1996). Modeling quality cost dynamics, *International Journal of Quality & Reliability Management,* 13(3), .8-26.

Carr, L.P. (1992). Applying cost of quality to a service business, *Sloan Management Reviews,* 33(4), 72-78.

Chang, S.J., Hyun, P.Y. and Park, E. H. (1996). Quality costs in multi-stage manufacturing systems, *Computers & Industrial Engineering,* 31(1-2), 115-118.

Chen, C.C. and Yang, C.C. (2002). Cost-effectiveness based performance evaluation for suppliers and operations, *Quality Management Journal,* 9(4), 59–73.

Chiadamrong, N. (2003). The development of an economic quality cost model, *Total Quality Management and Business Excellence,* 14(9), 999-1014.

Clark, H.J and Tannock, J.D.T. (1999), The development and implementation of a simulation tool for the assessment of quality economics within a cell-based manufacturing company, *International Journal of Production Research,* 37, 979-995.

Cooper, R. (1988). The rise of activity-based costing – Part I: What is an activity-based cost system?, *Journal of Cost Management,* 2(2),.45-54.

Cooper, R. and Kaplan, R. S. (1988). Measure costs right: Make the right decisions, *Harvard Business Review,* 66(5), .96-103.

Crosby, P. B. (1979). Quality is free, New York: McGraw-Hill.

Crossfield , R. T. and Dale, B. G. (1990). Mapping quality assurance systems: a methodology, *Quality and Reliability Engineering International,* 6(3), 167-178.

Dale, B.G. and Plunkett, J. J. (1999). Quality Costing, 3rd ed., Gower Press, Aldershot.

Dawes, E.W. (1989). Quality costs-new concepts and methods, quality costs: ideas & applications, in Campanella, J. (Ed.), *Quality Costs: Ideas and Applications,* Vol. 2(pp.440.), ASQC Quality Press, Milwaukee, WI.

Dawes, E.W. and Siff, W. (1993). Using quality costs for continuous improvement, *ASQC Annual Quality Congress Transactions,* 444-449.

Denton, D.K. and Kowalski, T.P. (1988). Measuring nonconforming costs reduced manufacturer's cost of quality in product by $200 000, *Industrial Engineering,* 20, 36-43.

Feigenbaum, A. V. (1956). Total quality control, *Harvard Business Review,* 34(6), 93-101.

Feigenbaum, A. V. (1961). Total Quality Control, McGraw-Hill Inc., New York, USA.

Gardner, L.L, Grant, M.E. and Rolston, L.J. (1995). Using simulation to assess costs of quality, *Proceedings of the Winter Simulation Conference.*

Giakatis, G., Enkawa, T. & Washitani, K. (2001). Hidden quality costs and the distinction between quality cost and quality loss, *Total Quality Management,* 12(2), 179-190.

Goulden, C. and Rawlins, L. (1995). A hybrid model for process quality costing, *International Journal of Quality & Reliability Management,* 12(8), 32-47.

Gupta, M. and Campbell, V. S. (1995). The cost of quality, *Productions and Inventory Management Journal,* 36(3), 43-49.

Harry, M.J. and Schroeder, R. (2000). Six Sigma: The breakthrough management strategy revolutionizing the world's top corporations, New York: Doubleday, Random House.

Heagy, C.D. (1991). Determining optimal quality costs by considering costs of loss sales, *Journal of Cost Management for the Manufacturing Industry,* Fall, 67-71.

Hester, W.F. (1993). True quality cost with activity based costing, *ASQC Annual Quality Congress Transactions,* 446-453.

Jorgenson, D.M. and Enkerlin, M.E. (1992). Managing quality costs with the help of activity based costing, *Journal of Electronics Manufacturing,* 2, 153-160.

Juran, J .M. (1952). Quality Control Handbook (1st ed.). New York: McGraw-Hill.

Juran, J. M. (1989). Juran on Leadership for Quality. New York: Free Press.

Juran, J.M., Gryna, F.M. and Bingham, R. (1975). Quality Control Handbook, 3rd edition, New York, McGraw-Hill.

Kent, R. (2005). Manufacturing strategy for window fabricators 14 – the cost of quality, Tanagram Technology, available at: www.tanagram.co.uk.

Krishnan, S.K., Agus, A. and Husain, N. (2000). Cost of quality: The hidden costs, *Total Quality Management,* 11(4, 5 & 6), 844–848.

Malchi, G. and McGurk, H. (2001). Increasing value through the measurement of the cost of quality (CoQ) – A practical approach, *Pharmaceutical Engineering,* 21(3), 92-95.

Marsh, J., (1989).Process modeling for quality improvement, *Proceedings of the Second International Conference on Total Quality Management,* IFS publication, Bedford, 111-121.

Merino, D. N. (1988). Economics of quality: Choosing among prevention alternatives, *International Journal of Quality & Reliability Management,* 5(7), 13-23.

Modarres, B. and Ansari, A. (1987). Two new dimensions in the cost of quality, *International Journal of Quality & Reliability Management,* 4(4), 9-20.

Morse, W. J. (1983). Consumer product quality control cost revisited, *Measuring Quality Costs, Cost and Management,* July/August, 16-20.

Moyer, D.R. and Gilmore, H.L. (1979) Product conformance in the steel foundry jobbing shop, *Quality Progress,* 12 (5),. 17-19.

Naylor, T.H., and Finger, J.M. (1967), Verification of computer models, *Management Science,* 14, pp92.

Omachonu, V.K., Suthummanon, S. And Einspruch, N. G. (2004). The Relationship Between Quality and Cost of Quality for a Manufacturing Company, *International of Quality & Reliability Management,* 21 (3), 277-290.

Plunkett, J.J. and Dale, B. G. (1988).Quality costs: A critique of some 'economic cost of quality' models, *International Journal of Production Research,* 26(11), 1713-1726.

Plunkett, J.J., Dale, B. G. (1987). A review of the literature on quality-related costs, *International Journal of Quality & Reliability Management,* 4 (1), .40-52.

Prickett, T.W. and Rapley, C. W. (2001). Quality costing: A study of manufacturing organizations Part 2: Main survey, *Total Quality Management,* 12(2), 211–222.

Purgslove, A.B. and Dale, B. G. (1995). Developing a quality costing system: Key features and outcomes, Omega*: International Journal of Management Science,* 23 (5), 567-575.

Ross, D.T. (1977). Structured analysis (SA): A language for communicating ideas, *IEEE Transactions on Software Engineering,* 3(1), 16-34.

Sandoval-Chavez, D.A. and Beruvides, M. G. (1998). Using opportunity costs to determine the cost of quality: A case study in a continuous-process industry, *Engineering Economist,* 43(2), 107-124.

Schiffauerova, A. and Thomson, V. (2006). A review of research on cost of quality models and best practices, *International Journal of Quality and Reliability Management,* 23 (6), 647-669.

Sharma, J. K. (2007). Business Statistics, 2nd edition, Pearson Education India.

Son, Y.K and Lie, F.H. (1991), A method of measuring quality costs, *International Journal of Production Research,* Vol. 29, pp. 1785-1794.

Sorqvist, L. (1997). Effective methods for measuring the cost of poor quality, *European Quality,* 4 (3), 40-42.

Sumanth, D. J., and Arora, D. P. S.(1992). State of the art on linkage between quality, quality costs and productivity, *International Journal of Materials and Product Technology,* 7(2), 150-169.

Suminsky Jr., L. T. (1994). Measuring cost of quality, *Quality Digest,* 14 (3), 26-32.

Tannock, J.D.T. (1995). Choice of inspection strategy using quality simulation, *International Journal of Quality & Reliability Management,* Vol. 12, No. 6, pp. 75-84

Tannock, J.D.T. (1997). An economic comparison of inspection and control charting using simulation, *International Journal of Quality & Reliability Management,* Vol. 14, No. 7, pp. 687-700.

Tatikonda, L.U. and Tatikonda, R.J. (1996). Measuring and reporting the cost of quality, *Production and Inventory Management Journal,* 37(2), 1-7.

Tsai, W. H. (1998). Quality cost measurement under activity-based costing, *International Journal of Quality and Reliability Management,* 15(7), 719-752.

Weheba, G. S. and. Elshennawy, A. K. (2004). A revised model for the cost of quality, *International Journal of Quality & Reliability Management,* 21(3), 291-308.

Yang, C.C. (2008). Improving the definition and quantification of quality costs, *Total Quality Management,* 19(3), March, 175 - 191.

3

GLP: Good Laboratory Practice

Isin Akyar
Acibadem University Faculty of Medicine Department of Medical Microbiology
Turkey

1. Introduction

In the early 70's FDA (United States Food and Drug administration) have realized cases of poor laboratory practice throughout the United States. FDA decided to check over 40 toxicology labs in-depth. They revealed lot dishonest activities and a lot of poor lab practices. Examples of some of these poor lab practices found were equipment not been calibrated to standard form, therefore giving wrong measurements, incorrect or inaccurate accounts of the actual lab study and incompetent test systems. Although the term "good laboratory practice" might have been used informal already for some time in many laboratories around the world GLP originated in the United States and it had a powerfull effect world wide.

2. History of Good Laboratory Practice (GLP)

GLP is an official regulation that was created by the FDA in 1978. The OECD (Organisation for Economic Co-operation and Development) Principles of Good Laboratory Practice were first created by an Expert Group on GLP set up in 1978 under the Special Programme on the Control of Chemicals. The GLP regulations that are accepted as international standards for non-clinical laboratory studies published by the US Food and Drug Administration in 1976 supplied the basis for the work of the Expert Group, which was guided by the United States and consisted experts from the following countries and organisations: Australia, Austria, Belgium, Canada, Denmark, France, the Federal Republic of Germany, Greece, Italy, Japan, the Netherlands, New Zealand, Norway, Sweden,Switzerland, the United Kingdom, the United States, the Commission of the European Communities, the World Health Organisation and the International Organisation for Standardisation. Eventually after United States other countries started making GLP regulations in their home countries. (Lori et al., 2009)

2.1 Those Principles of GLP were officially suggested for use in member countries by the OECD Council in 1981. They were set about as an essential part of the Council Decision on Mutual Acceptance of Data in the Assessment of Chemicals, which expresses that"data denoted in the testing of chemicals in an OECD member country in accordance with OECD Test Guidelines and OECD Principles of Good Laboratory Practice shall be accepted in other member countries for the aims of assessment and other uses relating to the protection of man and the environment".

2.1.1 The work of the OECD associated with chemical safety is fulfilled in the Environmental Health and Safety Division. The Environmental Health and Safety Division publishes free-off charge documents in six different series: Testing and Assessment; Principles on Good Laboratory Practice and Compliance Monitoring; Pesticides; Risk Management; Chemical Accidents and Harmonization of Regulatory Oversight in Biotechnology.

2.1.2 In spite of the fact that there are many national guidelines setting Good Laboratory Practice, the one guideline that is most universally accepted by the various national guidelines is the regulation of GLP through the Principles of Good Laboratory Practice of the Organisation of Economic Cooperation and Development (OECD), since these have been discussed by an international panel of experts and have been agreed on at an international level; they also form the basis for the OECD Council Decision/Recommendation on the Mutual Acceptance of Data in the Assessment of Chemicals which has to be regarded as one of the cornerstone agreements amongst the OECD member states with regard to trade in chemicals and to the removal of non-tariff barriers to trade. Besides the utilisation of the OECD Guidelines for the Testing of Chemicals, they restated the application of GLP Principles and the establishment of consorted national GLP compliance monitoring programmes as necessary parts of the mutual acceptability of data. The working group of experts who had createded the OECD Principles of Good Laboratory Practice also proceeded to inform and publish guidance for the Monitoring Authorities with regard to the introduction of procedures essential for the monitoring of industry's compliance with these Principles, as well as guidance with respect to the actual conduct of the necessary control activities such as laboratory inspections and study audits. (OECD, 1998).

2.1.3 Thus, the Principles of Good Laboratory Practice (GLP) have been developed to promote the quality and validity of test data used for determining the safety of chemicals and chemical products. Its principles are postulated to be followed by test facilities carrying out studies to be referred to national authorities for the purposes of assessment of chemicals and other uses in regards with the protection of man and the environment. Good laboratory practice might be used to detect collusion, but it could also serve to protect the researcher from unfounded allegations. In this manner, the application of the basic rules of GLP could be benefit even to a instution or laboratory.

2.1.4 Definition of GLP

The quality is the capability to systematically produce the same product to meet the same specifications time after time. GLP was altered to protect the integrity and quality of laboratory data used to back up a product application. The definition of the term "Good Laboratory Practice" itself, which identifies GLP as "a quality system related with the organisational process and the conditions under which non-clinical health and environmental safety studies are planned, performed, monitored, recorded, archived and reported." can be considered as an example of a brief and accurate definition. GLP describes good practices for non-clinical lab studies that support research or marketing approvals for FDA-regulated products(Seiler, 2005).

2.1.5 Purpose of GLP

Everyone makes mistakes that's why GLP is needed. GLP principles are a good idea even if you are not required to follow the standards. There are some simple rules such as: Say What You Do (with written standard operating procedures), do what you say (follow the procedures), be able to prove it (with good record keeping) (Jean Cobb, 2007).

2.1.6 The principles of good laboratory practice (GLP) is to support the development of quality and validity of test data used for determining the safety of chemicals and chemicals product (Clasby, 2005).

Hence GLP aims to decrease the occurrence of mistakes or mix-ups through large and specific labelling requirements. The registered information can be provided by demonstrating the application of the correct item in the stated amounts to the pertinent test systems.
2.1.7 GLP experience is important to employers in some cases. An employer may find it useful if you have: Practical experience with working on a study according to the GLP principles.
Good planning is the greater half of success. With a perfect propose in mind and a well figured out and defined testing procedure is it achiavable to acquire an evaluable outcome of a study. GLP places a high degree of reliance upon creating and following a pre-defined study plan.

2.2 The principles of good laboratory practice

Good Laboratory Practice is based on four principles:
The Management; The Quality Assurance; The Study Director; and The National Compliance Monitoring Authority. All of them serve important functions in the concordancy of performing and monitoring safety studies, and it should be kept in mind that all of them are required for GLP to achieve quality data.
2.2.1 Although GLP differs from other quality systems in aspects that are important not only for the traceability of data but especially for the full reconstructability of the study, there are certain co-occurances between GLP and other quality systems like accreditation schemes. (Seiler, 2005).
2.2.2 The aim of this chapter will be to give enough information about the GLP in details with the test facility organisation and personel, the facilities of quality assurance programme, test system, archive and waste disposal, apparatus, material, and reagents, physical, chemical, biological test systems, receipt, handling, sampling and storage and characterisation of the test and reference items, standard operating procedures, performance of the study, reporting of study results, storage and retention of records and materials.
2.2.3 The concerns of the chapter may be summarized as follows:

1. Test facility management
2. Quality assurance programme
3. Meeting the requirements of the test facility
4. Equipment
5. Receipt, handling, sampling and storage
6. Standard operating procedures.
7. Performance of the study.
8. Reporting of study results
9. Storage and retention of records and materials.

3. Test facility management

Test facility means the persons, premises and operational units that are necessary for conducting the non-clinical health and environmental safety study.
3.1 The term "test facility" may include several "test sites", at one or more geographical locations, where phases or components of a single overall study are conducted and does not only include buildings, rooms and other premises, but that it includes also the people who are working there and are liable for performing these studies (Seiler, 2005). For multi-site studies the test facility considers the site at which the Study Director is located and all

individual test sites, which individually or collectively can be considered to be test facilities. The test facility should be of appropriate size, construction and location to meet the requirements of the study. It should be designed safe enough to get the validation results confidently. Research laboratories where test/reference item characterisation considering determination of identity, purity/strength, stability, and other related activities is conducted, one or more agricultural or other in- or outdoor sites where the test or control item is applied to the test system are the different test sites in the test facility. And in some cases, a processing facility where collected commodities are treated to prepare other items where collected specimens are analysed for chemical or biological residues, or are otherwise evaluated. (OECD, 1998).

3.1.1 Properties of biological test systems are generally more complex and mutable than the ones of physical/chemical test systems. Hence biological test systems need very careful characterisation in order to guarantee the quality and integrity of the data derived from them. The outcome of a study may be influenced by the state and condition of the test system at the time of the study which has special importance with regard to the reconstructability. The GLP Principles, in uttering the requirements for the accomodation and siting of these systems, for their maintenance and utilization, and for the associating documentation, aims at supplying the essential basis for confidence into the results obtained from biological test systems. A test item should only be used in studies if it can safely be regarded as being in its pure, unspoilt and not decomposed. Any change in the properties of the test item may lead to spurious and erroneous results, and to wrong interpretations of the effects the test item is supposed to have produced. Stability testing will lead to the definition of a time interval within which the test item will stay in this state, and as a result "expiry" or "re-analysis" dates have to be mentioned on the label of the test item container. With this necessity GLP aims to reduce the possibility that an item will be used in a study which does no longer correspond to the item that had been intended for testing. The aim of any safety testing is to analyze possible effects of the test item on the test system. Therefore, the effects observed in any test system should be traceable to the application of the item which was the designated subject of the study.

3.1.2 After the conduct of the respective safety test, in order to find out this even retrospectively, the documentation on the test item has to fulfil a number of requirements:

3.1.3 There must be documented proof that one item that had been intended to be tested indeed reached the sensitive parts of the test system confirming that the effects observed had really been originated by the test item, and that the application of this item to man or the environment would therefore not be expected to result in any effects other than those which can be concluded from the observed ones in the test systems utilised. "Tidiness" is a crucial point with consideration to the general claims on the test facility. When the laboratory bench is filled up with clean and dirty instruments, glassware some of which are being used and some are not, it is not so easy to locate all the materials needed for a specific activity.

3.1.4 Tidiness therefore has both functions of inspiring trust into the quality of the work performed, and facilitate the performance of the daily activities according to the quality standards. Tidiness makes the life easier to survive a compliance monitoring inspection, if even under the stress the technician can find the folder with the SOPs at once like without trying to find a treasure.

3.1.5 A test facility needs a Management, a Study Director, a Quality Assurance Unit, study personnel and a person responsible for the archives (Seiler, 2005).

3.1.6 Test Facility Management should guarantee that these Principles of GLP are requested in its test facility. General Requirements for GLP consists of appropriately qualified personnel, adequate resources , appropriate procedures for: sanitation, health precautions, clothing, test protocol development, test methods, data analysis, report development, appropriately qualified study director, quality assurance function. Test site management should be aware of the fact that the test facility management may be liable to inspection by the national GLP compliance monitoring authority of the country in which the test site is located.

3.1.7 "The Study Director" has overall responsibility for the technical conduct of the study, as well as for the interpretation, analysis, documentation, and reporting of results, and represents the single point of study control." (OECD, 1998).

3.1.8 The GLP Principles are designed to avoid the factors that would endanger the reconstructability of a study, by giving the only and final responsibility for the GLP compliant conduct of a study to one single person, the Study Director. For each nonclinical laboratory study, a scientist or other professional of appropriate education, training, and experience should be identified as the study director.

3.1.9 The Study Director has to be aware of all possible circumstances that might affect the quality and integrity of a study. There should be communication between the Study Director and other personnel including all scientists involved in study conduct, in order to be kept at the forefront of developments in a study, and to be able to act, as considered appropriate, on unforeseen developments. All information has to be passed to the Study Director. He should make or at least acknowledge all the decisions. In such special circumstances where the Study Director cannot exercise his immediate control, the responsibilities of a Study Director may be extended to other individuals such as specialised scientists (Seiler, 2005).

3.1.10 When the Study Director cannot exercise immediate supervision, at each test site study procedures may be controlled by a member of the staff, called the Principal Investigator. The Principal Investigator means an individual responsible for the conduct of certain defined phases of the study, acting for the Study Director. The Study Director has the final responsibility of for the overall quality and integrity of the study. He cannot share this responsibility with any other individual involved in the study. Nonetheless, the Principal Investigator should take the responsibility for the defined, delegated part of the study, he is not responsible for the study plan, and he can not approve any improvements to it. The general management must have a stiff interpretation and working agreement with the test site management as to how and by whom the Quality Assurance Programme (QAP) will be carried out.(OECD, 1998).

3.1.11 Approved original and revised Standard Operating Procedures should be used in the studies. There should be a Quality Assurance Programme with assigned personel for each study an individual with the proper qualifications, training, and experience is designated by the management as the Study Director before the study is initiated. Personnel should clearly understand the functions that they are going to carry out, training should be provided when needed. Standard Operating Procedures should be established and followed. They should be appropriate and technically valid.

3.1.12 The GLP Compliance Statement signed by the Study Director in the final study report is the declaration that gives the Regulatory Authority the guarantee for a appropriately performed, valid study. The results and conclusions of the study can be trusted to reflect the real data obtained in the study (Seiler, 2005).

4. Quality assurance programme

Quality control is the process, procedures and authority used to accept or reject all components, drug product containers, closures, in-process materials, packaging material, labeling and drug products and the authority to review production records to assure that no errors have occurred, that they have been fully investigated. The quality and reliability of test data count on the state and condition of the test system which is used in its production. This is meant to be the control of a number of technical features and specifications which are needed to ensure the integrity of the system and the quality of the data generated. In a study for compliance with GLP, the most important aspects may be characterised as "suitability", "capacity" and "integrity" (OECD, 1998).

4.1 "Trust is Good, Control is Better" says an old proverb. The quality which is supposed to be achieved in GLP is not a quality which can be controlled by easy, numerical or other means, but it is the şcontrol over the intrinsic quality of a test facility and its studies. Only through this independence a reliable assurance of the studies inherent quality that can be achieved. (Seiler, 2005).

4.1.1 The test facility should have a documented Quality Assurance Programme to guarantee that studies performed comply with these Principles of Good Laboratory Practice. The Quality Assurance Programme should be performed by an individual or by individuals designated by. These staff should be familiar with the test procedures and directly responsible to management.This individual(s) should not be involved in the conduct of the study being assured(OECD, 1998). It must be clear that what the exact area of responsibility is for the defined individual, what exactly is to be done at those test sites where such "phases" are conducted in delegating parts or"phases" of a study through the terms of appointment for the Contributing Scientist or the Principal Investigator (Seiler, 2005).

4.1.2 As the person responsible for the overall conduct of the study, to the Study Director's management, and to the latter's Quality Assurance Programme, there should be a full, frank flow of information to the responsible test site management, to the responsible Principal Investigator(s) and to the Study Director. In the same way, for notification of critical activities it should be essential to assure effective communications from the Study Director and/or Principal Investigators to the quality assurance personel. Because of the complex nature of field studies, and the fact that the exact time of certain activities will depend upon local weather or other conditions flexible quality assurance procedures may be required. The geographical spread of test sites may mean that quality assurance personnel will also need to manage langquage differences in order to communicate with local study personnel, the Study Director, Principal Investigators and test site management. Independent from the test sites, the written reports of quality assurance personnel must reach both management and the Study Director.Those reports receipt by management and the Study Director should be documented in the raw data.

4.1.3 The Quality Assurance personnel should be responsible of maintaining copies of all approved study plans and Standard Operating Procedures in use in the test facility and have access to an up-to-date copy of the master Schedule, verifying that the study plan contains the information required for compliance with these Principles of Good Laboratory Practice, conducting inspections to determine if all studies are conducted in accordance with these Principles of Good Laboratory Practice. Inspections should also determine that study plans and Standard Operating Procedures have been made available to study personnel and are being followed. The study plan allows Quality Assurance: to monitor compliance of the study plan with GLP; to assess the clarity and consistency of the study plan; to identify the critical phases of the study; and to plan a monitoring programme in relation to the study (OECD, 1998).

4.1.4 Study plans and Standard Operating Procedures should be determined by the inspections and they should have been available to study personnel and are being followed. In the final reports it should be confirmed that the methods, procedures, and observations are accurately and completely described, and that the reported results accurately and completely reflect the raw data of the studies

4.1.5 Inspection of facilities and experimental activities is one of the tools of Quality Assurance for ascertaining and guaranteeing the continued obeyence to the rules of GLP in a test facility inside the studies performed. Since it is recognised that randomly conducted inspections will be sufficient to ensure compliance with, the GLP Principles do not necessitate a fixed supervision. These inspections should involve those parts of a study that have particular importance for the validity of the data and the conclusions to be drawn therefrom, or where deviations from the rules of GLP would most heavily have a powerfull effect on the integrity of the study. Quality Assurance thus has to find a balance in their inspectional activities, evaluating the study type and "critical phases", in order to achieve a well supported view of the GLP compliance at the test facility and within the studies conducted. It is clear that any deviations from the rules of GLP that are observed in these inspections should be corrected. The audit of the final report, hence serves to ascertain the quality and integrity of the specific study with its detailed assessment of GLP compliance throughout the study and with its concomitant review of all relevant information, records and data. It is the responsibility of management to provide policies, guidelines, or procedural descriptions to ensure that this statement reflects Quality Assurance's acceptance of the Study Director's GLP compliance statement. The Quality Assurance statement has two functions: Serving to demonstrate that Quality Assurance has adequately monitored the conduct and progress of the study, from the first check of the study plan for GLP conformity to the audit of the final report as a "second opinion" on the completeness of the reporting and the adequacy of raw data coverage and providing the study with the seal of approval by attesting to the GLP compliant conduct. Thus, the Quality Assurance statement has a particular importance for the assessment of the study's integrity and validity. The Quality Assurance statement should show that the study report accurately reflects the study's raw data.

4.1.6 Before signing the Quality Assurance statement, Quality Assurance should ensure that all issues raised in the Quality Assurance audit, i.e. in the audit report to the Study Director and to management, have been addressed through appropriate changes of the final report, that all agreed actions have been completed, and that no additional changes have been made to the report which would require a further report audit. Through management policy it should certainly be made clear that the Quality Assurance statement would only be completed if the Study Director's claim to GLP compliance can be supported(Seiler, 2005).

Laboratories use various supplied materials in studies conducted in compliance with the GLP Principles. Suppliers have attempted to produce products which satisfy users' obligations as set out in the GLP Principles.

4.1.7 Accreditation can be especially useful to suppliers. Often accreditation schemes monitor members' implementation of national and international standards thus, a supplier or manufacturer's accreditation certificate may signify to the customer the satisfactory implementation of a standard in addition to other aspects of accreditation.

4.1.8 It is recommended that suppliers seek membership, where feasible and/or appropriate, in national accreditation schemes. Although accreditation is a useful complementary tool to support compliance with the GLP Principles, it is not an acceptable alternative to GLP compliance nor will it lead to international recognition in the context of meeting the requirements for the mutual acceptance of data as set out in the OECD Council Acts. (OECD, 1998).

As an example ISO 17025 and GLP comparison can be considered (Table 1).

ISO Members	**OECD Members**
The same standard for all ISO	Different regulations in different countries
Designed for repetitive studies	Designed for single studies
Description of Quality System in Quality Manual	Description of Quality System in SOPs
General statements for responsibilities of personnel	Very specific responsibilities of personnel
No specific requirements for storage of records and reports	Specific requirements for storage, retention and archiving
No study plans required (standardized methods should be used)	Study plan required for each study
Written operating procedures without specific format	SOPs with detailed requirements for format and content
Analysis methods must be verified through inter-laboratory test (Proficiency testing)	Validation through inter-laboratory tests not required
Documented complaints procedures	In case of problems, only course of law
Storage of test samples and data until client accepts results	Storage of test samples according to local regulatory requirements

(Fox , 2011)

Table 1. ISO 17025 and GLP comparison.

5. Meeting the requirements of the test facility

The GLP principles do not address the question of the specific requirements for the location of an archive, except that it should be "of suitable size, construction and location to meet requirements". Therefore there is complete freedom for every test facility to define the location of its archives and to designate the proper locations for each type of materials to be stored (Seiler, 2005). Before they can be considered as GLP compliant General Requirements Facilities need to conform to a number of general rules. The facilities should be designed for the best suitability to the studies that are to be performed within. Some comfort for the employees comes of course with all the requirement of study quality, which means that the people working in a facility should certainly have sufficient room to move around in order to be able to perform the duties which the study calls for, and to perform them in a manner compatible with the quality, integrity and validity of the study. This is acknowledged absolutely in the general requirement that a test facility should be of appropriate size, construction and location, for both meeting the requirements of the study and minimising disturbance that would interfere with the validity of the study. jürgThe test facility should have a sufficient and suitable number of rooms or areas to assure the isolation of test systems and the isolation of individual projects, involving substances or organisms known to be or suspected of being biohazardous. There should be storage rooms or areas as needed for supplies and equipment and should provide adequate protection against infestation, contamination, and/or deterioration. Facilities for handling test and reference items should be planned. To prevent contamination or mix-ups, there should be separate rooms or areas for receipt and storage of the test and reference items, and mixing of the test items with a vehicle(Figure 1).

Fig. 1. There should be separate working areas in the laboratory.

5.1 Handling and disposal of wastes should be carried out in such a way as not to risk the integrity of studies. This includes provision for appropriate collection, storage and disposal facilities, and decontamination and transportation procedures. This policy is to assure that reagents used are specified in the standard operating procedure. Purchasing and testing should be handled by a quality assurance program. Reagents and solutions should be labeled, deteriorated or outdated reagents and solutions should not be used. The opening date should be recorded. They should be stored under ambient temperature and the expiration date should be considered(Lori et al , 2009).The equipments should be appropriately designed, adequate throughput capacity, appropriately located and routinely maintained & calibrated (Clasby, 2005).

5.1.1 In order to guarantee the quality of the data, appropriate conditions should be established and maintained for the storage, housing, handling and care of biological test systems. At the experimental starting date of a study, test systems should be free of any disease or condition that might interfere with the purpose or conduct of the study. If necessary to maintain the integrity of the study, test systems that become diseased or injured during the course of a study should be isolated and treated. Any diagnosis and treatment of any disease before or during a study should be recorded. Records of source, date of arrival, and arrival condition of test systems should be maintained. Biological test systems should be acclimatised to the test environment for an adequate period before the first administration/application of the test or reference item. All information needed to properly identify the test systems should appear on their housing or containers. Individual test systems that are to be removed from their housing or containers during the conduct of the study should bear appropriate identification, wherever possible. During use, housing or containers for test systems should be cleaned and sanitised at appropriate intervals. Any material that comes into contact with the test system should be free of contaminants at levels that would interfere with the study. Bedding for animals should be changed as required by sound husbandry practice. Use of pest control agents should be documented. Test systems used in field studies should be located so as to avoid interference in the study from spray drift and from past usage of pesticides (OECD, 1998).

5.1.2 The important principles can be summarised as follows:

There should be a unique identification for the study and all of its parts. All original observations in a study should be at once clearly and legibly recorded. The recording should be permanent and corrections should be made so as not to obscure the original entry; for all corrections the respective reasons have to be provided. All records should be in the form of bound notebooks or on continuously numbered sheets. All entries and corrections to them should be dated and initialled. Records related to the test system itself should be gathered and preserved. Specimens should be clearly identified so as to allow full traceability. At the end of a study, all raw data should be assembled, catalogued and archived. Archiving should support for secure storage of all raw data, samples and specimens, together with any other documents such as study plan and study report. (Jürg P. Seiler, 2005).

6. Equipment

6.1 Equipment, including validated computerised systems, used for the generation, storage and recovery of data, and for controlling environmental factors relevant to the study should be suitably located and of appropriate design and adequate capacity. Equipment records should include: name of the equipment and manufacturer, model or type for identification, serial number, date equipment was received in the laboratory, copy of manufacturers operating instruction(s). Equipment used in a study should be periodically inspected, cleaned, maintained, and calibrated according to Standard Operating Procedures. Records of these activities should be maintained. Calibration should be traceable to national or international standards of measurement. Instrumentation validation is a process necessary for any analytical laboratory. Data produced by "faulty" instruments may give the appearance of valid data. The frequency for calibration, re-validation and testing depends on the instrument and extent of its use in the laboratory. Chemicals, reagents, and solutions should be labelled to indicate identity, expiry date and specific storage instructions. Information concerning source, preparation date and stability should be available. The expiry date may be extended on the basis of documented evaluation or analysis. If a mistake is made, original data should not be obscured. Instead of this, a single strikeout should be drawn and a reason code should be added, later the date should be changed. Whenever an instrument's performance is outside the "control limits" reports must be discontinued

(Cobb, 2007). Equipment and materials used in a study should not interfere adversely with the test systems. (OECD, 1998).

6.1.1 Equipment used for the generation of physical/chemical data should be suitably located and of proper design and adequate capacity. The integrity of the physical/chemical test systems should be ensured. Appropriate conditions should be established and maintained for the storage, housing, handling and care of biological test systems, in order to ensure the quality of the data. Standardization, calibration, and verification are the definitions which have particular importance for the equipments. The difference between those should be well understood and performed by the laboratory personnel: Verification is the external check of equipment accuracy. It is the check balance accuracy against weights at laboratory. There is no adjustment.

6.1.2 In calibration equipment is adjusted based on comparison to certified or known reference materials. The balance is adjusted after comparison to certified weights by trained professional. Standardization is made by comparison with similar equipments, such as using two thermometers of similar design to compare readings.

6.1.3 While monitorizing the study laboratory staff should always have the following questions on mind: Was the equipment functioning properly? Who performed the work, what was the date, and what specific parameters did they use? Was there a problem? How was the problem fixed? Were there any problems with the reagents and solutions?

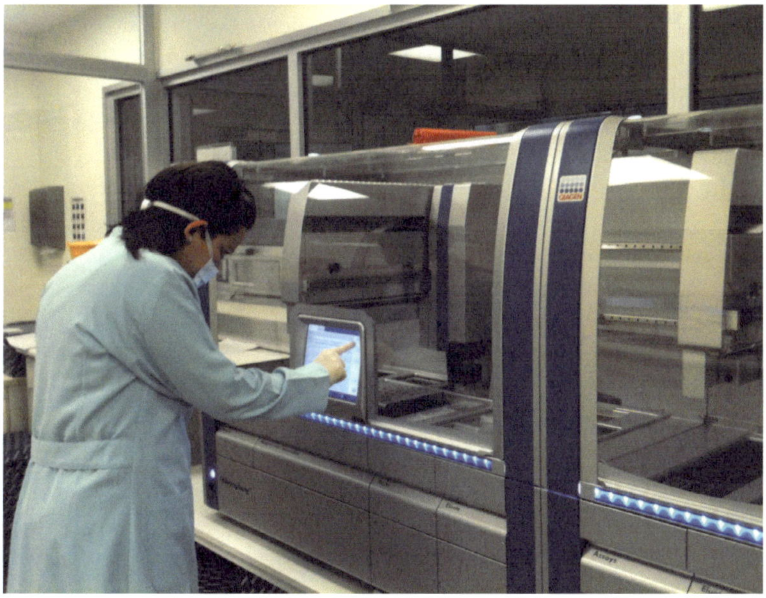

Fig. 2. Laboratory equipment should routinely be maintained and calibrated.

6.1.4 The GLP Principles do not suggest or require any specific time intervals for such activities. Cleaning and maintenance intervals may be different from one type of equipment to the other, and such intervals may as well depend on the frequency of use or the workload imposed on the respective equipment. On the other hand the question of the correct frequency of such activities should be considered as a scientific one, calling for the expert judgement of the responsible scientists.

6.1.5 Generally the manufacturer's manuals provide useful signs or suggestions for cleaning and maintenance intervals. These same aspects are valid also for calibration frequencies, where in some cases calibration is routinely performed before each measurement, while in other cases the respective frequencies may be set in an arbitrary manner. The key point in the consideration of maintenance and calibration frequencies is the necessary assurance of data validity. In some cases it would be necessary to ensure the traceability of the calibrations performed to "national or international standards of measurement". The results of a study can be relied on only as far as the study itself is being appropriately conducted. Suitability of apparatus, materials and reagents is thus one of the key points in this judgement. Computerised systems have taken over an ever increasing part of different tasks in various areas within our daily lives. They are used during the planning, conduct and reporting of studies for a variety of purposes, including the direct or indirect data capture from automated instruments, the recording, processing, reporting, general management and storage of data, as well as in controlling and steering functions in numerous kinds of equipment. For these different activities, computerised systems can be of varying complexity from a simple, microchip controlled instrument up to a complex laboratory information management system (LIMS) with multiple functions. Whatever the scale of computer involvement, the GLP Principles have to be followed. The correct application of the GLP Principles to ensure compliance of computerised systems with the GLP rules may, however, pose some problems, which might be regarded to stem at least in part from the very origins of GLP. All computerised systems used for the generation, measurement or assessment of data intended for regulatory submission should be developed, validated, operated and maintained in ways which are compliant with the GLP Principles. Appropriate controls for security and system integrity must also be adequately addressed during the whole life cycle of any computerised system(Seiler, 2005).

6.1.6 All equipment used in a GLP context have to satisfy the specified requirements of the users. For computerised systems the evidence of suitability is provided by the validation procedure. This has to start with the exact definition of the user requirements which have subsequently to be translated into proof of adequate operation of the system in the actual environment. With this prospective validation assurance it should be provided that the computerised system will perform the tasks designed to execute in a correct, reproducible and reconstructable way.

6.1.7 Computerised systems associated with the conduct of studies bound for regulatory submission should be of appropriate design, adequate capacity and suitable for their intended purposes. There should be appropriate procedures to control and maintain these systems, and the systems should be developed, validated and operated in a way which is in compliance with the GLP Principles. The demonstration that a computerised system is suitable for its intended purpose is of fundamental importance and is referred to as computer validation. The validation process provides a high degree of assurance that a computerised system meets its pre-determined specifications. Validation should be undertaken by means of a formal validation plan and performed prior to operational use. (OECD, 1998).

6.1.8 Whether any system has been fully and prospectively validated or has just been retrospectively evaluated and qualified, there is a need for continued maintenance of the validation status to be sure of the continuence of data validity. This is accomplished through formal procedures that require any changes to the system to be fully documented. Data integrity will, however, not only depend on the validation status of the system, but also, and to a very important extent, on the security measures developed for the utilisation of the system. Through the requirement of documented security procedures for the protection of hardware, software and data from corruption, unauthorised modification, or loss, GLP intends to provide for continuous data integrity. In general terms, security issues can be divided into measures of physical security, i.e. measures that can be instituted on the facility and apparatus level, and logical security, i.e. those that are related to software security at the access level (Seiler, 2005).

6.1.9 Physical location of computer hardware, peripheral components, communications equipment and electronic storage media should be considered. Extremes of temperature and humidity, dust, electromagnetic interference and proximity to high voltage cables should be avoided unless the equipment is specifically designed to operate under such conditions. Consideration must also be given to the electrical supply for computer equipment and, where appropriate, back-up or uninterruptable supplies for computerised systems, whose sudden failure would affect the results of a study. Adequate facilities should be provided for the secure retention of electronic storage media. (OECD, 1998).

6.1.10 Because of various reasons, in every test facility there may be computerised systems which have not been formally validated. Their use in a GLP environment should still be required, clear proof of their suitability can only be obtained through an evaluation of their past and actual performance. In order to get reconstructability and transparency, this proof has to be planned and documented, resulting in a final conclusion on the past, present and future suitability of the respective system. In this way GLP aims at providing evidence for the correct functioning of the computerised system and for estimating the extent of GLP compliance.

7. Receipt, handling, sampling and storage

Sampl e tracking vary among laboratories. Receipt, handling, sampling and storage should be prepared appropriately. Records including test item and reference item characterisation, date of receipt, expiry date, quantities received and used in studies should be maintained. Handling, sampling, and storage procedures should be identified in order that the homogeneity and stability are assured to the degree possible and contamination or mixup are precluded (Seiler, 2005). They should maintain the unmistakable connection between a set of analytical data and the samples from which they were obtained. Original source of samples must be recorded and unmistakably connected with the set of analytical data (Cobb, 2007). Records including test item and reference item characterisation, date of receipt, expiry date, quantities received and used in studies should be maintained. Handling, sampling, and storage procedures should be identified in order that the homogeneity and stability are assured to the degree possible and contamination or mix-up are precluded. Storage container(s) should carry identification information, expiry date, and specific storage instructions.

7.1 Receipt and storage areas for specimens must be separate from storage areas for pesticide formulations and other test or reference items. Areas used for specimen and sample preparation, instrumentation, calibration of sprays, reference Standard preparation, and for washing glassware should be adequately isolated from each other and from other functions of the laboratory which might introduce contamination. Storage areas for test and reference items at all test sites should be environmentally monitored, if required, to assure conformance with established stability limits for these materials. Test and reference items should not be placed in the same storage containers with collected test system specimens and other materials of low concentrations which are being stored for shipment to the analytical laboratory or to off-site archives. There should be adequate storage and disposal facilities available for pesticide and related wastes such that there is no potential for cross-contamination of test systems, of test or reference items or of collected specimens. (OECD, 1998). Storage container(s) should carry identification information, expiry date, and specific storage instructions Each test and reference item should be properly identified. For each study, the identity, including batch number, purity, composition, concentrations, or other characteristics to appropriately define each batch of the test or reference items should be known. In cases where the test item is supplied by the sponsor, there should be a mechanism, developed in co-operation between the sponsor and the test facility, to verify the identity of the test item subject to the study. The stability of test and reference items under storage and test conditions should be known for all studies. If the test item is administered or applied in a vehicle, the homogeneity, concentration and stability of the test item in that vehicle should be determined. A sample for analytical purposes from each batch of test item should be retained for all studies except short-term studies. A well thought-out concept of logistics is needed for receiving, storing, handling and disposing test items, together with provisions for the adequate documentation of all procedures connected with test item handling. One aspect in this area of test item logistics is the physical location of these activities, and the GLP Principles underline the importance of identifying adequate facilities for them.

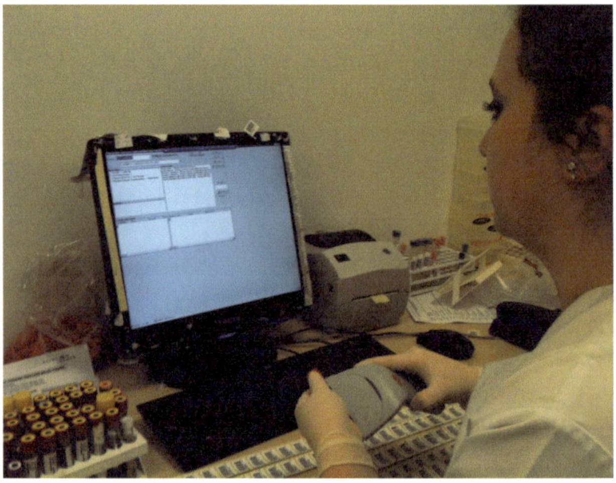

Fig. 3. Laboratory records of receipt, handling and storing should be carefully maintained.

7.1.1 While receipt and storage involves mainly the handling of closed containers, the opening of such a container exposes the test item to the facility environment and leads consequently to the possibility of contamination of either the test item or the environment. Moreover, the greater the number of different test items to be performed, the greater the danger that somebody would. Therefore, work in the special area where test items are prepared for application has to be carefully organised. For weighing of the test item and its mixing with the vehicle, it should be made compulsory that only one test item would be

present in that area at any one time. Special attention has to be given to such areas where test, control and reference items are prepared for in vitro studies.
7.1.2 In such studies, the term "contamination" does not only mean "contamination by traces of other items" but also contamination by microorganisms, etc.,hence necessitating areas where the preparation of these items for the application in the study could be performed under aseptic conditions. By the same reason, GLP mandates that the available test item storage locations should be separate from the rooms or areas containing test systems in order to prevent excessive exposure of the systems to test items other than the intended one.
7.1.3 Of course, the storage facilities should supply adequate conditions to save the identity, purity and stability of the test items. Therefore it is necessary that storage areas at different temperature levels, for storage at room temperature or in refrigerators and deep freezers. Also protection from light, humidity or oxygen may be necessary for special cases. Also, there are security aspects to be mentioned. A suitable limitation for access to the test items should be advisable. It is very important that a good, accurate accounting system should be in place, which could be used to reconstruct the course of test item utilisation. (Seiler, 2005).

8. Standard Operating Procedures (SOP)

According to EPA(Environmental Protection Agency) GLP regulations, "Raw data" means any laboratory worksheets, records, memoranda, notes, or exact copies thereof, that are the result of original observations and activities of a study and are necessary for the reconstruction and evaluation of the report of that study. Logbooks for recording temperatures or equipment use, repair, and maintenance, field or laboratory notebooks, forms for field or laboratory observations, training reports, computer printouts, recorded data from automated instrument are examples of raw data. It's so hard and not necessary for anyone remember all these details and that's one of the functions of the Standard Operating Procedures (SOPs).
8.1 In FDA it is said that :"If it is not documented..., it did not happen!" or, it's a rumor!" GLPs SOPs Can't do Guarantee "good science", guarantee good documentation, replace common sense, prevent all mistakes (Cobb, 2007). SOPs are written procedures for a laboratories program. They are approved protocols indicating test objectives and methods. Standard Operating Procedures are intended to ensure the quality and integrity of the data generated by the test facility. Revisions to Standard Operating Procedures should be approved by test facility management (OECD, 1998).
8.1.1 They define how to carry out protocol-specified activities. SOPs are most often written in a chronological listing of action steps. They are written to explain how the procedures are supposed to work SOP of routine inspection, cleaning, maintenance, testing and calibration, actions to be taken in response to equipment failure, analytical methods, definition of raw data, keeping records, reporting, storage, mixing, and recovery of data. (Standard Operating Procedures should have been written and approved by test facility management that are intended to ensure the quality and integrity of the data generated by that test facility. Revisions to Standard Operating Procedures should be approved by test facility management. Each separate test facility unit or area should have at once available current Standard Operating Procedures relevant to the activities being performed therein. Published text books, analytical methods, articles and manuals may be used as supplements to these Standard Operating Procedures. Deviations from Standard Operating Procedures related to

the study should be documented and should be acknowledged by the Study Director and the Principal Investigator(s). SOPs are written, approved procedures that describe routine activities that are specific for daily operations at each facility. SOPs should allow appropriately qualified personnel to perform a procedure once trained.
8.1.2 The details given under each heading are to be considered as illustrative examples. Room preparation and environmental room conditions for the test system, procedures for receipt, transfer, proper placement, characterisation, identification and care of the test system, test system preparation, observations and examinations, before, during and at the conclusion of the study, handling of test system individuals found in a severe position or dead during the study, collection, identification and handling of specimens ,siting and placement of test systems in test conspiracy should be reviewed. And also operation of Quality Assurance personnel in planning, scheduling, performing, documenting and reporting inspections should be examined. Personnel should perform the same tasks using the same procedures. SOPs should accurately reflect how routine tasks are performed written by each facility based on their specific field and/or laboratory operations. Laboratory management must be sure that the SOPs used in the laboratory are useful in daily operations. They should be scientifically sound. And they should always be updated as necessary, rewrites should be the part of the routine process. While writing SOP guidelines there must be some precautions such as avoiding restrictive language such as "vortex for exactly 1 minute" but include clear instructions such as "vortex until homogenized" if that satisfies the purpose. Unnecessary steps should not be added such as "consult the manual" unless personnel are required to follow this step (Cobb, 2007). Study personnel should easily access to the study plan and appropriate Standard Operating Procedures should be applicable to their involvement in the study. It is their responsibility to comply with the instructions given in these documents. Study personnel should exercise health precautions to minimise risk to themselves and to ensure the integrity of the study. Standard Operating Procedures (SOPs) are intended to describe procedures that are routinely employed in the performance of test facility operations. Indeed they are defined as "documented procedures which describe how to perform tests or activities normally not specified in detail in study plans or test guidelines." The definition moreover implies that SOPs should describe all steps in the performance of an activity in such a detailed way that somebody not familiar with this activity might be able to perform it correctly and without having to recourse to outside help (Seiler, 2005).
8.1.3 It is suggested that test site personnel should follow test site SOPs. When they are required to follow other procedures specified by the Study Director, for example SOPs provided by the test facility management, this necessity should be identified in the study plan (OECD, 1998).

9. Performance of the study

Performance of the Study should be monitorized carefully. All the standards supplied by the GLP should be followed from the beginning of the study to the end by the final report. For each study, a written plan should exist prior to the initiation of the study (Seiler, 2005). The study plan should contain the following information: Identification of the study, the test item and reference item, information concerning the sponsor and the test facility, dates, test methods, issues (where applicable)and records. (OECD, 1998).

9.1 The study plan should be approved by dated signature of the Study Director and verified for GLP compliance. Deviations from the study plan should be described, explained, recognized and dated in a timely fashion by the Study Director and/or Principal Investigator(s) and maintained with the study raw data.
9.1.1 In the study plan the identification of the study, the test item and reference item information should exist: A descriptive title; a statement which reveals the nature and purpose of the study; Identification of the test item by code or name; The reference item to be used. Information Concerning the Sponsor and the Test Facility should be declared. It should comprise: Name and address of the sponsor, any test facilities and test sites involved, Study Director, Principal Investigator(s), and the phase(s) of the study delegated by the Study Director and under the responsibility of the Principal Investigator(s) with the date of approval of the study plan by signature of the Study Director, of the study plan by signature of the test facility management and sponsor if required by national regulation or legislation in the country where the study is being performed, the proposed experimental starting and completion dates, reference to the OECD Test Guideline or other test guideline or method to be used, the justification for selection of the test system characterisation of the test system, such as the species, strain, substrain, source of supply, number, body weight range, sex, age and other pertinent information. It should also contain the method of administration and the reason for its choice; The dose levels and/or concentration(s), frequency, and duration of administration/application; detailed information on the experimental design, including a description of the chronological procedure of the study, all methods, materials and conditions, type and frequency of analysis, measurements, observations and examinations to be performed, and statistical methods to be used. Specimens from the study should be identified to confirm their origin. Such identification should enable traceability, as appropriate for the specimen and study. The study should be conducted in accordance with the study plan. All data generated during the conduct of the study should be recorded directly, punctually, correctly, and legibly by the individual entering the data. These entries should be signed or initialled and dated. Any change in the raw data should be made in order to understand the previous entry easily, should indicate the reason for change and should be dated and signed or initialled by the individual making the change.
9.1.2 Computerised system design should always supply for the retention of full audit trails to show all changes to the data without obscuring the original data. It should be possible to associate all changes to data with the persons having made those changes. Reason for changes should be given.

10. Reporting of study results

All studies generate raw data that are the original data gathered during the conduct of a procedure. They are essential for the reconstruction of studies and contribute to the traceability of the events of a study. Raw data are the results of the experiment upon which the conclusions of the study will be based. Some of the raw data may be used directly, and some of them will be treated statistically. The results and their interpretations provided by the scientist in the study report must be a true and accurate reflection of the raw data.
10.1 A final report should be prepared for each study. The study report, like all the other scientific aspects of the study, is the responsibility of the Study Director. He/she must ensure that it describes the study accurately. Reports of Principal Investigators or scientists involved in the study should be signed and dated by them. The final report should be signed and dated

by the Study Director to indicate acceptance of responsibility for the validity of the data.If necessary, corrections and additions to a final report should be in the form of amendments. Amendments should clearly specify the reason for the corrections or additions and should be signed and dated by the Study Director. The Study Director is responsible for the scientific interpretation included in the study report and is also responsible for declaring to what extent the study was conducted in compliance with the GLP Principles. The GLP Principles list the essential elements to be included in a final study report.

10.1.1 The final report should include, the following information: A descriptive title; identification of the test item by code or name, characterisation of the test item including purity, stability and homogeneity. Information concerning the sponsor and the test facility should imply; name and address of the sponsor, any test facilities and test sites involved, the study Director, the Principal Investigator(s) and the phase(s) of the study, delegated and scientists having contributed reports to the final report, experimental starting and completion dates.

10.1.2 A Quality Assurance Programme statement listing the types of inspections made and their dates, including the phase(s) inspected, and the dates any inspection results should be reported to management and to the Study Director and Principal Investigator(s). This statement should also serve to confirm that the final report reflects the raw data. It should contain the Description of Materials and Test Methods. A summary of results should be given. All information and data required by the study plan; A presentation of the results, including calculations and determinations of statistical significance; An evaluation and discussion of the results and, where appropriate, conclusions. It should imply the location(s) where the study plan, samples of test and reference items, specimens, raw data and the final report are to be stored.

10.1.3 A computerised system to be used in a GLP area should include both the dating and timing of the original entry and the retention of a full audit trail. Such identification could be possible either by the use of personal passwords recognised by the computer or by digital signatures. Furthermore, the system should not accept any changes to data without concomitant entry of a reason or justification. In manual recording the entries made on a sheet of paper can be dated and signed to bear witness to the validity of data and to accept responsibility.

10.1.4 Therefore GLP wants to ensure that data safety and integrity remains the same in electronically as in manually recorded data, irrespective of how they were recorded, and that reconstruction of the way in which the final results and conclusions were obtained remains fully possible (Seiler, 2005). The Study Director must sign and date the final report to indicate acceptance of responsibility for the validity of all the data. (OECD, 1998).

11. Storage and retention of records and materials

Storage and retention of records and materials should be prepared appropriately. The following should be retained in the archives for the period specified by the appropriate authorities : the study plan, raw data, samples of test and reference items, specimens, and the final report of each study records of all inspections performed by the Quality Assurance Programme, as well as master schedules, records of qualifications, training, experience and job descriptions of personnel; records and reports of the maintenance and calibration of apparatus; validation documentation for computerised systems. In the absence of a necessitated retention period, the final arrangement of any study materials should be documented.The necessary documents for GLP regulations are given in Table 2.

GLP Regulations (Rules)	Documentation (Tools)
Organization and Personnel	Training records, CVs, GLP training
Facilities	Maintain adequate space/separation of chemicals from office areas
Equipment	Calibration, logbooks of use, repair, and maintenance; check freezers
Facility Operation	Standard operating procedures
Test, Control and Reference Substances	Chemical and sample inventory, track expiration dates, labeling
Records and Reports	Timely reporting, storage of raw data and reports

(Cobb., 2007).

Table 2. Documentation for GLP rules.

11.1 When samples of test and reference items and specimens are disposed of before the expiry of the necessitated retention period for any reason, this should be justified and documented. Material preserved in the archives should be indexed so as to facilitate storage and retrieval in a tidy way. Safe storage should be provided for all of the samples, test materials and the reports produced. Figure 4 shows the storage of test material

11.1.1 Only personnel authorised by management should have access to the archives. Movement of material in and out of the archives should be recorded appropriately.

11.1.2 Documentation should not be accepted only written documents but also the material generally related to the test facility. Quality Assurance is obliged to retain the respective records in a special archive. Therefore, management is responsible for providing archive facilities for the safe storage and recovery of study plans, raw data, final reports, samples of test items and specimens. Storage should be safe , therefore the design of, and environmental conditions in, these facilities should protect the archived material from ill-timed deterioration. Although it may be enough to archive paper raw data, study plans and final reports to support the necessary space under dry conditions, protected from fire, water and corrosive gases, more stringent conditions will be essential for the storage of tissue specimens from toxicology studies. Samples of the test and reference items should to be stored under the original conditions which were applied during the testing phase. Reconstruction of a study could only be possible if all documents, records and materials from this study can be made available in an unadulterated and unspoiled condition. Traceability in GLP means that there has to be a perfect nonstop line of evidence, chaining together the test item with the effects displayed by the test systems. GLP aims to minimise mistakes or mix-ups through extensive and specific labelling requirements. Documented information can be provided evidencing the application of the correct item in the stated amounts to the relevant test system.

11.1.3 The storage of records must enable their safekeeping for long periods of time without loss or deterioration. In order to encourage safe storage of data, restricted access is used to archive facilities and record the documents logged in and out to a limited number of staff. (Seiler, 2005).

11.1.4 During the conduct of multi-site studies, the temporary storage of materials should be carefully made. Such storage facilities should be safe enough and protect the integrity of their contents. When test site storage facilities are not adequate to satisfy GLP requirements, records and materials should be transferred to a GLP compliant archive. Test site management should ensure that adequate records are available to demonstrate test site involvement in the study. OECD, 1998).

Fig. 4. Storage of the test material in an organized order.

12. Summary

GLP regulations are summarized in Table 3.

GLP	Describes good practices for non-clinical lab studies that support research or marketing approvals for FDA-regulated products
GLP General Requirements	Appropriately qualified personnel Adequate resources Appropriate procedures for: -Sanitation, health precautions, clothing -Test protocol development, test methods -Data analysis, report development Appropriately qualified study director Quality assurance function
GLP Facilities Requirements	Suitable size, construction, segregation -Animal care -Animal supplies -Test & control products maintained in a secure area -Operating "suite" -Specimen & data storage
Equipment Requirements	Appropriately designed Adequate thru-put capacity Appropriately located Routinely maintained & calibrated
Standard Operating Procedures	Animal room prep Animal care Receipt, ID, storage, handling, mixing & sampling of test & control articles Test system observations Lab tests Handling of moribund or dead animals Necropsy or postmortem exams of animals Collection & ID of specimens Histopathology Data handling, storage & retrieval Equipment maintenance & calibration Transfer, proper placement & ID of animals

Reagents & Solutions	Adequate labeling -Identity -Concentration -Storage requirements -Expiration date
Test & Control Articles	Adequate characterization Proper receipt, storage, distribution When mixed with a carrier, adequate methods to confirm -Mixture uniformity -Article concentration -Article stability
Study Implementation	Written, approved protocol indicating test objectives & methods Study conducted in accordance with protocol Study monitoring to confirm protocol compliance Appropriate labeling of specimens by test system, study, nature & collection date Records of gross findings from postmortems available to pathologist for specimen histopathology
Standard data capture/recording requirements	-Legibility -Permanence -Accountability -Changes
Final report of results Study records & data methodically archived to facilitate expedient retrieval	-Study documents -Raw data -Specimens -Protocols -QA inspections -Personnel training & qualifications -Calibration & maintenance records
Records retention (shortest of):	-≥ 2 yr after FDA marketing clearance -≥ 5 yr after data submitted to FDA in support of marketing application -≥ 2 yr after Sponsor decision not to proceed with marketing application -Wet specimens hold as long as viable Records transferable with written FDA notification
Facility Disqualification	Grounds for disqualification: -Failure to comply with regulations & -Noncompliance adversely affects study validity & -Previous regulatory actions have been unsuccessful in modifying facility operations

(Clasby, 2005)

Table 3. GLP regulations.

12.1 "Good laboratory practice" can be considered as " essentially tidiness, cleanliness, hygiene and common sense." (CWIS, 2000).
12.1.1 Quality combination with the GLP rules will be the way that the laboratories will tend to select more in the next years. This will be the leading way to the evidence based laboratory results with a more trustworthy approach.

13. References

Clasby Ginger (2005). Good Laboratory Practice CFR 21 Part 58. A Review for OCRA US RAC Study Group September 2005. Available at : http://www.google.com.tr/search?hl=tr&source=hp&q=A+Review+for+OCRA+US+RAC+Study+Group+September+2005+++&rlz=1W1ADFA_tr&aq=f&aqi=&aql=&oq

Cobb Jean (2007). GLP: Good Laboratory Practice for Field and Research. ALS 52 04 Available at: http://www.docstoc.com/docs/18191459/Good-Laboratory-Practices

CWIS (2000), L1 - Good Laboratory Practice, Liverpool John Moores University, Campus Wide Information Service, Health and Safety Unit, Laboratory and Workshop Safety Notes Dalton, R (1999), Roche's Taq patent "obtained by deceit", rules US court, Nature (news) 402, 16 December, p. 709. Available at: (www.cwis.livjm.ac.uk/hse)

Fox Arlene (2011). GLP Regulations vs. ISO 17025 Requirements: How do they differ? In Accreditation and Quality Assurance: Journal for Quality, Comparability, and Reliability in Chemical measurement. Volume 1/1996-volume 16/2011. DOI: 10.1007/s00769-011-0759-0. Available at: https://springerlink3.metapress.com/content/mr20ux0343141g4k/resource-secured/?target=fulltext.pdf&sid=sbx4al45ojtfu3vvjzteu045&sh=www.springerlink.com

Gladney Lori, Osakwe Izabella, Ford Endia (2009). Good Laboratory Practices. Available at: http://science.kennesaw.edu/~jhendrix/regs/GLP.ppt

OECD.(1998) OECD series on Principles of Good Laboratory Practice and Compliance Monitoring. Available at: http://www.oecd.org/officialdocuments/displaydocumentpdf/?cote=env/mc/chem(98)17&doclanguage=en

Seiler Jürg P (2005) Good Laboratory Practice. The why and the how. ISBN 3-540-25348-3, Springer-Verlag Berlin Heidelberg, Printed in the European Union.

Part 2

Evaluating Analytical Data

4

Partitioning Error Sources for Quality Control and Comparability Analysis in Biological Monitoring and Assessment

James B. Stribling
Tetra Tech, Inc., Center for Ecological Sciences, Owings Mills, Maryland USA

"...measurements are not passive accountings of an objective world but active interactions in which the thing measured and the way it is measured contribute inseparably to the outcome." (Lindley 2007: p. 154, attributing the concept to Neils Bohr)

"The experienced scientist has to learn to anticipate the possible sources of systematic error..." (Taylor 1997: p. 97)

"No simple theory tells us what to do about systematic errors. In fact, the only theory of systematic errors is that they must be identified and reduced...." (Taylor 1997: p. 106)

"...the only reason to carry out a test is to improve a process, to improve the quality..." (Deming 1986: p. i)

1. Introduction

Rationally, as scientists, we recognize that documented standard procedures constitute the first requirement for developing consistency within and among datasets; the second step is putting the procedures into practice. If the procedures were implemented as perfectly as they are written, there would be no need to question data. However, we are also cognizant of the fact that humans (a group of organisms to which we cannot deny holding membership) are called upon to use the procedures, and the consistency and rigor with which the procedures are applied are directly affected by an individual's skill, training, attention span, energy, and focus (Edwards, 2004). In fact, we fully expect inconsistency due to human foibles, and often substantial portions of careers are spent in efforts to recognize, isolate, correct, and minimize future occurrences of, error.

Many public and private organizations in the United States (US) and other countries collect aquatic biological data using a variety of sampling and analysis methods (Gurtz & Muir, 1994; ITFM, 1995a; Carter & Resh, 2001), often for meeting regulatory requirements, for example, by the United States' Clean Water Act (CWA) of 1972 (USGPO, 1989). While the information collected by an individual organization is usually directly applicable to a specific question or site-specific issue, the capacity for using it more broadly for comprehensive assessment has been problematic due to unknown data quality produced by different methods or programs (ITFM, 1995a; Diamond et al., 1996; NWQMC, 2001; The Heinz Center, 2002; GAO, 2004). If the occurrence and magnitude of error in datasets is unknown, a supportable conclusion based solely (or even in part) on those data is problematic at best. These datasets are more difficult to justify for analyses, communicate to broader audiences, base policy decisions on, and defend against potential misuse (Costanza et al., 1992; Edwards, 2004). To ensure the measurement system produces data that can be defended requires understanding the potential error sources that can affect variability of the data and approaches for monitoring the magnitude of error expression.

The purpose of this chapter is to communicate the concept of biological monitoring and assessment as a series of methods, each of which produces data and are as subject to error as any other measurement system. It will describe specific QC techniques and analyses that can be used to monitor variability (i.e., error), identify causes, and develop corrective actions to reduce or otherwise control error rates within acceptable limits. The chapter concludes by demonstrating that comparability analysis for biological data and assessment results is a two-step process, including 1) characterizing data quality, or the magnitude of error rates, associated with each method or dataset, and 2) determining acceptability. It should also be recognized that specific methods are not recommended in the chapter, but rather, emphasis is given that whatever methods are used, data quality and performance should be quantified. Additionally, special emphasis is given to biological monitoring where benthic macroinvertebrate sampling provides the primary data, but conceptually, this approach to QC is also applicable to other organism groups.

2. Quality control

Quality control (QC) is a process by which tests are designed and performed to document the existence and causes of error (=variability) in data, as well as helping determine what can be done to minimize or eliminate them, and developing, communicating, and monitoring corrective actions (CA). Further, it should also be possible to implement the QC process (Figure 1) in a routine manner such that, when those causes are not present, the cost of searching for them does not exceed budgetary constraints (Shewhart, 1939).

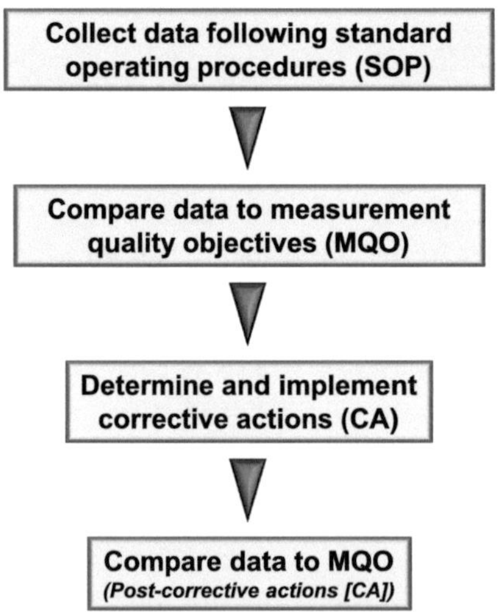

Fig. 1. Quality control (QC) process for determining the presence of and managing error rates, and thus, the acceptability of data quality.

The programmatic system that contains not only a series of QC tests and analyses, but also provides for organization and management of personnel, acquisition and maintenance of equipment and supplies essential to data collection, information management, information technology resources, safety protocols and facilities, enforcement of corrective actions, and budgetary support, is quality assurance (QA). It is acceptable to use the two terms jointly in reference to an overall quality program, as they often are, as QA/QC, but they should not be used interchangeably. The overall program is QA; the process for identifying and reducing error is QC.

Overall variability of data (= total uncertainty, or error) from any measurement system results from accumulation of error from multiple sources (Taylor 1988; Taylor & Kuyatt, 1994; Diamond et al., 1996; Taylor, 1997). Error can generally be divided into two types: systematic and random. Systematic error is the type of variability that results from a method and its application or mis-application; it is composed of bias that can, in part, be mediated by using an appropriate quality assurance program of training, audits, and documentation. Random error results from the sample itself or the population from which it is derived, and can only partly be controlled through a careful sampling design. It is often not possible to separate the effects of the two types of error, and they can directly influence each other (Taylor, 1988). The overall magnitude of error associated with a dataset is known as data quality; how statements of data quality are made and communicated are critical for data users and decision makers to properly evaluate the extent to which they should rely on technical, scientific, information (Keith, 1988; Peters, 1988; Costanza et al., 1992). Thus, an effective set of QC procedures helps not only reduce error in datasets, it provides tools for objective communication of uncertainty.

Biological assessment protocols are measurement systems consisting of a series of methods, each of which contribute to overall variability (Diamond et al., 1996; Cao et al., 2003; Brunialti et al., 2004; Flotemersch et al., 2006; Haase et al., 2006; Nichols et al., 2006; Blocksom & Flotemersch, 2008) (Figure 2). Our capacity as practitioners to control rates and magnitudes of error requires some attention be given to each component of the protocol. While it could be argued that error arising from any single component has only trivial effects on the overall indicator, lack of testing and documentation can substantially weaken that assertion, and opens the results to question. In fact, information without associated data quality characteristics might not even be considered data.

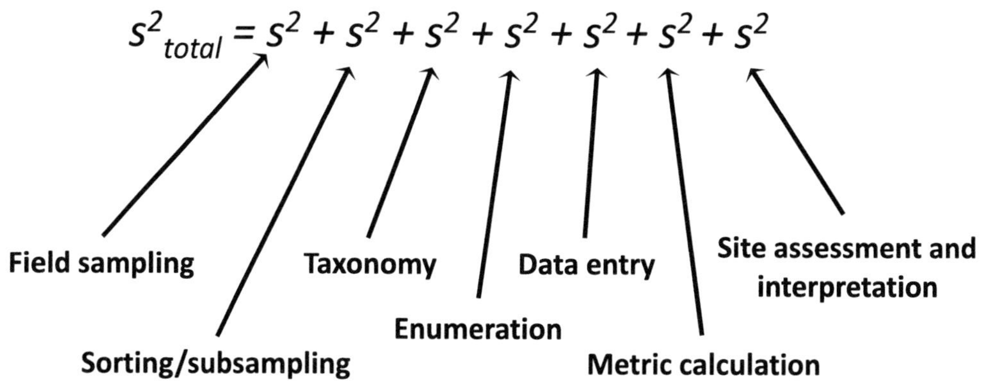

Fig. 2. Total error or variability (s^2) associated with a biological assessment is a combined result of that for each component of the process" (Flotemersch et al. 2006).

3. Indicators

All aquatic ecosystems are susceptible to cumulative impacts from human-induced disturbances including inorganic and organic chemical pollution, hydrologic alteration, channelization, overharvest, invasive species, and land cover conversion. Because they live in the presence of existing water chemistry and physical habitat conditions, the aquatic life of these systems (fish, insects, plants, shellfish, amphibians, reptiles, etc.) integrates cumulative effects of multiple stressors that are produced by both point and non-point source (NPS) pollution. The most common organism groups that are used by routine biological monitoring and assessment programs are benthic macroinvertebrates (aquatic insects, snails, mollusks, crustaceans, worms, and mites), fish, and/or algae, with indicators most often taking the form of a multimetric Index of Biological Integrity (IBI; Karr et al., 1986; Hughes et al., 1998; Barbour et al., 1999; Hill et al., 2000, 2003) or a predictive observed/expected (O/E) model based on the River Invertebrate Prediction and Classification System (RIVPACS; Clarke et al., 1996, 2003; Hawkins et al., 2000; Hawkins, 2006). Of these latter three groups, benthic macroinvertebrates (BM) are commonly used because the protocols are most well-established, the level of effort required for field sampling is reasonable (Barbour et al., 1999), and taxonomic expertise is relatively easily accessible. Thus, examples of QC tests and corrective actions discussed in this chapter are largely focused on benthic macroinvertebrates in the context of multimetric indexes, though, similar procedures for routine monitoring with algae and fish could be developed. Stribling et al. (2008) also used some of these procedures for documenting performance of O/E models.

4. Potential error sources in indicators

4.1 Field sampling

Whether the target assemblage is benthic macroinvertebrates, fish, or algae, the first step of biological assessment is to use standard field methods to gather a sample representing the taxonomic diversity and functional composition of a reach, zone, or other stratum of a waterbody. The actual dimensions of the sampling area ultimately depend on technical objectives and programmatic goals of the monitoring activity (Flotemersch et al., 2010). The spatial area from which the biological sample is drawn is that segment or portion of the waterbody the sample is intended to represent; for analyses and higher level interpretation, biological indicators are considered equivalent to the site. For its national surveys of lotic waters (streams and rivers), the U. S. Environmental Protection Agency defines a sample reach as 40x the mean wetted width (USEPA, 2004a); many individual states use a fixed 100m as the sampling reach.

Benthic macroinvertebrate samples are collected along 11 transects evenly distributed throughout the reach length, and a D-frame net with 500-μm mesh openings used to sample multiple habitats (Klemm et al., 1998; USEPA, 2004a; Flotemersch et al., 2006). An alternative approach to transects is to estimate the proportion of different habitat types in a defined reach (e.g., 100m), and distribute a fixed level of sampling effort in proportion to their frequency of occurrence throughout the reach (Barbour et al., 1999, 2006). For both approaches, organic and inorganic sample material (leaf litter, small woody twigs, silt, and sand) are composited in one or more containers, preserved with 95% denatured ethanol, and delivered to laboratories for processing. A composite sample over multiple habitats in a reach is a common protocol feature of many monitoring program throughout the US (Carter & Resh, 2001).

4.2 Laboratory processing

Processing of benthic macroinvertebrate samples is a 3-step process. Sorting and subsampling serves to 1) isolate individual organisms from nontarget material, such as leaf litter and other detritus, bits of woody material, silt, and sand, and 2) prepare the sample (or subsample) for taxonomic identification. Taxonomic identification serves to match nomenclature to specimens in the sample, and enumeration provides the actual counts, by taxon, of everything contained within the sample.

Although it is widely recognized that subsampling helps to manage the level of effort associated with bioassessment laboratory work (Carter & Resh, 2001), the practice has been the subject of much debate (Courtemanch, 1996; Barbour & Gerritsen, 1996; Vinson & Hawkins, 1996). Fixed organism counts vary among monitoring programs (Carter & Resh, 2001), with 100, 200, 300 and 500 counts being most often used (Barbour et al., 1999; Cao & Hawkins, 2005; Flotemersch et al., 2006). Flotemersch & Blocksom (2005) concluded that a 500-organism count was most appropriate for large/nonwadeable river systems, based on examination of the relative increase in richness metric values (< 2%) between successive 100-organism counts. However, they also suggested that 300-organism count is sufficient for most study needs. Others have recommended higher fixed counts, including a minimum of 600 in wadeable streams (Cao & Hawkins, 2005). The subsample count used for the USEPA national surveys is 500 organisms (USEPA, 2004b); many states use 200 or 300 counts.

If organisms are missed during the sorting process, bias is introduced in the resulting data. Thus, the primary goal of sorting is to completely separate organisms from organic and inorganic material (e.g., detritus, sediment) in the sample. A secondary goal of sorting is to provide the taxonomist with a sample for which the majority of specimens are identifiable. Note that the procedure described here assumes that the sorter and the taxonomist are different personnel. Although it is not the decision of the sorter whether an organism is identifiable, straightforward rules can be applied that minimize specimen loss. For example, "counting rules" can be part of the standard operating procedures (SOP) for both the sorting/subsampling and taxonomic identification, such as specifying what not to count:

- Non-benthic organisms, such as free-swimming gyrinid adults (Coleoptera) or surface-dwelling veliids (Heteroptera)
- Empty mollusk shells (Mollusca: Bivalvia and Gastropoda)
- Non-headed worm fragments
- Damaged insects and crustaceans that lack at least a head and thorax
- Incidental collections, such as terrestrial insects or aquatic vertebrates (fish, frogs or tadpoles, snakes, or other)
- Non-macroinvertebrates, such as copepods, cladocera, and ostracods
- Exuviae (molted "skins")
- Larvae or pupae where internal tissue has broken down to point of floppiness

If a sorter is uncertain about whether an organism is countable, the specimen should be placed in the vial and not added to the rough count total.

The sorting/subsampling process is based on randomly selecting portions of the sample detritus spread over a gridded Caton screen (Caton, 1991; Barbour et al., 1999; see also Figures 6-4a, b of Flotemersch et al., 2006 [note that an individual grid square is 6 cm x 6 cm, or 36 cm^2, *not* 6 cm^2 as indicated in Figure 6-4b]). Prior to beginning the sorting/subsampling process, it is important that the sample be mixed thoroughly and distributed evenly across the sorting tray to reduce the effect of organism clumping that may have occurred in the sample container. The grids are randomly selected, individually removed from the screen, placed in a sorting tray, and all organisms removed with forceps;

the process is completed until the rough count by the sorter exceeds the target subsample size. There should be at least three containers produced per sample, all of which should be clearly labeled: 1) subsample to be given to taxonomist, 2) sort residue to be checked for missed specimens, and 3) unsorted sample remains to be used for additional sorting, if necessary.

The next step of the laboratory process is identifying the organisms within the subsample. A major question associated with taxonomy for biological assessments is the hierarchical target levels required of the taxonomist, including order, family, genus, species or the lowest practical taxonomic level (LPTL). While family level is used effectively in some monitoring programs (Carter & Resh 2001), the taxonomic level primarily used in most routine monitoring programs is genus. However, even with genus as the target, many programs often treat selected groups differently, such as midges (Chironomidae) and worms (Oligochaeta), due to the need for slide-mounting. Slide-mounting specimens in these two groups is usually (though, not always) necessary to attain genus level nomenclature, and sometimes even tribal level for midges. Because taxonomy is a major potential source of error in any kind of biological monitoring data sets (Stribling et al., 2003, 2008a; Milberg et al., 2008; Bortolus, 2008), it is critical to define taxonomic expectations and to treat all samples consistently, both by a single taxonomist and among multiple taxonomists. This, in part, requires specifying both hierarchical targets and counting rules. An example list of taxonomic target levels is shown in Table 1. These target levels define the level of effort that should be applied to each specimen. If it is not possible to attain these levels for certain specimens due to, for example, the presence of early instars, damage, or poor slide mounts, the taxonomist provides a more coarse-level identification. When a taxonomist receives samples for identification, depending upon the rigor of the sorting process (see above), the samples may contain specimens that either cannot be identified, or non-target taxa that should not be included in the sample. The final screen of sample integrity is the responsibility of the taxonomist, who determines which specimens should remain unrecorded (for any of the reasons stated above). Beyond this, the principal responsibility of the taxonomist is to record and report the taxa in the sample and the number of individuals of each taxon. Programs should use the most current and accepted keys and nomenclature. *An Introduction to the Aquatic Insects of North America* (Merritt et al., 2008) is useful for identifying the majority of aquatic insects in North America to genus level. By their very nature, most taxonomic keys are obsolete soon after publication; however, research taxonomists do not discontinue research once keys are available. Thus, it is often necessary to have access to and be familiar with ongoing research in different taxonomic groups. Other keys are also necessary for non-insect benthic macroinvertebrates that will be encountered, such as Oligochaeta, Mollusca, Acari, Crustacea, Platyhelminthes, and others. Klemm et al. (1990) and Merritt et al. (2008) provide an exhaustive list of taxonomic literature for all major groups of freshwater benthic macroinvertebrates. Although it is not current for all taxa, the integrated taxonomic information system (ITIS; http://www.itis.usda.gov/) has served as a clearinghouse for accepted nomenclature, including validity, authorship and spelling.

4.3 Data entry

Taxonomic nomenclature and counts are usually entered into the data management system directly from handwritten bench or field sheets. Depending on the system used, there may be an autocomplete function that helps prevent misspellings, but which can also contribute to errors. For example, entering the letters 'hydro' could potentially autocomplete as either

Hydropsyche or *Hydrophilus*, and the data entry technician on autopilot might continue as normal. There are also, increasingly, uses of e-tablets for entering field observation data, or direct entry of laboratory data into spreadsheets, obviating the need for hardcopy paper backup.

4.4 Data reduction/indicator calculation

There is a large number of potential metrics that monitoring programs can use (Barbour et al., 1999; Blocksom & Flotemersch, 2005; Flotemersch et al., 2006), requiring testing, calibration, and final selection before being appropriate for routine application. Blocksom & Flotemersch (2005) tested 42 metrics relative to different sampling methods, mesh sizes, and habitat types, some of which are based on taxonomic information, as well as stressor tolerance, functional feeding group, and habit. Other workers and programs have tested more and different ones. For example, the US state of Montana calibrated a biological indicator for wadeable streams of the "mountains" site class (Montana DEQ 2006), resulting in a multimetric index comprised of seven metrics (Table 2).

Taxon	Target
Ceratopogonidae	Ceratopogoninae, leave at subfamily; all others, genus level
Dolichopodidae	(Dolichopodidae)
Phoridae	(Phoridae)
Scathophagidae	(Scathophagidae)
Syrphidae	(Syrphidae)
Decapoda	Family
Hirudinea	Family
Hydrobiidae	(Hydrobiidae)
Nematoda	(Nematoda)
Nematomorpha	(Nematomorpha)
Nemertea	(Nemertea)
Turbellaria	(Turbellaria)
Chironomidae, the following genera are combined under *Cricotopus/Orthocladius*	
Cricotopus	*Cricotopus/Orthocladius*
Orthocladius	*Cricotopus/Orthocladius*
Cricotopus/Orthocladius	*Cricotopus/Orthocladius*
Orthocladius/Cricotopus	*Cricotopus/Orthocladius*
Chironomidae, the following genera are combined under *Thienemannimyia genus group*	
Conchapelopia	*Thienemannimyia* genus group
Rheopelopia	*Thienemannimyia* genus group
Helopelopia	*Thienemannimyia* genus group
Telopelopia	*Thienemannimyia* genus group
Meropelopia	*Thienemannimyia* genus group
Hayesomia	*Thienemannimyia* genus group
Thienemannimyia	*Thienemannimyia* genus group
Hydropsychidae, the following genera are combined under *Hydropsyche*	
Hydropsyche	*Hydropsyche*
Ceratopsyche	*Hydropsyche*
Hydropsyche/Ceratopsyche	*Hydropsyche*
Ceratopsyche/Hydropsyche	*Hydropsyche*

Table 1. In this example list of hierarchical target levels, all taxa are targeted for identification to genus level, unless otherwise noted. Taxa with target levels in parentheses are left at that level.

This discussion assumes that the indicator terms have already been calibrated and selected, and deals specifically with their calculation. For this purpose, the raw data are taxa lists and counts; their conversion into metrics is data reduction usually performed with computer spreadsheets or in relational databases.

To ensure that database queries are correct and result in the intended metric values, a subset of values should be recalculated by hand. One metric is calculated for all samples, all metrics are calculated for one sample. When recalculated values differ from those values in the matrix, the reasons for the disagreement are determined and corrections are made. Reports on performance include the total number of reduced values as a percentage of the total, how many errors were found in the queries, and the corrective actions specifically documented.

4.5 Indicator reporting

Regardless of whether the indicator is based on a multimetric framework or multivariate predictive model, the ultimate goal is to translate the quantitative, numeric result, the score, into some kind of narrative that provides the capacity for broad communication. The final assessment for a site is usually determined based on a site score relative to the distribution of reference site scores to reflect degrees of biological degradation, the more similar a test site is to reference less degradation is being exhibited. Depending on the calibration process and how many condition categories are structured, narratives for individual sites can come from two categories (degraded, nondegraded), three (good, fair, poor), four (good, fair, poor, very poor), or five (very good, good, fair, poor, or very poor). There also may be other frameworks a program chooses to use, but the key is to have the individual categories quantitatively-defined.

Metric	Description
Number of Ephemeroptera taxa	Count of the number of distinct taxa of mayflies in sample
Number of Plecoptera taxa	Count of the number of distinct taxa of stoneflies in sample
% individuals as EPT	Percent of individuals in sample that is mayflies, stoneflies, or caddisflies (Ephemeroptera, Plecoptera, or Trichoptera, respectively)
% individuals as non-insects	Percent of individuals in sample as non-insects
% individuals as predators	Percent of individuals in sample as predators
% of taxa as burrowers	Percent of taxa in sample as burrower habit
Hilsenhoff Biotic Index	Abundance-weighted mean of stressor tolerance values for taxa in the sample

Table 2. Sample-based metrics calculated for benthic macroinvertebrates. Shown are those developed and calibrated for streams in the "mountains" site class of the state of Montana, USA (Montana DEQ 2006, Stribling et al. 2008b).

5. Measurement quality objectives (MQO)

For each step of the biological assessment process there are different performance characteristics that can be documented, some of which are quantitative and others that are qualitative (Table 3). Measurement quality objectives (MQO) are control points above (or below) which most observed values fall (Diamond et al., 2006; Flotemersch et al., 2006; Stribling et al., 2003, 2008a, b; Herbst & Silldorf, 2006), and are roughly analogous to the Shewhart (1939) concept of process control.

Component method or activity	Performance characteristics				
	Precision	**Accuracy**	**Bias**	**Representativeness**	**Completeness**
1. Field sampling	•	na	Δ	Δ	•
2. Laboratory sorting/subsampling	•	na	•	Δ	•
3. Taxonomy	•	•	na	na	•
4. Enumeration	•	•	Δ	na	•
5. Data entry	na	•	na	na	•
6. Data reduction (e. g., metric calculation)	na	•	Δ	na	na
7. Site assessment and interpretation	•	•	Δ	Δ	•

Table 3. Error partitioning framework for biological assessments and biological assessment protocols for benthic macroinvertebrates. There may be additional activities and performance characteristics, and they may be quantitative (•), qualitative (Δ) or not applicable (na).

Specific MQO should be selected based on the distribution of values attained, particularly the minima and maxima. Importantly, for environmental monitoring programs, special studies should never be the basis upon which a particular MQO is selected; rather, they should reflect performance expectations when *routine* techniques and monitoring personnel are used. Consider MQO that are established using data from the best field team, or the taxonomist with the most years of experience, or the dissolved oxygen measurements taken using the most expensive field probes. When those people or equipment are no longer available to the program, how useful would the database be to future or secondary users? Defensibility would potentially be diminished. Values that are >MQO are not automatically taken to be unacceptable data points; rather, such values are targeted for closer scrutiny to determine possible reasons for exceedence and might indicate a need for corrective actions (Stribling et al. 2003, Montana DEQ 2006). Simultaneously, they can be used to help quantify performance of the field teams in consistently applying the methods.

5.1 Field sampling

Quantitative performance characteristics for field sampling are *precision* and *completeness* (Table 3). Repeat samples for purposes of calculating precision of field sampling are obtained by sampling two adjacent reaches, shown as 500 m in this example (Figure 3), and can be done by the same field team for intra-team precision, or by different teams for inter-team precision. For benthic macroinvertebrates, samples from the adjacent reaches (also called duplicate or quality control [QC] samples) must be laboratory-processed prior to data being available for precision calculations. Assuming acceptable laboratory error, these precision values are statements of the consistency with which the sampling protocols 1) characterized the biology of the stream or river and 2) were applied by the field team, and thus, reflect a combination of natural variability and systematic error inherent in the dataset.

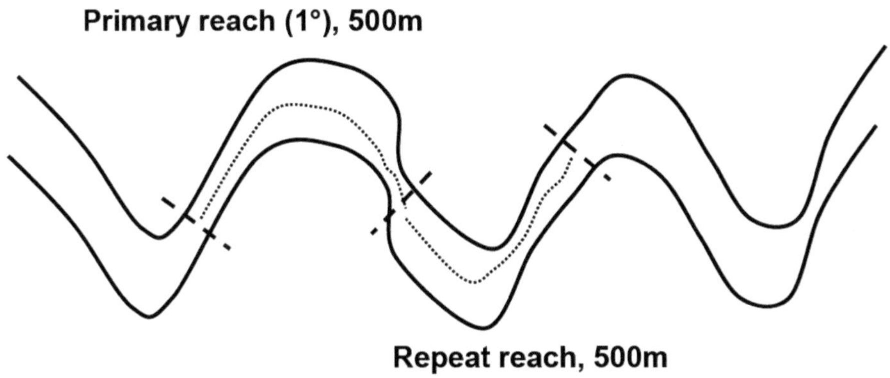

Fig. 3. Adjacent reaches (primary and repeat) for calculating precision estimates (Flotemersch et al. 2006).

The number of reaches for which repeat samples are taken varies, but a rule-of-thumb is 10%, randomly-selected from the total number of sampling reaches constituting a sampling effort (whether yearly, programmatic routine, or individual project). Because they are the ultimate indicators to be used in address the question of ecological conditions, the metric and index values are used to calculate different precision estimates. Root-mean square error (RMSE) (formula 1), coefficient of variability (CV) (formula 2), and confidence intervals (formula 3) (Table 4) are calculated on multiple sample pairs, and are meaningful in that context. Documented values for field sampling precision (Table 5) demonstrate differences among individual metrics and the overall multimetric index (Montana MMI; mountain site class). Relative percent difference (RPD) (formula 4) (Table 4) can have meaning for individual sample pairs. For example, for the composite index, median relative percent difference (RPD) was 8.0 based on 40 sample pairs (Stribling et al., 2008b). MQO recommendations for that routine field sampling for that biological monitoring program were a CV of 10% and a median RPD of 15.0. Sets of sample pairs having with CV>10% would be subjected to additional scrutiny to determine what might be the cause of increased variability. Similarly, individual RPD values for sample pairs would be more specifically examined.

Percent completeness (formula 5) (Table 3, 4) is calculated to communicate the number of valid samples collected as a proportion of those that were originally planned. This value serves as one summary of data quality over the dataset and it demonstrates an aspect of confidence in the overall dataset.

Also called standard error of estimate, **root mean square error (RMSE)** is an estimate of the standard deviation of a population of observations and is calculated by:

$$RMSE = \sqrt{\frac{\sum_{j=1}^{k}\sum_{i=1}^{n_j}(y_{ij} - \bar{y}_j)^2}{\sum df_{1...k}}} \quad (1)$$

where y_{ij} is the ith individual observation in group j, j = 1…k (Zar 1999). Lower values indicate better consistency; and are used in calculation of the **coefficient of variability (CV)**, a unit-less measure, by the formula:

$$CV = \frac{RMSE}{\bar{Y}} \times 100 \quad (2)$$

where $\bar{Y}$ is the mean of the dependent variable (e.g., metric, index across all sample pairs; Zar 1999). It is also known as relative standard deviation (RSD).

Confidence intervals (CI) (or detectable differences) are used to indicate the magnitude of separation of 2 values before the values can be considered different with statistical significance. A 90% significance level for the CI (i.e., the range around the observed value within which the true mean is likely to fall 90% of the time, or a 10% probability of type I error [α]). The 90% confidence interval (CI90) is calculated using RMSE by the formula:

$$CI90 = \pm([RMSE][z\alpha]) \quad (3)$$

where *zα* is the *z*-value for 90% confidence (i.e., p = 0.10) with degrees of freedom set at infinity. In this analysis, *zα* = 1.64 (appendix 17 in Zar 1999). For CI95, the *z*-value would be 1.96. As the number of sample repeats increases, CI becomes narrower; we provide CI that would be associated with 1, 2, and 3 samples per site.

Relative percent difference (RPD) is the proportional difference between 2 measures, and is calculated as:

$$RPD = \left(\frac{|A - B|}{(A + B)/2}\right) x100 \quad (4)$$

where A is the metric or index value of the 1st sample and B is the metric or index value of the 2nd sample (Keith, 1991; APHA, 2005; Smith, 2000). Lower RPD values indicate improved precision (as repeatability) over higher values.

Percent completeness (%C) is a measure of the number of valid samples that were obtained as a proportion of what was planned, and is calculated as:

$$\%C = {}^{v}\!/_{T}\, x100 \quad (5)$$

where *v* is the number of valid samples, and *T* is the total number of planned samples (Flotemersch et al., 2006).

Percent sorting efficiency (PSE) describes how well a sample sorter has done in finding and removing all specimens from isolated sample material, and is calculated as:

$$PSE = \frac{A}{A+B} x100 \quad (6)$$

where A is the number of organisms found by the original sorter, and B is the number of missed organisms recovered (specimen recoveries) by the QC laboratory sort checker.

Percent taxonomic disagreement (PTD) quantifies the sample-based precision of taxonomic identifications by comparing target level taxonomic results from two independent taxonomists, using the formula:

$$PTD = \left[1 - \left(\frac{a}{N}\right)\right] \times 100 \quad (7)$$

where a is the number of agreements, and N is the total number of organisms in the larger of the two counts (Stribling et al., 2003, 2008a).

Percent difference in enumeration (PDE) quantifies the consistency of specimen counts in samples, and is determined by calculating a comparison of results from two independent laboratories or taxonomists using the formula:

$$PDE = \frac{|n_1 - n_2|}{n_1 + n_2} x100 \quad (8)$$

where n_1 is the number of organisms in a sample counted by the first laboratory, and n_2, the second (Stribling et al. 2003).

Percent taxonomic completeness (PTC) describes the proportion of specimens in a sample that meet the target identification level (Stribling et al. 2008) and is calculated as:

$$PTC = {}^{x}\!/_{N} x100 \quad (9)$$

where x is the number of individuals in a sample for which the identification meets the target level, and N is the total number of individuals in the sample.

Discrimination efficiency (DE) is an estimate of the accuracy of multimetric indexes and individual metrics is characterized as their capacity to correctly identify stressor conditions (physical, chemical, hydrologic, and land use/land cover) and is quantified as discrimination efficiency using the formula:

$$DE = {}^{a}\!/_{b} x100 \quad (10)$$

where a is the number of *a priori* stressor sites identified as being below the quantified biological impairment threshold of the reference distribution (25th percentile, 10th, or other), and b is the total number of stressor sites (Flotemersch et al., 2006).

Table 4. Explanations and formulas for quantifying 10 different performance characteristics for different steps of the biological assessment process.

Qualitative performance characteristics for field sampling are *bias* and *representativeness* (Table 3). Programs that use multihabitat sampling, either transect-based similar to that used by the US national surveys (USEPA 2004a), or distributing sampling effort among different habitat types (Barbour et al., 1999, 2006), are attempting to minimize the bias through two components of the field method. First, the approaches are not limited to one or a few habitat types; they are focused on sampling stable undercut banks, macrophyte beds, root wads, snags, gravel, sand, and/or cobble. Second, allocation of the sampling effort is distributed throughout the entire reach, thus preventing the entire sample from being taken in a shortened portion of the reach. Further, if the predominant habitat in a sample reach is poor or degraded, that habitat would be sampled as well. These field sampling methods are intended to depict the benthic macroinvertebrate assemblage that the physical habitat in the streams and rivers has the capacity to support. Another note about representativeness is to be cognizant that, while a method might effectively depict the property it is intended to depict (Flotemersch et al., 2006), it could be interpreted differently at different spatial scales (Figure 4).

Metric	RMSE	Mean	CV	CI90		
				1 sample	2 samples	3 samples
Number of Ephemeroptera taxa	0.94	5.25	17.9	1.55	1.1	0.89
Number of Plecoptera taxa	0.9	2.42	37.3	1.48	1.05	0.85
% individuals as EPT	8.86	47.98	18.5	14.53	10.27	8.39
% individuals as non-insects	3	7.3	41.1	4.93	3.49	2.85
% individuals as predators	5.32	16.91	31.4	8.72	6.17	5.03
% of taxa as burrowers	3.93	28.91	13.6	6.45	4.56	3.72
Hilsenhoff Biotic Index	0.47	4.27	10.9	0.76	0.54	0.44
Multimetric index (7-metric composite)	3.80	55.6	6.8	6.23	4.41	3.60

Table 5. Precision estimates for sample-based benthic macroinvertebrate metrics, and composite multimetric index (Stribling et al., 2008b). Data shown are from the US state of Montana, and performance calculations are based on 40 sample pairs from the "mountain" site class (*abbreviations* - RMSE, root mean square error; CV, coefficient of variation; CI90, 90 percent confidence interval; EPT, Ephemeroptera, Plecoptera, Trichoptera).

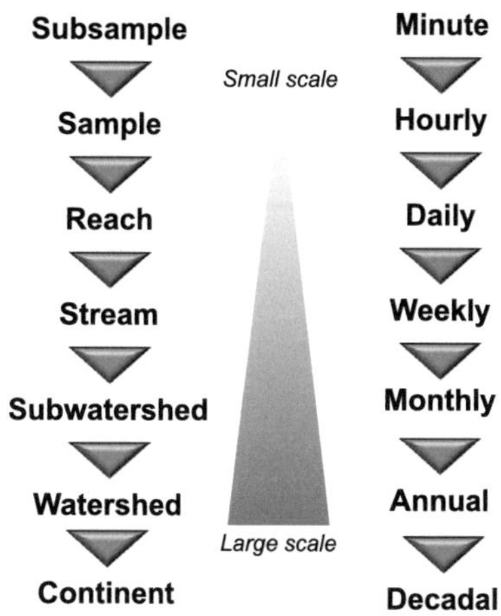

Fig. 4. Defining representativeness of a sample or datum first requires specifying the spatial and/or temporal scale of the feature it is intended to depict.

Accuracy is considered "not applicable" to field sampling (Table 3), because efforts to define analytical truth would necessitate a sampling effort excessive beyond any practicality. That is, the analytical truth would be all benthic macroinvertebrates that exist in the river (shore zone to 1-m depth). There is no sampling approach that will collect all individual benthic macroinvertebrate organisms.

5.2 Sorting/subsampling

Bias, precision, and, in part, *completeness,* are quantitative characteristics of performance for laboratory sorting and subsampling (Table 3). Bias is the most critical performance characteristic of the sorting process, and is evaluated by checking for specimens that may have been overlooked or otherwise missed by the primary sorter (Flotemersch et al., 2006). Checking of the sort residue is performed by an independent sort checker in a separate laboratory using the same procedures as primary, specifically, the same magnification and lighting as called for in the SOP. The number of specimens found by the checker as a proportion of the total number of originally found specimens is the percent sorting efficiency (PSE; formula 6) (Table 4), and quantifies sorting bias. This exercise is performed on a randomly-selected subset of sort residues (generally 10% of total sample lot), the selection of which is stratified by individual sorters, by projects, or by programs. As a rule-of-thumb, an MQO could be "less than 10% of all samples checked will have a PSE ≤90%". Table 6 shows PSE results from sort rechecks for a project within the state of Georgia (US). One sample (no. 8) exhibited a substantial failure with a PSE of 77.8, which became an immediate flag for a potential problem. Further evaluation of the results showed that the sample was fully sorted (100%), and still only 21 specimens were found by the original sorter, prior to the 6 recoveries by the re-check. Values for PSE become skewed when overall numbers are low, thus failure of this one sample did not indicate systematic error (bias) in the sorting process. Three additional samples fell slightly below the 90% MQO, but were only ≤ 0.2 percentage points low and were judged as passing by the QC analyst.

Precision of laboratory sorting is calculated by use of RPD with metrics and indexes as the input variables (Table 4). If, for example, the targeted subsample size is 200 organisms, and that size subsample is drawn twice from a sorting tray without re-mixing or re-spreading, metrics can be calculated from the two separate subsamples. RPD would be an indication of how well the sample was mixed and spread in the tray; the "serial subsampling" and RPD calculations should be done on two timeframes. First, these calculations should be done, and the results documented and reported to demonstrate what the laboratory (or individual sorter) is capable of in application of the subsampling method. Second, they should be done periodically to demonstrate that the program routinely continues to meet that level of precision. Representativeness of the sorting/subsampling process is addressed as part of the SOP that requires random selection of grid squares (Flotemersch et al., 2006) with complete sorting, until the target number is reached within the final grid. Percent completeness for subsampling is calculated as the proportion of samples with the target subsample size (±20%) in the rough sort. Considered as "not applicable", estimates of *accuracy* are not necessary for characterizing sorting performance.

5.3 Taxonomic precision (sample-based)

Precision and *completeness* are quantitative performance characteristics that are used for taxonomy (Table 3). Precision of taxonomic identifications is calculated using percent taxonomic

Sample no.	Number of specimens			PSE
	Original	Recovered	Total	
1	208	5	213	97.7
2	202	8	210	96.2
3	227	1	228	99.6
4	200	12	212	94.3
5	208	7	215	96.7
6	222	2	224	99.1
7	220	24	244	90.2
8	21	6	27	77.8[a]
9	215	22	237	90.7
10	220	25	245	89.8[b]
11	220	3	223	98.7
12	211	24	235	89.8[b]
13	205	12	217	94.5
14	213	24	237	89.9[b]
15	205	11	216	94.9
16	222	15	237	93.7
17	203	10	213	95.3
18	158	16	174	90.8

[a] Low PSE is due to there being small total number of specimens in the sample (n=27); this sample was also whole-pick (all 30 grid squares); [b] PSE values taken as passing, only ≤0.2 percentage points below MQO.

Table 6. Percent sorting efficiency (PSE) as laboratory sorting/ subsample quality control check. Results from 2006-2008 sampling for a routine monitoring program in north Georgia, USA.

disagreement (PTD) and percent difference in enumeration (PDE), both of which rely on the raw data (list of taxa and number of individuals) from whole-sample re-identifications (Stribling et al., 2003, 2008a). These two values are evaluated individually, and are used to indicate the overall quality of the taxonomic data. They can also be used to help identify the source of a problem. Percent taxonomic completeness (PTC) is calculated to document how consistently the taxonomist is able to attain the targeted taxonomic levels as specified in the SOP. It is important to note that the purpose of this evaluation approach is not to say that one taxonomist is correct over the other, but rather to make an effort to understand what is causing differences where they exist. The primary taxonomy is completed by one or more project taxonomists (T1); the re-identifications are completed as blind samples by one or more secondary, or QC taxonomists (T2) in a separate independent laboratory.

The number of samples for which this analysis is performed will vary, but 10% of the total sample lot (project, program, year, or other) is an acceptable rule-of-thumb. Exceptions are that large programs (>~500 samples) may not need to do >50 samples; small programs (<~30 samples) will likely still need to do at least 3 samples. In actuality, the number of re-identified samples will be program-specific and influenced by multiple factors, such as, how many taxonomists are doing the primary identification (there may be an interest in having 10% of the samples from each taxonomist re-identified), and how confident the ultimate data user is with the results. Mean values across all re-identified samples are estimates of taxonomic precision (consistency) for a dataset or a program.

5.3.1 Percent taxonomic disagreement (PTD)

The sample-based error rate for taxonomic identifications is quantified by calculation of percent taxonomic disagreement (PTD) (Table 4, formula 7). The key exercise performed by the QC analyst is determining the number of matches, or shared identifications between the two taxonomists (Table 7). Matches must be exact, that is, negative comparisons result even if the difference is *only* hierarchical (genus vs. family, or other), whether they have been assigned different names, or whether specimens are missing from the overall results of either T1 or T2. *Error typing* individual sample comparisons is the process of determining differences as either: a) straight disagreements, b) hierarchical differences, or c) missing specimens. While tedious, this QC exercise provides information that is extremely valuable in formulating corrective actions. An MQO of 15% has been found to be attainable by most programs, and is used for the USEPA national surveys. As testing continues and laboratories and taxonomists become more accustomed to the procedure, it is becoming apparent that potentially the national standard could eventually be set at 10%. A standard summary report for taxonomic identification QC (Table 8) can be effectively communicated to data users.

5.3.2 Percent difference in enumeration (PDE)

Another summary data quality indicator for performance in taxonomic identification is comparison of the total number of organisms counted and reported in the sample by the two taxonomists (not the sorters). There is some redundancy of this measure with PTD, but it has proven useful in helping highlight coarse differences immediately, and is calculated as percent difference in enumeration (PDE) (Table 4, formula 8). While sorters may be well-trained, experienced, and have substantial internal QC oversight, they may not always be able to determine identifiability, the final decision of which is the responsibility of the taxonomist. It is rare to find exact agreement on sample counts between two taxonomists but the differences are usually minimal, hence the low recommended MQO of 5%. When PDE>5, reasons are usually fairly obvious, and the QC analyst can turn attention directly to the error source to determine if it may be systematic, and the nature and necessity of corrective action(s).

5.3.3 Percent taxonomic completeness (PTC)

Percent taxonomic completeness (PTC) (Table 3, formula 9) quantifies the proportion of individuals in a sample that are identified to the specified target taxonomic level (Table 1). Results can be interpreted in a number of ways: the individuals in a sample are damaged or early instar, many are damaged with diagnostic characters missing (such as, gills, legs, antennae, etc.) or the taxonomist is inexperienced or unfamiliar with the particular taxon. MQO have not been used for this characteristic, but barring an excessively damaged sample, it is not uncommon to see PTC in excess of 97 or 98. For purposes of QC, it is more important to have the absolute difference (abs diff) of PTC between T1 and T2 to be a low number, as documentation of consistency of effort; those values are often typical at 5-6%, or below.

Sample no.	Count		No. matches	PDE	PTD	Target level (taxonomic completeness)				
	T1	T2				T1	PTC	T2	PTC	Abs diff
1	243	244	232	0.2	4.9	234	96.3	223	91.4	4.9
2	227	223	204	0.9	10.1	205	90.3	194	87.0	3.3
3	214	213	191	0.2	10.7	202	94.4	199	93.4	1.0
4	221	223	207	0.5	7.2	212	95.9	208	93.3	2.6
5	216	214	202	0.5	6.5	207	95.8	201	93.9	1.9
6	216	216	214	0	0.9	209	96.8	208	96.3	0.5
7	86	83	69	1.8	19.8	77	89.5	64	77.1	12.4
8	206	201	194	1.2	5.8	204	99	187	93.0	6.0
9	208	210	196	0.5	6.7	203	97.6	195	92.9	4.7
10	192	195	180	0.8	7.7	182	94.8	172	88.2	6.6

Table 7. Summary table for sample by sample taxonomic comparison results, from routine biological monitoring in US state of Mississippi. T1 and T2 are the primary and QC taxonomists, respectively. "No. matches" is the number of individual specimens counted and given the same identity by each taxonomist, and PDE, PTD, and PTC are explained in text. Target level is the number and percentage of specimens identified to the SOP-specified level of effort (see Table 3 as an example); "Abs diff" is the absolute difference between the PTC of T1 and T2.

A. Number of samples in lot	97
B. Number of samples used for taxonomic comparison	10
C. Percent of sample lot	10.3%
D. Percent taxonomic disagreement (PTD)	
1. MQO	15
2. No. samples exceeding	1
3. Average	7.9
4. Standard deviation	4.9
E. Percent difference in enumeration (PDE)	
1. MQO	5
2. No. samples exceeding	0
3. Average	0.6
4. Standard deviation	0.6
F. Percent taxonomic completeness (PTC, absolute difference)	
1. MQO	none
2. Average	4.3
3. Standard deviation	3.5

Table 8. Taxonomic comparison results from a bioassessment project in the US state of Mississippi.

5.4 Taxonomic accuracy (taxon-based)

Accuracy and *bias* (the inverse of accuracy) are quantitative performance characteristics for taxonomy (Table 3). Accuracy requires specification of an analytical truth, and for taxonomy that is 1) the museum-based type specimen (holotype, or other form of type specimen), 2) specimen(s) verified by recognized expert(s) in that particular taxon or 3) unique morphological characteristics specified in dichotomous identification keys. Determination of accuracy is considered "not applicable" for production taxonomy (most often used in routine monitoring programs) because that kind of taxonomy is focused on characterizing the sample; taxonomic accuracy, by definition, would be focused on individual specimens. Bias in taxonomy can result from use of obsolete nomenclature and keys, imperfect understanding of morphological characteristics, inadequate optical equipment, or poor training. Neither of these performance characteristics is considered necessary for production taxonomy, in that they are largely covered by the estimates of precision and completeness. For example, although it is possible that two taxonomists would put an incorrect name on an organism, it is considered low probability that they would put the *same incorrect name* on that organism.

5.5 Data entry accuracy

Recognition and correction of data entry errors (even the one mentioned in Section 4.3) could come from one of two methods for assuring accuracy in data entry; both do not need to be done. One is the double entry of all data by two separate individuals, and then performing a direct match between databases. Where there are differences, it is determined which database is in error, and corrections are made. The second approach is to perform a 100% comparison of all data entered to handwritten data sheets. Comparisons should be performed by someone other than the primary data entry person. When errors are found, they are hand-edited for documentation, and corrections are made electronically. The rates of data entry errors are recorded and segregated by data type (e.g., fish, benthic macroinvertebrates, periphyton, header information, latitude and longitude, physical habitat, and water chemistry). Issues could potentially arise when entering data directly into field e-tablets or laboratory computers. Because there would be no paper backup, QC checks of data entry are not possible.

5.6 Site assessment and interpretation

Quantitative performance characteristics for site assessment and interpretation are *precision*, *accuracy*, and *completeness* (Table 3). Site assessment precision is based on the narrative assessments from the associated index scores (good, fair, poor) from reach duplicates and quantifies the percentage of duplicate samples that are receiving the same narrative assessments. These comparisons are done for a randomly-selected 10% of the total sample lot. Table 9 shows this direct comparison that, for this dataset, 79% of the replicates returned assessments of the same category (23 out of 29); 17% were 1 category different (5 of 29); and 3% were 2 categories different (1 of 29). Assessment accuracy is expressed using discrimination efficiency (DE) (formula 10; Table 4), a value developed during the index calibration process, which relies upon, first, specifying magnitudes of physical, chemical, and/or hydrologic stressors that are unacceptable, and identifying those sites exhibiting those excessive stressor characteristics. The set of sites exhibiting unacceptable stressor levels constitute the analytical truth. The proportion of samples for which the biological index correctly identifies sites as impaired is DE. This is a performance characteristic that is directly suitable for expressing how well an indicator does what it is designed to do, detect stressor conditions, but it is not suitable for routine QC analyses. Percent completeness (%C) is the proportion of sites (of the total planned) for which valid final assessments were obtained.

6. Maintenance of data quality

The purpose of QC is to identify assignable causes of variation (error) so that the quality of the outcomes in future processes can be made, on average, less variable (Shewhart, 1939). For reducing error rates, it is first and foremost critical to know of the existence of error, and second, to know its causes. Once the causes are known, corrective actions can be designed to reduce or eliminate them. The procedures described in this chapter for gathering information that allow performance and data quality characteristics to be documented need to become a routine part of biological monitoring programs. If they are used only when "conditions are right", as part of special studies, or when there are additional resources, they are not serving their purpose and could ultimately be counter-productive. The counter-productivity would arise when monitoring staff begin to view QC samples and analyses as activities that are less than routine, and something for which to strive to do their best, that is, only when they are being tested. This perspective leads programs to work to meet a number, such as 15%, rather than using the information to maintain or improve.

Site	Replicate 1		Replicate 2		Categorical difference
	Narrative	Assessment category	Narrative	Assessment category	
A	Poor	3	Poor	3	0
B	Poor	3	Poor	3	0
C	Good	1	Good	1	0
D	Poor	3	Very Poor	4	1
E	Fair	2	Fair	2	0
F	Poor	3	Fair	2	1
G	Poor	3	Poor	3	0
H	Very Poor	4	Very Poor	4	0
I	Very Poor	4	Very Poor	4	0
J	Poor	3	Poor	3	0
K	Poor	3	Poor	3	0
L	Very Poor	4	Very Poor	4	0
M	Very Poor	4	Very Poor	4	0
N	Poor	3	Fair	2	1
O	Poor	3	Poor	3	0
P	Poor	3	Poor	3	0
Q	Poor	3	Very Poor	4	1
R	Poor	3	Poor	3	0
S	Fair	2	Very Poor	4	2
T	Fair	2	Fair	2	0
U	Good	1	Good	1	0
V	Poor	3	Fair	2	1
W	Fair	2	Fair	2	0
X	Poor	3	Poor	3	0
Y	Poor	3	Poor	3	0
Z	Very Poor	4	Very Poor	4	0
AA	Poor	3	Poor	3	0
BB	Fair	2	Fair	2	0
CC	Poor	1	Poor	1	0

Table 9. Assessment results shown for sample pairs taken from 29 sites, each pair representing two adjacent reaches (back to back (see Fig. 4). Assessment categories are 1 - good, 2 - fair, 3 - poor, and 4 - very poor.

Performance characteristic	MQO
Field sampling precision (multimetric index)	CV < 10%, for a sampling event (field season, watershed, or other strata)
Field sampling precision (multimetric index)	CI90 ≤ 15 index points, on a 100-point scale
Field sampling precision (multimetric index)	RPD < 15
Field sampling completeness	Completeness > 98%
Sorting/subsampling accuracy	PSE≥90, for ≥ 90% of externally QC'd sort residues
Taxonomic precision	Median PTD ≤ 15% for overall sample lot; samples with PTD ≥ 15% examined for patterns of error
Taxonomic precision	Median PDE ≤ 5%; samples with PDE ≥ 5% should be further examined for patterns of error
Taxonomic completeness	Median PTC ≥ 90%; samples with PTC ≤ 90% should be examined and those taxa not meeting targets isolated; mAbs diff ≤ 5%

Table 10. Key measurement quality objectives (MQO) that could be used to track maintenance of data quality at acceptable levels.

Key to maintaining data quality of known and acceptable levels is establishing performance standards based on MQO. Qualitative standards, such as some of the representativeness and accuracy factors (Table 3), can be evaluated by comparing SOP and SOP application to the goals and objectives of the monitoring program. However, a clear statement of data quality expectations, such as that shown in Table 10, will help to ensure consistency of success in implementing the procedures. As a program becomes more proficient and consistent in meeting the standards, efforts could be undertaken to "tighten up" the standards. With this comes necessary budgetary considerations; better precision can always be attained, but often at elevated costs.

7. Comparability analysis and acceptable data quality

All discussion to this point has been directed toward documenting data quality associated with monitoring programs, hopefully with sufficient emphasis that there are no data that are right or wrong, but just that they are acceptable or not. If data are acceptable for a decision (for example, in the context of biological assessment and monitoring), a defensible statement on the ecological condition of a site or an ecological system can be made. If they are not acceptable to support that decision, likewise, the decision not to use the data should also be defensible. Routine documentation and reporting of data quality within a monitoring program provides a statement of intra-programmatic consistency, that is, sample to sample comparability even if collected from different temporal or spatial scales. If there is an interest in or need to combine datasets from different programs (Figure 5), it is imperative for routinely documented performance characteristics be available for each. Lack of them will preclude any determination of acceptability for decision making by data users, whether scientists, policy-makers, or the public.

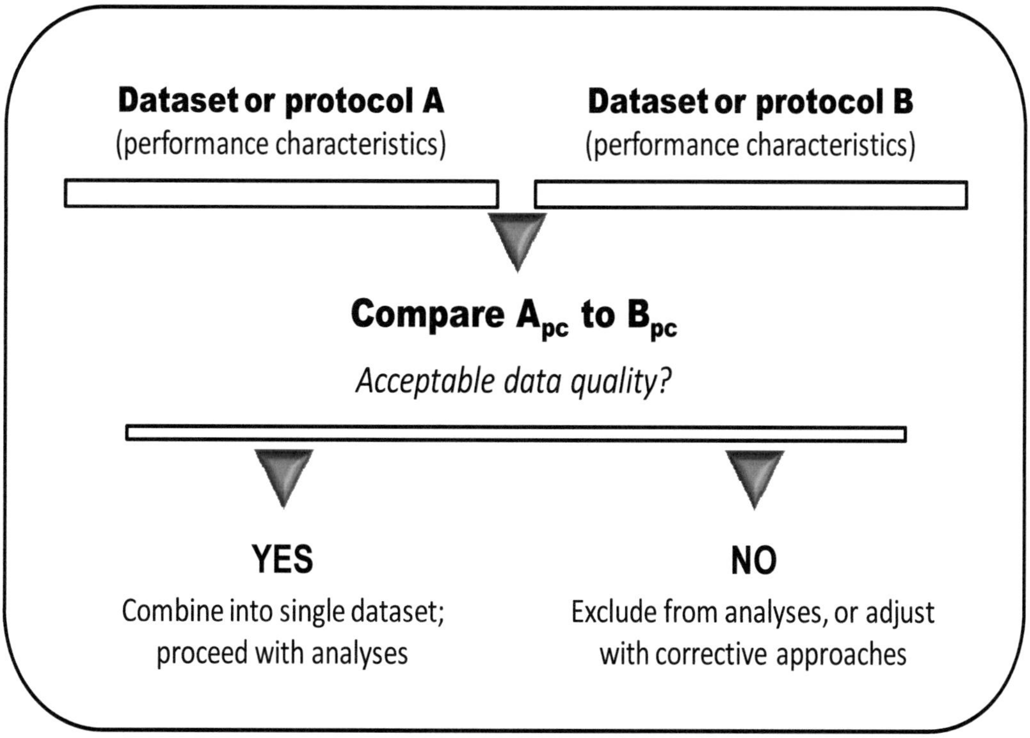

Fig. 5. Framework for analysis of comparability between or among monitoring datasets or protocols.

8. Conclusion

If data of unknown quality are used, whether by themselves or in combination with others, the assumption is implicit that they are acceptable, and hence, comparable. We must acknowledge the risk of incorrect decisions when using such data and be willing to communicate those risks to both data users and other decisionmakers. The primary message of this chapter is that appropriate and sufficient QC activities should be a routine component of any monitoring program, whether it is terrestrial or aquatic, focuses on physical, chemical, and/or biological indicators, and, if biological, whether it includes macroinvertebrates, algae/diatoms, fish, broad-leaf plants, or other organisms groups.

9. References

APHA. 2005. *Standard Methods for the Examination of Water and Wastewater.* 21st edition. American Public Health Association, American Water Works Association, and Water Environment Federation, Washington, DC.

Barbour, M.T., & J. Gerritsen. 1996. Subsampling of benthic samples: a defense of the fixed count method. *Journal of the North American Benthological Society* 15:386-391.

Barbour, M.T., J. Gerritsen, B.D. Snyder, J.B. Stribling. 1999. *Rapid Bioassessment Protocols for Streams and Wadeable Rivers: Periphyton, Benthic Macroinvertebrates and Fish.* Second edition. EPA/841-D-97-002. U.S. EPA, Office of Water, Washington, DC. *URL:* http://water.epa.gov/scitech/monitoring/rsl/bioassessment/index.cfm.

Barbour, M. T., J. B. Stribling, & P.F.M. Verdonschot. 2006. The multihabitat approach of USEPA's rapid bioassessment protocols: Benthic macroinvertebrates. *Limnetica* 25(3-4): 229-240.

Blocksom, K.A., & J.E. Flotemersch. 2005. Comparison of macroinvertebrate sampling methods for non-wadeable streams. *Environmental Monitoring and Assessment* 102:243-262.

Blocksom, K.A., & J.E. Flotemersch. 2008. Field and laboratory performance characteristics of a new protocol for sampling riverine macroinvertebrate assemblages. *River Research and Applications* 24: 373–387. DOI: 10.1002/rra.1073

Bortolus, A. 2008. Error cascades in the biological sciences: the unwanted consequences of using bad taxonomy in ecology. *Ambio* 37(2): 114-118.

Brunialti, G., P. Giordani, & M. Ferretti. 2004. Discriminating between the Good and the Bad: Quality Assurance Is Central in Biomonitoring Studies. Chapter 20, pp. 443-464, IN, G.B. Wiersma (editor), *Environmental Monitoring.* CRC Press.

Cao, Y., & C.P. Hawkins. 2005. Simulating biological impairment to evaluate the accuracy of ecological indicators. *Journal of Applied Ecology* 42:954-965.

Cao, Y., C.P. Hawkins, & M.R. Vinson. 2003. Measuring and controlling data quality in biological assemblage surveys with special reference to stream benthic macroinvertebrates. *Freshwater Biology* 48: 1898–1911.

Carter, J.L. & V.H. Resh. 2001. After site selection and before data analysis: sampling, sorting, and laboratory procedures used in stream benthic macroinvertebrate monitoring programs by USA state agencies. *Journal of the North American Benthological Society* 20: 658-676.

Caton, L. R. 1991. Improved subsampling methods for the EPA rapid bioassessment benthic protocols. *Bulletin of the North American Benthological Society* 8:317-319.

Clarke, R.T., M.T. Furse, J.F. Wright & D. Moss. 1996. Derivation of a biological quality index for river sites: comparison of the observed with the expected fauna. *Journal of Applied Statistics* 23:311–332.

Clarke, R.T., J.F. Wright & M.T. Furse. 2003. RIVPACS models for predicting the expected macroinvertebrate fauna and assessing the ecological quality of rivers. *Ecological Modeling* 160:219–233.

Costanza, R., S.O. Funtowicz & J.R. Ravetz. 1992. Assessing and communicating data quality in policy-relevant research. *Environmental Management* 16(1):121-131.

Courtemanch, D.L. 1996. Commentary on the subsampling procedures used for rapid bioassessments. *Journal of the North American Benthological Society* 15:381-385.

Deming, W.E. 1986. *Foreward.* In, Shewhart, W.A. 1939. *Statistical Methods from the Viewpoint of Quality Control.* The Graduate School, U.S. Department of Agriculture,

Washington, DC. 105 pp. Republished 1986, with a new Foreword by W.E. Deming. Dover Publications, Inc., 31 East 2nd Street, Mineola, NY.

Diamond, J.M., M.T. Barbour & J.B. Stribling. 1996. Characterizing and comparing bioassessment methods and their results: a perspective. *Journal of the North American Benthological Society* 15:713-727.

Edwards, P.N. 2004. "A vast-machine": Standards as social technology. *Science* 304 (7):827-828.

Flotemersch, J.E. & K.A. Blocksom. 2005 Electrofishing in boatable rivers: Does sampling design affect bioassessment *metrics? Environmental Monitoring and Assessment 102:263-283. DOI: 10.1007/s10661-005-6026-2*

Flotemersch J.E., J.B. Stribling, & M.J. Paul. 2006. *Concepts and Approaches for the Bioassessment of Non-Wadeable Streams and Rivers.* EPA/600/R-06/127. U.S. Environmental Protection Agency, Cincinnati, OH.

Flotemersch, J.E., J.B. Stribling, R.M. Hughes, L. Reynolds, M.J. Paul & C. Wolter. 2010. Site length for biological assessment of boatable rivers. *River Research and Applications.* Published online in Wiley InterScience (www.interscience.wiley.com) DOI: 10.1002/rra.1367.

General Accounting Office (GAO). 2004. *Watershed Management: Better Coordination of Data Collection Efforts.* GAO-04-382. Washington, DC , USA. Available from: <http://www.gao.gov/new.items/d04382.pdf>.

Gurtz, M.E. & T.A. Muir (editors). 1994. *Report of the Interagency Biological Methods Workshop.* U.S. Geological Survey, Open File Report 94-490, Reston, Virginia, USA.

Haase, P., J. Murray-Bligh, S. Lohse, S. Pauls, A. Sundermann, R. Gunn & R. Clarke. 2006. Assessing the impact of errors in sorting and identifying macroinvertebrate samples. *Hydrobiologia* 566:505–521. DOI 10.1007/s10750-006-0075-6

Hawkins, C.P. 2006. Quantifying biological integrity by taxonomic completeness: evaluation of a potential indicator for use in regional- and global-scale assessments. *Ecological Applications* 16:1277–1294.

Hawkins, C.P., R.H. Norris, J.N. Hogue & J.W. Feminella. 2000. Development and evaluation of predictive models for measuring the biological integrity of streams. *Ecological Applications* 10:1456–1477.

Heinz Center, The. 2002. *The state of the nation's ecosystems: measuring the lands, waters, and living resources of the United States.* The H. John Heinz III Center for Science, Economics, and the Environment, Washington, DC, USA. Cambridge University Press. Available from: <http://www.heinzctr.org/ecosystems/index.htm>.

Herbst, D.B. & E.L. Silldorf. 2006. Comparison of the performance of different bioassessment methods: similar evaluations of biotic integrity from separate programs and procedures. *Journal of the North American Benthological Society* 25:513–530.

Hill, B.H., A.T. Herlihy, P.R. Kaufmann, S.J. Decelles & M.A. Vander Borgh. 2003. Assessment of streams of the eastern United States using a periphyton index of biotic integrity. *Ecological Indicators* 2:325–338.

Hill, B.H., A.T. Herlihy, P.R. Kaufmann, R.J. Stevenson, F.H. McCormick & C.B. Johnson. 2000. Use of periphyton assemblage data as an index of biotic integrity. *Journal of the North American Benthological Society* 19:50–67.

Hughes, R.M., P.R. Kaufmann, A.T. Herlihy, T.M. Kincaid, L. Reynolds & D.P. Larsen. 1998. A process for developing and evaluating indices of fish assemblage integrity. *Canadian Journal of Fisheries and Aquatic Sciences* 55:1618–1631.

ITFM. 1995a. *The Strategy for Improving Water Quality Monitoring in the U.S.* Intergovernmental Task Force on Monitoring Water Quality. Report #OFR95-742, U.S. Geological Survey, Reston, Virginia, USA.

ITFM. 1995b. Performance-based approach to water quality monitoring. In: *Strategy for Improving Water Quality Monitoring in the U.S.*, Appendix M, Report #OFR95-742, Intergovernmental Task Force on Monitoring Water Quality, U.S. Geological Survey, Reston, Virginia, USA.

Karr, J.R., K.D. Fausch, P.L. Angermeier, P.R. Yant & I.J. Schlosser. 1986. *Assessing Biological Integrity in Running Waters: a Method and its Rationale.* Special publication 5. Illinois Natural History Survey, Champaign, Illinois, USA.

Keith, L.H. (editor). 1988. *Principles of Environmental Sampling.* ACS Professional Reference Book. American Chemical Society, Columbus, Ohio.

Keith, L.H. 1991. *Environmental Sampling and Analysis. A Practical Guide.* Lewis Publishers, Chelsea, Michigan.

Klemm, D.J., P.A. Lewis, F. Fulk & J.M. Lazorchak. 1990. *Macroinvertebrate Field and Laboratory Methods for Evaluating the Biological Integrity of Surface Waters.* EPA/600/4-90/030. Environmental Monitoring Systems Laboratory, U.S. Environmental Protection Agency, Cincinnati, OH. 256 pp.

Klemm, D.J., J.M. Lazorchak & P.A. Lewis. 1998. Benthic macroinvertebrates. Pages 147–182 in J. M. Lazorchak, D. J. Klemm, and D. V. Peck (editors). *Environmental Monitoring and Assessment Program – Surface Waters: Field Operations and Methods for Measuring the Ecological Condition of Wadeable Streams.* EPA/620/R-94/ 004F. U.S. Environmental Protection Agency, Washington, DC.

Lindley, D. 2007. *Uncertainty. Einstein, Heisenberg, Bohr, and the Struggle for the Soul of Science.* Anchor Books, a Division of Random House. ISBN: 978-1-4000-7996-4. New York, NY. 257 pp.

Merritt, R.W., K.W. Cummins & M.B. Berg (editors). 2008. *An Introduction to the Aquatic Insects of North America.* Fourth Edition. Kendall/Hunt Publishing Company, Dubuque, Iowa. ISBN 978-0-7575-5049-2. 1158 pp.

Milberg, P., J. Bergstedt, J. Fridman, G. Odell & L. Westerberg. 2008. Systematic and random variation in vegetation monitoring data. *Journal of Vegetation Science* 19: 633-644. http://dx.doi.org/10.3170/2008-8-18423.

Montana DEQ. 2006. *Sample collection, sorting, and taxonomic identification of benthic macroinvertebrates. Standard operation procedure. WQPBWQM-009. Revision no. 2.* Water Quality Planning Bureau, Montana Department of Environmental Quality, Helena, Montana. (Available from: http://www.deq.mt.gov/wqinfo/QAProgram/ WQPBWQM-009rev2_final_web.pdf)

NWQMC. 2001. *Towards a Definition of Performance-Based Laboratory Methods*. National Water Quality Monitoring Council Technical Report 01 - 02, U.S. Geological Survey, Reston, Virginia, USA.

Nichols, S.J., W.A. Robinson & R.H. Norris. 2006. Sample variability influences on the precision of predictive bioassessment. *Hydrobiologia* 572: 215–233. doi 10.1007/s10750-005-9003-4

Peters, J.A. 1988. Quality control infusion into stationary source sampling. Pages 317–333 in L. H. Keith (editor). *Principles of Environmental Sampling*. ACS Professional Reference Book. American Chemical Society, Columbus, Ohio.

Shewhart, W.A. 1939. *Statistical Methods from the Viewpoint of Quality Control.* The Graduate School, U.S. Department of Agriculture, Washington, DC. 105 pp. Republished 1986, with a new Foreword by W.E. Deming. Dover Publications, Inc., 31 East 2nd Street, Mineola, NY.

Smith, R.-K. 2000. *Interpretation of Organic Data.* ISBN 1-890911-19-4. Genium Publishing Corporation. Genium Group, Inc., Amsterdam, New York.

Stribling, J. B., S.R. Moulton II & G.T. Lester. 2003. Determining the quality of taxonomic data. *Journal of the North American Benthological Society* 22(4): 621-631.

Stribling, J.B., K.L. Pavlik, S.M. Holdsworth & E.W. Leppo. 2008a. Data quality, performance, and uncertainty in taxonomic identification for biological assessments. *Journal of the North American Benthological Society* 27(4): 906-919. doi: 10.1899/07-175.1

Stribling, J.B., B.K. Jessup & D.L. Feldman. 2008b. Precision of benthic macroinvertebrate indicators of stream condition in Montana. *Journal of the North American Benthological Society* 27(1):58-67. doi: 10.1899/07-037R.1

Taylor, J.K. 1988. *Defining the Accuracy, Precision, and Confidence Limits of Sample Data.* Chapter 6, pages 102-107, IN Lawrence H. Keith (editor), *Principles of Environmental Sampling.* ACS Professional Reference Book. ISBN 0-8412-1173-6. American Chemical Society. Columbus, Ohio.

Taylor, J.R. 1997. *An Introduction to Error Analysis. The Study of Uncertainties in Physical Measurements.* Second edition. University Science Books, Sausalito, California, USA.

Taylor, B.N. & C.E. Kuyatt. 1994. *Guidelines for Evaluating and Expressing the Uncertainty of NIST Measurement Results.* NIST Technical Note 1297. National Institute of Standards and Technology, U.S. Department of Commerce, Washington, DC. 24 pp.

USEPA. 2004a. *Wadeable Stream Assessment: Field Operations Manual.* EPA 841-B-04-004. Office of Water and Office of Research and Development, US Environmental Protection Agency, Washington, DC.

USEPA. 2004b. *Wadeable Stream Assessment: Benthic Laboratory Methods.* EPA 841-B-04007. Office of Water and Office of Research and Development, US Environmental Protection Agency, Washington, DC.

U. S. GPO (Government Printing Office). 1989. *Federal Water Pollution Control Act (33 U. S. C. 1251 et seq.)* as amended by a P. L. 92-500. In: Compilation of selected water resources and water pollution control laws. Printed for use of the Committee on Public Works and Transportation. Washington, DC, USA.

Vinson, M.R. & C.P. Hawkins. 1996. Effects of sampling area and subsampling procedure on comparisons of taxa richness among streams. *Journal of the North American Benthological Society* 15:392-399.

Zar, J.H. 1999. *Biostatistical Analysis. 4th edition.* Prentice Hall, Upper Saddle River, New Jersey, USA.

5

Patient Satisfaction with Primary Health Care Services in a Selected District Municipality of the Eastern Cape of South Africa

N. Phaswana-Mafuya[1,2], A. S. Davids[1], I. Senekal[3] and S. Munyaka[3]
[1]*Human Sciences Research Council, Port Elizabeth,*
[2]*Office of the Deputy Vice Chancellor: Research and Engagement, Nelson Mandela Metropolitan University, Port Elizabeth,*
[3]*University of Fort Hare,*
South Africa

1. Introduction

Traditionally, decisions about health services were made on the basis of health-provider and health authorities' views on what is in the best interest of the patient. This was based on a view that members of the general public lack the technical knowledge to make fully informed decisions themselves. Currently, the use of patient satisfaction surveys (PSS) in developing countries is advancing. Professionals have recognized that a systematic and consumer oriented perspective toward patient viewpoints about the level of care can result in feedback useful for promoting higher quality standards of patient care (Dağdeviren & Akturk 2004; Newman et al. 1998; Peltzer 2009).

Patient satisfaction surveys are seen as a means of determining patients' views on primary health care (PHC) (Ajayi, Olumide & Oyediran 2005; Andaleeb 2001; Campbell, Ramsay & Green 2001). These surveys are increasingly being promoted as a means of understanding health care service quality and the demand for these services in developing countries (Glick 2009) for various reasons. First, they highlight those aspects of care that need improvement in a health care setting (Ajayi, Olumide & Oyediran 2005; Muhondwa *et al.* 2008; Newman *et al.* 1998). Second, they are simple, quick and inexpensive to administer. Third, they are critical for developing measures to increase the utilization of PHC services. Fourth, they can help to educate medical staff about their achievements as well as their failures, assisting them to be more responsive to their patients' needs. Fifth, they allow managerial judgment to be exercised from a position of knowledge rather than guesswork in the important task of managing public expectations and resources (Glick 2009).

The South African government also endorses the centrality of consumers in service delivery. The White Paper on Transforming Public Services of 1997 (Department of Public Service and Administration 1997) and the Department of Health's policy on quality in health care (Department of Health 2007) state that public services need to respond to customers' needs, wants and expectations. Feedback from consumers is required in terms of experiences of health services – quality of care received. Feedback from customers will not only improve knowledge of decision makers, but will also facilitate more improved prioritization, improved strategic resource allocation and improved value for money. It will also serve as a platform for providing better services to citizens.

Against this background, a patient satisfaction survey with PHC services was conducted in a selected district of the Eastern Cape.

2. Methods

2.1 Design and setting

A cross-sectional descriptive design was employed to collect data among patients visiting 12 clinics in a selected health district of the Eastern Cape of South Africa in 2009. The majority of the South Africans are dependent on the public health sector, with only 15% of the citizenry belonging to a private medical aid scheme (McIntyre, 2010). In the Eastern Cape, private medical aid covers only 10.9% of the province's population and less than 7% of South Africa's private and independent hospitals, are located in the Eastern Cape (Hospital Association of South Africa, 2011). The current study focused on public health services (the main provider of health care) in a selected district in the Eastern Cape. We did not ask about private health care utilization.

The public health system of the Eastern Cape consists of 817 clinics, 81 hospitals and 18 community health care centres. The core norms, set by the National Department of Health in South Africa, for primary health care services are indicated in Table 1. Statistics South Africa estimated that the mid-year population for Eastern Cape in 2010 was 6743800, about 13,5% of the estimated total population of South Africa. Persons under the age of 15 years constitute 32.8% of the total population and the economically active population (15-64 years) is 61.2% of total population of the Eastern Cape. For the period 2006 to 2011 it is estimated that approximately 211 600 people will migrate from the Eastern Cape to other provinces of the country (Statistics South Africa, 2010). The burden of disease study for the year 2000, estimated that South Africans suffer from poverty-related diseases and conditions, emerging chronic diseases, injuries and HIV/AIDS, and differences in morbidity and mortality between socioeconomic groups (Bradshaw et al., 2003). The Eastern Cape had an estimated 23.6% unemployment rate at the end of the second quarter of 2009. This drives levels of poverty in the province, as those deemed to be living in poverty was 3564504, nearly 53% of the 2010 midyear population. The corresponding figure for South Africa is 38%, making the province one of the poorest in the country (ECSECC, 2011).

2.2 Sample and procedure

A purposive sample of 836 out of 939 patients (89% response rate), visiting 12 primary care facilities in a selected district of the Eastern Cape of South Africa were interviewed while exiting the clinic. Patients aged 18 years or above were considered eligible, provided that they were able to understand and respond to the interview questions. Patients were interviewed face to face by trained interviewers in their preferred language in five consecutive days per clinic. Two fieldworkers and one fieldwork coordinator (with at least a high school certificate) were trained per clinic. Two of the four local fieldworkers conducted the interviews in the respective clinics, while the other two served as a reserve. A clinic nurse supervised and coordinated the fieldwork process in the clinic where s/he was based in. Ethics approval for the study protocol was obtained from the University of Fort Hare's Research Ethics Committee and permission to conduct the study was received from the Eastern Cape Department of Health.

Core Norms	Core Services
• The clinic renders comprehensive integrated PHC services using a one-stop approach for at least 8 hours a day, five days a week. • Access, as measured by the proportion of people living within 5km of a clinic, is improved. • The clinic receives a supportive monitoring visit at least once a month to support personnel, monitor the quality of service and identify needs and priorities. • The clinic has at least one member of staff who has completed a recognised PHC course. • Doctors and other specialised professionals are accessible for consultation, support and referral and provide periodic visits. • Clinic managers receive training in facilitation skills and primary health care management. • There is an annual evaluation of the provision of the PHC services to reduce the gap between needs and service provision using a situation analysis of the community's health needs and the regular health information data collected at the clinic. • There is annual plan based on this evaluation. • The clinic has a mechanism for monitoring services and quality assurance and at least one annual service audit. • Community perception of services is tested at least twice a year through patient interviews or anonymous patient questionnaires.	• Women's Reproductive Health • Integrated Management of Childhood Illness • Diseases prevented by Immunisation • Adolescent and Youth Health • Management of Communicable Disease • Control of Cholera, diarrhoeal disease and dysentery • Helminths • Sexual Transmitted Diseases (STD)and HIV/AIDS • Malaria • Rabies • Tuberculosis • Leprosy • Prevention of Hearing Impairment due to Otitis Media • Rheumatic Fever and Haemolytic Streptococcal Infection • Trauma and Emergency • Oral and Mental Health • Treatment and support of victims of Sexual Offenses, Domestic Violence and Gender Violence • Substance Abuse • Chronic Diseases, Diabetes, Hypertension • Geriatric care • Rehabilitation Services

Source: Department of Health, South Africa

Table 1. The core norms and services for primary health care PHC set by the NDOH.

2.3 Data collection method

A patient satisfaction questionnaire adapted from the one developed by the Health Systems Trust in 2004 was used. Only slight changes were made to the questionnaire in collaboration with the Eastern Cape Department of Health to allow for cross-comparisons with earlier patient surveys that have been undertaken within the Eastern Cape Province using the same questionnaire. Further, some questions were asked on demographics, health status, reason for health visit, and health care utilization. The questionnaire was translated from English into Afrikaans and Xhosa. The Xhosa and Afrikaans versions were developed using back-translation methods (Brislin 1970). The procedure entailed having two native-speakers of the target languages independently do a back-translation. Discrepancies were arbitrated by a third consultant, and solutions were reached by consensus. The translated questionnaire underwent pilot-testing.

2.4 Measures

The questionnaire included demographics and eight domains, each having several items on a 5-point likert scale: Strongly Agree=5; Agree=4; Unsure=3; Disagree=2; and Strongly Disagree=1

2.5 Data analysis

Data was captured on SPSS version 17.0 and analysed. Frequency distributions of domain items were made and positive responses (Agree and Strongly agree) were grouped and are presented. Cross tabulations of domain items by gender were made. Chi square tests were performed to determine the relationship between each domain item and gender.

2.6 Limitations

Response biases introduced through the methodology of using exit interviews might act as filters and influence patient satisfaction ratings. For example, exit interviews automatically select out those who do not have access to public health facilities, but would otherwise have used services. In addition, using exit interviews in health facilities identified by the sub-district officials, means that respondents were purposively selected. Non-randomisation in the selection of respondents means that results are more difficult to generalise to a feeder population around a health facility. The study compensated for this limitation by collecting data from each facility over a week during a period of normal use and through achieving a high number of respondents. A further limitation is that the existing PSS methodology does not enable the relationship between aggregate satisfaction scores and changes in health status of populations to be explored.

3. Results

3.1 Sample characteristics

The majority of the respondents were African (50.9%), female (72.9) and unemployed (56.4%) with a mean age of 39.4 years. Only 5.5% of the respondents indicated that they had enough money to meet their basic needs for most of the time. Almost 85% had some form of formal education.

3.2 Utilization of health services

Most respondents visited clinics more frequently (i.e. about 7 times within 12 months) compared to private doctors (i.e. about 1.58 times), hospitals (i.e. about 1.14 times) and traditional healers were list visited (0.23 times). The main reason for visiting the health facility was to get treatment (41.8%), followed by suffering from non-communicable diseases (NCDs) (11.5%) and family planning (10.1%).

Mean Age: M (SD)	39 years (14.91)
Gender	**N (%)**
Male	230 (29.9)
Female	674 (72.9)
Race	**N (%)**
African	469 (50.9)
White	26 (2.8)
Indian	10 (1.1)
Coloured	375 (40.7)
Other	39 (4.2)
Occupation	**N (%)**
Employed	335 (37.1)
Not Employed	509 (56.4)
Other	59 (6.4)
Highest Level of Education	**N (%)**
None	139 (15.4)
Finished primary	325 (36.0)
Finished Grade 10	266 (29.4)
Finished Grade 12	108 (11.9)
Degree/Diploma	12 (1.3)
Other	54 (6.0)
Enough Money to meet own needs	**N (%)**
None	354 (39.3)
A little	318 (35.3)
Moderately	112 (12.4)
Mostly	40 (4.4)
Completely	10 (1.1)
Other	67 (7.4)

Table 2. Demographic Characteristics.

Items	**M (SD)**
Mean no. of clinic visits in 12 months	8.48 (6.798)
Mean no. of hospital visits in 12 months	1.14 (1.905)
Mean no. of private doctor visits in 12 months	1.58 (2.520)
Mean no. of traditional healer visits in 12 months	0.23 (1.121)
Main reason for visiting health facility	N (%)
Non-communicable diseases	108 (11.5)
Communicable diseases	**40 (4.3)**
Treatment	**391 (41.8)**
Treatment (for baby or child)	**57 (6.1)**
Bodily aches	**47 (5.0)**
Family Planning	**95 (10.1)**
Other/Unidentified	**198 (21.2)**

Table 3. Health Care Utilization.

3.3 Symptom reporting

More than two-thirds reported that coughing, headache, fever, and body/limb aches were the symptoms suffered in descending order.

Symptoms	N (%)
Coughing	183 (87.1)
Body/limb aches	110 (80.3)
Fever	153 (85.0)
Rash	46 (63.0)
Headache	178 (86.8)
Diarrhoea	23 (46.0)

Table 4. Symptom Reporting.

3.4 Prior diagnosis

More than 60% of the respondents indicated that they had prior diagnosis of other STIs (95%), other illnesses (88.9%), TB (78.4%), Diabetes (72.2%), High Blood Pressure (69.3%) and HIV (65.2%).

Prior Diagnosis	N (%)
TB	105 (78.4)
HIV	56 (65.1)
Diabetes	78 (72.2)
Other STI	891 (94.9)
High Blood Pressure	651 (69.3)
Pregnancy	38 (55.1)
Other illness	835 (88.9)

Table 5: Prior diagnosis.

4. Descriptions of patients evaluations: percentage of patients who used the most positive answering category by sex (N=836, percentages)

4.1 Access to PHC services

A larger portion of women respondents positively agreed with the items from the access domain, than did men. More than three quarters of women agreed that it was possible to get an appointment that suited them and about the same number indicated that no payment was required for treatment at that clinic. About three quarters of women also agreed that it was possible to get through to the clinic by telephone and that the clinic was disability-friendly. The same number of women also agreed that they were treated by nurses who spoke a language they could understand and that the clinic's opening hours were convenient. Nearly 75% of women further agreed that they are always treated and not asked to return on another day and that is was possible to speak to the nurse on the phone. More than seven in ten women did not think that nurses did not visit their places of residence often enough. In terms of time and financial costs, about seven in ten women respondents agreed that the journey there took longer than one hour and that it costs more than R10-00 (US$ 1.46) to get to the clinic.

Item	Men: N (%)	Women: N (%)	Total: N (%)	P
It takes longer than an hour to go to the clinic	70 (27.5)	184 (72.2)	255 (27.2)	0.067
It cost more than R10.00 to get to the clinic	50 (29.2)	120 (70.2)	171 (18.2)	0.218
The clinic has convenient opening hours	161 (25.8)	461 (74.0)	623 (66.3)	0.982
I don't think healthworkers/nurses come often enough to the place where I stay	76 (27.4)	201 (72.6)	277 (29.5)	0.689
I paid money to be treated in this clinic	28 (24.6)	86 (75.4)	144 (12.1)	0.088
The nurse who treated me spoke in a language I understood	191 (25.5)	556 (74.3)	748 (79.7)	0.07
When I come to this clinic I'm always treated & never told to return on another day	155 (25.8)	445 (74.0)	601 (64.0)	0.478
The clinic is user friendly to disabled persons	153 (25.4)	449 (74.6)	602 (64.1)	0.265
Getting an appointment to suit you	114 (23.7)	367 (76.3)	481 (51.2)	0.429
Getting through to the clinic on the phone	101 (25.0)	303 (75.0)	404 (43.1)	0.834
Being able to speak to the nurse practitioner on the telephone	93 (26.6)	256 (73.4)	349 (37.2)	0.588

Table 6. Perceived Access to PHC Services.

4.2 Perceived empathy

Women, when compared to men, were also more positive in their responses to items of the empathy domain. More than three quarters of women respondents agreed that their privacy was respected by all the staff involved in their treatment, that the nurse/doctor who treated them was polite and that they could answer all questions about their illness. The same number felt that this made it easy to tell the doctor/nurse about their problems. Just under three quarters of women respondents agreed that the nurse/doctor who treated them introduced themselves, that they gave their permission to be examined and treated and made them feel they had time during consultations.

4.3 General satisfaction

Larger proportions of women when compared to men had positive responses on items of this domain. Almost eight in ten women respondents positively agreed that patients do not usually appreciate all that the clinic staff does for them. More than three in four also agreed that staff do inform clients of changes in service, as well as any delays in services, on occasion. The same proportion of women agreed that their treatment is always better when an injection is administered and that they are pleased with the way they were treated at the clinic. Nearly three quarters agreed that they always get treatment when attending the clinic where they were interviewed and that they would attend the same facility again on another occasion. The same number will also recommend the clinic to friends and family when should they be sick. Despite these figures, just more than six in ten women agreed that the staff were helpful.

Item	Men: N (%)	Women: N (%)	Total: N (%)	P
The nurse/Doctor who treated me introduced him/herself	121 (25.9)	348 (73.9)	468 (49.8)	0.49
The nurse/Doctor who treated me answered all questions about my illness	162 (24.8)	490 (75.0)	653 (69.5)	0.474
I gave permission to be examined and treated	185 (25.4)	543 (74.5)	729 (77.6)	0.908
My privacy was respected by all the staff	172 (24.2)	538 (75.7)	711 (75.7)	0.022
The nurse/doctor who treated me was polite	162 (24.3)	503 (75.5)	666 (70.9)	0.251
The nurse in this clinic are very interested in their clients	165 (25.2)	490 (74.7)	656 (69.9)	0.883
Making you feel you had time during consultations	168 (26.4)	469 (73.6)	638 (67.9)	0.481
Interest in your personal situation	163 (26.1)	462 (74.0)	626 (66.6)	0.425
Making it easy for you to tell him or her about your problems	164 (24.8)	497 (75.2)	662 (70.5)	0.721

Table 7. Empathy.

Item	Men: N (%)	Women: N (%)	Total: N (%)	P
Next time I am ill I will come back here	184 (25.0)	551 (74.9)	736 (78.4)	0.034
I was pleased with the way I was treated at this clinic	172 (24.0)	544 (75.9)	717 (76.4)	0.034
If my friends/family are sick I will tell them to come to this facility	172 (25.0)	514 (74.8)	687 (73.2)	0.426
My treatment is always better if I have an injection	117 (23.6)	379 (76.4)	496 (52.8)	0.147
Patients don't usually appreciate all that staff in this clinic do for them	97 (21.6)	353 (78.4)	450 (47.9)	0.008
I always get treatment when I come here	172 (24.6)	526 (75.3)	699 (74.4)	0.397
Staff informs clients of delays in service from time to time	136 (23.7)	438 (76.3)	574 (61.1)	0.042
Staff informs clients of changes in service from time to time	128 (23.3)	421 (76.7)	549 (58.5)	0.019
The helpfulness of staff	277 (37.3)	465 (62.7)	624 (66.4)	0.715

Table 8. General Satisfaction.

4.4 Referral

The items in this domain also received a majority of positive responses from women respondents. For example, more than three quarters of women agreed that if they cannot be helped at the clinic they will be referred to the nearest hospital or doctor. They same number was also sure that nurses in this facility will call an ambulance if a client is very sick and that nurses in that facility ask patients to return to see how they are doing. The role of traditional healers is still an important aspect of health care, as more than three quarters of women responded that they usually visit a traditional healer before coming to the clinic.

Item	Men: N (%)	Women: N (%)	Total: N (%)	P
If I can't be helped here I will be referred to the nearest hospital/Doctor	163 (24.4)	503 (75.4)	667 (71.0)	0.216
Nurses in this facility call an ambulance if a client is very sick	167 (24.4)	517 (75.5)	685 (72.9)	0.358
Nurses in this facility ask patients to return to see how they are doing	152 (24.2)	475 (75.6)	628 (66.9)	0.321
When I'm sick I usually visit a traditional healer before I come to clinic	42 (23.3)	137 (76.1)	180 (19.2)	0.395

Table 9. Referral.

4.5 Service standards

Items in the service standards domain elicited more positive responses from women than from men. More than three in four women responded that they knew either the chairperson or a member of the clinic committee of that clinic that the health worker that assisted them had a name tag on her/him, that they knew where and to whom to raise complaints, and know of the availability of a suggestion box at the clinic. The same number also agreed that the registration procedures in the clinic were satisfactory, waiting time before examination was reasonable and that there were are fast queues in this clinic for certain services. Just fewer than three in four women agreed that when they had reason to complain, they received feedback and that such action improved service delivery.

Item	Men: N (%)	Women: N (%)	Total: N (%)	P
The registration procedures in this clinic are satisfactory	148 (24.7)	451 (75.2)	600 (63.9)	0.005
In this clinic the time I had to wait before I was examined was reasonable	127 (25.5)	372 (74.5)	499 (53.1)	0.449
There are fast queues in this clinic (e.g. under 5 Immunisation, TB clients, etc)	114 (24.9)	342 (74.8)	457 (48.7)	0.849
The health worker that assisted me had a name tag on him/her	150 (23.9)	476 (75.9)	627 (66.8)	0.248
I know where and to whom to raise my complaints	89 (26.0)	253 (74.0)	342 (36.4)	0.602
When I complain I write it and put it in the suggestion box provided	85 (24.7)	259 (75.3)	344 (36.6)	0.606
When I complained I received feedback	61 (24.6)	187 (75.4)	248 (26.4)	0.723
Raising complaints/suggestions improve service delivery	87 (27.0)	235 (73.0)	322 (34.3)	0.042
I know the chairperson/member of the clinic committee of this facility	48 (23.4)	157 (76.6)	205 (21.8)	0.515

Table 10. Service Standards.

4.6 Reliability

More than three quarters of women judged services as reliable as they did not wait long before receiving medication and that the clinic provided quick services for urgent health problems. Just under this figure regarded general waiting time in waiting rooms as positive.

Item	Men: N (%)	Women: N (%)	Total: N (%)	P
If I received medicines or pills I did not have to wait long for them	140 (24.2)	439 (75.8)	579 (61.7)	0.003
Waiting time in the waiting room	106 (25.2)	314 (74.8)	420 (44.7)	0.062
Providing quick services for urgent health problems	136 (24.2)	425 (75.8)	562 (59.8)	0.161

Table 11. Reliability.

4.7 Health promotion

A majority of women respondents were positive on items referring to health promotion at the clinic. For example, more than three quarters agreed that as patients are waiting to be seen, health workers in the clinic sometimes give talks on health related issues affecting the community. Also, nearly three in four replied that when they had to wait at the clinic, very useful things can be learnt from the posters and other IEC materials. The reason for this was that the posters and other IEC materials, the 'Batho Pele' (people first) principles and the patients' rights charter, were all in a language they could understand.

Item	Men: N (%)	Women: N (%)	Total: N (%)	P
I saw on the walls of this clinic a Patients Rights Charter in a language I could understand	148 (26.3)	413 (73.5)	562 (59.9)	0.887
I saw on the walls of this clinic Batho Pele Principles in a language I could understand	141 (26.3)	395 (73.6)	537 (57.2)	0.719
When I had to wait in this clinic I sometimes learn very useful things from the posters and other IEC (Information, Education & Communication) materials	138 (24.9)	415 (74.9)	554 (59.0)	0.417
The posters and other IEC material are in a language I understand	145 (25.0)	435 (74.9)	581 (61.9)	0.927
As patients are waiting to be seen, health workers in this facility sometimes talk to us about health related issues that affect our community	113 (22.9)	379 (76.9)	493 (52.5)	0.177

Table 12. Health Promotion.

4.8 Tangibles

Items under the tangibles domain also yielded positive responses from the majority of women respondents. More than three in four women agreed that the toilets were clean and in a good condition, that there were indeed toilets for patients in the clinic and that the clinic has enough consultation rooms and that there were enough benches for patients to sit while waiting to be seen by health workers. Just under three quarters agreed that there was clean drinking water for patients, that the building was in a good condition and the clinic and its surroundings are clean, that the services and hours of service displayed on the board outside the clinic was clear and in a language that could be understood.

Domain	Men: N (%)	Women: N (%)	Total: N (%)	P
The clinic building is in a good condition	174 (25.7)	501 (74.1)	676 (72.0)	0.976
The clinic and its surroundings are clean	182 (25.5)	532 (74.4)	715 (76.1)	0.903
There are toilets for patients in this clinic	167 (24.1)	526 (75.8)	694 (73.9)	0.026
The toilets are in a good condition	150 (23.7)	483 (76.2)	634 (67.5)	0
The toilets are clean	146 (23.7)	468 (76.1)	615 (65.5)	0.009
The clinic has enough consultation rooms	137 (24.5)	421 (75.3)	559 (59.5)	0.282
There are benches for patients to sit while waiting to be seen by health worker	165 (24.5)	507 (75.3)	673 (71.7)	0.615
There is clean water for patients in this clinic	171 (25.1)	509 (74.7)	681 (72.5)	0.432
The services rendered and hours of service are clearly displayed on a board outside the facility	149 (26.0)	424 (73.9)	574 (61.1)	0.463
The services and hours of service displayed on the board outside are in a language I can understand	142 (25.2)	421 (74.6)	564 (60.1)	0.378

Table 13. Tangibles.

4.9 Assurance

A greater percentage of women than men also responded positively to items of the assurance domain. More than three quarters of women agreed that the staff at the clinic had given preferential treatment to patients who looked more ill, that the nurses were able to tell them more about their illness and symptoms, and that were told how to store and self-administer their medication. The same proportion also agreed that health workers gave them help in dealing with the emotional problems related to your health status, that they felt comfortable to bring their partners to the facility when requested and that they felt assured that their treatment records remained confidential. Exactly three quarters agreed with their physical examination by health workers, and that they helped in making their patients understand the importance of following his or her medical advice and their preparation of patients as to what to expect from specialist or hospital care.

Just under three quarters of women respondents agreed that attending the health service meant quick relief of one's symptoms, that the explanation of the purpose of tests and treatments were clear and that they felt compelled to complete their treatment as was instructed. Slightly less than three in four women agreed that health workers at the facility listened to patients, that they can get them to always return when asked to do so and that they involved their patients in decisions affecting their medical care. The same proportion of women felt that health workers also helped patients to feel well enough to perform normal daily activities, were through, knew what advice were given to patients previously and were competent in offering advice on the prevention of diseases.

Domain	Men: N (%)	Women: N (%)	Total: N (%)	P
At the time I was waiting to be seen by a Health Worker there was a patient that looked more ill	89 (23.8)	285 (76.2)	374 (39.8)	0.172
I always return when asked by the nurse to come back	183 (25.4)	536 (74.4)	720 (76.7)	0.381
I finish all my treatment as instructed	190 (25.3)	559 (74.5)	750 (79.9)	0.885
I bring my partner(s) when requested to	157 (24.7)	478 (75.2)	636 (67.7)	0.852
I was told how to take my pills/medication	187 (24.6)	572 (75.3)	799 (80.9)	0.433
I was told how to store my pills/medication	171 (24.5)	527 (75.4)	699 (74.4)	0.136
Involving you in decisions about your medical care	168 (25.8)	483 (74.2)	652 (69.4)	0.608
Listening to you	182 (25.5)	533 (74.5)	716 (76.3)	0.934
Keeping your records and data confidential	175 (24.6)	529 (75.1)	705 (75.0)	0.956
Quick relief of your symptoms	167 (25.4)	490 (74.6)	658 (70.1	0.917
Helping you to feel well so that you can perform your normal daily activities	172 (25.9)	493 (74.1)	665 (70.9)	0.213
Thoroughness	150 (26.5)	416 (73.5)	567 (60.4)	0.499
Physical examination of you	155 (25.0)	466 (75.0)	622 (66.3)	0.202
Offering you services for preventing diseases	168 (26.2)	473 (73.8)	642 (68.4)	0.228
Explaining the purpose of tests and treatments	173 (25.4)	509 (74.6)	683 (72.7)	0.356
Telling you what you wanted to know about your symptoms and/or illness	162 (24.0)	514 (76.0)	678 (72.2)	0.604
Help in dealing with emotional problems related to your health status	161 (24.7)	490 (75.3)	651 (69.3)	0.237
Helping you understand the importance of following his or her advice	168 (25.1)	502 (75.0)	671 (71.4)	0.488
Knowing what s/he had done or told you during previous contacts	162 (26.0)	461 (74.0)	624 (66.5)	0.406
Preparing you for what to expect from specialist or hospital care	155 (25.1)	463 (75.0)	619 (66.0)	0.914

Table 14. Assurance.

5. Discussion

Seeking to understand patient perspectives is an important step in the efforts to improve the quality of health care. Research examining patient satisfaction with health care provision in South Africa and, more specifically, the perceived quality of care given by the health care providers is limited (Myburgh et al., 2005). In this study, there were consistently significant differences regarding patient satisfaction between male and female patients across selected items in the various domains.

Evidence from developed countries for gender differences in mean satisfaction levels is mixed. Some authors report that women are more satisfied than men with medical care received (Weiss, 1988), and some report that women are more critical of medical care than men #(Kaplan, 1996), whilst a 2005 Canadian study (Human Resources and Skills Development Canada, 2009) found almost similar satsifaction levels between male (86%) and female (84%) patients. However, a meta-analysis of 110 studies of patient satisfaction, using standard instruments, concluded that there was no average difference in satisfaction with medical care between women and men (Hall & Dornan, 1990). More recently, Sanmartin et al. (2002) suggested that user frequency might influence the descrepancies found betwee male and female patient statisfaction rates and that the type of service being assessed might be a further factor.

Wessels et al (2010) found that amongst oncology patients women rated care aspect of services more highly. A recent Ugandan study found some gender and age difference in patient satisfaction with TB services (Babikako et al. 2011). Past experience and consequently patient expectations, they argue might influence age and gender differences in patient satisfaction.

What is common across these studies is the importance of considering the influence of demographic variables on patient satisfaction. Simply controlling for demographic differences, might result in the needs of important demographic groupings being overlooked. In addition, demographic differences, such as gender are likely to shape patients' needs and preferences and might be a particularly important consideration in shaping specific health services to better meet needs and support treatment adherence. In the South African context, the role that gender plays in patient satisfaction and the gender differences in patient satisfaction need further exploration. We conclude that quality improvement and research in primary care could benefit from gender analysis of patient satisfaction data and from more gender-sensitive patient satisfaction measures.

6. Acknowledgements

We thank the Eastern Cape Department of Health for financially supporting the study.

We would like to extend our gratitude to the district manager and clinic managers of the 12 clinics for overseeing the study in their respective clinics.

Our gratitude is also due to the fieldworkers and patients who agreed to be interviewed.

7. References

Ajayi, I.O., Olumide, E.A. & Oyediran, O., 2005, 'Patient satisfaction with the services provided at a general outpatients' clinic, Ibadan, Oyo State, Nigeria', *African Journal of Medicine & Medical Science* 34(2), 133-140.

Andaleeb, S.S., 2001, 'Service quality perceptions and patient satisfaction: a study of hospitals in a developing country', *Social Science & Medicine* 52, 1359–1370.

Babikako, H.M., Neuhauser, D., Katamba, A., Mupere, E.(2011). Patient satisfaction, feasibility and reliability of satisfaction questionnaire among patients with pulmonary tuberculosis in urban Uganda: a cross-sectional study. *Health Research Policy and Systems* 2011, 9:6. Available from: http://www.health-policy-systems.com/content/9/1/6

Bediako, M.A., Nel, M. & Hiemstra, L.A., 2006, 'Patients' satisfaction with government health care and services in the Taung district, North West Province' *Curationis* 29(2), 12-15.

Bradshaw D, Groenewald P, Laubscher R, Nannan N, Nojilana B, Norman R, Pieterse D and Schneider M. *Initial Burden of Disease Estimates for South Africa*, 2000. Cape Town: South African Medical Research Council, 2003.

Brislin, R.W., 1970, 'Back translation for cross-cultural research', *Journal of Cross-Cultural Psychology* 1(3), 185-216.

Campbell, J.L., Ramsay, J. & Green, J., 2001, 'Age, gender, socioeconomic and ethnic differences in patients' assessments of primary health care', *Quality in Health Care* 10, 90-95.

Dağdeviren, N. & Akturk, Z., 2004, 'An evaluation of patient satisfaction in Turkey with the EUROPEP instrument', *Yonsei Medical Journal* 45(1), 23-28.

De Jager ,J., & Du Plooy, T., 2007, 'Service quality assurance and tangibility for public health care in South Africa' *Acta Commercii* 7, 96-117.

Department of Health, 2007, *A policy on quality in health care for South Africa*, Department of Health, Pretoria.

Department of Public Service and Administration, 1997, *Transforming Public Service Delivery*, Department of Public Service and Administration, Pretoria.

Eastern Cape Socio-Economic Consultative Council (ECSECC) 2011. Statistics at your fingertips *http://www.ecsecc.org/statistics-database* [Accessed 28 June 2011]

Glick, P., 2009, 'How reliable are surveys of client satisfaction with healthcare services? Evidence from matched facility and household data in Madagascar', *Social Science & Medicine* 68(2), 368-379.

Hall, J.A. & Dornan, M.C., 1990, 'Patient socio-demographic characteristics as predictors of satisfaction with medical care: A meta-analysis, *Social Science & Medicine*, 30(7), 811-818.

Hospital Association of South Africa, 2011. *http://www.hasa.co.za/hospitals/members/* [Accessed 28 June 2011]

Kaplan SH, Sullivan LM, Spetter D. *Gender and patterns of physician-patient communication*. In: Falik MM, Collins KS, eds. Women's health: The Commonwealth Fund Survey. Baltimore, MD: Johns Hopkins University Press; 1996.

McIntyre D, 2010, *Private sector involvement in funding and providing health services in South Africa: implications for equity and access to health care*, EQUINET Discussion Paper Series 84 Health Economics Unit (UCT), ISER Rhodes University, EQUINET: Harare.

Muhondwa, E.P., Leshabari, M.T., Mwangu, M., Mbembati, N. & Ezekiel, M.J., 2008, 'Patient satisfaction at the Muhimbili National Hospital in Dar es Salaam, Tanzania', *East African Journal of Public Health* 5(2), 67-73.

Myburgh, N.G., Solanki, G.C., Smith, M.J. & Lalloo, R., 2005, 'Patient satisfaction with health care providers in South Africa: the influences of race and socioeconomic status', *International Journal for Quality in Health Care* 17(6), 473-477.

Peltzer, K., 2009, 'Patient experiences and health system responsiveness in South Africa' , *BMC Health Services Research* 9, 117, DOI:10.1186/1472-6963-9-117.

Peltzer, K., 2000, 'Community perceptions of biomedical health care in a rural area in the Limpopo Province South Africa', *Health SA Gesondheid* 5 (1), 55-63.

Newman, R.D., Gloyd, S., Nyangezi, J.M., Machobo, F. & Muiser, J., 1998, 'Satisfaction with Outpatient Health Care Services in Manica Province, Mozambique', *Health Policy & Planning* 13(2), 174-180.

Sanmartin, C., Houle, C., Berthelot, J. & White, K., 2002, *Access to Health Care Services in Canada, 2001;* Statistics Canada, Ottawa.

Weiss, G.L., 1988, 'Patient Satisfaction with Primary Medical Care Evaluation of Sociodemographic and Predispositional Factors', *Medical Care* 26(4), 383-392.

Wessels, H., De Graff, A., Wynia, K., De Heus, M., Kruitwagen, C.L.J.J., Woltjer, G.T.G.J., Teunissen, S.C.C.M., Voest, E. (2010). Gender-Related Needs and Preferences in Cancer Care Indicate the Need for an Individualized Approach to Cancer Patients. *The Oncologist*,15:648–655 doi: 10.1634/ theoncologist.2009-0337

6

Application of Sampling Strategies for Hot-Mix Asphalt Infrastructure: Quality Control-Quality Assurance Sampling; Specification for Performance Test Requirements

Bor-Wen Tsai[1], Jiangmiao Yu[1,2] and Carl L. Monismith[1]
[1]*University of California at Berkeley,*
[2]*South China University of Technology,*
[1]*U.S.A.*
[2]*P.R.C.*

1. Introduction

Due to the lack of a rational, effective, and systematic quality control-quality assurance (QC/QA) methodology, the nonconformity of construction quality with design requirements for public works, especially for civil engineering infrastructure systems, can result in increased expenditures over time. Thus, development of a rational QC/QA methodology to ensure that the construction quality complies with the design requirements should have a high priority. The limited sample size constrained by the consideration of cost and time may result in the misjudgement that the construction quality does not meet the design requirements.

In this chapter, the effects of sampling size, sampling strategies, and acceptance/rejection criteria for QC/QA projects using statistically based decision making in hot-mix asphalt (HMA) construction are presented. Also, there has developed an increased interest recently in ensuring that the HMA as placed will meet certain performance requirements by measuring the actual performance parameters on test specimens prepared from in situ samples rather than from surrogate values such as asphalt content and aggregate gradation. Examples include direct measures of mix permanent deformation characteristics and fatigue characteristics, mix stiffness, and degree of compaction as measured by air-void content.

Determination of sample size is primarily based on an acceptable error level for a performance parameter specified by the agency. It is not uncommon to base quality assurance by many agencies on three samples. Through the t distributions, discussion is presented as to why it is not appropriate to take only this number of samples for quality-assurance. Based only on three samples in a large project, the agency will have insufficient power to reject the null hypothesis given that the null hypothesis is false unless the project quality delivered by the contractor is extremely poor so that the agency is confident enough to reject the project.

In addition to providing a general introduction to fundamental statistics and hypothesis testing, two case studies are used to clarify the relationships among sampling size, sample strategies, and performance specifications (or acceptance/rejection criterion). These include the following:

(1) A QC/QA case study is used to illustrate a methodology to determine strategies for a sampling scheme and selection of sample size for QC/QA for HMA construction to ensure that the acceptable level of a mix parameter is obtained with the same risk to the contractor and the agency. A sampling scheme and sampling size based on statistical simulation of a fixed length of a one-lane-width placement of HMA are discussed. Sample size is based on the combination of the sample size of the contractor and that of the agency to balance the risk to both organizations which will result in a mix that will meet the minimum performance requirement. An example is presented for the placement of 15,000 tons of HMA according to the California Department of Transportation (Caltrans) QC/QA requirements. For this total tonnage, the contractor and agency are assumed to perform a specific number of performance tests using the California stabilometer methodology for QC and QA.
(2) A QA case study is used to illustrate the application of the use of uniform design (UD) as a sampling strategy to ensure that the most representative sampling scheme can be achieved with a specified sample size. A sampling scheme using uniform design and sampling size through statistical simulation of a fixed length of a two-lane-width placement of HMA with several segregation data patterns is discussed. Based on the simulation, a QA guideline is developed by inspecting the accuracy of sample mean and the precision of sample standard deviation criteria combined with the application of the UD table is proposed and verified with two full scale pavement sections by measured air-void contents (measure of degree of compaction).

2. Case I: quality control-quality assurance sampling strategies for hot-mix asphalt construction

The effects of sampling strategies and size on statistically based decision making in hot-mix asphalt (HMA) construction are presented. For sample sizes agreed upon by the agency and the contractor, an acceptable level for an HMA mix parameter is determined with risk balanced between the two organizations. With increased emphasis on specific performance requirements, the use of performance tests on HMA specimens prepared from in situ samples is developing. Examples include direct measures of mix stiffness and permanent deformation characteristics. A measure of rutting resistance, the stabilometer S-value, is used by the California Department of Transportation (Caltrans) for quality control-quality assurance (QC/QA) projects. Although the S-value was used for this simulation because extensive tests were available, this approach is applicable to any performance measures already in use, such as HMA thickness or compacted air-void content. A sampling scheme and sampling size through statistical simulation of a fixed length of a one-lane-width placement of HMA are discussed. Sample size is based on the combination of the sample size of the contractor and that of the agency to balance the risk to both organizations and results in a mix that meets the minimum performance requirement.

2.1 Hypothesis testing of inequality

The acceptance or rejection of the null hypothesis, H_0, is referred to as a decision. Therefore, a correct decision is made in situations where (1) the H_0 is correctly accepted if H_0 is true and (2) the H_0 is correctly rejected if the H_0 is not true. As shown in the following for a decision based on a sample, when the null hypothesis is valid, the probability α of erroneously rejecting it is designated as the Type I error (or seller's risk), i.e., α = P{Type I error} =

P\{reject H_0 | H_0 is true\} ; when the null hypothesis is not true, the probability β of erroneously accepting it is named the Type II error (or buyer's risk), i.e., β = P\{Type II error\} = P\{fail to reject H_0 | H_0 is false\}.

	Truth about the population	
	H_0 True	H_0 Not True
Reject H_0	Type I Error (α)	Correct Decision
Accept H_0	Correct Decision	Type II Error (β)

The power is defined as the probability 1 - β of correctly rejecting H_0 if H_0 is not true, i.e., 1 - β = P\{reject H_0 | H_0 is false\}. Hence, from the viewpoint of the agency (the buyer), it is necessary to have the power as high as possible; likely, from the perspective of the contractor (the seller), the Type I error should be as minimum as possible.

2.1.1 Testing inequality μ ≥ C_s and size of test α

The objective is to test the null hypothesis that $H_0 : \mu \geq C_S$ from the viewpoint of the contractor. The relevant t statistic is given by $t = \dfrac{\hat{\mu} - C_S}{SE(\hat{\mu})}$, where $\hat{\mu}$ is the sample mean of the stabilometer tests, C_S is the minimum specification limit for the stabilometer test, and SE the standard error. The critical region for the t test of size α of the null hypothesis is given by $t \leq -t_{1-\alpha,n-p}$, where $n = n_1 + \cdots + n_p$ and p the number of laboratories. In other words, the t statistic lies in the acceptance region $t > -t_{1-\alpha,n-p}$ if and only if

$$(\hat{\mu} - C_S) / SE(\hat{\mu}) \; > \; -t_{1-\alpha,n-p} \Rightarrow \hat{\mu} > C_S - t_{1-\alpha,n-p} SE(\hat{\mu}) \tag{1}$$

Note that the critical region for the t test of size $\alpha = 0.05$ of the null hypothesis $H_0 : \mu \geq C_S$ can be given by $t \leq -t_{1-\alpha,n-p} \cong -\Phi(1-\alpha) = -\Phi(0.95) = -1.64485$, if $n - p >> 0$, where Φ is the distribution function of a standard normal distribution. The size of test $\alpha = 0.05$ represents that at most a 5% chance is allowed to erroneously reject a valid null hypothesis; that is, there is a 95% chance that H_0 is accepted if H_0 is valid.

2.1.2 Test power, sample size, and operating-characteristic curve

Suppose that the hypothesis is not true, that is, $\mu < C_S$ (the opposite of $\mu \geq C_S$). Then the power as shown by Stone (Stone, 1996) is:

$$\begin{aligned} 1 - \beta &\cong P\left(\frac{\hat{\mu} - C_S}{SD(\hat{\mu})} \leq -z_{1-\alpha} \right) \\ &= P\left(\frac{\hat{\mu} - \mu}{SD(\hat{\mu})} \leq -z_{1-\alpha} + \frac{C_S - \mu}{SD(\hat{\mu})} \right) \\ &= \Phi\left(-z_{1-\alpha} + \delta\right) \end{aligned}$$

where $\delta = \frac{|\mu - C_S|}{SD(\hat{\mu})}$ and $z_{1-\alpha}$ is quantile of a standard normal distribution. For the specified α and β levels under the null hypothesis that $H_0 : \mu \geq C_S$, Table 1 lists the $SD(\hat{\mu})$, test power, and required sample sizes for the case with the agency and the contractor.

The test power equation shown in Table 1 indicates that the power of testing a null hypothesis is actually a standard normal distribution function in terms of the test of size α, d ($d = \frac{|\mu - C_S|}{S_p}$), and the number of tests. Figure 1 plots power versus d with $\alpha = 0.05$ at various numbers of tests, designated as the operating-characteristic curves. Several observations can be addressed in the following:

1. With the same number of tests and power level, increasing α will decrease the value of *d*; alternatively, at the same number of tests and a fixed value of *d*, increasing α will increase the power.
2. At fixed levels of α and β, increasing the number of tests will reduce the value of *d*.
3. For $n = 4$ to ensure that the test power is greater than 0.95, d > 1.645, i.e., $d \cong \frac{|\hat{\mu} - C_S|}{S_S} > d_{0.95}$. In other words, if the sample mean of tests $\hat{\mu}$ is either $\hat{\mu} > C_S + d_{0.95} \cdot S_S$ or $\hat{\mu} < C_S - d_{0.95} \cdot S_S$, then the agency has enough power to confidently accept or reject the null hypothesis $H_0 : \mu \geq C_S$. If $\hat{\mu}$ lies in the range of $(C_S - d_{0.95} \cdot S_S, C_S + d_{0.95} \cdot S_S)$, then the agency does not have enough power with $n = 4$. Thus the number of tests has to be increased to reach the same level of power.
4. The test power approaches the test of size α as $d \to 0$.

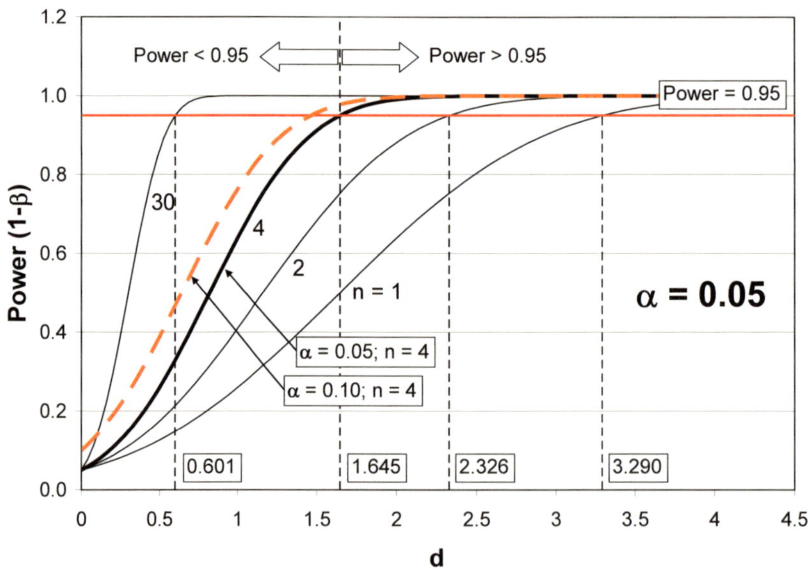

Fig. 1. Operating-characteristic curves with α = 0.05.

The agency and the contractor: $n_1 = k \cdot n_2$, $0 < k \le 1$ (n_1 **: number of QA samples;** n_2 **: number of QC samples.)**	
$SD(\hat{\mu})$	$\frac{S_p}{p} \cdot \sqrt{\frac{1}{n_1} + \frac{1}{n_2}} = \frac{S_2}{2} \cdot \sqrt{\frac{n_1 + n_2}{n_1 \cdot n_2}}$
Test Power $1-\beta$	$\Phi\left(-z_{1-\alpha} + 2 \cdot d \cdot \sqrt{\frac{k \cdot n_2}{1+k}}\right)$
Sample size	$n_2 \approx \frac{1+k}{k} \cdot \left(\frac{z_{1-\beta} + z_{1-\alpha}}{2 \cdot d}\right)^2$
Minimum Requirement of contractor	$\hat{\mu} \ge C_S - \frac{t_{1-\alpha, n_1+n_2-2}}{2} \cdot \sqrt{\frac{1+k}{k \cdot n_2}} \cdot S_P$
Upper and Lower bounds of Agency	$\hat{\mu} \ge C_S + \frac{z_{1-\alpha} + z_{1-\beta}}{2} \cdot \sqrt{\frac{1+k}{k \cdot n_2}} \cdot S_P$ (upper bound)
	$\hat{\mu} \le C_S - \frac{z_{1-\alpha} + z_{1-\beta}}{2} \cdot \sqrt{\frac{1+k}{k \cdot n_2}} \cdot S_P$ (lower bound)

Note:

The pooled sample variance, S_p^2, is defined as,

$$S_p^2 = \frac{\sum_{i=1}^{n_1}\left(x_{1,i} - \bar{x}_1\right)^2 + \sum_{i=1}^{n_2}\left(x_{2,i} - \bar{x}_2\right)^2 + \cdots + \sum_{i=1}^{n_p}\left(x_{p,i} - \bar{x}_p\right)^2}{n_1 + n_2 + \cdots + n_p - p}; \text{ if } p = 1, \text{ then } S_P = S_S.$$

$z_{1-\alpha}$ and $z_{1-\beta}$ are quantiles of a standard normal distribution; Φ is the distribution function of a standard normal distribution.

$$d = \frac{|\mu - C_S|}{S_2}; \ \mu = \frac{\mu_1 + \mu_2}{2}$$

Table 1. Test power, required sample size, minimum requirement of contractor, and upper and lower bounds of agency.

2.1.3 Size of test α and power 1 - β

For the contractor, under the null hypothesis $H_0 : \mu \ge C_S$, the acceptance region for the *t* test of size α is given by Equation 1, that is, $\hat{\mu} \ge C_S - t_{1-\alpha, n-p} \cdot SE(\hat{\mu})$.
For the agency, as noted earlier, the power of a test under the null hypothesis is given by

$$1 - \beta \cong \Phi\left(-z_{1-\alpha} + \delta\right) \Rightarrow z_{1-\beta} \cong -z_{1-\alpha} + \delta$$

where $\delta = \frac{|\mu - C_S|}{SD(\hat{\mu})} \cong \frac{|\hat{\mu} - C_S|}{SE(\hat{\mu})}$; $z_{1-\alpha}$ and $z_{1-\beta}$ are quantiles of a standard normal distribution.

Therefore, to satisfy the power requirement of the agency, $\delta \ge z_{1-\beta} - (-z_{1-\alpha})$, i.e.,

$$\frac{|\hat{\mu}-C_S|}{SE(\hat{\mu})} \geq z_{1-\alpha}+z_{1-\beta}$$

$$\Rightarrow \quad \hat{\mu} \geq C_S+\left(z_{1-\alpha}+z_{1-\beta}\right)\cdot SE(\hat{\mu}) \text{ , or } \hat{\mu} \leq C_S-\left(z_{1-\alpha}+z_{1-\beta}\right)\cdot SE(\hat{\mu}) \qquad (2)$$

The $C_S+\left(z_{1-\alpha}+z_{1-\beta}\right)\cdot SE(\hat{\mu})$ will be designated as the upper bound and $C_S-\left(z_{1-\alpha}+z_{1-\beta}\right)\cdot SE(\hat{\mu})$ the lower bound of $1-\beta$ power.

It should be noted that (1) if $\beta=0.5$, then $z_{1-\beta}=0$ and (2) $z_{1-\alpha} \cong t_{1-\alpha,n-p}$, if $n-p>>0$. Thus, Equation 1 is equivalent to the lower bound of Equation 2. Based on Equation 1, the minimum requirement of the contractor, and Equation 2, the upper and lower bounds of power requirement of the agency, the case of the agency and the contractor is defined in Table 1.

Figure 2 illustrates plots of the upper and lower bounds at various power levels of the agency and the minimum requirement of the contractor under $H_0: \mu \geq 37$ in terms of μ and sample size, n_2. The minimum requirements of the contractor in Figure 2 are plotted based on the *t*- distribution and standard normal distribution. It will be noted that the two curves coincide after $n_2=10$. From Table 1 and Figure 2, two observations can be made:

1. It is very important to recognize that the minimum requirement of the contractor actually matches the lower bound of 0.5 power of the agency.
2. The distance enclosed by the upper and lower bounds at a specified power level decreases with smaller S_P, larger α and β, larger k $(0<k\leq 1)$, and, more importantly, larger sample size.

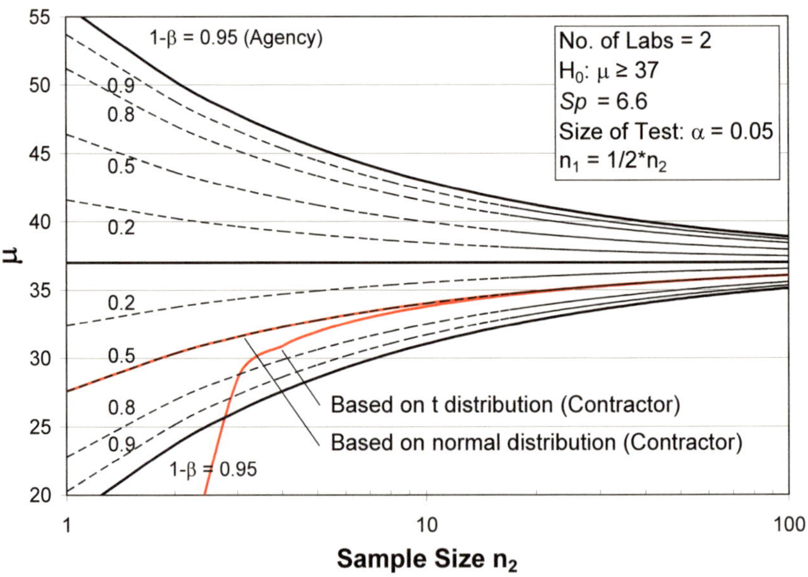

Fig. 2. Minimum stability requirements of the contractor and power requirement of the agency under the same null hypothesis.

2.2 QC/QA demonstration example

In this demonstration example 15,000 tons of HMA will be placed on 20 sublots (750 tons per sublot). The contractor is required to conduct 20 tests (n_2), i.e., one test per sublot. The number of tests conducted by the agency ($n_1 = k \cdot n_2$) will include the minimum required by the agency according to Caltrans specifications, i.e, k = 0.1 (2 tests in this case); in addition, determinations will be made for four tests (k = 0.2), six tests (k = 0.3), and eight tests (k = 0.4). The minimum stabilometer S-value has been set at 37 (Type A HMA) (California Department of Transportation [CALTRANS], 2007), and a standard deviation S_P is used for the S-value for tests between two laboratories of 6.6 (Paul Benson, private communication transmitting analyses of stabilometer test results for periods 1967– 1970 and 1995-1999). The demonstration example will include sampling consistency between QC and QA, sampling stabilization of S_P, and minimum requirements for both the agency and the contractor.

To conduct the sampling size simulation, several assumptions were made:

1. Lane width: 12 ft (3.66 m),
2. Unit weight of HMA - 145 lb/ft^3 (2,323 kg/m^3),
3. HMA layer thickness - 8 in. (20 cm), and
4. One stability sample is represented by a 4 × 4-in (10 × 10-cm) square with each square assigned a normalized stability value.

For these assumptions, the 15,000 tons of HMA will produce a section ~26,000 ft (7,925 m). long and 12 ft (3.66 m) wide. This results in a N(0,1) stability population of 12 x 3 x 26,000 x 3 = 2,808,000 data points to generate three types of data patterns as schematically shown in Figure 3: (1) random pattern, (2) transverse strip pattern with 40 vertical strips, and (3) longitudinal strip pattern with 6 horizontal strips. The *N(0,1)* distribution is separated by the points of quantiles into several intervals, e.g., 6 intervals for transverse strip pattern or 4 intervals for longitudinal strip pattern as shown in Figure 3. These intervals are then permuted to vary randomly across the x-direction or the y-direction of a lane of HMA paving. Those points within the interval are also randomly distributed over the transverse strip or the longitudinal strip.

The sampling scheme used was illustrated in Figure 4 with cases of $M \times N$ cells (N [y-direction] = 1; M [x-direction] = 10, 20, 30, 40, 50, 100, 200, and 500). That is, one random QC sample from each cell and one random QA sample from one random cell of $n_1 = k \cdot n_2$ random transverse strips. A total of 8 cases were simulated over three data patterns. Each case, per data pattern, was simulated 200 times.

To verify the minimum sampling size for an HMA paving strip is to show (1) no apparent difference of sampling consistency between the contractor (QC) and the agency (QA) and (2) stabilization of the pooled sample estimate of standard deviation of stability value, S_P. (Tsai & Monismith, 2009).

In each sampling simulation, the normalized stability values form a distribution with mean and standard deviation; hence, when repeated 200 times, the standard deviations will form another distribution. For each case, the standard deviations of the standard deviation distributions (SDSD) were calculated for QC and QA respectively. The difference of SDSD between QA and QC were used as an index to represent the sampling consistency between the agency (QA) and the contractor (QC).

Likewise, for each simulation, the S_p was calculated based on the equation in Table 1; hence, when repeated 200 times, the standard deviation of the S_p distribution will be used to inspect its stability over the $M \times N$ domain.

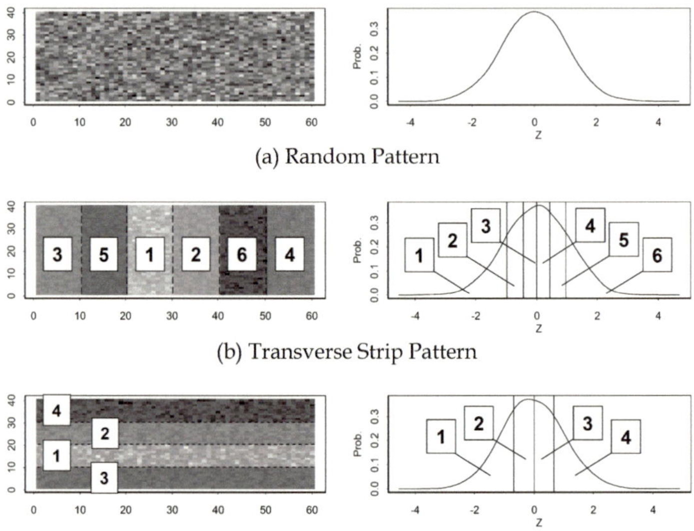

(a) Random Pattern

(b) Transverse Strip Pattern

(c) Longitudinal Strip Pattern

Fig. 3. Schemetic illustration of three data patterns: (a) random pattern, (b) transverse strip pattern, and (c) longitudinal strip pattern.

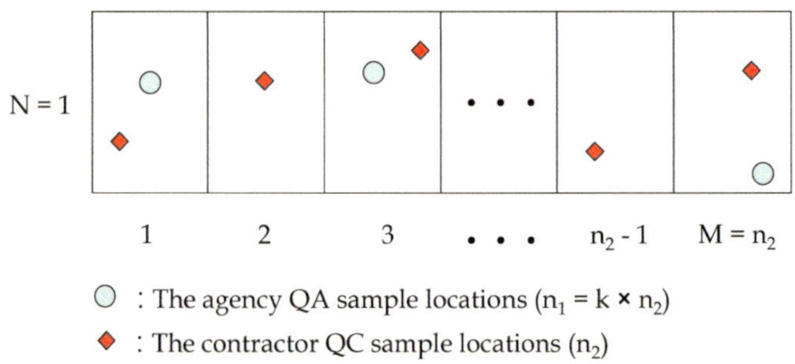

Fig. 4. Sampling scheme.

Figure 5a illustrates the simulation results for sampling consistency between QC and QA at various k values in terms of global smoothed line over three different data patterns. As would be expected, the sampling consistency between QC and QA increases as the k value increases. Figure 5b indicates that sampling stabilization of S_P depends only on the contractor's sampling size, n_2, rather than the k value.

From a series of operating-characteristic curves for the four k values and two α values (5% and 10%), the values in Table 2 were determined for the required minimum value of S, termed $\mu_{\min}$. With Figure 6a as an example, under the condition that $\alpha = 5\%$, n_2 = 20, k = 0.2, and power = 0.95, d has to be greater than 0.902 to satisfy the agency's power requirement; that is, $\hat{\mu}$ has to be greater than 42.95 so that the agency has power 0.95 to clearly accept the contractor's mix. Figure 6b shows a smaller d (0.803) will be obtained when the α value is increased to 10%.

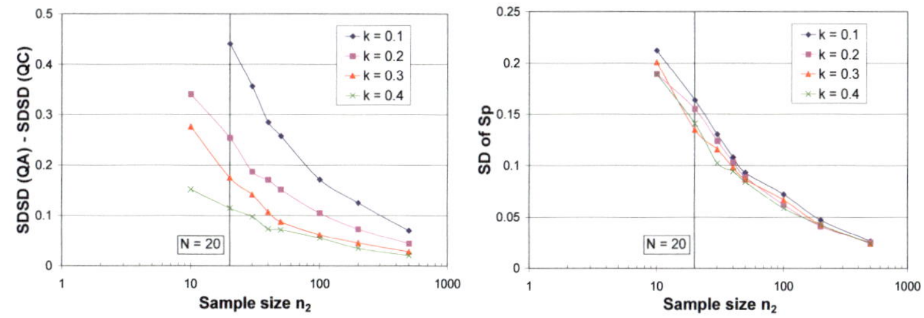

(a) Sampling Consistency (b) Sampling Stabilization

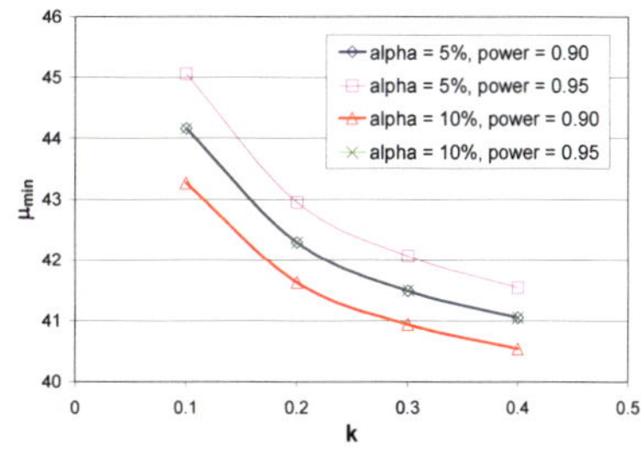

(c) Relation between μ_{min} and k

Fig. 5. Summary of simulation results: (a) sampling consistency; (b) sampling stabilization; and (c) relationship between k and μ_{min}.

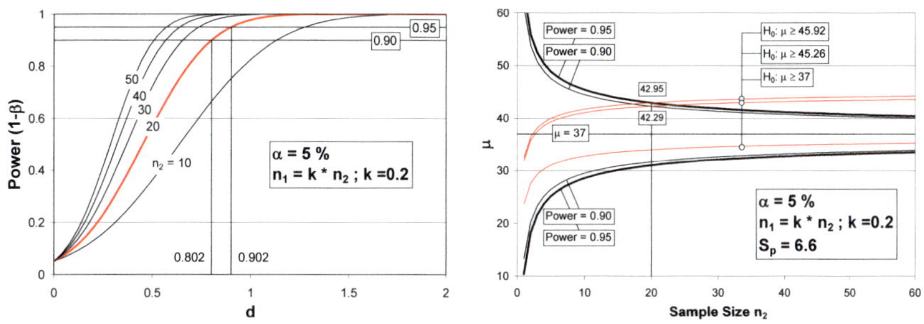

(a) α = 5%; n_2 = 20, k = 0.2

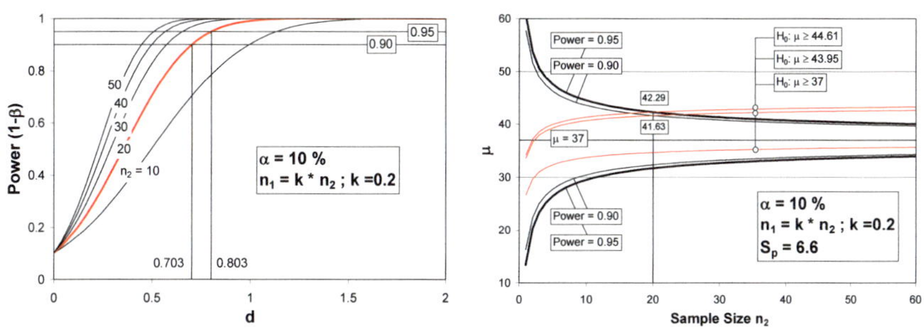

(b) α = 10%; n_2 = 20, k = 0.2

Fig. 6. Examples of operating-characteristic curves and μ_{min} required to meet the agency's power requirement and the contractor's minimum requirement: (a) α = 5% and (b) α = 10%.

Figure 5c illustrates the relationship between k and the minimum S-value. It is apparent that an increase of k-value reduces the value of $\mu_{\min}$. It is interesting to observe that the curve of $\alpha = 5\%$ and power = 0.90 is exactly the same as the curve of $\alpha = 10\%$ and power = 0.95. From Figure 5c, it is also shown that the higher $\mu_{\min}$-criterion is needed if both the agency and the contractor require a high power level and a low α-level, whereas if both the agency and the contractor require a low power level and a high α level, then the $\mu_{\min}$ criterion can be much smaller.

α (Contractor)	Power 1-β (Agency)	k	d	Acceptance $\mu_{\min}$	Target Hypothesis (Contractor)
5%	0.90	0.1	1.085	44.16	$H_0^*: \mu \geq 48.19$
		0.2	0.802	42.29	$H_0^*: \mu \geq 45.26$
		0.3	0.680	41.49	$H_0^*: \mu \geq 44.02$
		0.4	0.613	41.05	$H_0^*: \mu \geq 43.31$
	0.95	0.1	1.220	45.05	$H_0^*: \mu \geq 49.07$
		0.2	0.902	42.95	$H_0^*: \mu \geq 45.92$
		0.3	0.766	42.06	$H_0^*: \mu \geq 44.58$
		0.4	0.688	41.54	$H_0^*: \mu \geq 43.81$
10%	0.90	0.1	0.951	43.27	$H_0^*: \mu \geq 46.41$
		0.2	0.703	41.63	$H_0^*: \mu \geq 43.95$
		0.3	0.597	40.94	$H_0^*: \mu \geq 42.91$
		0.4	0.537	40.54	$H_0^*: \mu \geq 42.31$
	0.95	0.1	1.086	44.16	$H_0^*: \mu \geq 47.30$
		0.2	0.803	42.29	$H_0^*: \mu \geq 44.61$
		0.3	0.682	41.50	$H_0^*: \mu \geq 43.46$
		0.4	0.613	41.05	$H_0^*: \mu \geq 42.81$

Note:

Null hypothesis: $H_0: \mu \geq 37$.

$n_1 = k \cdot n_2\ (0 \leq k \leq 1)$, where n_1 is the number of tests of agency; n_2 the number of tests of contractor.

$d = \frac{|\mu - C_S|}{S_P}$, where $\mu = \frac{\mu_1 + \mu_2}{2}$, μ_1 the average stabilometer value from agency; μ_2 the average stabilometer value from contractor; $C_S = 37$; $S_P = 6.6$.

$\mu_{\min} = d \cdot S_P + C_S$.

Table 2. Acceptance $\mu_{\min}$ values and target hypotheses for contractor with combinations of various α levels, power levels, and k- values.

3. Case II: HMA sampling strategies using uniform experimental design for quality assurance

The application of using uniform design (UD) as a sampling strategy for quality assurance (QA) ensures that the most representative and unbiased sampling scheme can be achieved with the sample size based on an acceptable error level of a hot-mix asphalt (HMA) parameter specified by the agency. Through statistical simulations and demonstration of air-void measurements of two field pavement sections, a QA guideline combined with the UD sampling scheme was developed to justify construction quality using the sample mean and sample standard deviation criteria. This approach can also be applied to any performance measure already in use.

3.1 Uniform experimental design

Statisticians have developed a variety of experimental design methods for different purposes, with the expectation that use of these methods will result in increased yields from experiments, quality improvements, and reduced development time or overall costs. Popular experimental design methods include full factorial designs, fractional factorial designs, block designs, orthogonal arrays, Latin square, supersaturated designs, etc. One relatively new design method is called Unifrom Design (UD). Since it was proposed by Fang and Wang in the 1980s (Fang, 1980; Fang et al., 2000; Wang & Fang, 1981), UD has been successfully used in various fields, such as chemistry and chemical engineering, quality and system engineering, computer sciences, survey design, pharmaceuticals, and natural sciences, etc.

Generally speaking, uniform design is a space-filling experimental design that allocates experimental points uniformly scattered in the domain. The fundamental concept of UD is to choose a set of experimental points with the smallest discrepancy among all the possible designs for a given number of factors and experimental runs.

Suppose that there are s factors in an experiment. Without loss of generality we can assume that the experimental domain is the unit cube $C^s = [0,1]^s$ (after making a suitable linear transformation). The aim is to choose a set of n experiment points $P = \{x_1, \ldots, x_n\} \subset C^s$ that is uniformly scattered on C^s. Let M be a measure of uniformity of P such that the smaller M corresponds to better uniformity. Let $Z(n,s)$ be the set of sets of n points on C^s. A set $P^* \in Z(n,s)$ is called a uniform design if it has the minimum M-value over $Z(n,s)$, i.e., $M(P^*) = \min_{P \in Z(n,s)} M(P)$.

Many different measures of uniformity have been defined. However, the centered L_2-discrepancy (CD_2) is regarded as one of the most commonly used measures in constructing the UD tables, the reason is that the CD_2 considers the uniformity not only of P over C^s, but also of all the projection uniformity of P over C^u which is the u-dimensional unit cube involving the coordinates in u, P_u is the projection of P on C^u. Hickernell gave an analytical expression of CD_2 as follows (Fang & Lin, 2003):

$$CD_2(P) = \left[\left(\frac{13}{12}\right)^s - \frac{2}{n}\sum_{k=1}^{n}\prod_{j=1}^{s}\left(1 + \frac{1}{2}\left|x_{kj} - 0.5\right| - \frac{1}{2}\left|x_{kj} - 0.5\right|^2\right)\right.$$
$$\left. + \frac{1}{n^2}\sum_{k=1}^{n}\sum_{j=1}^{n}\prod_{i=1}^{s}\left(1 + \frac{1}{2}\left|x_{ki} - 0.5\right| + \frac{1}{2}\left|x_{ji} - 0.5\right| - \frac{1}{2}\left|x_{ki} - x_{ji}\right|\right)\right]^{\frac{1}{2}}$$

where $x_k=(x_{k1},\ldots,x_{ks})$ is the k-th experimental point, s is the number of factors in an experiment, n is the number of runs.

One of the most noteworthy advantages of the uniform design is that it allows an experiment strategy to be conducted in a relatively small number of runs. It is very useful when the levels of the factors are large, especially in some situations in which the number of runs is strictly limited to circumstances when factorial designs and orthogonal arrays can not be realized in practice.

Given that the strength of uniform design is that it provides a series of uniformly scattered experiment points over the domain, this homogeneity in two factors has physically become the spatial uniformity of sampling from a pavement section in x and y directions. The application of uniform design resulted in the generation of a sampling scheme with a UD table consisting of pairs of (x, y) coordinates.

3.2 Fundamental statistics

If $\bar{x}$ is the sample mean of a random sample of size n from a normal population, $X \sim N(\mu,\sigma^2)$, then $Z=(\bar{x}-\mu)/(\sigma/\sqrt{n})$ has a standard normal distribution. A $100(1-\alpha)\%$ confidence interval (CI) can be defined as (Figure 7a), $p\left(-z_{\alpha/2} \le \dfrac{\bar{x}-\mu}{\sigma/\sqrt{n}} \le z_{\alpha/2}\right)=1-\alpha$.

Hence, If μ and σ are specified, a $100(1-\alpha)\%$ confidence interval on $\bar{x}$ can be then given by

$$\mu - z_{\alpha/2}\cdot\frac{\sigma}{\sqrt{n}} \le \bar{x} \le \mu + z_{\alpha/2}\cdot\frac{\sigma}{\sqrt{n}} \tag{3}$$

It can be assumed that the error $E=|\bar{x}-\mu|$ is equivalent to $z_{\alpha/2}\cdot\dfrac{\sigma}{\sqrt{n}}$ (Figure 7b). Then the required sample size will be

$$n=\left(\frac{z_{\alpha/2}\cdot\sigma}{E}\right)^2 \tag{4}$$

That is to say, if $\bar{x}$ is used as an estimate of μ, we can be $100(1-\alpha)\%$ confident that the error $|\bar{x}-\mu|$ will not exceed a specified amount E when the sample size is $n=\left(z_{\alpha/2}\cdot\sigma/E\right)^2$ (Montgomery & Runger, 2010). If the specified error level is selected as the fraction of standard deviation of $N(\mu,\sigma^2)$ distribution, i.e., $E=|\bar{x}-\mu|=k\sigma$, where $k\ge 0$, then the Equation 4 can be simplified as $n=\left(z_{\alpha/2}\cdot\sigma/E\right)^2=\left(z_{\alpha/2}/k\right)^2$. It should be noted that $z_{\alpha/2}=1.9600$ if $\alpha=0.05$; $z_{\alpha/2}=1.6449$ if $\alpha=0.10$.

The same argument of sample mean can also be applied to sample standard deviation s. Let $X_1, X_2, \ldots, X_n$ be a random sample of size n from a normal distribution $N(\mu,\sigma^2)$, and let s^2 be the sample variance. Then the random variable $X^2=\dfrac{(n-1)s^2}{\sigma^2}$ has a chi-square (χ^2) distribution with $n-1$ degrees of freedom. As shown in Figure 7c, we may write

$$p\left(\chi^2_{\alpha/2,n-1} \le \frac{(n-1)s^2}{\sigma^2} \le \chi^2_{1-\alpha/2,n-1}\right)=1-\alpha$$

k $\left(\begin{array}{l}E=\|\bar{x}-\mu\| \\ =k\sigma\end{array}\right)$	Sample[1] Size	Mean Two-Sided[2]	Standard Deviation Two-Sided[3]		Standard Deviation One-Sided[4]
		$\frac{z_{\alpha/2}}{\sqrt{n}}$	$\sqrt{\frac{\chi^2_{\alpha/2,n-1}}{n-1}}$	$\sqrt{\frac{\chi^2_{1-\alpha/2,n-1}}{n-1}}$	$\sqrt{\frac{\chi^2_{1-\alpha,n-1}}{n-1}}$
1.0	4	0.9800	0.2682	1.7653	1.6140
0.9	5	0.8765	0.3480	1.6691	1.5401
0.8	7	0.7408	0.4541	1.5518	1.4487
0.7	8	0.6930	0.4913	1.5125	1.4176
0.6	11	0.5910	0.5698	1.4312	1.3530
0.5	16	0.4900	0.6461	1.3537	1.2909
0.4	25	0.3920	0.7188	1.2807	1.2318
0.3	43	0.2989	0.7868	1.2128	1.1764
0.2	97	0.1990	0.8587	1.1411	1.1174
0.1	385	0.0999	0.9293	1.0707	1.0591
0.62	10	0.6198	0.5478	1.4538	1.3711
0.44	20	0.4383	0.6847	1.3149	1.2596
0.36	30	0.3578	0.7439	1.2556	1.2114
0.31	40	0.3099	0.7788	1.2208	1.1829
0.28	50	0.2772	0.8025	1.1971	1.1636
0.25	60	0.2530	0.8199	1.1798	1.1493

Note:

Sample size is calculated by $n=\left(z_{\alpha/2}\cdot\sigma/E\right)^2=\left(z_{\alpha/2}/k\right)^2$.

The two-sided 100(1-α)% confidence interval of sample mean is calculated by

$$\mu-z_{\alpha/2}\cdot\sigma/\sqrt{n}\leq\bar{x}\leq\mu+z_{\alpha/2}\cdot\sigma/\sqrt{n}.$$

The two-sided 100(1-α)% confidence interval of sample standard deviation is calculated by

$$\sqrt{\chi^2_{\alpha/2,n-1}/(n-1)}\cdot\sigma\leq s\leq\sqrt{\chi^2_{1-\alpha/2,n-1}/(n-1)}\cdot\sigma.$$

The one-sided 100(1-α)% confidence interval of sample standard deviation is calculated by

$$s\leq\sqrt{\chi^2_{1-\alpha,n-1}/(n-1)}\cdot\sigma.$$

Table 3. Summary of 95% confidence intervals of sample mean and sample standard deviation at various error levels and sample sizes for a *N*(0, 1) distribution.

If s^2 is the sample variance from a random sample of n observations from a normal distribution with known or specified variance σ^2, then a two-sided $100(1-\alpha)\%$ confidence interval on s is

$$\sqrt{\frac{\chi^2_{\alpha/2,n-1}}{n-1}}\cdot\sigma\leq s\leq\sqrt{\frac{\chi^2_{1-\alpha/2,n-1}}{n-1}}\cdot\sigma$$

As for the one-sided $100(1-\alpha)\%$ upper confidence bound as shown in Figure 7d, we may write

$$p\left(\frac{(n-1)s^2}{\sigma^2}\leq\chi^2_{1-\alpha,n-1}\right)=1-\alpha$$

then the confidence upper bound on s is

$$s \le \sqrt{\frac{\chi^2_{1-\alpha,n-1}}{n-1}} \cdot \sigma \tag{5}$$

Table 3 summarizes the 95% confidence interval of sample mean and sample standard deviation at various error levels and sample sizes. Notice that the sample size listed in Table 3 was rounded to its ceiling value.

Figure 8a plots the sample size versus the specified error ($E = |\bar{x} - \mu|$) in terms of standard error (σ) with 95% confidence interval. The two-sided 95% confidence interval on the sample mean and the one-sided 95% upper confidence bound on the sample standard deviation of a $N(0, 1)$ distribution, as a function of sample size, can be illustrated as shown in Figures 8b and 8c, respectively.

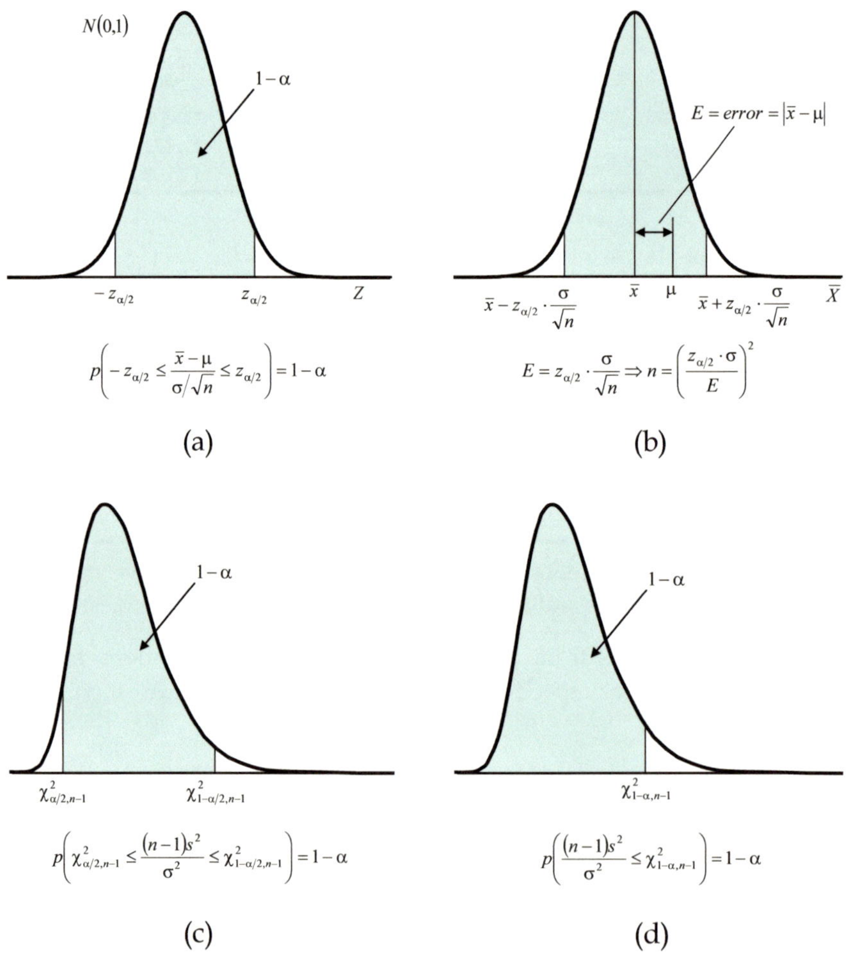

Fig. 7. (a) 100(1-α)% confidence interval of $N(0, 1)$ distribution, (b) sample size determination with a specified error level, (c) 100(1-α)% two-sided confidence interval of χ^2 distribution, and (d) 100(1-α)% one-sided confidence interval of χ^2 distribution.

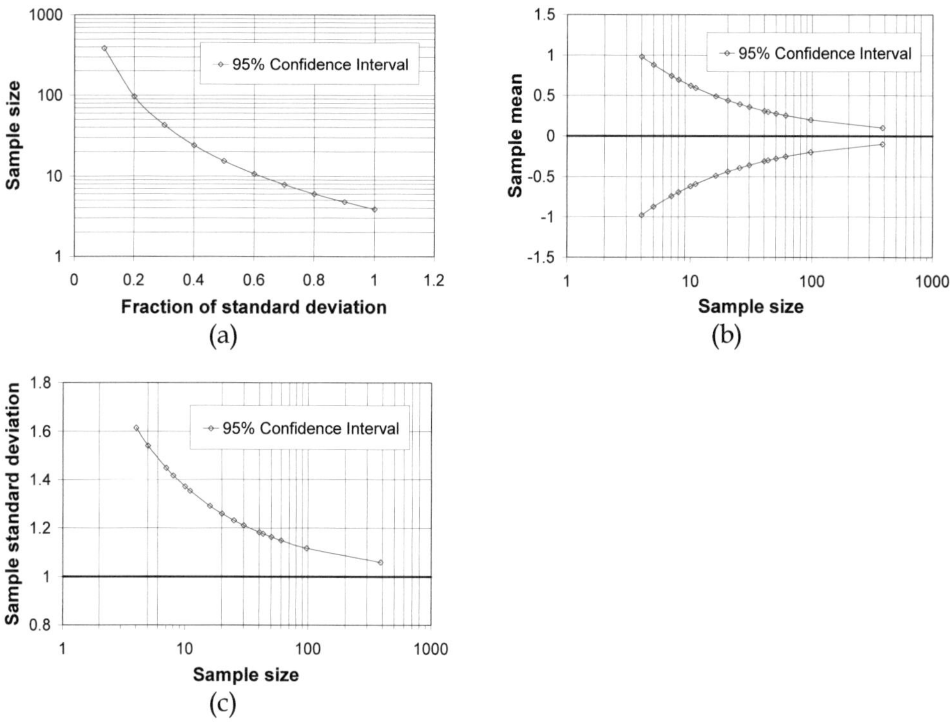

Fig. 8. (a) Sample size versus fraction of standard deviation, (b) 95% two-sided confidence interval of sample mean, and (c) 95% one-sided upper confidence bound of sample standard deviation.

3.3 Sampling scheme and size simulation

In this approach, it was assumed that the air-void contents on a project can be represented by a standard normal $N(0, 1)$ distribution. The data from the $N(0, 1)$ distribution were used to generate five data patterns: random pattern, central segregation pattern, bilateral segregation pattern, central-bilateral segregation pattern, and block segregation pattern (Figure 9). The reasons for selecting these pattern types are as follows:

1. Random pattern: non-segregation, with ideal construction quality.
2. Central segregation pattern: the gap between two augers of an asphalt paver makes coarse aggregate concentrated near the center of the paved area.
3. Bilateral segregation pattern: the gap between the auger and the lateral board of the asphalt paver makes coarse aggregate concentrated near the bilateral regions of the paved area, or provides less compaction of the side area.
4. Central-bilateral segregation pattern: a combined situation of patterns 2 and 3.
5. Block segregation pattern: as demonstrated in gradation segregation, temperature segregation, uneven compaction, etc.

The segregation horizontal strips as shown in Figures 9b, 9c, and 9d were randomly generated using the data in the shaded area of the $N(0, 1)$ distribution, which represent higher air-void contents. In the block segregation pattern (Figure 9e), the $N(0, 1)$ distribution was divided into 6 intervals and the data of each interval were randomly distributed into blocks of pavement sections.

The prospective road section was divided into $n(X)$ (x-direction) × $n(Y)$ (y-direction) cells. The $n(X)$ represents the number of intervals in the x-direction. N points were then assigned to these $n(X)$ × $n(Y)$ cells. Hence, a sampling scheme was defined by $n(X)$, $n(Y)$, and N. For instance, *x30y6n30* represents 30 runs that were assigned to 30 cells of the 30 × 6 cells based on the UD table. The sampling schemes considered in this study were combinations of various numbers of $n(X)$ and $n(Y)$ —that is, $n(X)$ = 3, 5, 10, 15, 20, 25, 30, 35, 40, 45, 55, 60 and $n(Y)$ = 1, 2, 3, 4, 6—and N = $n(X)$; however, the cases with $n(Y) > n(X)$ were excluded, resulting in a total of 62 cases. Each case was assigned a UD table with minimum CD_2 value. Figures 10a through 10c respectively illustrate the example sampling schemes (i.e., UD tables), *x10y6n10*, *x30y6n30*, and *x60y6n60*, from the uniform design. These sampling schemes are on the same scales of a 900 ft × 24 ft (274 m × 7.32 m) pavement section. The black rectangle cell physically represents the area of which one measure should be sampled randomly.

For this sampling simulation, a total of 2700 × 72 points with a standard normal distribution of air-void contents were used to generate five data patterns with the following assumptions:

1. Lane width: 24 ft (7.32 m).
2. Time frame of construction: 1 hour with 900 ft (274 m) of HMA placed, i.e., paver speed = 15 ft/min. (4.57 m/min.).
3. One air-void sample is represented by a 4 × 4-in. (10 × 10-cm) square with each square assigned a normalized air-void value.

Each type of sampling scheme per data pattern was simulated 200 times. For each simulation, the sample mean and sample standard deviation were calculated. It should be noted that the data of each simulation were randomly drawn from the cells specified in the UD table with replacement. Consequently, the distributions of the sample mean and standard deviation were generated after 200 simulations. The boxplot was then utilized to characterize the location and dispersion of sample means and standard deviations.

The boxplot illustrates a measure of location (the median [solid black dot or white strip]), a measure of dispersion (the interquartile range IQR [lower quartile: left or bottom-edge of box; upper quartile: right or top-edge of box]), and the possible outliers (data points with light circle or horizontal line outside the 1.5 IQR distance from the edges of box; the most extreme data points within 1.5 IQR distance are marked with square brackets) and also gives an indication of the symmetry or skewness of the distribution.

The Trellis graph introduced by Cleveland in 1993 (Cleveland, 1993) is a graphical way of examining high-dimensional data structure by means of conditional one-, two-, and three-dimensional graphs. As an example, we would like to determine how the sample mean distribution depends on $n(X)$, $n(Y)$, and the data pattern. To inspect this graphically, the simulation results can be split up into groups and can be plot separately as opposed to blurring the effects in a single graph. The Trellis graph of boxplots presented in Figures 11 and 12 was arranged in such a way that each panel consists of all the $n(Y)$ = 1, 2, 3, 4, 6 cases (i.e., 5 boxplots in each panel), each row is made by all the N = $n(X)$ = 3, 5, 10, 15, 20, 25, 30, 35, 40, 45, 55, 60 cases (i.e., 13 panels in a row) with the same data pattern, and each column has 5 panels (i.e., 5 data patterns) with the same $n(X)$. Thus, for each individual column, the effects of data pattern and $n(Y)$ can be examined at the specified $n(X)$; for each individual row, the effects of $n(X)$ and $n(Y)$ can be inspected at the specified data pattern. The Trellis graph was categorized by $n(X)$, $n(Y)$, and five data patterns.

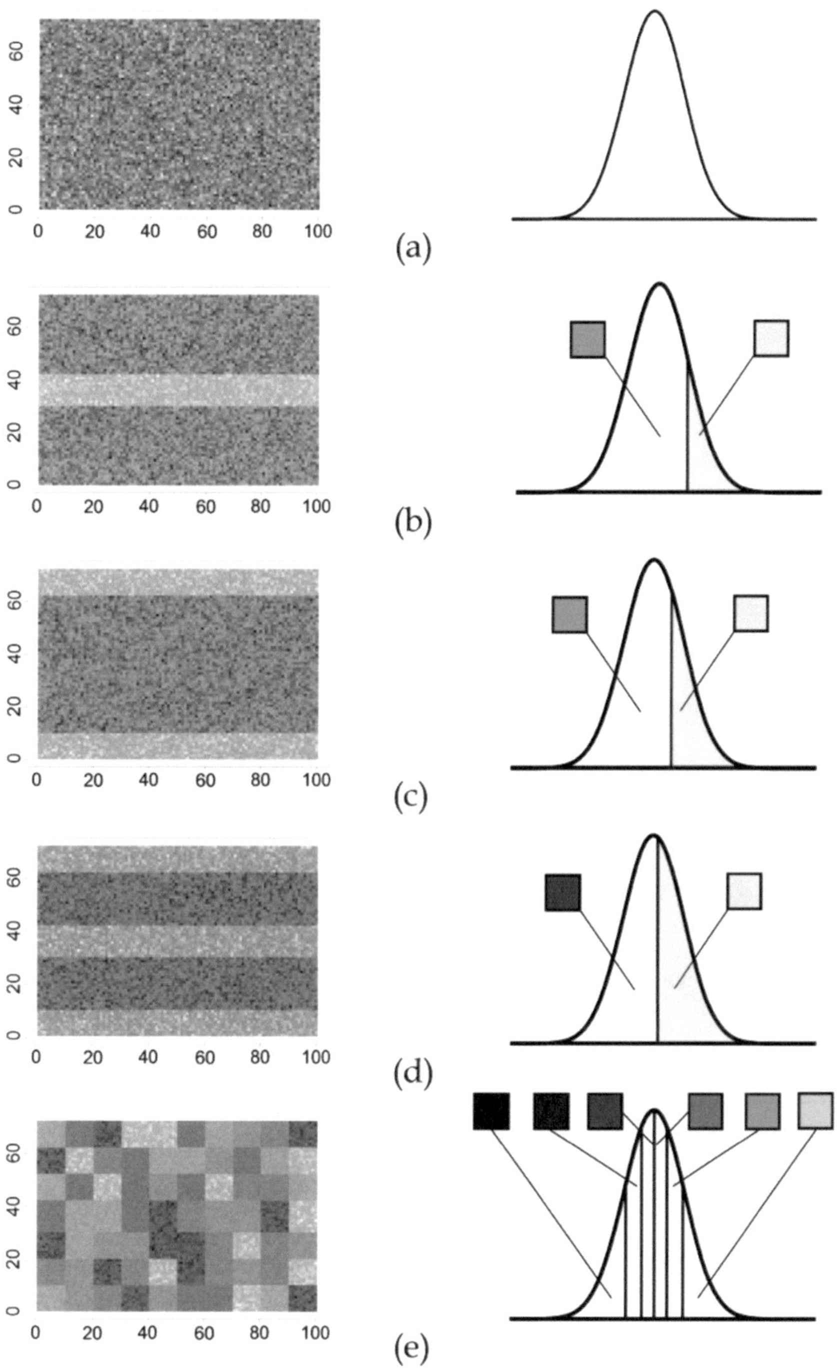

Fig. 9. Schematic illustration of five data patterns: (a) random pattern, (b) central segregation pattern, (c) bilateral segregation pattern, (d) central-bilateral segregation pattern, and (e) block segregation pattern.

The Trellis graphs of the boxplots shown in Figures 11 and 12 summarize respectively the simulation results of sample means and sample standard deviations. Several observations from the Trellis graphs can be made:

1. As $n(X)$ increases, i.e., increase of N, the variations of sample mean and standard deviation reduce exponentially regardless of data patterns.
2. For the segregation data patterns 2, 3, and 4, the increase of $n(Y)$ does benefit the decrease of variation per $n(X)$ and per data pattern. However, no apparent decrease of variation on random or block segregation patterns was perceived. This implies that the UD table provides a uniform sampling strategy. From the perspective of practice, it is suggested that $n(Y)$ should be as large as possible to include all the possible data patterns.
3. It should be noted that the distributions of sample standard deviation at small $n(X)$s exhibit unsymmetrical and skewed distributions due to the intrinsic properties of χ^2 (chi-square) distribution. From the point of view of HMA construction, the $100(1-\alpha)\%$ one-sided upper bound is suggested for the judgment of sample standard deviation. As a result, the smaller the sample standard deviation, the more uniform the construction quality of HMA. Also, from Figures 8b, 8c, 11, and 12, it is apparent that the change of variation decreases sharply at the beginning and the rate of change of variation stabilizes after $N = 20 \sim 30$.

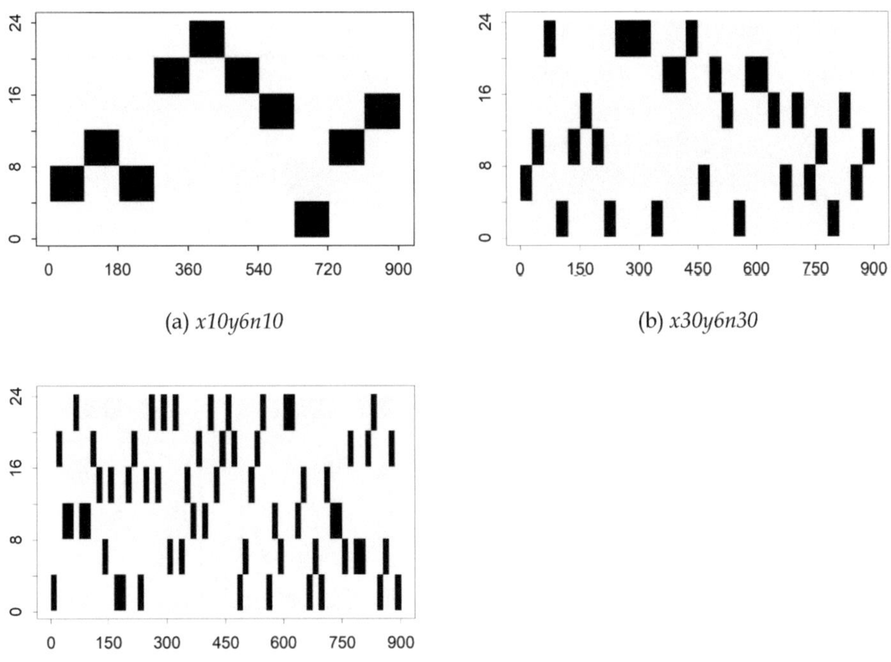

(c) *x60y6n60*

Fig. 10. Examples of UD tables (a) *x10y6n10*, (b) *x30y6n30*, and (c) *x60y6n60*.

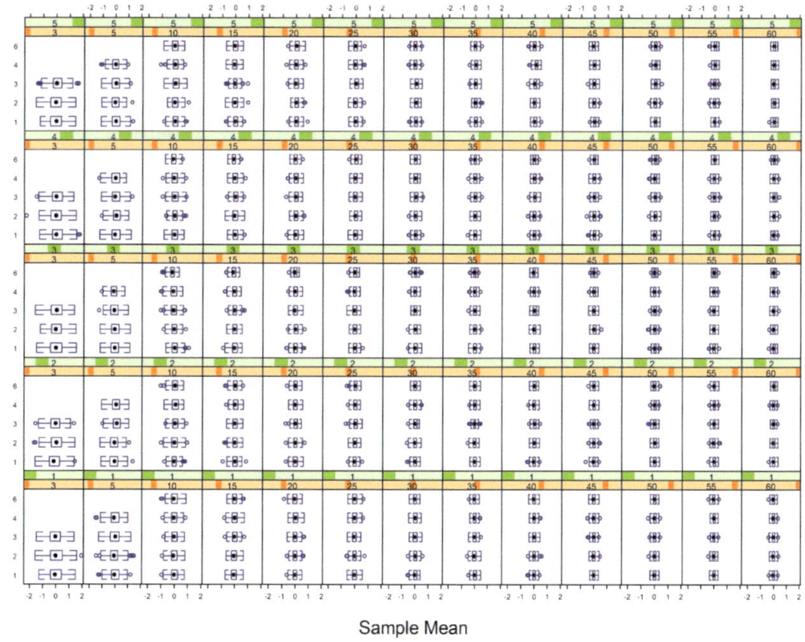

Fig. 11. Trellis graph of boxplots of sample mean categorized by *n(X)*, *n(Y)*, and five data patterns with *N* = *n(X)*.

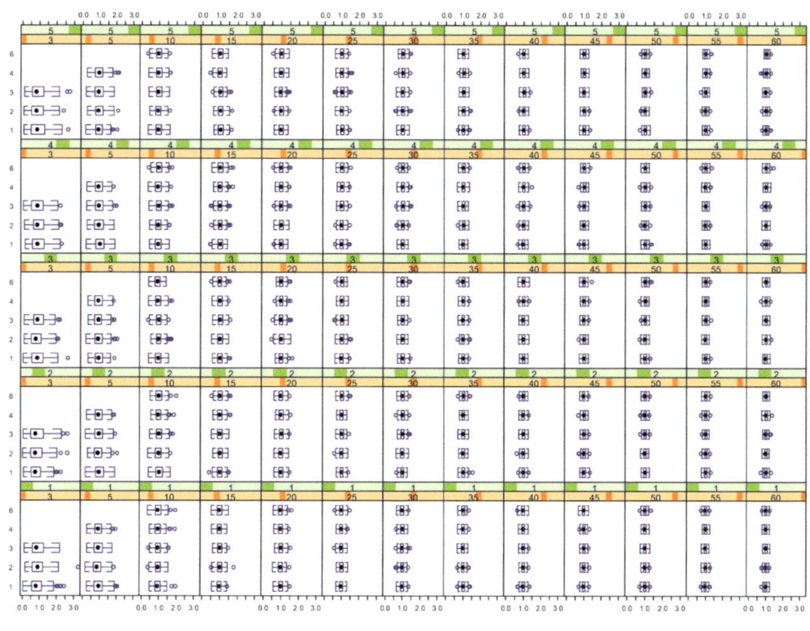

Fig. 12. Trellis graph of boxplots of sample standard deviation categorized by *n(X)*, *n(Y)*, and five data patterns with *N* = *n(X)*.

3.4 UD demonstration example using two field sections

In this demonstration example, the percent air-void content data of two field pavement sections each with 164 ft (50 m) in length and 36 ft (11 m) in width were acquired by the Pavement Quality Indicator (PQI), which is a non-nuclear density measurement device calibrated with core samples. The percent air-void content was taken by a 3.3 × 3.3-ft (1 × 1-m) square. These two pavement sections served as the "testing sections" of which the paving operation, compaction pattern/effort, and other construction details were verified and corrected (if necessary) by the contractor. Several performance tests were comprehensively conducted by the agency to guarantee that the pavement quality of the whole project met the specifications afterwards. The material properties of two pavement sections, AC-13 and AC-20, are as follows.

	Pavement Section AC-13	Pavement Section AC-20
Asphalt mix type	Asphalt concrete with nominal maximum aggregate size (NMAS) 13 mm	Asphalt concrete with NMAS 20 mm
Binder type	SBS modified binder (equivalent to PG76-22)	
Aggregate type	Diabase (fully crushed, hard rock)	Granite (fully crushed)
Design binder content	5.6%	4.8%
Target air-void content	$N(\mu, \sigma^2) = N(5, 1)$: mean 5%; standard deviation 1%.	
Acceptable air-void content range	5±2 %, i.e., $P(3 \le AV \le 7) = 0.95$ of a $N(5, 1)$ distribution	

The measured percent air-void contents are illustrated in Figures 13a and 13b respectively for the AC-13 and AC-20 pavement sections. As can be seen from the figures, the AC-13 section presents high air-void content on the section edges and seems to have a wide variation of air-void content. The AC-20 section appears to have more uniform distribution of air-void content.

To illustrate the proposed QA approach, it was decided that 20 points (20 runs) will be sampled to ensure that the agency is 95% confident that the error $|\bar{x} - \mu|$ will not exceed 0.44σ, i.e., 0.44 percent (Table 3). Two UD tables (Figures 13c and 13d) were generated for both sections which are subdivided into 10 (x-direction) by 11 (y-direction), i.e., *x10y11n20*. In this case study, the sampling for each UD table was conducted only once. Figures 13e and 13f summarized the sampled, measured, and specified distributions of air-void content. Several findings can be addressed in the following:

1. The sampled distribution based on the UD table matches the measured distribution reasonably well: AC-13 sampled $N(6.29, 1.40^2)$ versus AC-13 measured $N(6.18, 1.43^2)$; AC-20 sampled $N(5.41, 1.22^2)$ versus AC-20 measured $N(5.12, 1.24^2)$.
2. The sample mean, 6.29, of AC-13 section is outside the 95% CI (4.56, 5.44) (Table 3); therefore, it is identified as an "inaccurate" distribution. The sample standard deviation, 1.40 exceeds the 95% one-sided upper bound 1.26 (Table 3); thus, it is designated as an "imprecise" distribution. As a result, the construction quality of AC-13 section is not acceptable because of its "inaccurate" and "imprecise" distribution.
3. On the contrary, the construction quality of AC-20 section is not rejected because of its "accurate" and "precise" distribution: the sample mean 5.41 lies in the 95% CI although

on the high side; the sample standard deviation 1.22 is a slightly less than the 95% one-sided upper bound 1.26.

Accordingly, a proposed QA sampling guideline can be provided by the agency in the following steps:

1. Specify an error level (E) of sample mean in terms of standard deviation of the specified distribution $N\left(\mu,\sigma^2\right)$, i.e., $E=\left|\bar{x}-\mu\right|=k\sigma$.
2. Specify the α value to construct a $100(1-\alpha)\%$ confidence interval.
3. Determine the sample size based on the Equation 4, $n=\left(\dfrac{z_{\alpha/2}\cdot\sigma}{E}\right)^2$.
4. Generate a uniform design table (UD table) as the sampling scheme, the X factor should have n (sample size) levels, i.e., $N = n(X) = n$. It is suggested the Y factor should have at least 3 levels per lane, i.e., $n(Y) \geq 3$ per lane.
5. Randomly take the measurement from each (x,y) cell specified in UD table.
6. Check the sample mean $\bar{x}$ from n observations. If $\mu-\dfrac{z_{\alpha/2}}{\sqrt{n}}\cdot\sigma\leq\bar{x}\leq\mu+\dfrac{z_{\alpha/2}}{\sqrt{n}}\cdot\sigma$ (Equation 3), then we accept the sample mean; otherwise, the sample mean has been rejected and the agency has to reject the project.
7. Check the sample standard deviation s if step 6 has been satisfied. If $s\leq\sqrt{\dfrac{\chi^2_{1-\alpha,n-1}}{n-1}}\cdot\sigma$ (Equation 5), then we accept the sample standard deviation; otherwise the project should be rejected because of non-uniformity of construction quality.

4. Why is it inappropriate to take only three samples?

It is not uncommon for agencies to base QA on three samples. However, the following discussion using t distribution is presented showing why it is inappropriate to take only this number of samples for quality assurance. When estimating the mean of a normally distributed population with unknown mean μ and unknown variance σ^2, the t distribution should be applied especially with small sample size. Let $X_1, X_2, \ldots, X_n$ be a random sample from a normal distribution with unknown mean μ and unknown variance σ^2. The random variable $T=\dfrac{\bar{X}-\mu}{S/\sqrt{n}}$ has a t_{n-1} distribution with $n-1$ degrees of freedom, where $\bar{X}$ is the sample mean and S the sample standard deviation. Now if $\mu=\mu_0$, then

$$T_0=\frac{\bar{X}-\mu_0}{S/\sqrt{n}}=\frac{\bar{X}-\mu_0-\delta+\delta}{S/\sqrt{n}}=\frac{\bar{X}-\left(\mu_0+\delta\right)}{S/\sqrt{n}}+\frac{\delta\sqrt{n}}{S}$$

If $\delta=0$, then it is called the central t_{n-1} distribution. When the true value of the mean is $\mu=\mu_0+\delta$, the distribution of T_0 is termed the noncentral t_{n-1} distribution with the noncentrality parameter $\delta\sqrt{n}/S$. Based on the definition of type II error (β) : P{fail to reject H_0 | H_0 is false} under the hypothesis testing $H_0:\mu=\mu_0$, $H_1:\mu<\mu_0$, the Type II error is

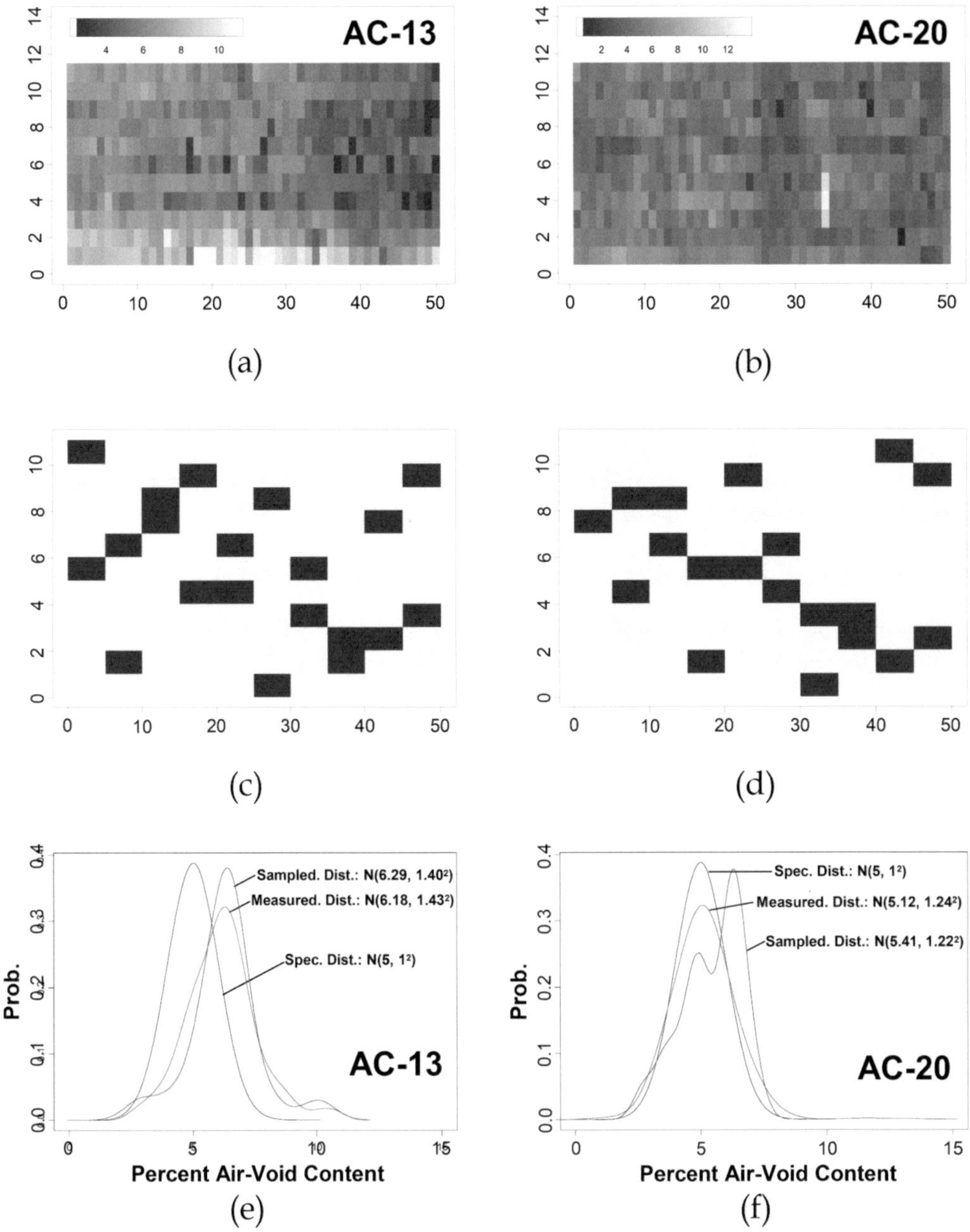

Fig. 13. Image plots of air-void measures for sections (a) AC-13 and (b) AC-20; UD tables *x10y11n20* for sections (c) AC-13 and (d) AC-20; the specification, measured, and sampled distributions for sections (e) AC-13 and (f) AC-20

made only if $T_0 \geq -t_{\alpha,n-1}$ [shown in Figures 14a and 14b respectively for the situations that (a) δ > 0 and (b) δ < 0] where T_0 is noncentral t_{n-1} distribution. Hence, we have the probability of Type II error

$$\beta = 1 - \mathrm{T}_{n-1}\left(-t_{\alpha,n-1} - \frac{\delta\sqrt{n}}{\sigma}\right), \text{ i.e., } power = 1 - \beta = \mathrm{T}_{n-1}\left(-t_{\alpha,n-1} - \frac{\delta\sqrt{n}}{\sigma}\right), \quad (6)$$

where T_{n-1} is the distribution function with $n-1$ degrees of freedom. From Figure 14, it is apparent that the more positive δ value the larger the β value, i.e., the smaller the power; on the contrary, the more negative δ value the smaller the β value, i.e., the larger the power.

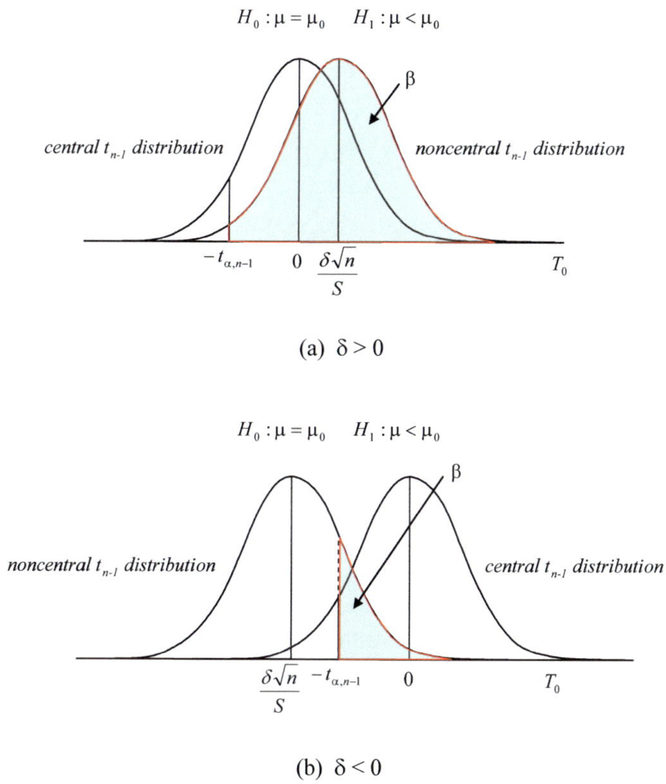

Fig. 14. Definition of type II error (β) of a *t* distribution under the hypothesis testing: $H_0 : \mu = \mu_0$, $H_1 : \mu < \mu_0$ for the situations that (a) δ > 0 and (b) δ < 0.

Equation 6 indicates that power is a function of α, *n*, and δ/S. Figure 15 plots power versus δ/S at various sample sizes. Under the hypothesis testing $H_0 : \mu = \mu_0$, $H_1 : \mu < \mu_0$, α = 0.05 and *n* = 3, the interpretation of Figure 15 is that one will have power greater than 0.8 to reject the null hypothesis if $\delta/S \leq -2.30$; on the other hand, if $\delta/S \geq -2.30$, then the agency has insufficient power to reject the null hypothesis that $H_0 : \mu = \mu_0$. It should be noted that, to increase sample size from three to five, the agency will have power greater than 0.8 if $\delta/S \leq -1.37$; that is, the agency can detect smaller mean difference from 2.30*S* down to 1.37*S* by increasing two samples. In sum, by taking only three samples out of a project, the agency will have insufficient power to reject $H_0 : \mu = \mu_0$ given that H_0 is false unless the quality of the project delivered by the contractor is so poor that the agency is confident enough to reject the project.

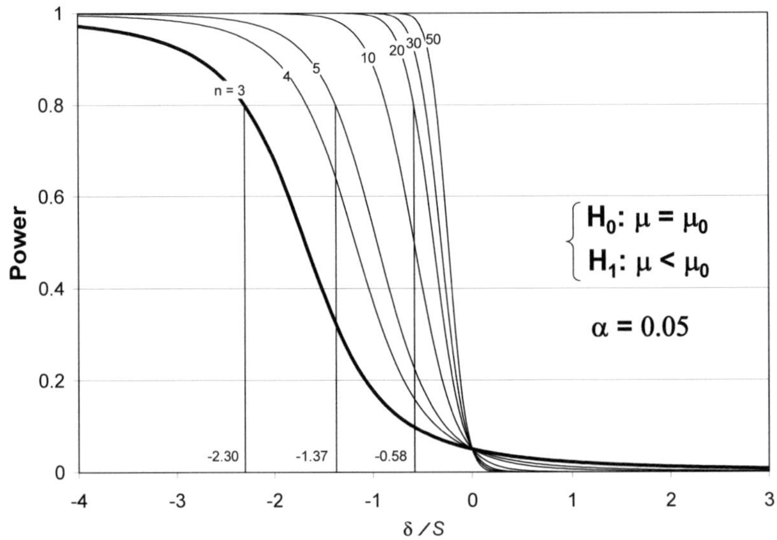

Fig. 15. Power versus δ/S curves at different sample sizes for the one-sided *t*-test at a significance level α = 0.05.

5. Findings and conclusions

For the Case I study, an attempt has been made to illustrate an approach and the extent of testing required using a performance test to insure reasonable quality in as-placed HMA. Stabilometer S-value test results were used in this example since extensive data were available. It should be emphasized that the same approach could be applied using other test parameters to control the quality of the as-constructed mix.

Based on stabilometer test results, the brief discussion of hypothesis testing, and the simulation results of sampling scheme and size, the following observations and suggestions are offered:

1. Cooperation between the agency and the contractor is essential. It is necessary to have the testing process, test equipment, test results, and specimen preparation as consistent as possible between the two organizations.
2. The sampling simulation of the Case I demonstration example suggests that the sample size required to stabilize the sampling consistency and sampling stabilization is around 50 ~ 70 for the placement of 15,000 tons HMA.
3. Likely, sampling as noted (2) is perhaps impractical. However, increasing the sample size is actually beneficial for both the agency and the contractor since it reduces the potential for dispute and guarantees the quality of the constructed mix. By extension, it is advisable for the agency to provide incentives to encourage the contractor to increase sampling size and testing.

4. To ensure the success of the proposed QC/QA guidelines, the contractor's minimum value of the testing null hypothesis must exceed that required by the agency.
5. From the Caltrans case study, the μ_{min} criterion depended not only on the contractor's α value and the agency's power level as expected but also on the *k* value that the agency would select for use. The μ_{min} criterion can be smaller if both the agency and the contractor require low power level and high α level and/or the agency increases the *k* value.

A concluding general observation relates to the concern for developing longer lasting pavement at this period of time because of increased costs of both pavement materials and increased traffic that must be accommodated. The added costs of testing by both the contractor and the agency are a very small proportion of the total costs associated with long lasting pavements. Accordingly an "attitude adjustment" for both parties relative to QC and QA testing would enhance long-term pavement performance.

From above discussion of Case II for determining sample size, simulation results of the sampling size and sampling scheme using UD tables, along with a demonstration example, the following observations and suggestions are offered:

1. It is important to recognize that the agency can be $100(1-\alpha)\%$ confident that the error $|\bar{x}-\mu|$ will not exceed a specified amount *E* if and only if the sample size is $n=\left(z_{\alpha/2}\cdot\sigma/E\right)^2$. The variations of sample mean and sample standard deviation for the 900 ft HMA paving simulation (Figures 8, 11, and 12) suggests that the minimum sample size required to stabilize the variation is around 20 ~30.
2. The UD table not only provides the most representative sampling scheme with the sample size for a given specified error level by the agency but also minimizes the possible effect of the underlying data pattern. Moreover, the UD table gives the agency a more unbiased "random" sampling scheme that can be followed in the quality assurance process.
3. The sample mean and sample standard deviation criteria proposed in the QA guideline demonstrates the accurate/inaccurate and precise/imprecise concept of sampling outcomes. If the sample mean is located in the range of $100(1-\alpha)\%$ confidence interval, then it is accurate. Precision is a term to describe the degree of data dispersion; if the sample standard deviation is less than the $100(1-\alpha)\%$ one-sided upper bound, then it is precise. The case study presents a very good example of an inaccurate/ imprecise case of the AC-13 field section and an accurate/precise case of the AC-20 field section. The quality of a project can only be accepted if and only if these criteria have been fulfilled simultaneously.
4. The proposed QA guideline with the introduction of the UD table is relatively simple, practical, and robust. The sample mean and sample standard deviation criteria are rational enough for both the agency and the contractor to agree upon.
5. It should be emphasized that the proposed QA approach could be applied with other performance measurement parameters to control the quality of the as-constructed mix, such as thickness, stabilometer testing as used in California, performance testing of fatigue and rutting, etc. Moreover, the decision-making based on this proposed QA approach can also be a basis for pay factor determination.
6. By taking only three samples out of a project, the agency will have insufficient power to reject $H_0:\mu=\mu_0$ given that H_0 is false unless the quality of the project delivered by the

contractor is so poor that the angency is confident enough to reject the project. However, by increasing sample size from three to five, the agency can detect smaller mean difference from 2.30*S* down to 1.37*S* by simply increasing two samples.

7. It is likely that the proposed sampling size is impractical. In this regard, the alternative is to establish a "testing section" similar to those in the case study and follow the proposed QA approach with the minimum sampling size (at least greater than 20) to ensure that the compaction pattern/effort, paving operation, and other construction details are appropriate to guarantee that the pavement quality meets the specifications.

6. Acknowledgments

The research associated with the first case study was conducted as a part of the Partnered Pavement Research Program supported by the California Department of Transportation (Caltrans) Division of Research and Innovation. Special thanks go to Mr. Kee Foo of Caltrans who provided the stability data from Caltrans projects. The contents of this paper reflect the views of the authors who are responsible for the facts and accuracy of the information presented and do not reflect the official views of the policies of the State of California or the Federal Highway Administration.

The field data associated with the second case study was sponsored by the Ministry of Transport of the People's Republic of China. The contents of this paper reflect the views of the authors who are responsible for the facts and accuracy of the information presented and do not reflect the official views of the policies of the Ministry of Transport of the People's Republic of China.

7. References

California Department of Transportation (Nov. 2007). *Standard Specifications,* Sacramento, Calif., USA

Cleveland, W.S. (1993). *Visualizing Data,* Hobart Press, ISBN 978-0963488404, Summit, NJ, USA

Fang, K.T. (1980). The Uniform Design: Application of Number Theoretical Methods in Experimental Design. *Acta Mathematicae Applagatae Sinica,* Vol.3, pp. 353-372, ISSN 1618-3832

Fang, K.T.; Lin, D.K.J, Winker, P. & Zhang, Y. (2000). Uniform Design: Theory and Application. *Technometrics,* Vol.42, No.3, pp. 237-248, ISSN 1537-2723

Fang, K.T. & Lin, D.K.J (2003). Uniform Experimental Designs and Their Applications in Industry. In: *Handbook of Statistics 22,* Khattree, R. & Rao, C.R., pp. 131-170, ISBN 0444-506144

Montgomery, D.C. & Runger, G.C. (2010). *Applied Statistics and Probability for Engineers,* John Wiley & Sons, Inc., ISBN 978-0-470-05304-1, USA

Stone, C.J.A. (1996). *Course in Probability and Statistics,* Duxbury Press, ISBN 0-534-23328-7, Pacific Grove, Calif., USA

Tsai, B.-W. & Monismith, C.L. (2009). Quality Control – Quality Assurance Sampling Strategies for Hot-Mix Asphalt Construction. *In Transportation Research Board: Journal of the Transportation Research Board,* No.2098, pp. 51-62, ISSN 0361-1981

Wang, Y. & Fang, K.T. (1981). A Note on Uniform Distribution and Experimental Design. *KeXue TongBao,* Vol.26, pp. 485-489, ISSN 0250-7862

7

Analysis of MLPA Data Using Novel Software Coffalyser.NET by MRC-Holland

Jordy Coffa[1,2] and Joost van den Berg[3]
[1]MRC-Holland
[2]Free University Amsterdam
[3]Berg IT solutions
The Netherlands

1. Introduction

Genetic knowledge has increased tremendously in the last years, filling gaps and giving answers that were inaccessible before. Medical genetics seeks to understand how genetic variation relates to human health and disease (National Center for Biotechnology Information, 2008). Although genetics plays a larger role in general, the knowledge of the genetic origins of disease has increased our understanding of illnesses caused by abnormalities in the genes or chromosomes, offering the potential to improve the diagnosis and treatment of patients. Normally, every person carries two copies of every gene (with the exception of genes related to sex-linked traits), which cells can translate into a functional protein. The presence of mutant forms of genes (mutations, copy number changes, insertion/deletions and chromosomal alterations) may affect several processes concerning the production of these proteins often resulting in the development of genetic disorders. Genetic disease is either caused by changes in the DNA of somatic cells in the body or it is inherited, e.g. by mutations in the germ cells of the parents.

Genetic testing is "the analysis of, chromosomes (DNA), proteins, and certain metabolites in order to detect heritable disease-related genotypes, mutations, phenotypes, or karyotypes for clinical purposes (Holtzman et al, 2002). In order to make this suitable for routine diagnostics dedicated, affordable, fast, easy-to-interpret and simple-to-use genetic tests are necessary. This allows scientists to easily access information that for instance can be used to: confirm or rule out a suspected genetic condition or help determine a person's chance of developing or passing on a genetic disorder. Several hundred genetic tests are currently in use, and more are being developed (Sequeiros et al, 2008). The Multiplex Ligation-dependent Probe Amplification (MLPA) is a PCR-based technique, which allows the detecting of copy number changes in DNA or RNA. MLPA can quantify up to 50 nucleic acid sequences or genes in one simple reaction, with a resolution down to the single nucleotide level (Schouten et al., 2002) needing only 20 ng of DNA. The MLPA procedure itself needs little hands on work allowing up to 96 samples to be handled simultaneously while results can be obtained within 24 hours. These properties make it a very efficient technique for medium-throughput screening of many different diseases in both a research and diagnostic settings (Ahn et al, 2007).

Over a million of MLPA reactions were performed last year worldwide but researchers are still concerned with the application of tools to facilitate and improve MLPA data analysis on large, complex data sets. MLPA kits contain oligo-nucleotide probes that through a biochemical reaction can produce signals that are proportional to the amount of the target sequences present in a sample. These signals are detected and quantified on a capillary electrophoresis device producing a fragment profile. The signals of an unknown sample need to be compared to a reference in order to assess the copy number. Profile comparison is a matter of professional judgment and expertise. Diverse effects may furthermore systematically bias the probe measurements such as: quality of DNA extraction, PCR efficiency, label incorporation, exposure, scanning, spot detection, etc., making data analysis even more challenging. To make data more intelligible, the detected probe measurements of different samples need to be normalized thereby removing the systematic effects and bringing data of different samples onto a common scale.

Although several normalization methods have been proposed, they frequently fail to take into account the variability of systematic error within and between MLPA experiments. Each MLPA study is different in design, scope, number of replicates and technical considerations. Data normalization is therefore often context dependent and a general method that provides reliable results in all situations is hard to define. The most used normalization strategy therefore remains the use of in-house brew analysis spreadsheets that often cannot provide the reliability required for results with clinical purposes. These sheets furthermore do not provide easy handling of large amounts of data and file retrieval, storage and archival needs to be handled by simple file management systems. We therefore set out to develop software that could tackle all of these problems, and provide users with reliable results that are easy to interpreter.

In this chapter we show the features and integrated analysis methods of our novel MLPA analysis software called Coffalyser.NET. Our software uses an analysis strategy that can adapt to fit the researcher objectives while considering both the biological context and the technical limitations of the overall study. We use statistical parameters appropriate to the situation, and apply the most robust normalization method based on the biology and quality of the data. Most information required for the analysis is extracted directly from the MRC-Holland database, producer of the MLPA technology, needing only little user input about the experimental design to define an optimal analysis strategy. In the next section we review the MLPA technology in more detail and explain the principles of MLPA data normalization. Then in section 3, we describe the main features of our software and their significance. The database behind our software is reviewed in section 4 and section 5 explains the exact workflow of our program reviewing the importance and methodology of each analysis step in detail. In the final section, we summarize our paper and present the future directions of our research.

2. Background

MLPA data is commonly used for sophisticated genomic studies and research to develop clinically validated molecular diagnostic tests, which e.g. can provide individualized information on response to certain types of therapy and the likelihood of disease recurrence. The most common application for MLPA is the detection of small genomic aberrations, often accounting for 10 to 30% of all disease-causing mutations (Redeker et al., 2008). In case of the very long DMD gene -involved in Duchenne Muscular Dystrophy— exon deletions and duplications even account for 65-70% of all mutations (Janssen et al., 2005). Since MLPA can detect sequences that differ only a single nucleotide, the technique is also widely used for the analysis of complicated diseases such as congenital adrenal hyperplasia and spinal muscular atrophy, where pseudo-genes and gene conversion complicate the analysis (Huang et al., 2007). Methylation-specific MLPA has also proven to be a very useful method for the detection of aberrant methylation patterns in imprinted regions such as can be found with the Prader-Willi/Angelman syndrome and Beckwith-Wiedemann syndrome (Scott et al., 2008). The MS-MLPA method can also be used for the analysis of aberrant methylation of CpG islands in tumour samples using e.g. DNA derived from formalin-fixed, paraffin-embedded tissues.

MLPA kits generally contain about 40-50 oligo-nucleotide probes targeted to mainly the exonic regions of a single or multiple genes. The number of genes that each kit contains is dependent on the purpose of the designed kit. Each oligo-probe consists of two hemi-probes, which after denaturation of the sample DNA hybridize to adjacent sites of the target sequence during an overnight incubation. For each probe oligo-nucleotide in a MLPA kit there are about 600.000.000 copies present during the overnight incubation. An average MLPA reaction contains 60 ng of human DNA sample, which correlates to about 20.000 haploid genomes. This abundance of probes as compared to the sample DNA allows all target sequences in the sample to be covered. After the overnight hybridization adjacent hybridized hemi-probe oligo-nucleotides are then ligated using a ligase enzyme and the ligase cofactor NAD at a slightly lower temperature than the hybridization reaction (54 °C instead of 60 °C). The ligase enzyme used, Ligase-65, is heat-inactivated after the ligation reaction. Afterwards the non-ligated probe oligonucleotides do not have to be removed since the ionic conditions during the ligation reaction resemble those of an ordinary 1x PCR buffer. The PCR reaction can therefore be started directly after the ligation reaction by adding the PCR primers, polymerase and dNTPs. All ligated probes have identical end sequences, permitting simultaneous PCR amplification using only one primer pair. In the PCR reaction, one of the two primers is fluorescently labeled, enabling the detection and quantification of the probe products.

The different length of every probe in the MLPA kit then allows these products to be separated and measured using standard capillary fragment electrophoresis. The unique length of every probe in the probe mix is used to associate the detected signals back to the original probe sequences. These probe product measurements are proportional to the amount of the target sequences present in a sample but cannot simply be translated to copy numbers or methylation percentages. To make the data intelligible, data of a probe originating from an unknown sample needs to be compared with a reference sample. This reference sample is usually performed on a sample that has a normal (diploid) DNA copy number for all target sequences. In case the signal strengths of the probes are compared with those obtained from a reference DNA sample known to have two copies of the chromosome, the signals are expected to be 1.5 times the intensities of the respective probes from the reference if an extra copy is present. If only one copy is present the proportion is expected to be 0.5. If the sample has two copies, the relative probe strengths are expected to be equal. In some circumstances reliable results can be obtained by comparing unknown samples can to reference samples by visual assessment, simply by overlaying two fragment profiles and comparing relative intensities of fragments (figure 1).

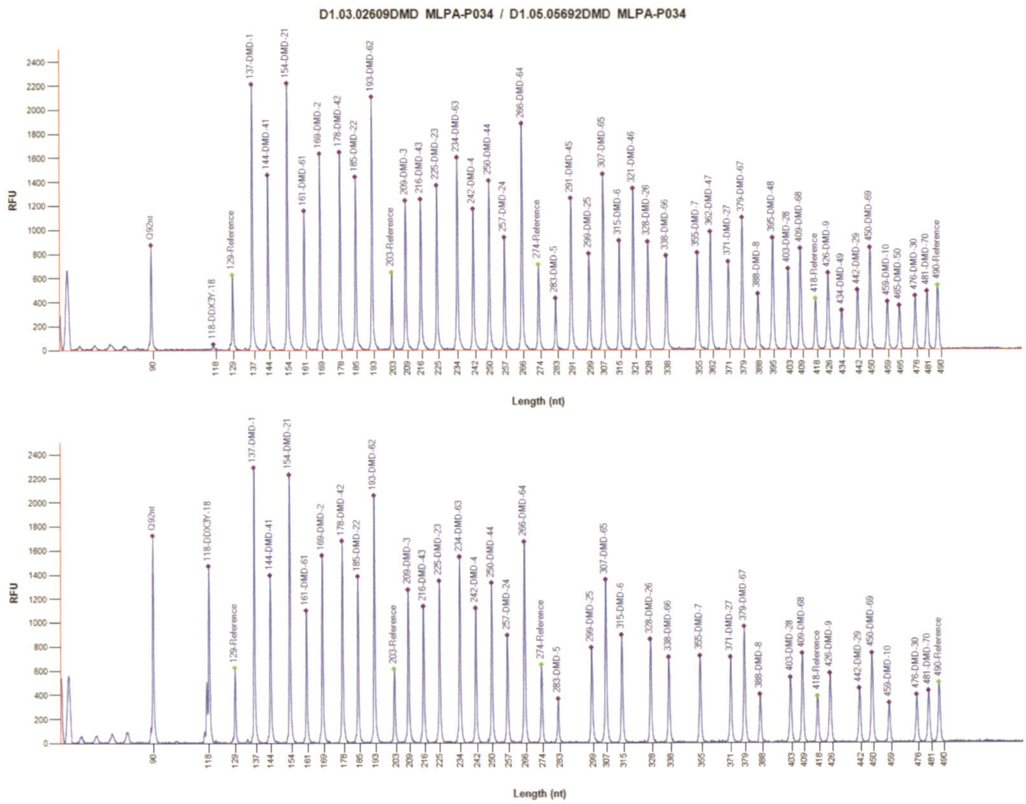

Fig. 1. MLPA fragment profile of a patient sample with Duchenne disease (bottom) and that of a reference sample (top). Duchenne muscular dystrophy is the result of a defect in the DMD gene on chromosome Xp21. The fragment profile shows that the probe signals targeted to exon 45-50 of the DMD gene have a 100% decrease as compared to the reference, which may be the result of a homozygous deletion.

It may however not be feasible to obtain reliable results out of such a visual comparison if:

1. The DNA quality of the samples and references is incomparable.
2. The MLPA kit contains probes targeted to a number of different genes or different chromosomal regions, resulting in complex fragment profiles
3. The data set is very large, making visual assessment very laborious.
4. The DNA was isolated tumor tissue, which often shows DNA profiles with altered reference probes

To make (complex) MLPA data easier understandable unknown and reference samples have to be brought on a common scale. This can be done by normalization, the division of multiple sets of data by a common variable in order to cancel out that variable's effect on the data. In MLPA kits, so called reference probes are usually added, which may be used in multiple ways in order to comprise a common variable. Reference probes are usually are targeted to chromosomal regions that are assumed to remain normal (diploid) in DNA of applicable samples. The results of data normalization are probe ratios, which display the balance of the measured signal intensities between sample and reference. In most MLPA studies, comparing the calculated MLPA probe ratios to a set of arbitrary borders is used to recognize gains and losses (González, 2008). Probe ratios of below 0.7 or above 1.3 are for instance regarded as indicative of a heterozygous deletion (copy number change from two to one) or duplication (copy number change from two to three), respectively. A delta value of 0.3 is a commonly accepted empirically derived threshold value for genetic dosage quotient analysis (Bunyan et al. 2004). To get more conclusive results probes may be arranged according to chromosomal location as this may reveal more subtle changes such as those observed in mosaic cases.

3. Key features

3.1 Support wide range of file format

Our software is compatible with binary data files produced by all major capillary electrophoresis systems including: ABIF files (*.FSA, *.AB1, *.ABI) produced by Applied Biosystems devices, SCF and RSD files produced by MegaBACE™ systems (Amersham) and SCF and ESD files produced by CEQ systems (Beckman). We can also import fragment lists in text or comma separate format, produced by different fragment analysis software programs such as Genescan (Applied Biosystems), Genemapper (Applied Biosystems), CEQ Fragment analysis software (Beckman) and Genetools. Raw data files are however preferred since they allow more troubleshooting and quality check options as compared to size called fragment lists. Next to this, raw and analyzed data are then stored in a single database and more advanced reports can be created.

3.2 Optimized peak detection / quantification method for MLPA

All applied algorithms in our software are specifically designed to suit MLPA or MLPA-like applications. We designed an algorithm for peak detection and quantification specifically for MLPA peak patterns. Most peak detection algorithms simply identify peaks based on amplitude, ignoring the additional information in the shape of the peaks. In our experience, 'true' peaks have characteristic shapes, and including fluorescence of artifacts may introduce ambiguity into the analysis and interpretation process. Our algorithm has the ability to differentiate most spurious peaks and artifacts from peaks that originate from a probe product. We differentiate a number of different peak artifacts, such as: shoulder peaks, printout spikes, dye artifacts, split peaks, pull-up peaks, stutter peaks and non-template additions. It is often difficult to identify the correct peaks due to appearance of nonspecific peaks in the vicinity of the main allele peak. Our algorithm is therefore optimized to discriminate the different artifacts from the probe signals by usage of minimum and maximum threshold values on the peak -amplitude, -area, -width and -length. Next to this, it may also recognize split and shoulder -peaks by means of shape recognition, making correct identification of probe signals even more reliable. Following peak detection, quantification and size calling, our software allows one or more peaks to be linked to the original MLPA probe target sequence. This pattern matching is greatly simplified as compared to other genotyping programs and additionally provides a powerful technique for identifying and separating signal due to capillary electrophoresis artifacts. Our software may employ three different metrics to reflect the amount of probe fluorescence: peak height, peak area and peak area including its siblings. Peak siblings are the peak artifacts that are created during the amplification of the true MLPA products but have received an alternative length. To determine which metric should be used for data normalization, our program uses an algorithm that compares the signal level of each metric over the reference probes in all samples, and compares this to the amount of noise over the same signals. The metric that has the largest level signal to noise is then used in the following normalization steps.

3.3 Performances and throughput

After a user logs in, analysis of a complete experiment can be performed in two simple steps: the processing of raw data and the comparison of different samples. Depending on the analysis setup and type of computer, the complete analysis may be completed in less than a minute for 24 samples. Our software can also make use of extra cores running in a computer, multiplying the speed of the analysis almost by two for each core. Because of problems arising from poor sample preparations, presence of PCR artifacts, irregular stutter bands, and incomplete fragment separations, a typical MLPA project requires manual examination of almost all sample data. Our software was designed to eliminate this bottleneck by substantially minimizing the need to review data. By creating a series of quality scores to the different processes users can easily pinpoint the basis for the failed analysis. These scores include quality assessment related to: the sample DNA, MLPA reaction, capillary separation and normalization steps (figure 6). The quality of each step can fall roughly into three categories.

1. High-quality or green. The results of these analysis steps can be accepted without reviewing.
2. Low-quality or red. These steps represent samples with contamination and other failures, which render the resulted data unsuitable to continue with. This data can quickly be rejected without reviewing; recommendations can be reviewed in Coffalyser.NET and used for troubleshooting.
3. Intermediate-quality or yellow. The results of these steps fall between high- and low-quality. The related data and additional recommendations can be reviewed in Coffalyser.NET and used to optimize the obtained results.

When the analysis is finished the results can be visualized in a range of different display and reporting options designed to meet the requirement of modern research and diagnostic facilities. Results effortlessly can be exported to all commonly used medical report formats such as: pdf, xls, txt, csv, jpg, gif, png etc.

3.4 Reliable recognition of aberrant probes

Results interpretation of clinically relevant tests can be one of the most difficult aspects of MLPA analysis and is a matter of professional judgment and expertise. In practice, most users only consider the magnitude of a sample test probe ratio, comparing the ratio against a threshold value. This criterion alone may often not provide the conclusive results required for diagnosing disease. MLPA probes all have their own characteristics and the level of increase or decrease that a probe ratio displays that was targeted to a region that contains a heterozygous gain or loss, may differ for each probe. Interpretation of normalized data may even be more complicated due to shifts in ratios caused by sample-to-sample variation such as: dissimilarities in PCR efficiency and size to signal sloping. Other reasons for fluctuations in probe ratios may be: poor amplification, misinterpretation of an artifact peak/band as a true probe signal, incorrect interpretation of stutter patterns or artifact peaks, contamination, mislabeling or data entry errors (Bonin et al., 2004). To make result interpretation more reliable our software combines effect-size statistics and statistical interference allowing users

to evaluate the magnitude of each probe ratio in combination with it's significance in the population. The significance of each ratio can be estimated by the quality of the performed normalization, which can be assessed two factors: the robustness of the normalization factor and the reproducibility of the sample reactions.
During the analysis our software estimates the reproducibility of each sample type in a performed experiment by calculating the standard deviation of each probe ratio in that sample type population. Since reference samples are assumed to be genetically equal, the effect of sample-to-sample variation on probe ratios of test probes is estimated by the reproducibility of these probes in the reference sample population. These calculations may be more accurate under circumstances where reference samples are randomly distributed across the performed experiment. Our program therefore provides an option to create a specific experimental setup following these criteria, thereby producing a worksheet for the wet analysis and a setup file for capillary electrophoresis devices. DNA sample names can be selected from the database and may be typed as a reference or test sample, positive control or negative control. This setup file replaces the need for filling in the sample names in the capillary electrophoresis run software thereby minimizing data entry errors.
To evaluate the robustness of the normalization factor our algorithm calculates the discrepancies computed between the probe ratios of the reference probes within each sample. Our normalization makes use of each reference probe for normalization of each test probe; thereby producing as many dosage quotients (DQ) as there are references probes. The median of these DQ's will then be used as the definite ratio. The median of absolute deviations between the computed dosage quotients may reflects the introduced mathematical imprecision of the used normalization factor. Next, our software calculates the effect of both types of variation on each test sample probe ratio and determines a 95% confidence range. By comparing each sample's test probe ratio and its 95% confidence range to the available data of each sample type population in the experiment, we can conclude if the found results are significantly different from e.g. the reference sample population or equal to a positive sample population. The algorithm then completes the analysis by evaluating these results in combination with the familiar set of arbitrary borders used to recognize gains and losses. A probe signal in concluded to be aberrant to the reference samples; if a probe signal is significantly different as from that reference sample populations and if the extent of this change meets certain criteria. The results are finally translated into easy to understand bar charts (figure 2) and sample reports allowing users to make a reliable and astute interpretation of the results.

3.5 Advanced data mining options

The database behind our software is designed in SQL and is based on a relational database management system (RDBMS). In short this means that data is stored in the form of tables and the relationship among the data is also stored in the form of tables. Our database setup contains a large number of subtraction levels, not only allowing users to efficiently store and review experimental sample data, but also allowing users to get integrative view on comprehensive data collections as well as supplying an integrated platform for comparative genomics and systems biology. While all data normalization occurs per experiment, experiments can be organized in projects, allowing advanced data-mining options enabling users to retrieve and review data in many different ways. Users can for instance review multiple MLPA sample runs from a single patient in a single report view. Results of multiple MLPA mixese may be clustered together, allowing users gain more confidence on any found results. The database can further handle an almost unlimited number of specimens for each patient, and each specimen can additionally handle an almost unlimited number of MLPA sample runs. To each specimen additional information can be related such as sample type, tissue type, DNA extraction method, and other clinical relevant data, which can be used for a wide range of data mining operations for discovery purposes. Some of these operations include:

1. Segmenting patients accurately into groups with similar health patterns.
2. Evidence based medicine, where the information extracted from the medical literature and the corresponding medical decisions are key information to leverage the decision made by the professional.

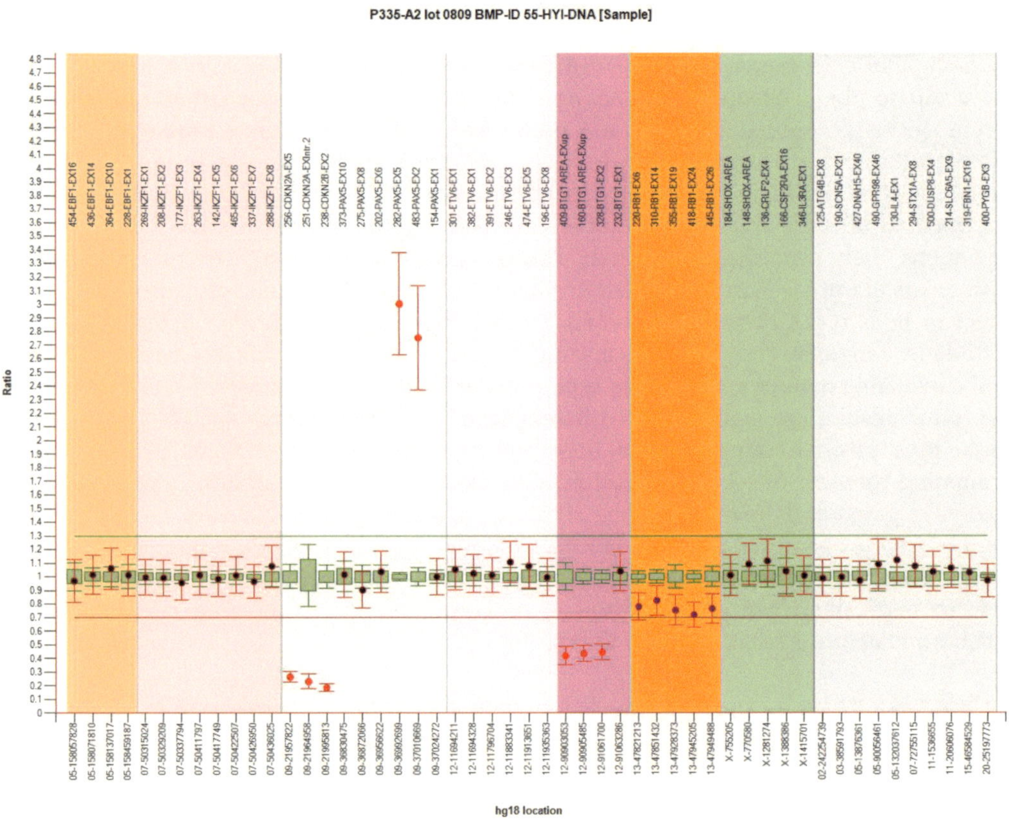

Fig. 2. Ratio chart of the results of a tumor sample analyzed with the P335 MLPA kit. Red dots display the probe ratios and the error bars the 95% confidence ranges. The orange box plots in the background show the 95% confidence range of the used reference samples. Map view locations are displayed on the x-axis and ratio results on the Y-axis. The red and green lines at ratio 0.7 and 1.3 indicate the arbitrary borders for loss and gain respectively. The displayed sample contains several aberrations and extra caution with interpretation is needed due to normal cell contamination.

3. Non-parametric tests (distribution-free) used to compare two or ore independent groups of data.
4. Classification methods that can be used for predicting medical diagnosis.

4. About the database

4.1 Client server database model

Our software uses a SQL client–server database model to store all project/experiment-related data. The client-server model has one main application (server) that deals with one or several slave applications (clients). Clients may communicate to a server over the network, allowing data sharing within and even beyond their institutions. Even though this system may provide great convenience e.g. for people who are working on a single project but are working on different locations, both client and server may also reside in the same system. Having both client and server on the same system has some advances over running both separately: the database is better protected and both client and server will always have the same version number. In case an older client will try to connect to a server that has a newer version number, the client needs to be updated first. A client does not share any of its resources, but requests a server's content or service function. Clients therefore initiate communication sessions with servers that await incoming requests. When a new client is installed on a computer it will implement a discovery protocol in order to search for a server by means of broadcasting. The server application will then answer with its dynamic address that resolves any issues with dynamic IP addresses.

4.2 User access

In addition to serving as a common data archive, the database provides user authentication, robust and scalable data management, and flexible archive capabilities via the utilities provided within Software. Our database model acts in accordance with a simple legal system, linking users to one or multiple organizations. Each user receives a certain role within each organization to which certain right are linked. These rights may for instance include denial of access to certain data but may also be used to deny access to certain parts of the program. These same levels may also be applied on project level. Projects will have project administrators and project members. The initial project creators will also be the project administrators who are responsible for user management of that project.

4.3 Sessions

As soon as a user makes a connection with the server a session will be started with a unique identifier. Subsequent made changes by any user will be held to this identifier, in order to keep track of the made changes. This number is also used to secure experiment data when in use; this ensures no two users try to edit essential data simultaneously (data concurrency). When a user logs in on a certain system, all previously open session of that user will be closed. Every user can thus only be active on a single system. On closing a session, either by logout or by double login all old user locks will disappear.

4.4 Data retrieval and updates

In our software is equipped with MLPA sheet manager software, allowing users to obtain information about commercial MLPA kits and size markers directly from the MRC-Holland database. Next to this, the sheet manager also allows users to create custom MLPA mixes.

The sheet manager software can be used to check if updates to any of the MLPA mixes are available. The sheet manager can further carry out automatic checks for updates at the frequency you choose, or it can be used to make manual checks whenever you wish. It can display scheduled update checks and can work completely in the background if you choose. With just one click, you can check to see if there are new versions of the program, or updated MLPA mix sheets. If updates are available, you can download them quickly and easily. In case some MLPA mixes are already in use, users may choose to hold on to both the older version and updated versions of the mix or replace the older version.

5. Coffalyser.NET workflow

Figure 3 shows the graphical representation of the workflow of our software. After creating an empty solution, users can add new or existing items to the empty solution by using the "add new project" or "add new experiment" command from the client software context menu. By creating projects, users can collect data of different experiments in one collection. Next, data files can then be imported to the database and linked to an experiment. Users then need to define for each used channel or dye stream of each capillary (sample run) what the contents are. Each detectable dye channel can be set as a sample (MLPA kit) or a size marker. Samples may further be typed as: MLPA test sample, MLPA reference sample, MLPA positive control, or MLPA digested sample. The complete analysis of each MLPA experiment can be divided in 2 steps: raw data analysis and comparative analysis. Raw data analysis includes all independent sample processes such as: the recognition and signal determination of peaks in the raw data streams of imported data files, the determination of the sizes of these peaks in nucleotides and the process of linking these peaks to their original probe target sequences. After raw data analysis is finished, users can evaluate a number of quality scores (figure 6), allowing users to easily assess the quality of the produced fragment data for each sample. Users may now reject, accept and adjust sample types before starting the comparative analysis. During the comparative part of the analysis several normalization and regression analysis methods are applied in order to isolate and correct the amount of variation that was introduced over the repeated measured data. Found variation that could not be normalized out of the equation is measured and used to define confidence ranges. The software finally calculates the variation of the probes over samples of the same types, allowing subsequent by classification of unknown samples. After the comparative analysis is finished, users may again evaluate a number of quality scores this time concerning the quality of different properties related to the normalization. The users can finally evaluate the results by means of reporting and visualization methods.

5.1 Import / export of capillary data

Importing data is the process of retrieving data from files to the SQL Server™ (for example, an ABIF file) and inserting it into SQL Server tables. Importing data from an external data source is likely to be the first step you perform after setting up your database. Our software contains several algorithms to decode binary files from the most commonly used capillary electrophoresis devices (see paragraph 2.1). Capillary devices usually store measurements of relative fluorescent units (RFU) and other related data that is collected during fragment separation in computer files encoded in binary form. Binary

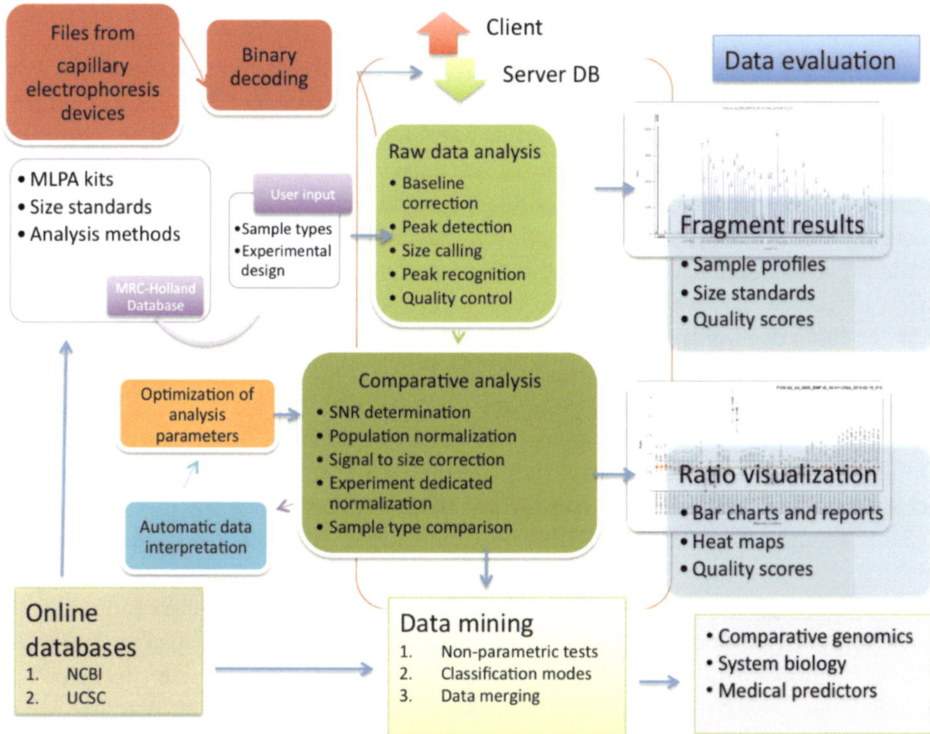

Fig. 3. Schematic overview of the Coffalyser.NET software workflow.

files are made up as a sequence of bytes, which our program decodes back into lists of the different measurements. The most important measurement being the laser induced fluorescence of the covalently bound fluorescent tags on the probe products and the size marker. The frequency at which these measurements occur depends on the type of system. A complete scan will always check all filters (or channels) and result in one data point. Almost all capillary systems are able to detect multicolor dyes permitting the usage of an internal size marker providing a more accurate size call than the usage of external size marker. Multicolor dyes may also permit the analysis of loci with overlapping size ranges, thus allowing multiple MLPA mixes to be run simultaneously in different dye colors.

After data has been imported into your SQL Server database, users can start the analysis. Users can choose to analyze the currently imported data or data that was imported in the past or a combination of both. Due to the relative nature of all MLPA data, it is recommended to analyze data within the confinements of each experiment. There do exist circumstances in which better results may be obtained by applying older collected reference data but one should use these options with caution.

Exporting data is usually a less frequent occurrence. Coffalyser.NET therefore does not have standard tools to export raw capillary data but rather depends on the provided tools and features of the SQL server. The data may be exported to a text file and then be read by third party applications such as Access or Microsoft Excel, which can then be used to view or manipulate the data.

5.2 Raw data analysis
5.2.1 Baseline correction

When performing detection of fluorescence in capillary electrophoresis devices it is some times the case that spectra can be contaminated by fluorescence. Baseline curvature and offset are generally caused by the sample itself and little can be designed in an instrument to avoid these interferences (Nancy T. Kawai, 2000). Non-specific fluorescence or background auto fluorescence should be subtracted from the fluorescence obtained from the probe products to obtain the relative fluorescence as a result of the incorporation of the fluorophore. The baseline wander of the fluorescence signals may cause problems in the detection of peaks and should be removed before starting peak detection. Our software corrects for this baseline by applying two times a median signal filter on the raw signals. First, the signals of the first 200 data points of each dye channel are extracted and its median was calculated. Then for every 200 subsequent data points till the end of the data stream, the same procedure is carried out. These median values are then subtracted from the signal of the original data stream to remove the baseline wander, resulting in baseline 1. This corrected baseline 1 is then fed as input for a filter that calculates the median signal over every 50 subsequent data points. These median values are then subtracted from all the signals that are below 300 RFU (for ABI-devices) on baseline 1, resulting in baseline 2. This second baseline is often necessary due to the relatively short distance between the peaks that derive from probe products with only a few nucleotides difference. By applying this second baseline correction solely on the signals that are in the lower range of detection, even peaks that reside close to each other may reside back to zero-signal, without subtracting too much fluorescence that originates from the probe products. Program administrators can modulate the default baseline correction settings, and also may store different defaults for each used capillary system.

5.2.2 Peak detection

In capillary-based MLPA data analysis, peak detection is an essential step for subsequent analysis. Even though various peak detection algorithms for capillary electrophoresis data exist, most of them are designed for detection of peaks in sequencing profiles. While peak detection and peak size calling are very important processes for sequencing applications, peak quantification is not so important. Due to the relatively nature of the MLPA data, peak quantification is particularly important and has a large influence on the final results. Our peak detection algorithm exists of two separate steps; the first step exists of peak detection by comparison of the intensities of fluorescent units to set arbitrary thresholds and shape recognition, the second step exist of filtering of the generated peak list by relative comparison. Program administrators can modulate the peak detection algorithm thresholds, which make use of the following criteria:

1. Detection/Intensity threshold:
   This threshold is used to filter out small peaks in flat regions. The minimal and maximal peak amplitudes are arbitrary units and default values are provided for each different capillary system.
2. Peak area ratio percentage:
   Peak area is computed as the area under the curve within the distance of a peak candidate. Peak area ratio percentage is computed as the peak area divided by the total

amount of fluorescence times one hundred. The peak area ratio percentage of a peak must be larger than the minimum threshold and lower than the maximum set threshold.

3. Model-based criterion:
   The application of this criterion can consists of 3-4 steps:
   - Locate the start point for each peak: a candidate peak is recognized as soon as the signal increases above zero fluorescence.
   - Check if the candidate peak meets minimal requirements: the peak signal intensity is first expected to increase, if the top of the peak is reached and the candidate peak meets the set thresholds for peak intensity and peak area ratio percentage, then the peak is recognized as a true peak.
   - Discarding peak candidates: if the median signal of the previous 20 data points is smaller then the current peak intensity or if the current peak intensity returns to zero.
   - Detect the peak end: the signal is usually expected to drop back to zero designating the peak end. In some cases the signal does not return to zero, a peak end will therefore also be designated if the signal drops at least below half the intensity of the peak top and if the median signal of the 14 last data points is lower than the current signal.
4. Median signal peak filter:
   The median peak signal is calculated by the percentage of intensity of each peak as opposed to the median peak signal intensity of all detected peaks. Since the minimum and maximum thresholds are dependent on detected peaks, this filter will be applied after an initial peak detection procedure based on the criteria point 1-3.
5. Peak width filter:
   After peak end points have been identified, the peak width is computed as the difference of right end point and left end point. The peak width should be within a given range. This filter is also applied after an initial peak detection procedure.
6. Peak pattern recognition:
   This method is only applied for the size marker channel, and involves the calculation of the correlation between the data point of the peak top of the detected peak list (based on the criteria point 1-5) and the expected lengths of the set size marker. In case the correlation is less than 0.999, the previous thresholds will be automatically adapted and peak detected will be restarted. These adaptations mainly include adjustment of minimal and maximal threshold values.

5.2.3 Peak size calling

Size calling is a method that compares the detected peaks of a MLPA sample channels against a selected size standard. Lengths of unknown (probe) peaks can then be predicted using a regression curve between the data points and the expected fragment lengths of the used size standard, resulting in a fragment profile (figure 4). Coffalyser.NET allows the use of 2 different size-calling algorithms:

1. Local least squares method
2. 1st, 2nd or 3rd order least squares

The local least squares method is the default size calling method for our software. It determines the sizes of fragments (nucleotides) by using the local linear relationship between fragment length and mobility (data points). Local linearity is a property of functions that have graph that appear smooth, but they need not to be smooth in a mathematical sense. The local linear least squares method makes use of a function that is only once differentiable at a point where it is locally linear. Different from the other methods, this function is not differentiable, because the slope of the tangent line is undefined. To solve the local linear function our algorithm first calculates the intercept and coefficient for each size marker point of the curve by use of a moving predictor. A local linear size of 3 points provides three predictions for each point along its curve that is surrounded by at least 2 points. The average intercept and coefficient are then stored for that point. Points at the beginning and the end of the curve will receive a single prediction, since they do not have any surrounding known values. The coefficient (β) and intercept (α) are calculated by solving the following equations 1 and 2.

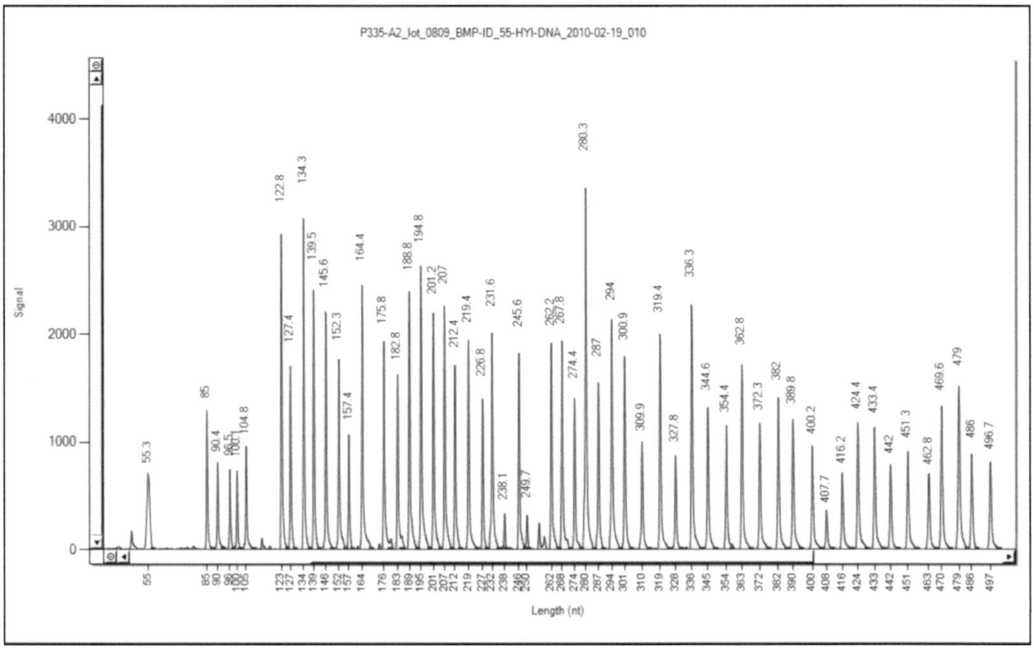

Fig. 4. MLPA fragment length profile displaying the lengths of all detected peaks from a sample. Peak lengths were determined by comparison of the data against a GS500-ABI size marker and determination of the length using the local least squares method.

$$\beta = \left(\frac{\sum X_i Y_i - \frac{1}{n} \sum X_i \sum Y_i}{\sum X_i^2 - \frac{1}{n} (\sum X_i)^2} \right) \tag{1}$$

$$\alpha = \left(\overline{Y} - \left(\beta * \overline{X} \right) \right) \tag{2}$$

E.g. if we use a size marker that has 15 known points and a local linear size of 3 points, the coefficient and intercept of point 5 will be calculated by equation 3 and 4.

$$\beta_5 = \frac{1}{3}\sum \frac{\sum X_{3-5}Y_{3-5} - \frac{1}{3}\sum X_{3-5}\sum Y_{3-5}}{\sum X_{3-5}^2 - \frac{1}{3}\left(\sum X_{3-5}\right)^2}; \frac{\sum X_{4-6}Y_{4-6} - \frac{1}{3}\sum X_{4-6}\sum Y_{4-6}}{\sum X_{4-6}^2 - \frac{1}{3}\left(\sum X_{4-6}\right)^2}; \frac{\sum X_{5-7}Y_{5-7} - \frac{1}{3}\sum X_{5-7}\sum Y_{5-7}}{\sum X_{5-7}^2 - \frac{1}{3}\left(\sum X_{5-7}\right)^2} \quad (3)$$

$$\alpha_5 = \frac{1}{3}\sum\left(\overline{Y} - \left(\beta_{3-5} * \overline{X}\right)\right);\left(\overline{Y} - \left(\beta_{4-6} * \overline{X}\right)\right);\left(\overline{Y} - \left(\beta_{5-7} * \overline{X}\right)\right) \quad (4)$$

To calculate the length of an unknown fragment our algorithm uses the calculated coefficient and intercepts calculated over the surrounded size marker peaks above and one below its peak. Each unknown point will be predicted twice where after the average value will be stored for that peak. If we wish to predict the length (Y) of an unknown fragment (X) of which the data point of the peak top is in between the data points of known fragments 5 and 6, we need to solve equation 5.

$$Y = \frac{1}{2}\sum \alpha_5 + \beta_5 * X; \alpha_6 + \beta_6 * X \quad (5)$$

5.2.4 Peak identification

Once all peaks have been size called, the profiles must be aligned to compare the fluorescence of the different targets across samples, an operation that is perhaps the single most difficult task in raw data analysis. Peaks corresponding to similar lengths of nucleotides may still be reported with slight differences or drifts due to secondary structures or bound dye compounds. These shifts in length make a direct numerical alignment based on the original probe lengths all but impossible. Our software uses an algorithm that automatically considers what the same peaks are between different samples, allowing easy peak to probe linkage. This procedure follows a window-based peak binning approach, whereby all peaks within a given window across different samples are considered to be the same peak (figure 5). Our software algorithm follows four steps: reference profile analysis, applying and prediction of new probe lengths, reiteration of profile analysis and data filtering of all samples.

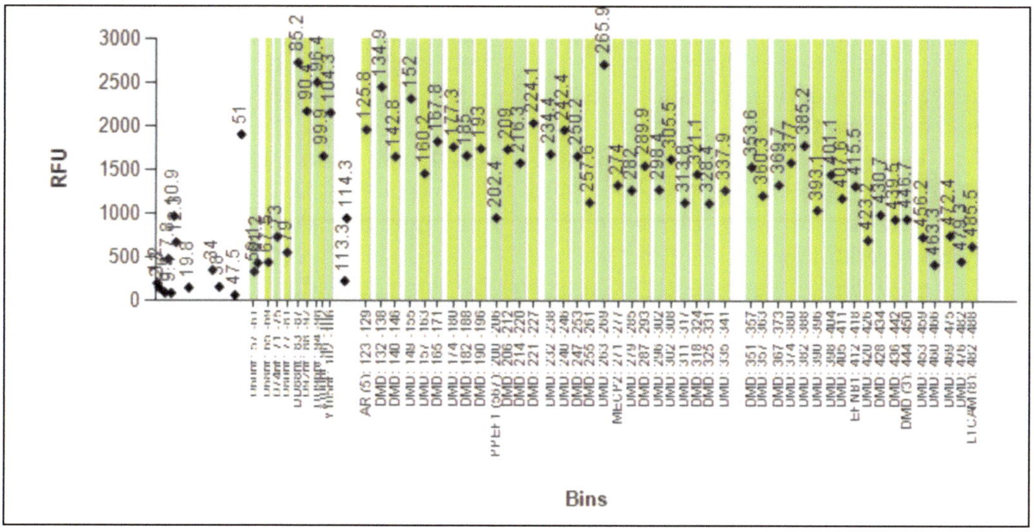

Fig. 5. Visualization of the collection of bins for a MLPA mix (x-axis) and the signal intensities in relative fluorescent units for detected peaks of a sample (y-axis).

The crucial task in data binning is to create a common probe length reference vector (or bin). In the first step our algorithm applies a bin set that searches for all peaks with a length closely resembling that of the design length of that probe. Next, the largest peak in each temporary bin is assumed to be the real peak descending from the related probe product. To create a stable bin, we calculate the average length over all real peaks of all used reference samples. If no reference samples exist, the median length over all collected real peak from all samples will be used. Since some probes may have a large difference between their original and detected length the previously created results may often not suffice. We therefore check if the length that we have related to each probe is applicable in our sample set. We do this by calculating how much variation exists over collected peaks length in each of the previous bins. If the variation was too large (standard deviation > 0.2) or no peak at all was found in any of the bins, the expected peak length for that probe will be estimated by prediction. The expected probe peak lengths may be predicted by using a second-order polynomial regression on using the available data of the probes for which reproducible data was found. Even though a full collection of bins is now available, the lengths of the probe products that were predicted may not be very accurate. The set of bins for each probe in the selected MLPA mix will therefore be improved by iteration of the previous steps. The lengths provided for the bins are now based on the previously detected or predicted probe product lengths allowing a more accurate detection of the real probe peaks. Probes that were not found are again predicted and a final length reference vector or bin is constructed for each probe. This final bin set can be used directly for data filtering but may also be edited manually in case the automatically created bin set may not suffice.

Data filtering is the actual process where the detected fragments of each sample are linked with gene information to a probe target or control fragment. Our algorithm assumes that peaks within each sample that fall within the same provided window or bin and have sufficient fluorescence intensity are the same probe (figure 4). Our algorithm is also able to link more than one peak to a probe within one sample. The amount of fluorescence of each probe product may then be expresses the peak height, peak area of the main peak and the summarized peak area of all peaks in a bin. An algorithm can then be used to compare these metrics and decide which should optimally be used as described at 3.2, alternatively users may set a default metric. The summarized peak area may reflect the amount of fluorescence best if peaks are observed that show multiple tops which all originate from the amplification of the same ligation product. Such peaks may be observed if:

1. Too much input DNA is added the amplification reaction and the polymerase was unable to complete the extension for all amplicons (Clark, J. M. 1988).
2. Peaks were discovered which are one base pair longer than the actual target due to non-template addition.
3. The polymerase was unable to complete the adenine addition on all products that resulted in the presence of shoulder peaks or +A/-A peaks (Applied Biosystems, 1988).

5.2.5 Raw data quality control

In the final step of the raw data analysis the software performs several quality checks and translates this into simple scores (figure 6).

These quality checks are the result of a comparison of sample specific properties such as: baseline height, peak signal intensity, signal to size drop, incorporated percentage of primer etc., to expected standards specific for each capillary system. Several quality checks are furthermore performed using the control fragments providing information about the used

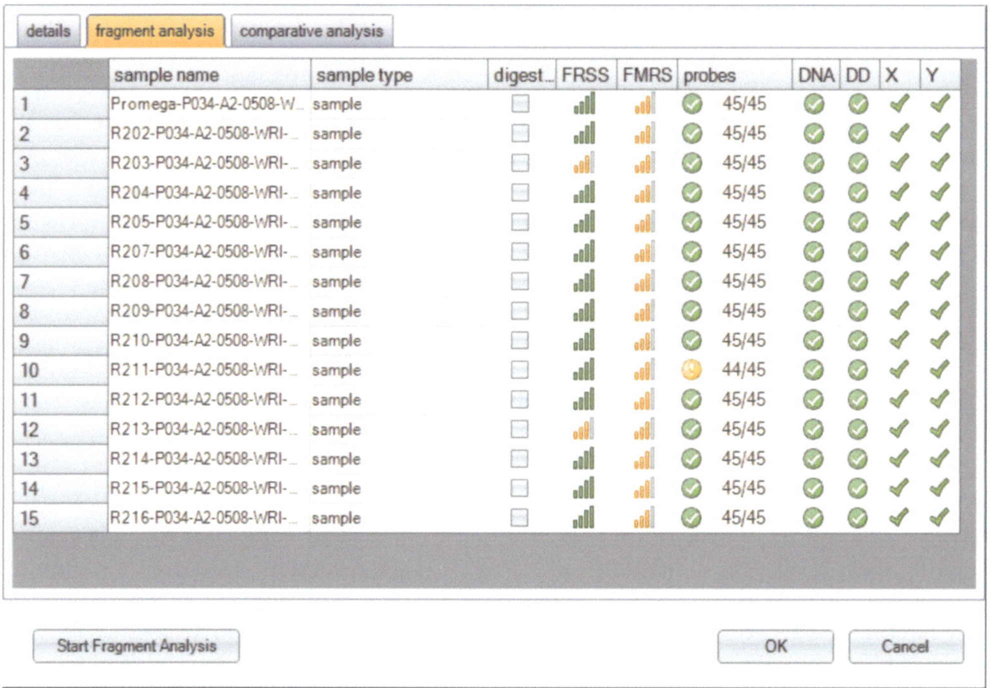

Fig. 6. Coffalyser.NET screenshot. FRSS means fragment run separation score. FMRS means fragment MLPA reaction score. Probes, displays the number of found signals to the number of expected signals. The last columns display the quality of the DNA concentration and denaturation and the presence of the X and Y- fragments.

DNA itself as described before (Coffa, 2008). The quality scores allow users to easily find problems due to: the fragment separation process, MLPA reaction, DNA concentration or DNA denaturation. Users may then reject, accept and adjust sample types before starting the comparative analysis.

5.3 Comparative analysis

During the comparative part of the analysis we aim to isolate the amount of variation that was introduced over the repeated measured data and provide the user with meaningful data by means of reporting and visualization methods. The program is equipped with several normalization strategies in order to allow underlying characteristics of the different types of data sets to be compared. During normalization we bring MLPA data (probe peak signals) of unknown and reference samples to a common scale allowing easier understandable data to be generated. In MLPA, normalization refers to the division of multiple sets of data by a common variable or normalization constant in order to cancel out that variable's effect on the data. In MLPA kits, so called reference probes are usually added, which are targeted to chromosomal regions that are assumed to remain normal (diploid) in the DNA of all used samples.

Our algorithm is able to make use of the reference probes in multiple ways in order to comprise a common variable. In case a MLPA kit does not contain any reference probes, the common variable can be made out of probes selected by the user or the program will make an auto-selection. After normalization the relative amount fluorescence related to each probe can be expressed in dosage quotients, which is the usual method of interpreting MLPA data (Yau SC, 1996). This dosage quotient or ratio is a measure for the ratio in which the target sequence is present in the sample DNA as compared to the reference DNA, or relative ploidy. To make the normalization more robust our algorithm makes use of every MLPA probe signal, set as a reference probe for normalization to produce an independent ratio ($DQ_{i,\,h,\,j,\,z}$). The median of all produced ratios is then taken as the final probe ratio ($DQ_{i,\,h\,j}$). This allows for the presence of aberrant reference signals without profoundly changing the outcome. If we want to calculate the dosage quotient for test Probe J of unknown Sample I as compared to t reference Sample H, by making use of reference Probes Z (1-n), we need to solve the equation 6.

$$DQ_{i,h,j} = med\left(\frac{\left[S_iP_j / S_iP_{z=1}\right]}{\left[S_hP_j / S_hP_{z=1}\right]}, \frac{\left[S_iP_j / S_iP_{z=2}\right]}{\left[S_hP_j / S_hP_{z=2}\right]}, \ldots \frac{\left[S_iP_j / S_iP_{z=n}\right]}{\left[S_hP_j / S_hP_{z=n}\right]}\right) \tag{6}$$

The data for each test probe of each sample ($DQ_{i,\,h,\,j}$) will be compared to each available reference sample (S_h=n), producing as many dosage quotients as there are reference samples. The final ratio ($DQ_{i,\,j}$) will then estimated by calculating the average over these dosage quotients. In case no reference samples are set, each sample will be used as reference and the median over the ratios be calculated.

5.3.1 Dealing with sample to sample variation

Each MLPA probe is multiplied during the amplification reaction with a probe specific efficiency that is mainly determined by the sequence of the probe, resulting in a probe specific bias. Even though the relative difference of these probes in signal intensity between different samples can be determined by normalization or visual assessment (figure 1), the calculated ratio results may not always be easy to understand by employing arbitrary thresholds only. This is mainly due to sample-to-sample variation or more specific, a difference in the amplification efficiency of probe targets between reference and sample targets. Chemical remnants from the DNA extraction procedure and other treatments sample tissue was subjected to, may allot to impurities that influence the *Taq* DNA polymerase fidelity. Alternatively target DNA sequences may have been modified by external factors, e.g. by aggressive chemical reactants and/or UV irradiation which may result in differences in amplification rate or extensive secondary structures of the template DNA that may prevent access to region of the target DNA by the polymerase enzyme (Elizatbeth van Pelt-Verkuil, 2008). An effect that is commonly seen with MLPA data is a drop of signal intensity that is proportional with the length of the MLPA product fragments (figure 7). This signal to size drop is caused by a decreasing efficiency of amplification of the larger MLPA probes and may be intensified by sample contaminants or evaporation during the hybridization reaction. Signal to size drop may further be influenced by injection bias of the capillary system and diffusion of the MLPA products within the capillaries.
In order to minimize the amount of variation in and between reference and sample data and create a robust normalization strategy our algorithm follows 7 steps. By automatic interpretation of results after each step our algorithm can adjust the parameters used for the next step thereby minimizing the amount of error that may be introduced by the use of

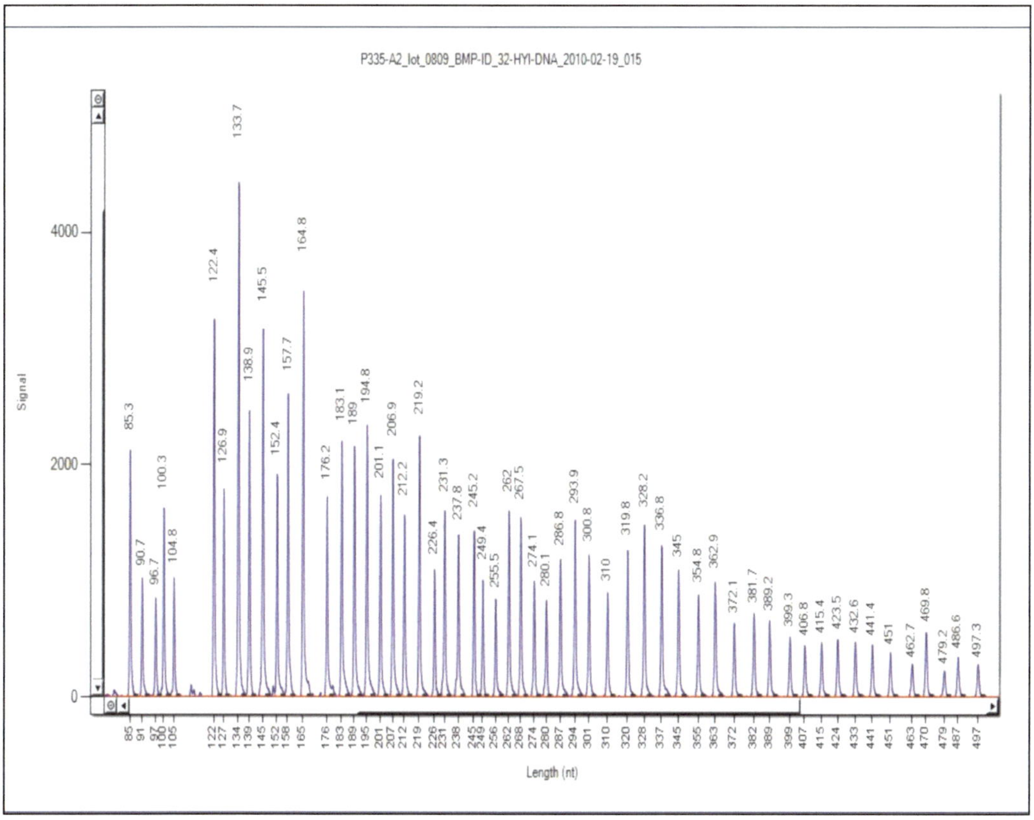

Fig. 7. MLPA fragment profile of a sample with a large drop in signal as size. This effect may have a similar result on the dosage quotients if not corrected for.

aberrant reference signals. The following 7 steps are performed in a single comparative analysis round:

1. Normalization of all data in population mode. Each sample will be applied as a reference sample and each probe will be applied as a reference probe.
2. Determination of significance of the found results by automatic evaluation using effect-size statistics and comparison of samples to the available sample type populations.
3. Measure of the relative amount of signal to size drop. If the relative drop is less than 12% a direct normalization will suffice, any larger drop will automatically be corrected by means of regression analysis (step 4-5).
4. Before correction of the actual amount of signal to size drop, samples are corrected for the MLPA mix specific probe signal bias. This can be done by calculating the extent of this bias in each reference run by regressing the probe signals and probe lengths using a local median least squares method. Correction factors for these probe specific biases are then computed by dividing the actual probe signal through its predicted signal. The final probe-wise correction factor is then determined by taking a median of the calculated values over all reference runs. This correction factor is then applied to all runs to reduce the effect of probe bias due to particular probe properties on the forthcoming regression normalization.

5. Next we calculate the amount of signal to size drop for every sample by using a function where the log-transformed probe bias corrected signals are regressed with the probe lengths using a 2nd order least squares method. Signals from aberrant targets are left out of this function, by applying an outlier detection method that makes use of the results found at step 2 as well as correlation measurements of the predicted line. The signal to size corrected values can then be obtained by calculating the distance of each log transformed pre-normalized signal to its predicted signal.
6. Normalization of signal to size corrected data in the user selected mode and determination of significance of the found results.

Our algorithm then measures the amount variation that could not be resolved in the final normalization to aid in results interpretation and automatic sample classification. To measure the imprecision of the normalization constant, each time a sample is normalized against a reference, the median of absolute deviations ($MAD_{i,\,h,\,j}$) is calculated between the final probe ratio ($DQ_{i,\,h,\,j}$) and the independent dosage quotients using each reference probe ($DQ_{i,\,h,\,j,\,z}$). The average of all collected $MAD_{i,\,j}$ values over the samples are then average to estimate the final amount of variation introduced by the imprecision of reference probes. Our algorithm estimates the final $MAD_{i,\,j}$ for each probe J in sample I and by equation 7.

$$MAD_{i,j} = \frac{1}{N}\sum_{z=1}^{N} med^{m}{}_{z=1}\left(\left|DQ_{i,h,j,z} - DQ_{i,h,j}\right|\right) \tag{7}$$

Since the final probe ratio ($DQ_{i,\,j}$) for each probe in each sample is estimated by the average over the dosage quotients ($DQ_{i,\,h,\,j}$) that were calculated using each reference sample (equation 8), the amount variation that was introduced over the different samples is estimated by calculating the standard deviation over these probe ratios (equation 9).

$$\sigma_{i,j} = \frac{1}{N}\sum_{h=1}^{N}\left(DQ_{i,h,j} - DQ_{i,j}\right)^{2} \tag{8}$$

$$DQ_{i,j} = \frac{1}{N}\sum_{h=1}^{N}\underset{z=1}{\overset{M}{Med}}\left(\frac{\left[S_iP_j / S_iP_z\right]}{\left[S_hP_j / S_hP_z\right]}\right) \tag{9}$$

Our algorithm then estimates the 95% confidence range of each probe ratio ($DQ_{i,\,j}$) of each sample by following 3 steps:

1. Conversion of the MAD values to standard deviations by multiplying with 1.4826 Albert, J. (2007)
2. Calculation of a single standard deviation for each probe ratio by combining the calculated value of step 1 with the standard deviation calculated over the reference samples by equation 9. This can be done by first converting both standard variations to variations by converting the values to the power of two. Then we sum up the outcome of both and take the square root.
3. Defining the limits of the confidence range by adding and subtracting a number of standard deviations of the final probe ratio ($DQ_{i,\,j}$) from equation 8.

Discrepancies on estimated dosage quotient by the used reference probes and/or reference samples may lead to an increase of the width of this confidence range, indicating a poor normalization. Since 95% is commonly taken as a threshold indicating virtual certainty (ZAR, J.H., 1984), our algorithm on default uses 1.96 standard deviations (equation 10) to calculate the confidence ranges for probe ratios.

$$\overset{95\%}{DQ}_{i,j} = +/-1.96 * \left(\sqrt{\left(\left(1.4826 * \left(MAD_{i,j}\right)\right)^2 + \left(\sigma_{i,j}\right)^2 \right)} \right) \tag{10}$$

5.3.2 Interpretation of the calculated dosage quotients

The previous sections explained how probe ratio are calculated and how our algorithm estimates the amount of introduced variation. In this section, we reflect on what those results mean for empirical comparison of users. To make data interpretation easier our program allows the use advanced visualization methods but also contains an algorithm allowing automatic data interpretation. Our algorithm compares the ratio and standard deviation of a test probe from a single sample to the behavior of that probe within a sub-collection of samples. This allows the program for instance to recognize if a result from an unknown sample is significantly different from the results found in the reference sample population. Alternatively, it may find if a sample is equal to a sample population, for instance a group of positive control samples. To make an estimation of the behavior of a probe ratio within a sample population, we calculate the average value and standard deviation for each probe over samples with the same sample type. In order to calculate the confidence range of probe J in for instance the reference sample population, we need to solve equation 11. N in this case refers to all probe ratio results ($DQ_{i,\,j}$) from samples that were defined in the normalization setup with the sample type: reference sample (h).

$$\overset{95\%}{DQ}_{ref\ j} = \overline{DQ_{i,j}} +/-1.96 * \left(\frac{1}{N} \sum_{i=h}^{N} \left(DQ_{i,j} - \overline{DQ_{i,j}} \right)^2 \right) \tag{11}$$

Probe result of each sample are then classified in three categories, by comparison to the confidence ranges of available sample types. A probe result of a sample is either significantly different to a sample population, equal to a sample population or the result is ambiguous. To define if a probe result of an unknown sample is significantly different (>>*) to sample population, our algorithm employs 2 criteria:

1. The difference in the magnitude of the probe ratio, as compared to the average of that probe calculated over samples with the same sample type, needs to exceed a delta value of 0.3. In case an unknown sample is compared to the reference sample population, the average ratio for each probe is always approaches 1.
2. The confidence range of the probe of the unknown sample (equation 10) cannot overlap with the confidence range of that probe in a sample population (equation 11).

An unknown sample in classified to be equal (=) to the population of a certain sample type if:

1. The difference in the magnitude of the probe ratio, as compared to the average of that probe calculated over samples with the same sample type, is less than 0.3.
2. The probe ratio of the unknown sample falls within the confidence range of that probe in a sample population (equation 11).

Probe results that are ambiguous, consequently only meet one of the two criteria in order to characterize the result to be different or equal. Ambiguous probe results that do show a difference in the magnitude of the probe ratio, as compared to the average of that probe calculated over samples with the same sample type, but have overlapping 95% confidence ranges will be marked with an asterisk (*). In case the overlap of the confidence ranges is less than 50% the probe results will be marked with a smaller or greater than symbol plus asterisk (<* or >*). Ambiguous probe results that do not show a difference in the magnitude of the probe ratio, but do show a difference in confidence ranges may be displayed with a single and double smaller or greater than symbols, depending on the size of the difference.

COFFALYSER

Coffalyser MLPA sample report: P335-A2 lot 0809 BMP-ID 55-HYI-DNA

Sample type: Sample | Project: p335 | Experiment: p335 bmp

Performed by: Administrator | Machine: ABI | Report date: 10-8-2011 | Run date: 2-8-2011

D [nt]	Gene-Exon	Chr.band	Hg18 loc.	Probe Nr.	Height	Area	T-Area	Pre	Ratio	IRatio	[RPS]	[RPQ]	Stdev	[REF]	[Sam]	[Pos]	Width	E [nt]	d[nt]
59	Q60nt	-	-	-	0	0	0	0.00	0.00	0.00	0.00	0.00	0.00	-	-	-	0	0	0.0
66	Q68nt	-	-	-	0	0	0	0.00	0.00	0.00	0.00	0.00	0.00	-	-	-	0	0	0.0
73	Q74nt	-	-	-	0	0	0	0.00	0.00	0.00	0.00	0.00	0.00	-	-	-	0	0	0.0
79	Q80nt	-	-	-	0	0	0	0.00	0.00	0.00	0.00	0.00	0.00	-	-	-	0	0	0.0
85	DD88nt	-	-	-	1285	12957	12957	0.00	0.00	1.63	0.00	0.00	0.00	-	-	-	50	85.2	-0.1
91	Q92nt	-	-	-	789	8014	8014	0.00	0.00	1.00	0.00	0.00	0.00	-	-	-	49	90.4	-0.1
96	DD96nt	-	-	-	746	7663	7663	0.00	0.00	0.95	0.00	0.00	0.00	-	-	-	41	96.7	0.1
100	X100nt	-	-	-	733	7344	7344	0.00	0.00	0.93	0.00	0.00	0.00	-	-	-	41	100.1	-0.1
105	Y105nt	-	-	-	937	9527	9527	0.00	0.00	1.19	0.00	0.00	0.00	-	-	-	50	104.6	-0.1
Control fragments																			
454	EBF1-16	05q33.3	05-158057828	12527-13577	872	11996	11996	0.83	0.94	0.94	0.02	0.06	0.05	=	=	=	50	451.1	0.1
436	EBF1-14	05q33.3	05-158071810	13868-15386	1088	14794	14794	0.88	0.99	0.99	0.01	0.06	0.05	=	=	=	50	432.8	0.1
364	EBF1-10	05q33.3	05-158137017	14059-15657	1672	21110	21110	1.00	1.03	1.02	0.01	0.06	0.04	=	=	=	50	362.9	0.0
228	EBF1-1	05q33.3	05-158459187	12509-14269	1404	14904	14904	1.10	0.99	1.00	0.01	0.06	0.04	=	=	=	43	226.3	-0.1
Region: 05q	*Chr. band: 05q33.3 Size: 401359 nt ; 401 kb*				**Region analysis:**			Average 0.99	Median 1.00	Stdev 0.03	Min 0.94	Max 1.02							
269	IKZF1-1	07p12.2	07-050315024	13877-15918	1866	20281	20281	1.04	0.97	0.95	0.01	0.06	0.04	=	=	=	50	267.3	0.0
208	IKZF1-2	07p12.2	07-050329206	14056-15654	2219	23996	23996	1.10	0.96	0.97	0.00	0.07	0.05	=	=	=	50	206.7	0.0
177	IKZF1-3	07p12.2	07-050337794	12536-13586	1879	19141	19141	1.10	0.93	0.93	0.03	0.06	0.05	=	=	=	48	176.1	0.0
263	IKZF1-4	07p12.2	07-050411797	13873-15917	1896	21431	21431	1.06	0.98	0.97	0.00	0.06	0.04	=	=	=	50	261.7	-0.1
142	IKZF1-5	07p12.2	07-050417749	13872-15390	2400	24318	24318	1.18	0.96	0.96	0.02	0.06	0.05	=	=	=	50	138.8	-0.1
465	IKZF1-6	07p12.2	07-050422507	14061-15923	678	9996	9996	0.85	0.98	1.00	0.01	0.06	0.05	=	=	=	50	462.9	0.2
337	IKZF1-7	07p12.2	07-050426950	13869-15387	2217	26815	26815	0.93	0.94	0.95	0.02	0.06	0.05	=	=	=	50	336.6	-0.2
288	IKZF1-8	07p12.2	07-050436025	12549-15921	1537	18376	18376	1.10	1.05	1.04	0.03	0.06	0.06	=	=	=	50	286.5	-0.1
Region: 07p	*Chr. band: 07p12.2 Size: 121001 nt ; 121 kb*				**Region analysis:**			Average 0.97	Median 0.96	Stdev 0.03	Min 0.93	Max 1.04							
256	CDKN2A-5	09p21.3	09-021957822	10334-10871	225	2161	3289	0.18	0.17	0.17	0.00	0.01	0.01	<<*	<<*	?	26	255.3	0.0
251	CDKN2A-2a	09p21.3	09-021964958	10333-15916	297	2734	2734	0.24	0.22	0.23	0.01	0.01	0.01	<<*	<<*	?	21	249.2	0.0
238	CDKN2B-2	09p21.3	09-021995813	10337-15914	336	3298	3298	0.20	0.18	0.18	0.00	0.01	0.01	<<*	<<*	?	33	237.7	0.1
373	PAX5-10	09p13.2	09-036830475	12521-13571	1130	14355	14355	0.94	0.99	1.02	0.01	0.06	0.05	=	=	=	50	372.1	0.0
275	PAX5-8	09p13.2	09-036872066	14057-15919	1378	15004	15004	0.93	0.88	0.86	0.02	0.05	0.04	<<	=	=	50	273.9	0.0
202	PAX5-6	09p13.2	09-036956622	14055-15653	2158	23552	23552	1.16	1.01	1.00	0.03	0.06	0.05	=	=	=	50	201.1	0.1
282	PAX5-5	09p13.2	09-036992699	13870-15920	3261	40183	40183	3.08	2.92	2.93	0.05	0.18	0.14	>>*	>>*	>>*	50	280	0.1
483	PAX5-2	09p13.2	09-037010669	13876-15924	1448	21743	21743	2.27	2.67	2.69	0.07	0.17	0.14	>>*	>>*	>>*	49	479.9	0.6
154	PAX5-1	09p13.2	09-037024272	12501-13551	1719	17812	17812	1.18	0.97	0.97	0.03	0.06	0.05	=	=	=	50	152.3	-0.1
Region: 09p	*Chr. band: 09p13.2 - 09p21.3 Size: 15066 kb ; 15 Mb*				**Region analysis:**			Average 1.12	Median 0.97	Stdev 0.97	Min 0.17	Max 2.93							
301	ETV6-1	12p13.2	12-011694211	14058-15656	1766	21049	21049	1.06	1.02	1.05	0.01	0.06	0.05	=	=	=	47	300.5	-0.1
382	ETV6-1	12p13.2	12-011694328	14060-15658	1367	17495	17495	0.94	0.99	1.01	0.01	0.06	0.04	=	=	=	50	381.6	0.0
391	ETV6-2	12p13.2	12-011796704	13875-15393	1171	15572	15572	0.92	0.98	1.00	0.01	0.06	0.04	=	=	=	50	389.3	0.1
246	ETV6-3	12p13.2	12-011883341	13874-15915	1803	19440	19440	1.18	1.08	1.05	0.01	0.07	0.05	=	=	=	49	245.1	0.0
474	ETV6-5	12p13.2	12-011913651	13871-15389	1291	18192	18192	0.90	1.05	1.08	0.03	0.07	0.06	=	=	=	50	469.9	0.2
196	ETV6-8	12p13.2	12-011935363	14054-15652	2557	27619	27619	1.12	0.97	0.98	0.02	0.06	0.05	=	=	=	50	194.6	-0.1
Region: 12p	*Chr. band: 12p13.2 Size: 241152 nt ; 241 kb*				**Region analysis:**			Average 1.03	Median 1.03	Stdev 0.04	Min 0.98	Max 1.08							
409	BTG1 AREA-up	12q21.33	12-090903053	14117-15718	357	4900	4900	0.37	0.41	0.41	0.01	0.03	0.02	<<*	<<*	<<*	44	407	0.1
160	BTG1 AREA-up	12q21.33	12-090905485	14116-15717	1071	11192	11192	0.51	0.42	0.43	0.01	0.03	0.02	<<*	<<*	<<*	45	157.5	-0.2
328	BTG1-2	12q21.33	12-091061700	12553-13603	862	10513	10513	0.43	0.43	0.43	0.00	0.03	0.02	<<*	<<*	<<*	50	327.8	-0.3
232	BTG1-1	12q21.33	12-091063286	12542-15913	1946	22529	22529	1.13	1.01	0.98	0.02	0.06	0.05	=	=	=	50	231.3	0.1
Region: 12q	*Chr. band: 12q21.33 Size: 160233 nt ; 160 kb*				**Region analysis:**			Average 0.56	Median 0.43	Stdev 0.24	Min 0.41	Max 0.98							
220	RB1-6	13q14.2	13-047821213	01782-01346	1900	20475	20475	0.86	0.76	0.76	0.01	0.05	0.04	<<	<<	>>	50	219	-0.1
310	RB1-14	13q14.2	13-047851432	01789-01353	975	12043	12043	0.82	0.80	0.82	0.02	0.05	0.04	<<	=	>>	48	309.8	-0.1
355	RB1-19	13q14.2	13-047928373	01792-15922	1115	13537	13537	0.71	0.73	0.73	0.02	0.05	0.04	<<	<<	=	45	354.6	0.0
418	RB1-24	13q14.2	13-047945205	01797-01360	684	9239	9239	0.64	0.70	0.70	0.01	0.04	0.03	<<	<<	>>	50	415.5	0.2
445	RB1-26	13q14.2	13-047949488	01799-01362	773	10665	10665	0.66	0.74	0.75	0.02	0.04	0.04	<<	<<	=	50	441.7	0.2
Region: 13q	*Chr. band: 13q14.2 Size: 128275 nt ; 128 kb*				**Region analysis:**			Average 0.75	Median 0.75	Stdev 0.04	Min 0.70	Max 0.82							
184	SHOX-AREA-down	Xp22.33P	X-000755205	06293-06219	1577	16612	17063	1.12	0.96	0.96	0.02	0.06	0.05	=	=	=	38	183	-0.1
148	SHOX-AREA-down	Xp22.33P	X-000770580	05648-06218	2171	22014	22014	1.30	1.06	1.07	0.03	0.06	0.05	=	=	=	50	145.3	-0.1
136	CRLF2-4	Xp22.33P	X-001281274	13889-15427	3065	31230	31230	1.34	1.08	1.09	0.03	0.07	0.05	=	=	=	50	133.6	-0.1
166	CSF2RA-16	Xp22.33P	X-001388386	14052-15650	2381	24386	24386	1.21	1.01	0.99	0.03	0.06	0.05	=	=	=	49	164.7	-0.2
346	IL3RA-1	Xp22.33P	X-001415701	13907-15445	1279	15854	15854	0.96	0.98	0.96	0.01	0.06	0.04	=	=	=	50	344.9	-0.1
Region: Xp	*Chr. band: Xp22.33PAR Size: 660496 nt ; 660 kb*				**Region analysis:**			Average 1.01	Median 0.99	Stdev 0.05	Min 0.96	Max 1.09							

Fig. 8. Part of a pdf report from a tumor sample analyzed with the P335 MLPA kit. The report shows clear aberrations at 9p21.3, 9p13.2 and 12q21.33. Less clear is the ratio of RB1, which displays a slight decrease in signal as opposed to the reference population, but doesn't surpass the threshold value, due to sample mosaicism.

5.3.3 Reporting and visualization

Automatic data interpretation cannot replace the specialist judgment of a researcher. Knowledge about the expected genetic defect of the target DNA and other sample information may be crucial. To assist the user with data interpretation, our software automatically sorts all probe results based on the last updated map view locations of the probes. Chromosomal aberrations often-span larger regions (M. Hermsen, 2002), which allow probes targeted to that region to cluster together by sorting. Our software can then make a single page PDF reports, containing a summary of all relevant data, probe ratios (figure 8), statistics, quality controls and charts (figure 2 & 4) of a single sample.

	Gene	Positi...	Length	Order	P335-A2_lo...	P335-A2_lo...	P335-A2_lo...	P335-A2_lo...
46	EBF1 - 16	05-158...	454	001	0.96	0.97	1.02	0.99
44	EBF1 - 14	05-158...	436	002	0.98	0.96	1	0.94
36	EBF1 - 10	05-158...	364	003	1.13	1.1	1.08	1.08
24	IKZF1 - 1	07-050...	269	004	1.08	1.08	0.55	1.11
14	IKZF1 - 2	07-050...	208	005	0.99	0.99	0.51	1.03
9	IKZF1 - 3	07-050...	177	006	0.94	0.8	0.48	0.86
23	IKZF1 - 4	07-050...	263	007	1.04	1.02	0.57	1.11
4	IKZF1 - 5	07-050...	142	008	1	0.95	0.49	0.9
47	IKZF1 - 6	07-050...	465	009	1.03	0.88	0.49	0.98
33	IKZF1 - 7	07-050...	337	010	0.91	0.92	0.56	0.92
27	IKZF1 - 8	07-050...	288	011	1.13	1.14	0.54	1.14
22	CDKN2A - 4	09-021...	256	012	0.61	0.88	0.99	0.98
21	CDKN2A - 2a	09-021...	251	013	0.64	1.06	0.94	1.01
19	CDKN2B - 2	09-021...	238	014	0.14	0.9	0.95	1.04
37	PAX5 - 10	09-036...	373	015	1.24	1.11	1	0.96
25	PAX5 - 8	09-036...	275	016	1.05	0.94	1.09	0.88
13	PAX5 - 6	09-036...	202	017	1.01	0.88	1.05	0.87
26	PAX5 - 5	09-036...	282	018	1.11	0.94	1.17	0.95
49	PAX5 - 2	09-037...	483	019	1.13	0.99	1.09	1.09
6	PAX5 - 1	09-037...	154	020	1.01	0.93	0.97	1.03
29	ETV6 - 1	12-011...	301	021	0.46	0.93	1.12	1
38	ETV6 - 1	12-011...	382	022	0.55	1.02	1.06	0.98
39	ETV6 - 2	12-011...	391	023	0.59	0.95	1.01	1.02
48	ETV6 - 5	12-011...	474	024	0.53	1.02	1.01	1.08
12	ETV6 - 8	12-011...	196	025	1.07	1.01	1.07	1.01
41	BTG1 AREA	12-090...	409	026	1.13	1.05	0.98	1
7	BTG1 AREA - 29	12-090...	160	027	0.99	1.01	1	0.96
32	BTG1 - 2	12-091...	328	028	1.08	1.01	1.04	1.11
18	BTG1 - 1	12-091...	232	029	1.08	0.96	1.04	0.96
16	RB1 - 6	13-047...	220	030	0.97	0.94	0.97	1
30	RB1 - 14	13-047...	310	031	0.97	0.95	0.97	0.97
35	RB1 - 19	13-047...	355	032	0.99	0.93	0.95	0.94
42	RB1 - 24	13-047...	418	033	1.01	0.94	1.02	1
45	RB1 - 26	13-047...	445	034	1	0.99	0.97	0.95
10	SHOX-AREA	23-755...	184	035	1.01	0.97	1.04	1.49
5	SHOX-AREA	23-770...	148	036	1.07	1.05	1.05	1.5
3	CRLF2 - 4	23-001...	136	037	1.06	0.99	1.07	1.46
8	CSF2RA - 16	23-001...	166	038	1.06	1.05	1.11	1.66
34	IL3RA - 1	23-001...	346	039	0.93	0.92	1.13	1.43
17	EBF1 - 1	05-158...	226	040	1.01	0.84	0.96	0.86
20	ETV6 - 3	12-011...	244	041	0.53	1.03	0.94	0.98
1	REFERENCE (125 nt)	02-242...	125	REF001	0.9	0.9	0.93	0.98
11	REFERENCE (190 nt)	03-038...	190	REF002	0.91	0.96	1.02	0.98
43	REFERENCE (427 nt)	05-013...	427	REF003	0.93	0.92	0.95	0.87
50	REFERENCE (490 nt)	05-090...	490	REF004	1.11	1.08	1.05	1.11
2	REFERENCE (130 nt)	05-132...	130	REF005	1.07	1.04	0.95	1.02
28	REFERENCE (294 nt)	07-072...	294	REF006	1.01	1.15	1.62	1.09
51	REFERENCE (500 nt)	11-001...	500	REF007	1	1	0.94	0.95
15	REFERENCE (214 nt)	11-020...	214	REF008	1.03	1.04	1.04	1.09
31	REFERENCE (319 nt)	15-046...	319	REF009	0.98	0.9	0.99	0.95
40	REFERENCE (400 nt)	20-025...	400	REF010	1.02	1.01	1.1	1.01

Fig. 9. Screen shot of from a tumor sample analyzed with the P335 MLPA kit. Probe ratio results of targets estimated as significantly increased as opposed to the reference population are marker green; those estimated as significantly decreased are marked red.

Our software enables users further, to display MLPA sample results in large array of different chart types figure 2 & 4). Charts may all be exported to different formats such as: jpg, gif, tiff, png, bmp. The results of a complete experiment may be plot together in grids and heat map algorithms may be applied to provide users a simple overview (figure 9). These grids may be exported to file formats (XML, txt, csv) that may be opened in Microsoft Excel. Alternatively these grids may also be exported to PDF files or several imaging formats.

6. Conclusions and future research

In this chapter we showed the options and applied algorithms of our MLPA analysis software, called Coffalyser.NET. Our software integrates new technologies enhancing the speed, accuracy and ease of MLPA analysis. Recognition of aberrations is improved by companioning effect-size statistics with statistical interference allowing users to interpreter units of measurement that are meaningful on a practical level (L. Wilkinson, 1999), while also being able to draw conclusions from data that are subject to random variation, for example, sampling variation (Bickel, Peter J.; Doksum, Kjell A., 2001). Our software contains extensive methods for results reporting and interpretation. It may also provide an alternative to software such as: Applied BioSystems Genotyper® and GeneScan® or GeneMapper®software; LiCor's SAGA, MegaBACE® Genetic Profiler and Fragment Profiler. Compatible with outputs from all major sequencing systems i.e. ABI Prism®, Beckman CEQ and MegaBACE® platforms. Coffalyser.NET is public freeware and can be downloaded from the MRC-Holland website.

Using data-mining techniques such as support vector machines in the large volumes of data obtained by large-scale MLPA experiments, may serve as a powerful and promising mechanism for recognizing of results patterns, which can be used for classification. Our future directions therefore concentrate on developing novel methods and algorithms that can improve recognition of disease related probe ratio patterns optimizing results in terms of validity, integrity and verification.

7. References

Ahn, J.W. (2007). Detection of subtelomere imbalance using MLPA: validation, development of an analysis protocol, and application in a diagnostic centre, BMC Medical Genetics, 8:9

Albert, J. (2007) Bayesian Computation with R. Springer, New York

Applied Biosystems. (1988). AmpFℓSTR® Profiler Plus™ PCR Amplification Kit user's manual.

Bickel, Peter J.; Doksum, et al. (2001). Mathematical statistics: Basic and selected topics. 1

Clark, J. M. (1988). Novel non-templated nucleotide addition reactions catalyzed by procaryotic and eucaryotic DNA polymerases. *Nucleic Acids Res* 16 (20): 9677–86.

Coffa, J. (2008). MLPAnalyzer: data analysis tool for reliable automated normalization of MLPA fragment data, Cellular oncology, 30(4): 323-35

Ellis, Paul D. (2010). The Essential Guide to Effect Sizes: An Introduction to Statistical Power, Meta-Analysis and the Interpretation of Research Results. United Kingdom: Cambridge University Press.

Elizatbeth van Pelt-Verkuil, Alex Van Belkum, John P. Hays (2008). Principles and technical aspects of PCR amplification.

González J. 2008. Probe-specific mixed model approach to detect copy number differences using multiplex ligation dependent probe amplification (MLPA), BMC bioinformatics, 9:261

Hermsen M., Postma C. (2002). Colorectal adenoma to carcinoma progression follows multiple pathways of chromosomal instability, Gastroenterology, 123 (1109-1119)

Holtzman NA, Murphy PD, Watson MS, Barr PA (1997). "Predictive genetic testing: from basic research to clinical practice". *Science (journal)* 278 (5338): 602–5.

Huang, C.H., Chang, Y.Y., Chen, C.H., Kuo, Y.S., Hwu, W.L., Gerdes, T. and Ko, T.M. (2007). Copy number analysis of survival motor neuron genes by multiplex ligation-dependent probe amplification. *Genet Med.* 4, 241-248.

Janssen, B., Hartmann, C., Scholz, V., Jauch, A. and Zschocke, J. (2005). MLPA analysis for the detection of deletions, duplications and complex rearrangements in the dystrophin gene: potential and pitfalls. *Neurogenetics.* 1, 29-35.

Kluwe, L., Nygren, A.O., Errami, A., Heinrich, B., Matthies, C., Tatagiba, M. and Mautner, V. (2005). Screening for large mutations of the NF2 gene. *Genes Chromosomes Cancer.* 42, 384-391.

Michils, G., Tejpar, S., Thoelen, R., van Cutsem, E., Vermeesch, J.R., Fryns, J.P., Legius, E. and Matthijs, G. (2005). Large deletions of the APC gene in 15% of mutation-negative patients with classical polyposis (FAP): a Belgian study. *Hum Mutat.* 2, 125-34.

Nakagawa, Shinichi; Cuthill, Innes C (2007). "Effect size, confidence interval and statistical significance: a practical guide for biologists". *Biological Reviews Cambridge Philosophical Society* 82 (4): 591–605

"NCBI: Genes and Disease". NIH: National Center for Biotechnology Information (2008).

Redeker, E.J., de Visser, A.S., Bergen, A.A. and Mannens, M.M. (2008). Multiplex ligation-dependent probe amplification (MLPA) enhances the molecular diagnosis of aniridia and related disorders. *Mol Vis.* 14, 836-840.

Schouten, J.P. (2002), Relative quantification of 40 nucleic acid sequences by multiplex ligation-dependent probe amplification. Nucleic Acids Research, 20 (12): e57

Scott, R.H., Douglas, J., Baskcomb, L., Nygren, A.O., Birch, J.M., Cole, T.R., Cormier-Daire, V., Eastwood, D.M., Garcia-Minaur, S., Lupunzina, P., Tatton-Brown, K., Bliek, J., Maher, E.R. and Rahman, N. (2008). Methylation-specific multiplex ligation-dependent probe amplification (MS-MLPA) robustly detects and distinguishes 11p15 abnormalities associated with overgrowth and growth retardation. *J Med Genet.* 45, 106-13.

Sequeiros, Jorge; Guimarães, Bárbara (2008). Definitions of Genetic Testing EuroGentest Network of Excellence Project.

Taylor, C.F., Charlton, R.S., Burn, J., Sheridan, E. and Taylor, GR. (2003). Genomic deletions in MSH2 or MLH1 are a frequent cause of hereditary non-polyposis colorectal cancer: identification of novel and recurrent deletions by MLPA. *Hum Mutat.* 6, 428-33.

Wilkinson, Leland; APA Task Force on Statistical Inference (1999). "Statistical methods in psychology journals: Guidelines and explanations". *American Psychologist* 54: 594–604. doi:10.1037/0003-066X.54.8.594.

Yau SC, Bobrow M, Mathew CG, Abbs SJ (1996). "Accurate diagnosis of carriers of deletions and duplications in Duchenne/Becker muscular dystrophy by fluorescent dosage analysis". *J. Med. Genet.* 33 (7): 550–558. doi:10.1136/jmg.33.7.550.

Zar, J.H. (1984) *Biostatistical Analysis.* Prentice Hall International, New Jersey. pp 43–45

Part 3

Quality Control for Biotechnology

8

Establishment and Quality Control Criteria for Population Culture Collection - Promising Strategy for Animal Genetic Resource Preservation

Yuehui Ma et al.[1]
Institute of Animal Sciences, Chinese Academy of Agricultural Sciences, Beijing PR China

1. Introduction

Animal genetic resources top the list of the most fundamental properties for animal husbandry and medical sciences, play an irreplaceable role in human survival and social prosperity, constitute an invaluable substrate for scientific researches, and have an enormous bearing on sustainable development of economy. Biodiversity takes an underlying part in ecological balance. The havoc people wreaked on nature had been aggravating ever since agricultural civilization, accelerating the extinction of animal species and breed incomparable than gradual natural loss, which became more and more apparent upon the advent of industrial age. The statistics of Food and Agriculture Organization (FAO) in 1995 revealed that approximately 15% of the total 738 registered livestock and poultry breeds in Sub-Saharan Africa were on the brink of extinction. The situation has been aggravating ever since. To date, the proportion of livestock breed in danger has increased from 8% to 19%, while that of poultry has risen from 20% to 34%. Among the 1251 registered breeds in Asia, 10% are severely endangered. From 1995 to 1999, livestock breeds about to be extinct grew from 11% to 14%, and the proportion of poultry was 32% to 37%. Owing to economic pressure, some low yield breeds are being subjected to marketing elimination and shrinkage in population, for instance, the production of poultry and swine depends heavily on only a few breeds. The trend is extremely obvious in Eastern Europe, which is further worsened by political unstability. Similarly, sustaining intensification of animal husbandry makes the food production rely more and more on a small number of high yield breeds, thereby exacerbating the animal diversity crisis. In Latin America, the number of endangered breeds accounts for 20% of the whole. As was reported by the FAO in 2000, livestock and poultry throughout the earth are disappearing at the rate of 2 breeds per week. Worse still, 1350 breeds are next to imminent extinction. For all that matter, animal genetic resources are confronted with a progressive narrowing in diversity. Accordingly, it's absolutely exigent to protect and preserve them with effective measures.

---

[1] Weijun Guan, Xiangchen Li, Xiaohong He, Yabin Pu, Qianjun Zhao, Dapeng Jin, Shen Wu, Taofeng Lu, Xiaohua Su, Chunyu Bai
Institute of Animal Sciences, Chinese Academy of Agricultural Sciences, Beijing, PR China

In the context of biodiversity crisis, no country around the world stands indifferent regarding the preservation of animal genetic resources. The contest for genetic resources, the basis for animal husbandry and sustainable development, is nearly incandescent.
Scientists from all around the world have been endeavouring to preserve and to make use of animal genetic resources, which are now stored in terms of individual animals, semen, embryos, genomic libraries, cDNA libraries, etc. Unfortunately, these alternative methods remain problematic for several reasons: i) endangered species and breeds are incredibly diversified, rendering the costs for preservation of individual animals unaffordable; ii) some core techniques for semen and embryos are still immature; iii) genomic DNA or organ preservation are not applicable in long term because of their finite proliferative capabilities; iv) genomic libraries and cDNA libraries are not the basic unit of cellular activities, moreover, their biological function can only be represented in transgenic techniques. Accordingly, preservation of animal genetic resources in terms of somatic cells is essentially an effective and appealing procedure to protect vulnerable mammalian and avian species, as well as all other kinds of animals. In comparison, somatic cell line, by virtue of its low costs, large capacity, convenient application, proliferative potential and so on, is supposed to be a promising strategy for storage of animal genetic resources.
Consistent with this notion, culture collections of animal materials, mainly identified cell lines, have been established and developed over time. American Type Culture Collection (ATCC), for instance, endeavours to isolate, collect, preserve and supply reliable cell lines, with its all identified type culture applicable for register, preservation, instant use, and even commercial provision. European Collection of Animal Cell Culture (ECACC), jointly run by England and Switzerland, has collected and identified some 1600 cell lines. Established in 1986, Kunming Institute of Zoology, Chinese Academy of Sciences has collected cell stains, tissues and germ cells of numerous precious species of wild life.
The lab of Animal Genetic Resources, Institute of Animal Sciences, Chinese Academy of Agricultural Sciences, by virtue of its solid technical platform, advantages in animal genetic resources, and persistent efforts in the past decade or so, has established Animal Population Culture Collection of China (APCCC), the most massive animal population culture collection in the world, comprising somatic cell lines with different tissue origins, including ear margin, blood, kidney, heart, brain, muscle, stomach and so on, from 88 animal breeds primarily in China, *e.g.* chicken (*Gallus gallus*), duck (*Anseriformes Anatidae*), goose (*Anser cygnoides orientalis*), sandpiper (*Scolopacidae*), cattle (*Bos taurus*), sheep (*Ovis aries*), goat (*Capra hircus*), pig (*Suidae*), ferret-polecat (*Mustela Pulourius Furot*), raccoon dog (*Nyctereutes procyonoides*), horse (*Equus caballus*), mule, red deer (*Cervus elaphus*), sika deer (*Cervus nippon*), fox (*Vulpinae*), wolf (*Canis lupus*), bactrian camel (*Camelus bactrianus*), tiger (*Panthera tigris*), *etc.*, and further endeavours to conserve other animal species and breeds in the world. A well-orchestrated series of standardized technical lines and quality control criteria is steadily ameliorated in this process.
This chapter will introduce the preservation of animal genetic resources in terms of somatic cells and the quality control criteria by detailed experimental description and technical line.

2. Isolation, *in vitro* culture and identification of somatic cell lines

2.1 Sampling and cell culture

Tissue pieces (about 1 cm^3 in size) were sampled from animals and placed into sterile tubes containing Dulbecco's modified Eagle's medium (DMEM, Gibco) (for livestock breeds)/

Modified Eagle's Medium (MEM, Gibco) (for poultry breeds) medium supplemented with ampicillin (100 U/ml) and streptomycin (100 µg/ml). The samples were immediately brought back to the laboratory for further experiments.

The tissues were rinsed and chopped to 1 mm^3 pieces, which were then plated onto the bottom of a tissue culture flask in an incubator at 37°C with 5% CO_2 for 2 h until the tissue pieces spontaneously adhered to the flask surface, and then DMEM/MEM containing 10% fetal bovine serum (FBS, Gibco) was added. Cells were harvested when they reached 80%-90% confluence and were passaged into more flasks at the ratio of 1:2 or 1:3 (Freshney 2000).

After three passages, the cells in logarithmic phase were harvested and resuspended at the concentration of 4×10^6/ml in cryogenic media containing 40% DMEM/MEM, 50% FBS and 10% DMSO (Sigma), aliquoted into cryovials, and kept at 4°C for 20-30 min to equilibrize the DMSO. Then they were put into a programmed cryopreservation system with controllable temperature dropping rate, and finally transferred to liquid nitrogen for long-term storage (Jenkins, 1999).

2.2 Trypan Blue exclusion test

Viabilities before cryopreservation and after resuscitation were determined using Trypan blue exclusion test. The cells were plated in 6-well plates at 10^4/well and counted with a hemocytometer (Qi et al., 2007).

2.3 Growth dynamics

According to the method of Gu et al. (Gu et al., 2006) and Ikeda Y et al. (Ikeda Y, 1990), cells at the concentration of 1.5×10^4/ml were plated into 24-well plates. Three wells were counted each day until the plateau phase. Based on the numbers, the mean values of cell density were then calculated and plotted against the culture time. The population doubling time (PDT) was calculated accordingly.

2.4 Microbial detection

- Tests for contamination with bacteria, fungi and yeasts: the cells were cultured in antibiotic free media. Bacterial, fungal and yeast contamination was assessed within 3 days as described by Doyle et al. (1990).
- Test for viruses: the cells were subjected to Hay's hemadsorption protocol using phase-contrast microscopy to detect cytopathogenesis (Hay, 1992).
- Test for mycoplasmas: cells were cultured in antibiotic free media for at least 1 week, and then fixed and stained with Hoechst 33258 (Sigma) according to the method of Masover and Becker (1998) and Freshney (2000). The ELISA Mycoplasma Detection kit (Roche, Lewes, East Sussex, UK) was used to confirm the results of the DNA fluorescent staining.

2.5 Karyotyping and chromosomal indices

The cells were harvested upon 80%-90% confluence. Microslide preparation and chromosome staining were performed as described by Suemori et al. (2006). Fifty to 100 spreads were sampled for counting chromosome numbers of diploid cells. There are three important parameters for chromosomal analysis, i.e. relative length, arm ratio, and centromere index, which were determined according to the protocol of Kawarai et al. (2006).

2.6 Isoenzyme analysis

Isoenzyme profiles of lactic dehydrogenase (LDH) and malic dehydrogenase (MDH) were identified by vertical slab non-continuous polyacrylamide gel electrophoresis (PAGE). The cells were harvested, pelleted and resuspended in protein extraction solution (0.9% Triton X-100, 0.06 mmol/L NaCl:EDTA in volume ratio of 1:15) at the density of 5×10^7 cells/ml. Then the suspension was centrifuged and the supernatant was stored in aliquots at -70°C. Isovolumic 40% (m/v) sucrose and 2.5 ml loading buffer were added to each sample to get loading solution (Zhongxiao and Shuzheng, 1999). Subsequently the electrophoresis was performed at the voltage of 120 V. When the bromophenol blue migrated into the separation gel, the electrophoresis voltage was modified to 220 V. The electrophoresis was terminated when the bromophenol blue migrated to the bottom (0.5 - 1 cm to the margin). Different mobility patterns were differentiated by the relative mobility front (RFs), which was calculated as the ratio of the migration distances of the isozyme bands to that of the bromophenol blue.

2.7 Expression of exogenous genes

According to the method of Tsuchiya et al. (2002), the fluorescent plasmids pEGFP-C1, pEGFP-N3, pEYFP-N1, pDsRed1-N1,pECFP-N1 and pECFP-mito were transfected into the cells with Lipofectamine™ 2000 transfection reagent (Invitrogen Corp., Carlsbad, CA). The plasmid DNA (μg) to Lipofectamine 2000 (μl) ratio was 1:3. After 8 h, the serum-free transfection media were replaced with complete media. To evaluate the transfection efficiency, the cells were observed under a confocal microscope (Nikon TE-2000-E, Japan)at 24 h, 48 h and 72 h after transfection, respectively. The test data were subjected to multiple comparisons to analyze statistical difference. For each sample, images were captured from ten visual fields, and the total and positive cells were counted in each field to determine the transfection efficiency.

3. Results

3.1 Morphological observation

The somatic cells sprouting from tissue explant pieces grew rapidly and migrated from the tissues with a different time and speed according to the species origin. Fibroblasts were initially mingled with epithelial cells, but the fibroblasts, by virtue of their proliferative superiority, would outgrow the epithelial cells gradually after 2-3 passages, and prevail in the population (Fig. 1). Then, purified fibroblast lines were obtained. The cells had fibrous contour with plump cytoplasm, and during growth they were morphologically fibroblast-like with radiating, flame-like or whirlpool migrating patterns. The cells were then subjected to programmed cryopreservation. Trypan blue exclusion test showed non-significant difference ($P>0.05$) in viability upon proper freezing procedures, and resuscitated cells displayed good morphology and proliferative activities.

3.2 Growth dynamics

The growth curves of the somatic cell lines in APCCC before cryopreservation and after cryopreservation displayed a typical "S" shape (Fig. 2) and the PDT was approximately 24 h to 48 h, which varies from species to species or even between subspecies. There is usually a lag time or latency phase of about 24 h to 48 h after plating, corresponding to the adaptation and recovery of the cells from trypsinization, and then the cells proliferate rapidly and enter exponential phase. As the cell density increased, proliferation slows down due to contact inhibition. The cells subsequently enter the plateau phase and begin to degenerate. The growth curves before cryopreservation is generally consistent with that after resuscitation.

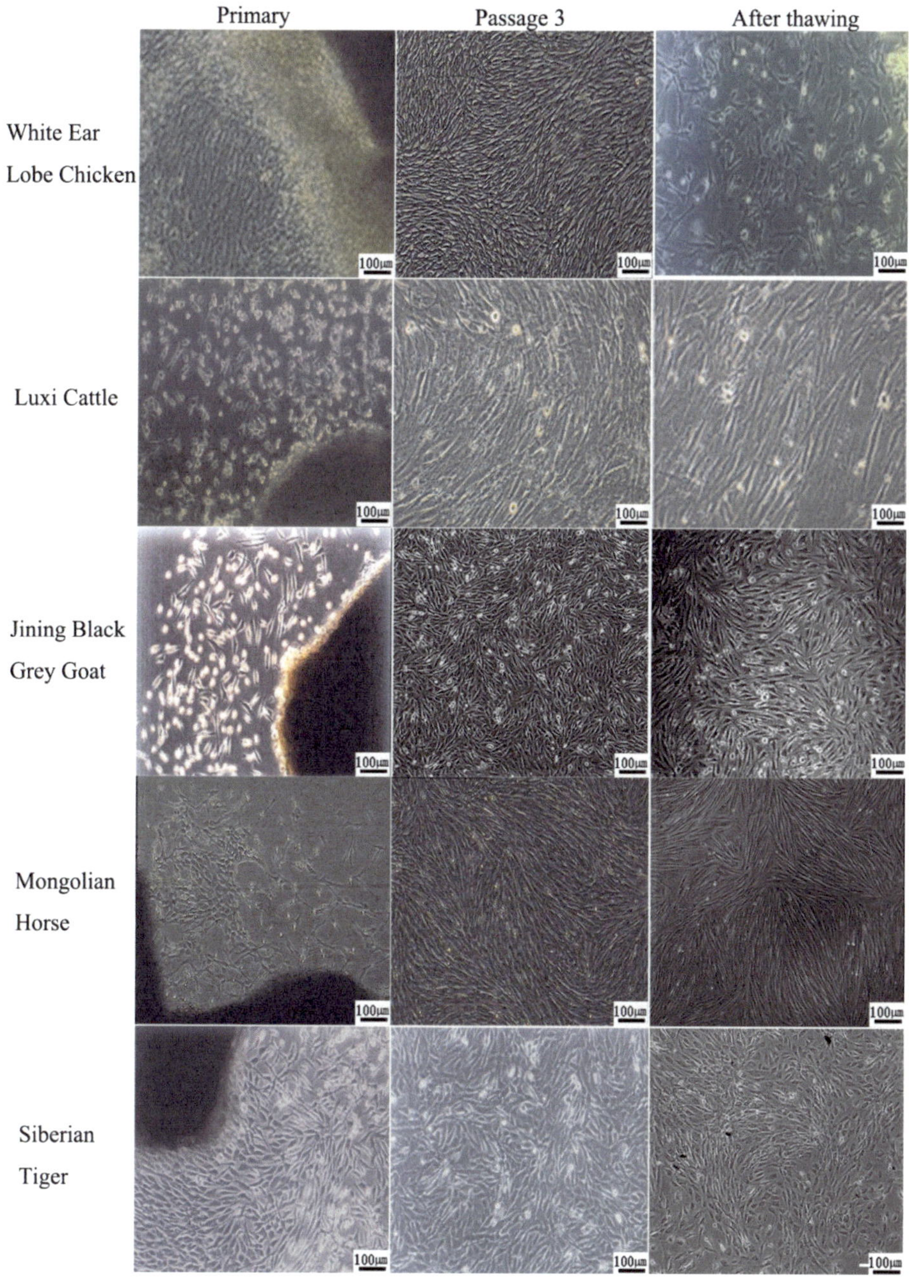

Fig. 1. Morphology of somatic cells of White Ear Lobe chicken, Luxi cattle, Jingning Black Grey goat, Mongolian horse and Siberian tiger in primary culture, before cryopreservation and after resuscitation.

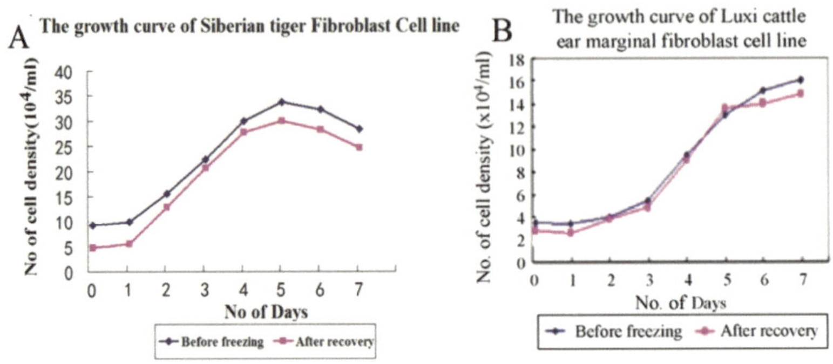

Fig. 2. Growth dynamics. Growth curves of (A) Siberian Tiger fibroblast line, and (B) Luxi Cattle ear marginal fibroblast line before cryopreservation and after resuscitation. A representative growth curve consist of latency phase, exponential phase, plateau phase and decline phase.

3.3 Microbial detection

In a sharp contrast with infections by bacteria, fungi and yeasts, characterized by turbidity, colony or hypha, which can be observed by unaided eyes, the mycoplasma contamination, usually undistinguishable, is only accompanied with slightly slower growth and increased cell fragmentation. As a result, Hoechst 33258 staining or molecular assays are required further. Would there be abundant punctiform and filiform blue fluorescence in the nucleoli, it could be concluded that the cells were contaminated by mycoplasmas (Fig. 3B).

In APCCC, all the somatic cells are subjected to microbial detection prior to cryopreservation to ensure they are free of bacterial, fungal and yeast contamination. No microorganisms were observed in the culture media. No viruses were indicated either by the hemadsorption test. Mycoplasma testing by both the ELISA detection kit and Hoechst 33258 staining (Fig. 3A) was negative. Through microbial detection, the safety and reliability of future commercial and experimental applications of the cell lines are to a large extent ensured.

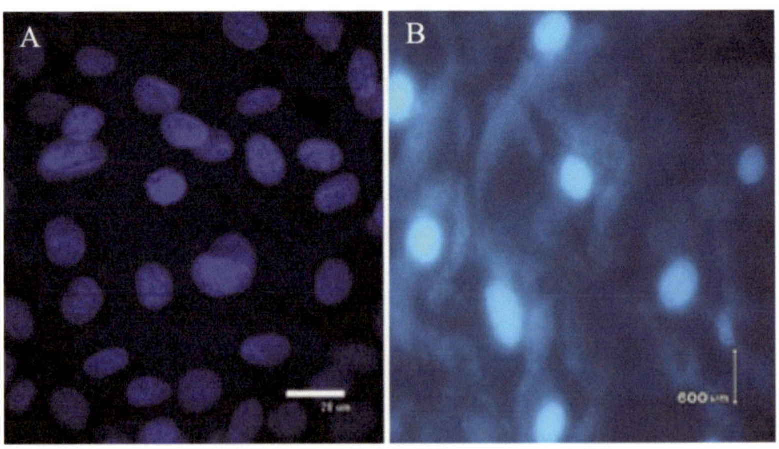

Fig. 3. Hoechst 33258 staining for (A) the detection of mycoplasma in the Siberian tiger fibroblasts; and (B) a positive control of mycoplasma contamination.

3.4 Karyotyping and chromosomal indices

Diploid cells of a given species possess a characteristic chromosome number, shape and structure, which remain very stable in normal cells (Fig. 4). Therefore, karyotype analysis is a major method for distinguishing normal cells from mutants. The percentage of diploid cells tends to decrease with increasing passage number. However, the fact that the diploid proportion is normally higher than 90% warrants the hereditary stability.

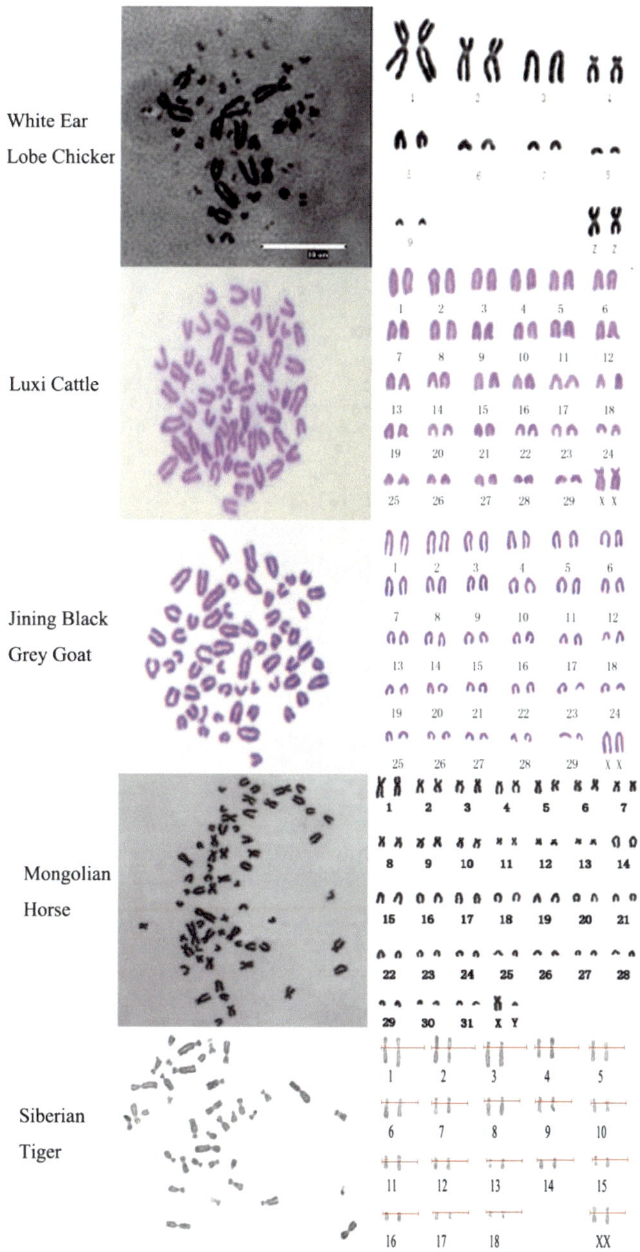

Fig. 4. Chromosome at metaphase (left) and karyotype (right).

Chromosome number	Relative length (%)	Centromere type
1	5.58 ± 0.26	T
2	5.12 ± 0.16	T
3	4.68 ± 0.34	T
4	4.49 ± 0.41	T
5	4.23 ± 012	T
6	4.05 ± 0.45	T
7	3.87 ± 0.38	T
8	3.86 ± 0.57	T
9	3.81 ± 0.04	T
10	3.76 ± 0.22	T
11	3.61 ± 0.11	T
12	3.56 ± 0.19	T
13	3.41 ± 0.33	T
14	3.36 ± 0.20	T
15	3.27 ± 0.41	T
16	3.26 ± 0.32	T
17	3.01 ± 0.09	T
18	2.97 ± 0.19	T
19	2.97 ± 0.06	T
20	2.71 ± 0.31	T
21	2.70 ± 0.24	T
22	2.60 ± 0.12	T
23	2.58 ± 0.27	T
24	2.21 ± 0.19	T
25	2.14 ± 0.22	T
26	2.09 ± 0.53	T
27	2.07 ± 0.10	T
28	1.85 ± 0.35	T
29	1.75 ± 0.32	T
X	4.47 ± 0.11	SM

Note: Relative lenth, 1.0-1.6, Metacentric chromosome (M); 1.7-2.9, Submetacentric chromosome (SM); 3.0-6.0 Subtelocentric chromosome (ST); ≥7.0 Telocentric chromosome (T).

Table 1. Chromosomal parameters of White ear lobe chicken(♀).

The chromosome number of Luxi cattle was 2n = 60, comprising 58 autosomes and two sex chromosomes, XY or XX. All the autosomes are acrocentric, and only the two sex chromosomes (XY) were submetacentric (Table 1). The chromosome numbers were counted for 100 spreads of passages 1, 3 and 4 respectively, and the frequencies of cells with 2n = 60 were 92.2%, 91.6% and 90.7% accordingly.

The chromosome number of Siberian tiger is 2n=38, consisting of 36 autosomes and two sex chromosomes, XY or XX. The karyotype composition of the Siberian tiger is 12 (M) + 16 (SM) + 4 (ST) + 4 (T), XY (M, M) (Table 2). The chromosome numbers were counted for 100 spreads of passages 1, 3 and 4 respectively, and the frequencies of cells with 2n=38 were 91.6%, 91.2% and 90.2% accordingly.

Chromosome No.	Relative lenth (%)	Centromere type
1	10.13±0.93	SM
2	9.48±1.05	M
3	8.43±0.92	ST
4	6.65±0.85	M
5	6.31±0.81	SM
6	5.85±0.75	ST
7	5.66±0.70	M
8	5.34±0.67	SM
9	5.22±0.71	SM
10	4.47±0.61	SM
11	4.11±0.75	M
12	3.51±0.66	SM
13	3.54±0.44	M
14	3.34±0.69	T
15	3.18±0.67	T
16	2.84±0.25	SM
17	2.43±0.32	SM
18	2.25±0.64	M
X	5.54±0.62	M

Table 2. Chromosomal parameters of Siberian tiger.

3.5 Isoenzyme analysis

Isoenzyme profiles of at least 5 kinds of animals were analysed simultaneously. Each kind of animal has its specific bands. The LDH bands obtained from Siberian tiger, for instance, were compared with those of other species or breeds, and five breed-specific isoenzyme bands (LDH-1, -2, -3, -4, -5) were observed (Fig. 5A). Enzymatic activities were in the order of LDH-3, LDH-2, LDH-4, LDH-5, LDH-1. LDH-2, LDH-3 and LDH-4 were dominant, while LDH-1 and LDH-5 were scarcely observable. In the MDH patterns of Siberian tiger and other breeds, two MDH bands (s-MDH, m-MDH) were observed (Fig. 5B), with the m-MDH band near the cathode and the s-MDH band (comprise two subbands but hardly identified) near the anode (Fig. 5B). Similar activity was seen from both m-MDH and s-MDH. There were significant differences in the isoenzyme patterns of LDH and MDH between the Siberian tiger fibroblasts and other cell lines in APCCC. These animals have their distinctive bands with different relative mobility. These results showed that there was no cross-contamination between different cell lines.

3.6 Expression of exogenous genes

Six fluorescent protein genes with stable structures, high expression levels and species-independent efficiency (Baird et al., 2000) have been used as marker genes to observe the expression, distribution and function of target proteins in live cells and organisms (Heim et al., 1995; Genyang et al., 2003). At APCCC, the 6 kinds of fluorescent genes were introduced into the preserved cells to evaluate the expressibility of exogenous genes. Positive cells were usually the most abundant and with the strongest fluorescence at 24 h-48 h after transfection. While the transfection efficiency decreased, strong expression levels were observed after a week, indicating that the exogenous genes can be replicated, transcribed,

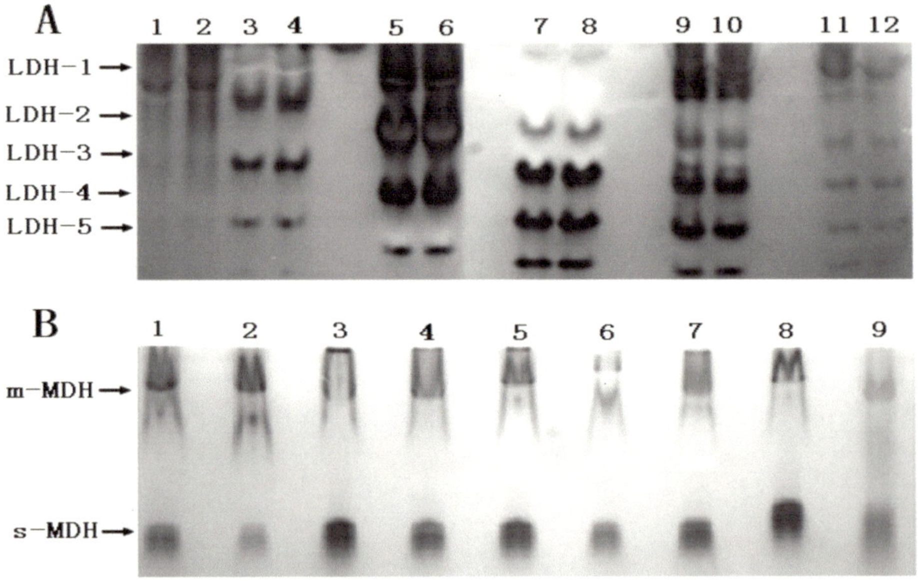

Fig. 5. LDH zymotype and MDH zymotype of several cell lines. A, SDS–PAGE electrophoresis of LDHs, from up to down, there were LDHs-1, 2, 3, 4 and 5. Panel A: 1,2 Simmental cattle, 3,4 Zhiwei goat, 5,6 Jining black goat, 7,8 Mongolian horse, 9,10 Bengal tiger, and 11,12 Siberian tiger; Panel B, MDHs from up to down were mMDH and sMDH. 1 Siberian tiger, 2 Bengal tiger, 3 Large white pig, 4 Songliao Black pig, 5 Jining Black goat, 6 Mongolian sheep, 7 Saf sheep, 8 Simmental cattle.

translated and modified within the cells. The transfected cells were not significantly less viable than the control cells ($P>0.05$), showing that the expression of fluorescent proteins had no obvious effect on the growth and proliferation of the transfected cells.

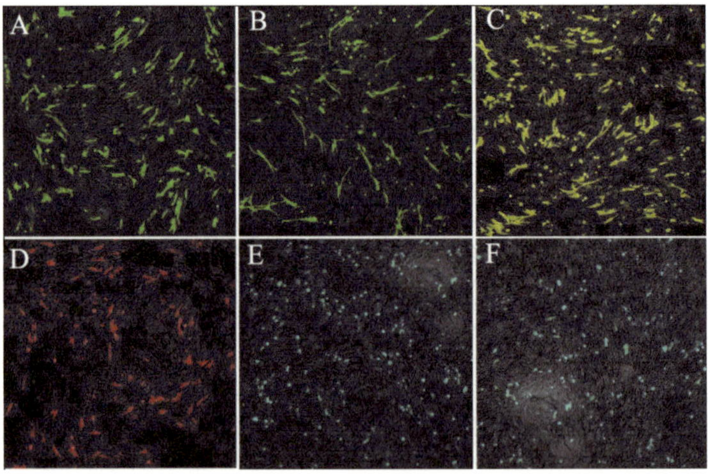

Fig. 6. Comparative figures of six fluorescent proteins in White Ear Lobe chicken fibroblasts at 24 h after transfection (×10). A, B, C, D, E and F were the transfection results of pEGFP-C1, pEGFP-N3, pEYFP-N1, pDsRed1-N1, pECFP-N1 and pECFP-mito, respectively.

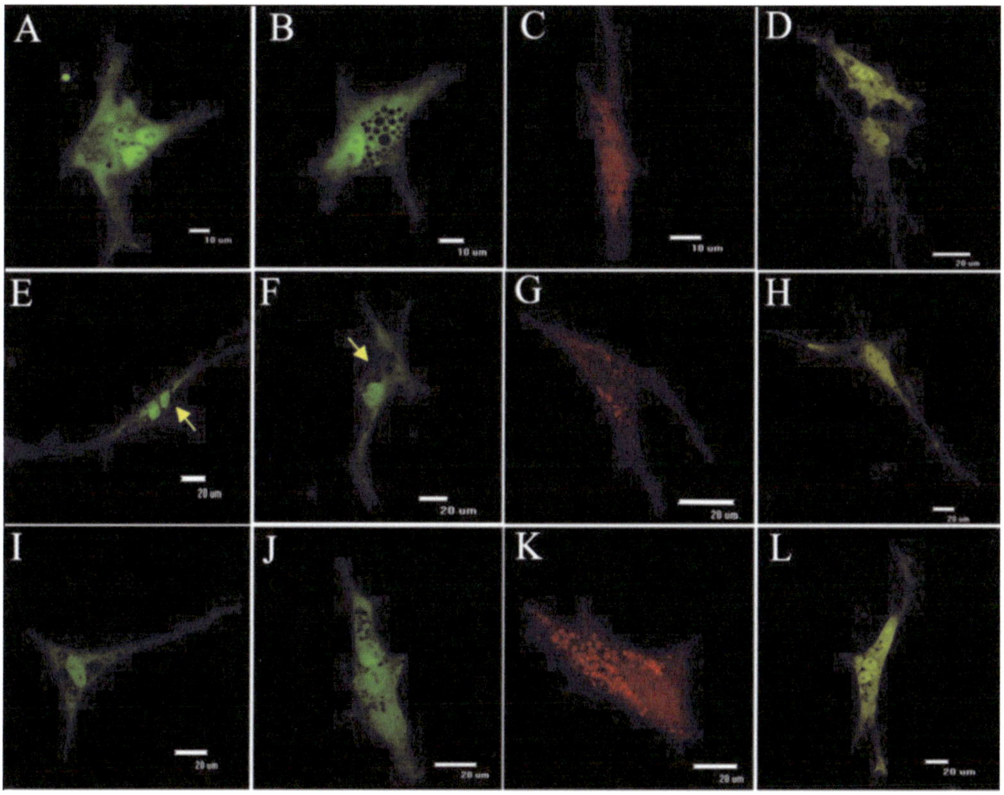

Fig. 7. The expression and distribution of pEGFP-C1, pEGFP-N3, pDsRed1-N1 and pEYFP-N1 in White Ear Lobe chicken fibroblasts (×40). A, B, C and D are the expression of pEGFP-C1, pEGFP-N3, pDsRed1-N1 and pEYFP-N1 at 24 h; E, F, G and H at 48 h; and I, J, K and L at 72 h after transfection.

4. Conclusion

Animal resources, a fundamental respect of agriculture and industry in close correlation with production and social stability, supply human beings with meat, eggs, milk, furs, medicinal materials, products for athletic and ornamental purposes, etc. In most developed countries, scalization of animal husbandry has restricted animal feeding to high yield breeds or crossbreeds with an intensified operating system, greatly compromising the diversity of local animal breeds. In the meanwhile, despite the abundance of animal genetic resources in developing countries, the lack of efficient preservation strategies and blind introduction of exotic breeds for hybridization also has reduced the animal variety. Emerging evidence has revealed that owing to the interference from human activities, species extinction has sped up for about 1000 fold, 100 million times faster than speciation, or in other words, 1 species per day. The total 7176 livestock and poultry breeds throughout the earth are disappearing at the rate of 2 per week, and 690 are on the edge of extinction.

Species extinction signifies a perpetual loss of the precious hereditary information, and will be an irreparable defeat of world genetic resources and biological theoretical repositories. Haven't the genetic resources been preserved in any forms before their extinction, not only the genetic resources will be lost evermore, but also it becomes impossible to investigate the unknown cell and molecular mechanisms, let alone to regenerate corresponding species via cloning technique. Therefore, it is exigent to employ practical measures to conserve endangered animal species.

Therefore, the APCCC has as yet preserved, for each cell line, 45-1250 cryovials of somatic cells from 30-212 individuals using primary explantation, serial passage and programmed cryopreservation. Each vial contains approximately 1.5×10^6 cells. The cells are cryopreserved within 3 passages, and are subjected to evaluation in terms of morphology, growth kinetics, viability, microbial detection, karyogram and isoenzyme analyses according to quality control standards of ATCC. The purified fibroblasts are fusiform, displaying flame-like or swirl-like patterns. The growth curves are sigmoidal with characteristic PDTs. Trypan blue exclusion test suggests that programmed preservation exerts a non-significant effect ($P>0.05$) on cell viability compared with that before freezing. Tests for bacteria, fungi, viruses and mycoplasmas are unanimously negative. Karyograms of peripheral blood lymphocytes and the in vitro cultured cells are photographed, according to which the mode of chromosome numbers are determined as that of the diploid cells, and indices including relative length, arm ratio and centromeric index and kinetochore type are calculated or dertermined. Comparison between peripheral blood lymphocytes and the in vitro cultured cells in respects of chromosome number and non-banding karyotype reveals no perceptible differences, manifesting the genetic stability of the cell lines established. Isoenzyme patterns of LDH and MDH are detected using vertical slab non-continuous PAGE assay, the breed specific bands of which rule out cross-contamination amongst the cell lines, and in the meanwhile further evince the hereditary stability.

Aforementioned results indicated that the APCCC conforms to all the ATCC criteria for somatic cell lines. In addition, plasmids of pEGFP-N3, pEGFP-C1, pECFP-N1, pECFP-mito, pDsRed1-N1, and pEYFP-N1 encoding the corresponding fluorescent proteins are transfected into the cells using lipofectin mediated protocol to study the expression of exogenous genes. By observation or detection of spatiotemporal expression of the fluorescent proteins, proliferation and growth of positive cells, apoptotic rate and viability, the ability and characteristics to accommodate exogenous genes are initiatively adopted as a constitutional index for cell line quality control.

The establishment of the APCCC is technically and theoretically conducive to preserve genetic resources of animals at somatic cell level, and definitively has a profound and long-lasting influence on biological and biomedical research in the future. The quality control standards it's been adopting will definitely provide insights for future development of culture collections.

5. Abbreviations

APCCC - Animal Population Culture Collection of China
ATCC - American Type Culture Collection
DMEM - Dulbecco's modified Eagle's medium
ECACC - European Collection of Animal Cell Culture
FAO - Food and Agriculture Organization
LDH - lactic dehydrogenase
MDH.- malic dehydrogenase
MEM - modified Eagle's medium

PAGE - polyacrylamide gel electrophoresis
PDT - population doubling time

6. References

Baird GS, Zacharias DA, & Tsien RY. (2000). Biochemistry, mutagenesis, and oligenerization of DsRed, a red fluorescent protein from coral. Proc Natl Acad Sci. Vol. 97, No. 22, pp. (11984-9). ISSN 0027-8424

Doyle A., Hay R., & Kirsop B.E. (1990). Animal Cells (Living Resources for Biotechnology). Cambridge University Press, ISBN 0521352231, Cambridge, UK.

Freshney R.I. (2000). Culture of animal cells: a manual of basic technique(4th ed). Wiley-Liss,Inc., ISBN 0471348899, New York.

Genyang Cheng, Xiangmei Cheng, & Xueyuan Bai. (2003). The gene construction and location in tubular epithelial cells of fused by green fluorescence protein and human kidney gene NaDC3. J Cell Biol. Vol. 25, No. 3, pp.(170-3). ISSN 1540-8140.

Gu Y.P., Li H.Z., & Mik J. (2006). Phenotypic characterization of telomeraseimmortalized primary non-malignant and malignant tumor-derived human prostate epithelial cell lines. Experimental Cell Research. Vol. 312, No. 6. Pp. (841–843). ISSN 0014-4827.

Hay R.I., (1992). Cell line preservation and characterization. In: Animal Cell Culture: A Practical Approach (2nd ed). Freshney R.I. pp. (104–135), Oxford University Press,. ISBN 0199637962,Oxford.

Heim R, Cubitt AB, & Tsien RY. (1995). Improved green fluorescence. Natur. Vol. 373, No. 6516, pp. (663-4).ISSN 1476-4687.

Jenkins N. (1999). Animal cell biotechnology methods and protocols. Humana Press, ISBN 0896035476, New Jersey.

Kawarai S., Hashizaki K., & Kitao S. (2006). Establishment and characterization of primary canine hepatocellular carcinoma cell lines producing alpha-fetoprotein. Vet. Immunol. Immunopathol. Vol. 113, No. 1-2, pp. (30–36), ISSN 0165-2427.

Ikeda Y, Ezaki M, Hayashi I, Yasuda D, Nakayama K, & Kono A. (1990). Establishment and characterization of human pancreatic adenocarcinoma cell line in tissue culture and the nude mouse. Jpn J Cancer Res. Vol. 81, No. 10, pp. (987-93). ISSN 0910-5050.

Masover G.K., Becker F.A., (1998). Detection of mycoplasmas in cell cultures by cultural methods. In: Methods in Molecular Biology, Miles R.J., Nicholas R.A.J., et al. pp. (207–215, 217–226). Humana Press Inc., Totawa NJ.

Qi Yitao, Tu Yiding, Yang Di, Chen Qian, Xiao Jun, & Chen Yiqiang. (2007). Cyclin A but not cyclin D1 is essential in c-myc-modulated cell cycle progression. J Cell Physiol. Vol. 210, No. 1, pp. (63-71). ISSN 1097-4652.

Suemori H., Yasuchika K., Hasegawa K., Fujioka T., Tsuneyoshi N., & Nakatsuji N. (2006). Efficient establishment of human embryonic stem cell lines and long-term maintenance with stable karyotype by enzymatic bulk passage. Biochem. Biophys. Res. Commun. Vol. 345, No. 3, pp. (926–932), ISSN 0006-291X.

Tsuchiya R., Yoshiki F., Kudo Y., & Morita M. (2002). Cell type-selective expression of green fluorescent protein and the calcium indicating protein, yellow cameleon, in rat cortical primary cultures. Brain Res. Vol. 956, No. 2, pp. (221–229), ISSN . 0006-8993.

Zhongxiao He, & Shuzheng Zhang. (1999). Electrophoresis. Scientific Press; Beijing.

9

Genomic Microarray Quality Assurance

Catherine D. Kashork, Lisa G. Shaffer and Kyle S. Sundin
Signature Genomics Laboratories
USA

1. Introduction

The use of microarray technology is revolutionizing the field of clinical cytogenetics. This new technology has transformed the cytogenetics laboratory by the adaptation of techniques that had previously been the province of molecular geneticists. Proficiency with these techniques is now a must for the modern cytogeneticist. This chapter will focus on quality assurance principles associated with microarray analysis for the diagnosis of copy number changes associated with genetic disease.

Microarrays consist of a glass slide or other solid support on which small amounts of DNA ("probes" or "targets") are deposited and immobilized in an ordered fashion (DeRisi et al., 1996; Schena et al., 1995). Probes vary in size from oligonucleotides manufactured to represent genomic regions of interest (25-85 base pairs [bp] of DNA) to large genomic clones such as bacterial artificial chromosomes (BACs, 80-200 thousand base pairs [kb]). Analysis methodology for microarray-based comparative genomic hybridization (aCGH) is consistent regardless of the probe content. First, DNA is extracted from a test sample (e.g., blood, skin, cells from pregnancy). The patient DNA is labeled with a fluorescent dye, while a DNA from a normal control (reference) sample or pooled control samples is labeled with a different-colored fluorescent dye. The two genomic DNAs, test and reference, are then mixed together and applied to the array. Because the DNAs have been denatured, they are single strands; when applied to the array, the single-strand DNAs hybridize with the arrayed single-strand probes. Using a dual-color scanner, digital images are captured, and the relative fluorescence intensities of the hybridized labeled DNA probes are quantified. The fluorescence ratio of the test and reference hybridization signals is determined at different positions along the genome and provides information on the relative copy number of sequences in the test genome compared to the normal diploid genome, enabling the detection of submicroscopic chromosomal deletions and duplications at an unprecedented level (Beaudet & Belmont, 2008; Shaffer & Bejjani, 2009).

Launching a new assay in the clinical setting requires an effective validation of the assay, clear protocols for use at the bench and clearly defined quality assurance (QA) and quality control (QC) procedures prior to the launch. Every laboratory must develop a strong Quality Management System (QMS) that is coordinated with the defined policies under regulatory bodies, such as CLIA '88 (Schwartz, 1999), College of American Pathologists and state regulating agencies. These agencies perform rigorous inspections and verify that a diagnostic laboratory follows defined principles to ensure quality patient care and correct diagnosis. This chapter covers many of the QA and QC principles identified and monitored for laboratories offering microarray-based diagnostics.

2. Quality systems with strong monitoring for quality metrics

Regulatory bodies require diagnostic laboratories to build a strong QMS (Deming, 2000). A robust QMS integrates the organization's processes, policies and procedures for total quality management. In the diagnostic laboratory industry, CAP and other accrediting bodies require defined metrics throughout all phases of testing, including pre-analytical, analytical and post-analytical.

For microarray technology, pre-analytical metrics may include assessment of DNA quality and yield. Each laboratory must define the ideal quality of DNA prior to implementing the assay into clinical testing. In our experience, a gel assessment that indicates clean genomic DNA free of RNase and degradation should lead to quality microarray results. If a DNA specimen has artifacts or appears to have degraded (Fig. 1), the laboratory should inform the client that results may be compromised because of DNA quality or obtain a new specimen from which to perform the analysis. In addition to a visual assessment of the DNA via gel electrophoresis, the laboratory should assess the DNA yield post-extraction. The quantity of DNA required in the analytic labeling phase of microarray analysis determines DNA yield requirements. Spectrophotometric assessment of DNA offers two indicators of quality DNA, including quantity and purity. A 260/280 nm reading indicates quantity, and a 260/230 nm reading indicates purity. These measurements are imperative for the downstream labeling process. Insufficient DNA quantity and quality (purity) will compromise successful microarray analysis. The spectrophotometer measures optical density (OD), which is the physical process of absorbing light. The OD, or absorbance, is calculated as a mathematical quantity. OD readings for pure DNA should measure at 1.8 (Sambrook & Russell, 2001). Our laboratory uses OD measurements from 1.4 to 1.8, although quality of labeling product can be compromised at the lower OD readings. There are many causes of poor yield, including compromised technique during extraction and poor sample quality (e.g., from increased age or exposure).

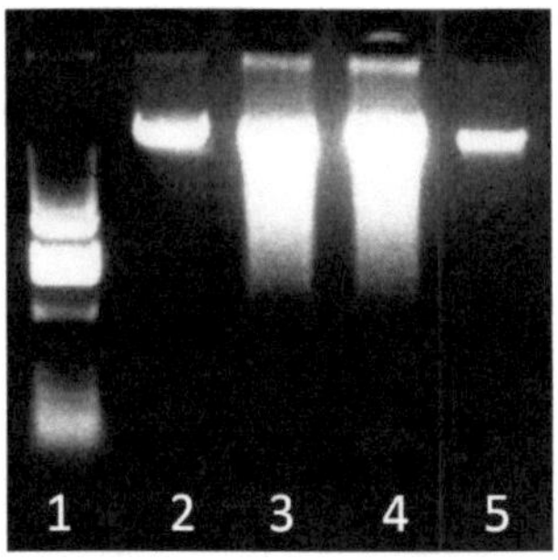

Fig. 1. Gel electrophoresis for the assessment of DNA degradation. Lane 1 has the molecular mass standard. Lanes 2 & 5 show high molecular weight samples that do not exhibit any signs of degradation. The two samples in lanes 3 & 4 show lower molecular weight DNA below the main high molecular weight bands in the other lanes. Degraded DNA typically leads to compromised array results.

Assessment of quality should be implemented throughout all phases of testing including the analytic phase. For aCGH, analytical metrics may include, but are not limited to, spectrophotometric assessment of the labeling product and the identification of labeling efficiency, which has an impact on results. Microarray analysis usually requires a dye incorporation using a random priming method. The dyes are tagged to a promoter molecule, and the promoter molecule is incorporated into the genomic DNA. As with pre-analytic assessment of DNA, the quantity assessment of the labeling product is performed by assessing the 260/280 nm readings from the spectrophotometer. Laboratories should define post-labeling quantity requirements that indicate labeling efficiency. In addition to labeling efficiency as a quality indicator, visual assessment of the set-up of labeling product on the array should be considered. In the event that there are air bubbles or non-complete contact of array product to the hybridization area (Fig. 2), the quality of the microarray result may be compromised and should be documented.

Fig. 2. Microarray after hybridization with an air bubble. The air bubble creates an area of incomplete contact of array product to the hybridization area, which compromises the array.

Post-analytic assessments may include average standard deviation (SD), intensity and background values. The SD value is the standard deviation of the normalized $\log_2$ intensity ratios for autosomal regions (excluding large copy number imbalances) and provides a measure of quality for aCGH experiments (Vermeesch et al., 2005). The SD value provides a quantitative metric that is relative to the overall noise on an array. As the overall noise of an aCGH experiment increases, so does the SD value. Our laboratory has established SD values that indicate whether an aCGH experiment is optimal, suboptimal, or failed. In our laboratory the average SD is used daily to monitor the collective SD values for all patients. Daily monitoring of the average SD value allows for establishment of a system to monitor the average SD over time. If a shift in the average SD value is observed, the laboratory processes can be evaluated to determine the potential cause and potentially prevent a system-wide failure. Each laboratory will need to define and validate a quality metric to measure the quality of the array and implement a system to track the performance of the metric.

In addition to monitoring the SD value of the array, the signal intensity of the two fluorescent dyes relative to the background can be monitored. These two values can be tracked independently or together by monitoring the signal-to-noise ratio (SNR). The SNR is the signal intensity divided by the background noise. Low signal intensity or high background noise will result in a low SNR value. A low SNR is an indicator for poor-quality array data (Basarsky et al., 2000). Low signal intensities can result from several factors, including poor fluorescent dye incorporation in labeling, inadequate denaturation of the probe, inadequate quantity of the probe, and suboptimal hybridization. Several factors can result in high background noise, including labeling reaction impurities, drying of the array during hybridization or during the post-hybridization washes, and incorrect Cot-1 to probe ratio. Constant monitoring of these metrics allows the laboratory staff to anticipate potential system failures leading to failed or inaccurate findings.

3. Verification of array results

In addition to the microarray assessments, there are other post-analytical assessments of quality that validate the microarray findings and lead to a complete, quality result used by the clinician for the diagnosis of the patient. Methods for confirmation of array results may include fluorescence *in situ* hybridization (FISH), multiplex ligation-dependent probe amplification (MLPA), quantitative PCR (Q-PCR) and other PCR techniques.

FISH is an established technique that is used to identify numerical and structural chromosome abnormalities by using fluorescently labeled DNA probes to detect the presence or absence of the DNA in the interphase nucleus or in metaphase, the stage of active cell division when the chromosomes are visibly condensed and can be observed in a microscope (Kashork et al., 2010). FISH commonly uses unique-sequence BAC probes; depending on the specific probe used, the resolution of metaphase FISH is ~80-200 kb (Shaffer et al., 2001). In our experience, deletions are easy to visualize by FISH. However, tandem duplications represent a challenge to any laboratory using confirmatory FISH because the duplicated material is not of sufficient distance from the original genomic location to generate a distinct fluorescent signal to allow detection by interphase or metaphase analysis. In some cases the intensity of the signal may be twice as intense on the duplicated homolog compared to the normal homolog, but this is not always the case.

In addition, although microarray analysis can detect DNA copy number changes, it does not identify the provenance of the abnormality; seemingly identical array results may be caused by distinct molecular mechanisms. Complete understanding of the rearrangement so that accurate genetic counseling can be provided requires visualization of the rearrangement, which can be accomplished with FISH. For example, a copy-number gain identified by microarray analysis (Fig. 3) may be a duplication, an insertion, a marker chromosome or an unbalanced translocation (Fig. 4).

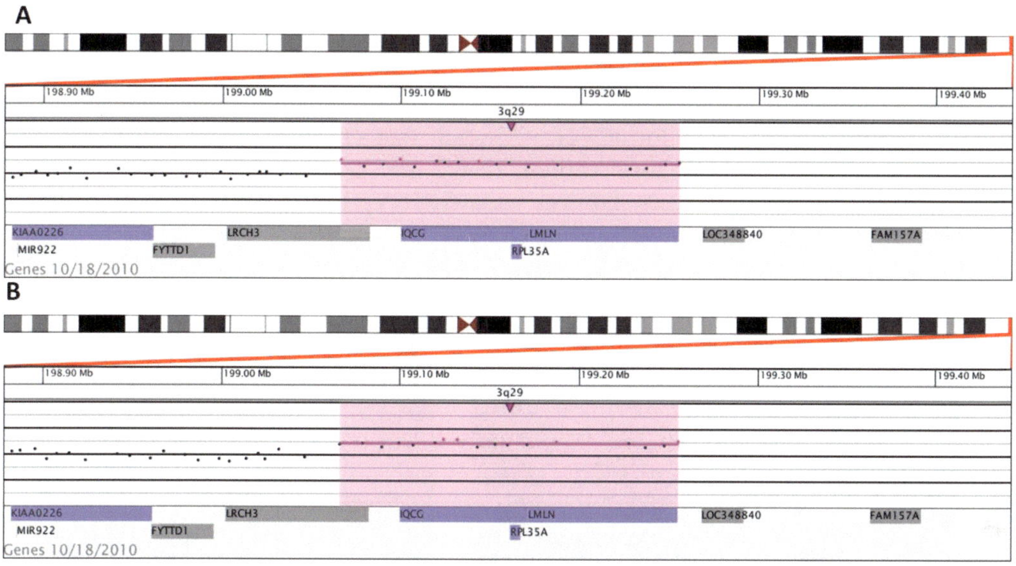

Fig. 3. (A-B) Microarray plots from two subjects showing identical single-copy gains of 17 oligonucleotide probes from the terminal long arm of chromosome 3, approximately 189 kb in size (chr3:199,067,024-199,255,755, hg18 coordinates). Probes are ordered on the x-axis according to physical mapping positions, with the most proximal 3q29 probes to the left and the most distal 3q29 probes to the right. Values along the y-axis represent $\log_2$ ratios of patient:control signal intensities. Results are visualized using Genoglyphix (Signature Genomics, Spokane, WA).

One disadvantage of FISH visualization of microarray results, particularly when using high-density oligonucleotide arrays, is that high-density oligonucleotide arrays can detect abnormalities well below the size of the smallest FISH probes, which are 100-350 kb in size. However, PCR-based methodologies such as MLPA and Q-PCR can visualize small gains and losses. MLPA targets the region of interest with two oligonucleotide probes, one probe containing a forward primer sequence and the second probe containing the reverse primer sequence (Schouten et al., 2002). The oligonucleotide probes are allowed to hybridize to the DNA followed by a ligation step. If the two probes are adjacent to each other the ligation will combine the two probes into a single probe with a fluorescently tagged forward primer on one end and reverse primer on the other end. The probe is amplified by PCR and only the ligated probe is amplified. The amplified product is dependent on the number of target sites present in the DNA. The forward primer is fluorescently labeled, which allows a comparison of the ratio of the fluorescent intensity between reference sample and the test subject to determine the relative quantity of the probe.

Q-PCR amplifies and simultaneously quantifies the relative amount of DNA when compared against a reference. Two Q-PCR methods have been developed. The first method uses fluorescent dyes that intercalate nonspecifically with the double-stranded DNA which produces fluorescent signals relative the quantity of DNA present (VanGuilder et al., 2008). The ratio is compared against a normal reference to confirm the relative quantity of the sample to the control. The second method uses a fluorescently labeled probe that is targeted to the region of interest. The fluorescently labeled probe has a fluorescent reporter and a quencher to hide the fluorescent signal until the region is amplified (Udvardi et al., 2008). During each round of the PCR process, the exonuclease activity of the polymerase releases the fluorescent reporter, unquenching the signal and allowing detection.

Additional PCR-based methods such as polymorphic microsatellite analysis have also been used as a confirmatory assay for aCGH. Although these molecular assays can confirm a copy number gain or loss, they cannot reveal the chromosomal rearrangement or mechanism giving rise to the copy number variant (CNV). Each laboratory must determine the appropriate confirmatory assay to meet its needs.

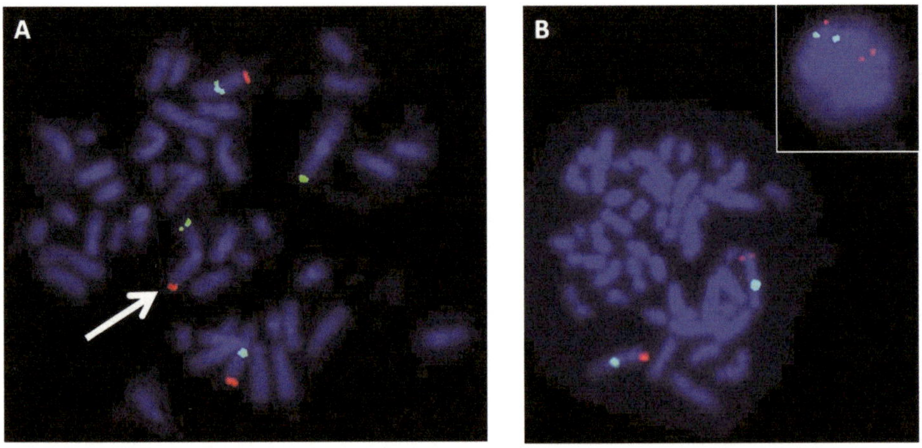

Fig. 4. (A) FISH visualization of the gain shown in Fig. 3A revealed an unbalanced translocation of the 3q29 material to 1p. BAC clone RP11-23M2 from 3q29 is labeled in red, and BAC clones RP11-9A9 from 3q11.2 and RP11-438F14 from 1q44 are labeled in green as controls. The presence of one red signal on one of the chromosome 1 homologues indicates translocation of 3q29 onto 1p (arrow). (B) FISH of the gain shown in Fig. 3B. BAC clone RP11-159K3 from 3q29 is labeled in red, and chromosome 3 centromere probe D3Z1 is labeled in green as a control. The presence of two red signals on metaphase FISH rules out an unbalanced translocation, while the additional finding of three red signals on interphase FISH (inset) suggests a duplication. The patient shown in Fig. 3A may have inherited the unbalanced translocation from a parent with a balanced translocation, whereas the duplication in the patient shown in Fig. 3B may have arisen *de novo*, although parental testing is necessary to confirm the inheritance.

4. Environmental controls

Over time, specific environmental controls have been developed and implemented within the laboratory to ensure quality microarray diagnostics. For aCGH, ozone is an important environmental factor to control. Ozone is a common pollutant found in the lower atmosphere and is the primary component of smog. Ozone is formed when nitric oxides and volatile organic compounds (VOCs) react in the presence of sunlight (US Environmental Protection Agency [EPA], http://www.epa.gov/air/ozonepollution, last accessed May 9, 2008). Nitric oxides and VOCs are emitted by motor vehicle exhaust, industrial emissions, gasoline vapors, chemical solvents, and natural sources. Consequently, ozone levels are higher in urban and industrial areas, especially during the summer months.

National ozone standards have been established by the EPA to protect public health. The established standard peak ozone level set by the EPA is 80 ppb, which is based on the annual fourth maximum 8-hour average (EPA, http://www.epa.gov/air/ozonepollution, last accessed May 9, 2008). The EPA has also established an air quality index system for monitoring the daily pollution levels. The "good" air quality range is 0 to 60 ppb. The fluorescent dyes commonly used in aCGH are sensitive to ozone levels as low as 5 to 10 ppb (Branham et al., 2007; Fare et al., 2003). Thus, ozone levels considered normal for environmental standards are well above those ranges, demonstrating sensitivity of the dyes.

Ozone has been shown to strongly affect dyes that are commonly used in aCGH, including cyanine 5 (Cy5) and Alexa dyes (Alexa Fluor 647 and Alexa Fluor 5) and, to a lesser extent, cyanine 3 (Cy3) and the Alexa equivalents (Alexa Fluor 555 and Alexa Fluor 3) (Branham et al., 2007; Byerly et al., 2009; Fare et al., 2003). Several studies have identified the post-hybridization washes as the most sensitive period for exposure to ozone (Branham et al., 2007; Byerly et al., 2009; Fare et al., 2003). These studies have demonstrated the difficulty with which laboratories identify the source of ozone-related problems, especially considering the extremely low levels of ozone (5 to 10 ppb) that cause these problems, the duration of exposure (as little as 10 to 30 seconds), and the seasonal emergence of ozone itself. The effects of ozone must be addressed when aCGH is performed, particularly in a clinical diagnostic setting, where it is critical to have consistent high quality and reproducible results. Failure to protect the fluorescent dyes from ozone during the post-hybridization washes will result in considerable negative impact on the array data. The implementation of quality control measures such as ozone reduction and monitoring to ensure high-quality aCGH results is mandatory for any aCGH laboratory. There are many commercially available enclosures and scrubbers designed to protect the dyes during the post-hybridizations washes. Some laboratories have gone as far as developing ozone-free rooms where post-hybridization washes and the subsequent scanning and analysis are performed. The latter is most desirable but may not always be feasible.

In addition to ozone degradation, the dyes are also photosensitive and often must be used in a reduced-light environment. Systems should be implemented to prevent photobleaching of the fluorescent dyes. To mitigate against the effect of photobleaching, the dyes should be protected from the light whenever possible. This can be done by using indirect lighting in the work area, using amber tubes, covering the samples with tin foil or placing the samples in areas with little or no light when not being directly handled.

5. Normalization

Normalization, which aims to separate biologically relevant signal from experimental artifacts, is a crucial step of microarray analysis (Neuvial et al., 2006). Each laboratory must identify a system for normalization. Most microarray vendors offer software with built-in normalization methods optimized for their own platforms. Laboratories can use a normalization package that is developed by the microarray vendor or can develop their own package. One normalization method that is used in the laboratory is the locally weighted polynomial regression (LOESS) (Cleveland, 1979). This normalization applies a spatial correction to correct position-dependent non-uniformity of signals across the array.

Another normalization system used in the laboratory is the Qspline fit normalization (Workman et al., 2002). This normalization compensates for the inherent differences in signal intensities between the two dyes. There are many normalization methodologies available; each laboratory will have to define its method of choice.

6. Automation

One of the key elements for any clinical assay is reproducibility. By replacing manual processes with automation, a laboratory can substantially improve the consistency and reproducibility of its daily operations. In addition, automation can increase throughput, which is often an advantage for a growing laboratory and helps to reduce the dependency on staffing levels. The laboratory protocols that have been successfully automated include DNA isolation, labeling and hybridization, washing and analysis.

However, automation can present several challenges. Because it is based on a plate format, if the input materials are flawed or a technical issue occurs (e.g., labeling master mix or an automation failure) it will impact the entire plate, which may consist of 48 or more patients. The consequences of a failure of this magnitude are substantial in terms of cost and the potential loss of the sample.

7. Multiplexing of microarray platforms

As laboratories become more accustomed to using microarrays, the demand for the assay may increase. In addition to automation, multiplexed array formats can help the laboratory satisfy the increased demand. Multiplexed array formats allow for the simultaneous hybridization of 2 to 12 or more samples depending on the probe coverage of the array and the array manufacturer. The multiplex design (Fig. 5) has many positive features including decreased costs and higher throughput. As this technology continues to advance, higher multiplexed formats are likely to be developed.

Although increased throughput has its advantages, it also creates challenges for any quality system. The laboratory must ensure that there is no cross contamination between each sub-array, which would affect patient samples. Some manufacturers have included QA/QC features into the development of multiplexed arrays, such as tracking controls that can be spiked into the experiments that identify unique positions on the array such that each position of the array can have a unique tracking control. Unique tracking controls are added to each sample before they are introduced to the array, which allows the laboratory to monitor each sub-array for cross contamination. In the event of cross contamination or leaking between the sub-arrays, the laboratory can determine the sample involved in the cross-contamination event based on the unique tracking controls involved.

8. Validation

As laboratories begin to adopt microarray technology within their facility, they should identify the validation criteria which they are responsible for meeting. The American College of Medical Genetics (ACMG; Shaffer et al., 2007) and other US state guidelines have been developed to ensure laboratories have thoroughly tested and reviewed the capacity and expectation of the assay prior to clinical release. These validations include testing known abnormal specimens to verify the expected outcomes. Section E13.2 of the ACMG

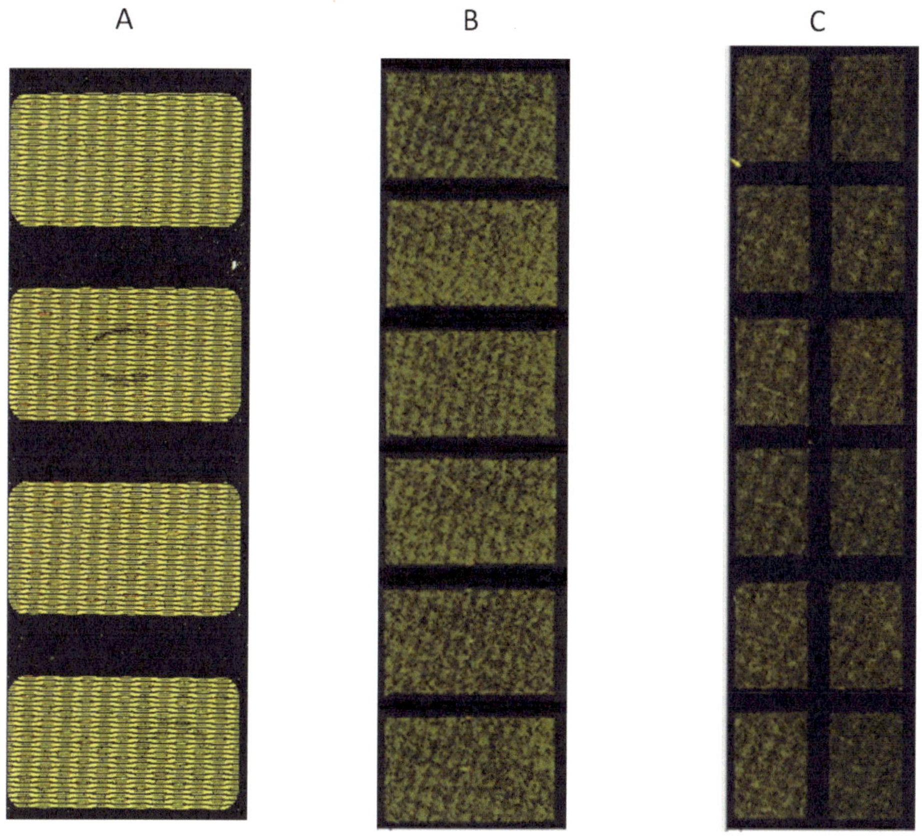

Fig. 5. Different multiplex array formats. (A) Agilent 4-plex, (B) NimbleGen 6-plex, and (C) NimbleGen 12-Plex.

guidelines distinguishes between different levels of validation depending on platform type (e.g., FDA-approved, investigation-use-only/research-use-only, or home-brew microarrays) and requires a demonstration of expertise of array performance and analysis through defined validation criteria for new microarray platforms, new versions of existing microarrays, and new lots of the same microarray. Some states have additional validation requirements. For example, New York has the Clinical Laboratory Evaluation Program (CLEP). Accreditation through CLEP requires additional quality assurances and validation criteria to which laboratories must adhere. Outside of accreditations, laboratories need to consider state regulations when considering offering their testing services nationwide. Some states have regulations that not only impact laboratories that reside in that state but also impact laboratories that test samples from that particular state.

9. Control samples

When defining the control specimen of choice, a laboratory can choose to use same-sex or sex-mismatched controls from the patient. Same-sex controls offer detection of autosomal gains and losses, and complex sex chromosome abnormalities are more easily visualized. Sex-mismatched controls offer the same detection of autosomal gains and losses but can be more challenging when assessing sex chromosome changes. However, sex-mismatched controls offer the laboratory an internal assessment of hybridization success because of the expected deviations when comparing a male against a female. The deviations that are inherent to sex-mismatched controls are a result of the copy number variation of the X/Y chromosome ratio. When a female genome is compared to a male genome, there is an apparent gain of chromosome X (two copies in the female against the single copy in males) and a loss of chromosome Y (Fig. 6).

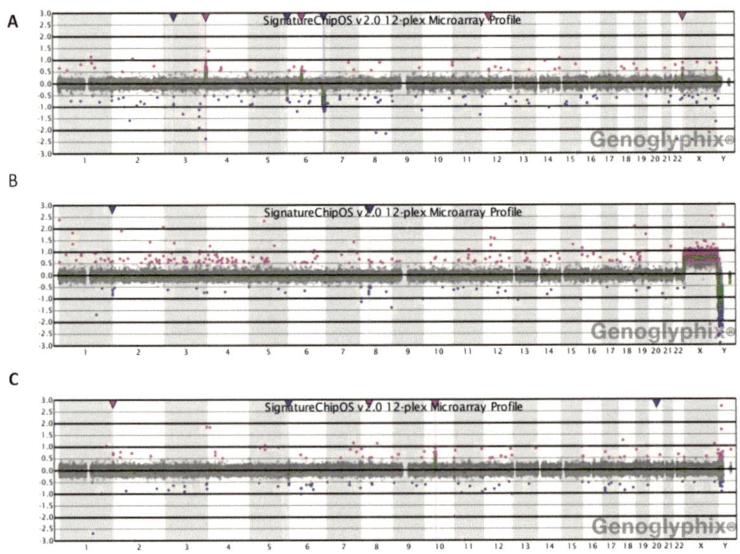

Fig. 6. Microarray plots from three different gender parings: male/male, female/male, and female/female. The probes are ordered on the x-axis according to physical mapping positions starting with chromosome 1 on the left and the X and Y chromosomes on the right. Values along the y-axis represent $\log_2$ ratios of patient:control signal intensities. (A) A same-sex male/male comparison showing identical dosage at the X and Y regions. (B) An opposite-sex female/male comparison showing a gain of the X chromosome (two copies in the female versus a single copy in the male) and a loss of Y (no copies in the female versus one copy in the male). (C) A same-sex female/female comparison showing identical dosage at the X region and no hybridization at the Y region. Results are visualized using Genoglyphix (Signature Genomics, Spokane, WA).

In addition to determining the ideal sex of the control, the laboratory must decide if a single control or pooled controls will be used in the facility. Often, these decisions are made based on available controls. Some institutions have used a consistent male and female control for easy identification of known CNVs and for monitoring the performance of arrays. However, in the absence of consistent candidates for controls, laboratories can create or purchase a pooled DNA control. All variations of these controls must be assessed with any change in the pool so that they will not significantly impact reporting. In addition, CNVs present in the control DNA can be used as a positive indicator of assay performance when using same-sex controls. However, pooled control samples will have diluted CNVs, which may not be apparent on the microarray results or may appear as mosaics or background noise.

10. Conclusion

This chapter highlights many of the quality assurance principles that impact a laboratory setting up or using aCGH. This is not an exhaustive set of challenges to implementation as there may be lab-, region-, environment- and vendor-specific variations. Each laboratory should perform initial quality verification at the time of test development. Post development, a thorough validation must be performed, which may uncover variation that should be controlled prior to launching the clinical assay. A laboratory's role in developing new assays should include an established, documented and maintained quality system that ensures that the test conforms to specified requirements and ultimately leads to accurate results.

11. Acknowledgments

We thank Aaron Theisen (Signature Genomics, Spokane, WA) for his critical review of our manuscript.

12. References

Basarsky T., Verdnik D., Willis D. & Zhai J. (2000). An overview of a DNA microarray scanner: design essentials for an integrated acquisition and analysis platform, In: *Microarray biochip technology*, Schena M. (Ed.). Eaton Publishing, Natick, MA

Beaudet A.L. & Belmont J.W. (2008). Array-based DNA diagnostics: let the revolution begin. *Annu Rev Med*, Vol. 59. pp. 113-129, ISSN 0066-4219

Branham W.S., Melvin C.D., Han T., Desai V.G., Moland C.L., Scully A.T. & Fuscoe J.C. (2007). Elimination of laboratory ozone leads to a dramatic improvement in the reproducibility of microarray gene expression measurements. *BMC Biotechnol*, Vol. 7. pp. 8, ISSN 1472-6750

Byerly S., Sundin K., Raja R., Stanchfield J., Bejjani B.A. & Shaffer L.G. (2009). Effects of ozone exposure during microarray posthybridization washes and scanning. *J Mol Diagn*, Vol. 11. No. 6, (Nov), pp. 590-597, ISSN 1943-7811

Cleveland W.S. (1979). Robust locally weighted regression and smoothing scatterplots. *J Amer Statist Assoc*, Vol. 74. pp. 829-836

Deming W.E. (2000). *The New Economics For Industry, Government & Education* (2nd), The MIT Press, Cambridge, MA

DeRisi J., Penland L., Brown P.O., Bittner M.L., Meltzer P.S., Ray M., Chen Y., Su Y.A. & Trent J.M. (1996). Use of a cDNA microarray to analyse gene expression patterns in human cancer. *Nat Genet*, Vol. 14. No. 4, (Dec), pp. 457-460

Fare T.L., Coffey E.M., Dai H., He Y.D., Kessler D.A., Kilian K.A., Koch J.E., LeProust E., Marton M.J., Meyer M.R., Stoughton R.B., Tokiwa G.Y. & Wang Y. (2003). Effects of atmospheric ozone on microarray data quality. *Anal Chem*, Vol. 75. No. 17, (Sep 1), pp. 4672-4675, ISSN 0003-2700

Kashork C.D., Theisen A. & Shaffer L.G. (2010). Diagnosis of cryptic chromosomal syndromes by fluorescence in situ hybridization (FISH). *Curr Protoc Hum Genet*, Vol. Chapter 8. Unit 8.10 (Oct), pp. 11-20, ISSN 1934-8258

Neuvial P., Hupe P., Brito I., Liva S., Manie E., Brennetot C., Radvanyi F., Aurias A. & Barillot E. (2006). Spatial normalization of array-CGH data. *BMC Bioinformatics*, Vol. 7. pp. 264, ISSN 1471-2105

Sambrook J. & Russell D. (2001). *Molecular Cloning: A Laboratory Manual* (3rd), Cold Spring Harbor Laboratory Press, Woodbury, NY

Schena M., Shalon D., Davis R.W. & Brown P.O. (1995). Quantitative monitoring of gene expression patterns with a complementary DNA microarray. *Science*, Vol. 270. No. 5235, (Oct 20), pp. 467-470

Schouten J.P., McElgunn C.J., Waaijer R., Zwijnenburg D., Diepvens F. & Pals G. (2002). Relative quantification of 40 nucleic acid sequences by multiplex ligation-dependent probe amplification. *Nucleic Acids Res*, Vol. 30. No. 12, (Jun 15), pp. e57

Schwartz M.K. (1999). Genetic testing and the clinical laboratory improvement amendments of 1988: present and future. *Clin Chem*, Vol. 45. No. 5, (May), pp. 739-745, ISSN 0009-9147

Shaffer L.G., Ledbetter D.H. & Lupski J.R. (2001). Molecular cytogenetics of contiguous gene syndromes: mechanisms and consequences of gene dosage imbalance. In: *Metabolic and Molecular Basis of Inherited Disease*, Scriver C.R., Beaudet A.L., Sly W.S., Valle D., Childs B., Kinzler K.W. & Vogelstein B. (Eds.). pp. 1291-1324, McGraw Hill, New York

Shaffer L.G., Beaudet A.L., Brothman A.R., Hirsch B., Levy B., Martin C.L., Mascarello J.T. & Rao K.W. (2007). Microarray analysis for constitutional cytogenetic abnormalities. *Genet Med*, Vol. 9. No. 9, (Sep), pp. 654-662

Shaffer L.G. & Bejjani B.A. (2009). Using microarray-based molecular cytogenetic methods to identify chromosome abnormalities. *Pediatr Ann*, Vol. 38. No. 8, (Aug), pp. 440-447, ISSN 0090-4481

Udvardi M.K., Czechowski T. & Scheible W.R. (2008). Eleven golden rules of quantitative RT-PCR. *Plant Cell*, Vol. 20. No. 7, (Jul), pp. 1736-1737, ISSN 1040-4651

VanGuilder H.D., Vrana K.E. & Freeman W.M. (2008). Twenty-five years of quantitative PCR for gene expression analysis. *Biotechniques*, Vol. 44. No. 5, (Apr), pp. 619-626, ISSN 0736-6205

Vermeesch J.R., Melotte C., Froyen G., Van Vooren S., Dutta B., Maas N., Vermeulen S., Menten B., Speleman F., De Moor B., Van Hummelen P., Marynen P., Fryns J.P. & Devriendt K. (2005). Molecular karyotyping: array CGH quality criteria for constitutional genetic diagnosis. *J Histochem Cytochem*, Vol. 53. No. 3, (Mar), pp. 413-422, ISSN 0022-1554

Workman C., Jensen L.J., Jarmer H., Berka R., Gautier L., Nielser H.B., Saxild H.H., Nielsen C., Brunak S. & Knudsen S. (2002). A new non-linear normalization method for reducing variability in DNA microarray experiments. *Genome Biol*, Vol. 3. No. 9, (Aug 30), pp. 40-48, ISSN 1465-6914

Part 4

Planning for Quality Control

10

QA/QC in Transport Infrastructures: Issues and Perspectives

Filippo G. Praticò
University Mediterranea at Reggio Calabria
Italy

1. Introduction

Transport infrastructures (road and highway networks, railways, terminal facilities, airports, mass transit systems, bicycle paths and pedestrian walkways, etc.) have significant impacts on the development of regional and national economies. However, the effectiveness of these impacts over the time has to be established based on the actual quality of all the pertaining components of the infrastructure asset (pavement, safety barriers, signals, illumination, embankment, drainage, etc.). Quality can be interpreted as the degree of excellence of a product or service, or as the degree to which a product or service satisfies the needs of a specific customer or, finally, as the degree to which a product or service conforms with a given requirement. In more detail, **quality assurance** (QA) refers to all those planned and systematic actions necessary to provide confidence that a product or facility will perform satisfactorily in service. At the same time, **quality control** (QC), also called process control, relates to those QA actions and considerations necessary to assess and adjust production and construction processes so as to control the level of quality being produced in the end product (Fig.1). Note that **QA (which includes QC) is an infrastructure (e.g. highway) agency responsibility** and involves all the process (planning, design, plan and specifications, construction, etc.), while **QC is a producer–contractor** responsibility which mainly affects construction. Furthermore, QC is not simply QA in construction, due to the fact that both independent assurance and acceptance procedures refer to QA in construction but they are NOT a part of QC. The entire QA/QC process includes: i) setting up the initial data collection or experimentation to determine typical parameters of current construction; ii) designing the acceptance plan itself, including selecting quality characteristics (and corresponding specification limits), statistical quality measure (and corresponding quality levels), buyer's and seller's risks, lot size, number of samples (sample size), specification and/or acceptance limits, and payment–adjustment provisions. As is well known (Burati et al, 2003), traditionally, highway specifications spelled out in detail the work that was to be done by the contractor under the form of **materials and methods specifications** (also called method specifications, recipe specifications, or prescriptive specifications). In this case, specifications direct the contractor to use specified materials in definite proportions and specific types of equipment and methods to place the material. On the contrary, **end result specifications** require the contractor to take the entire responsibility for supplying a product or an item of construction.

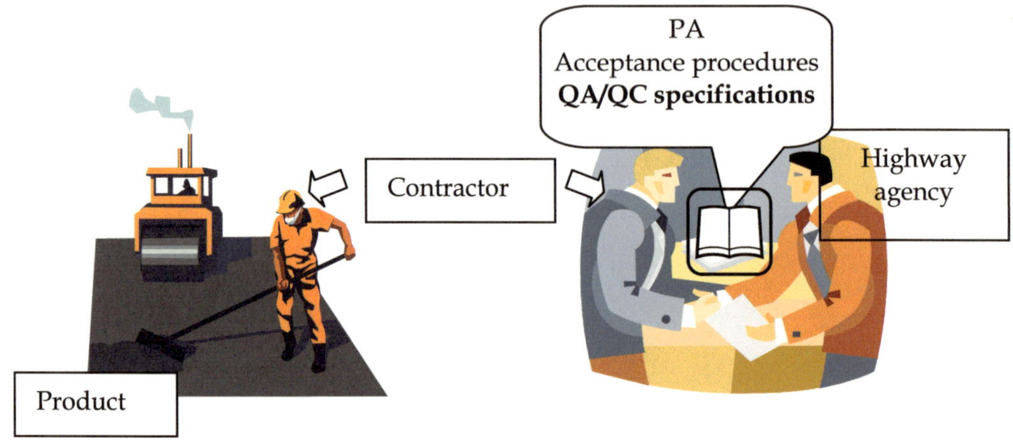

Fig. 1. Contractor vs. agency relationship.

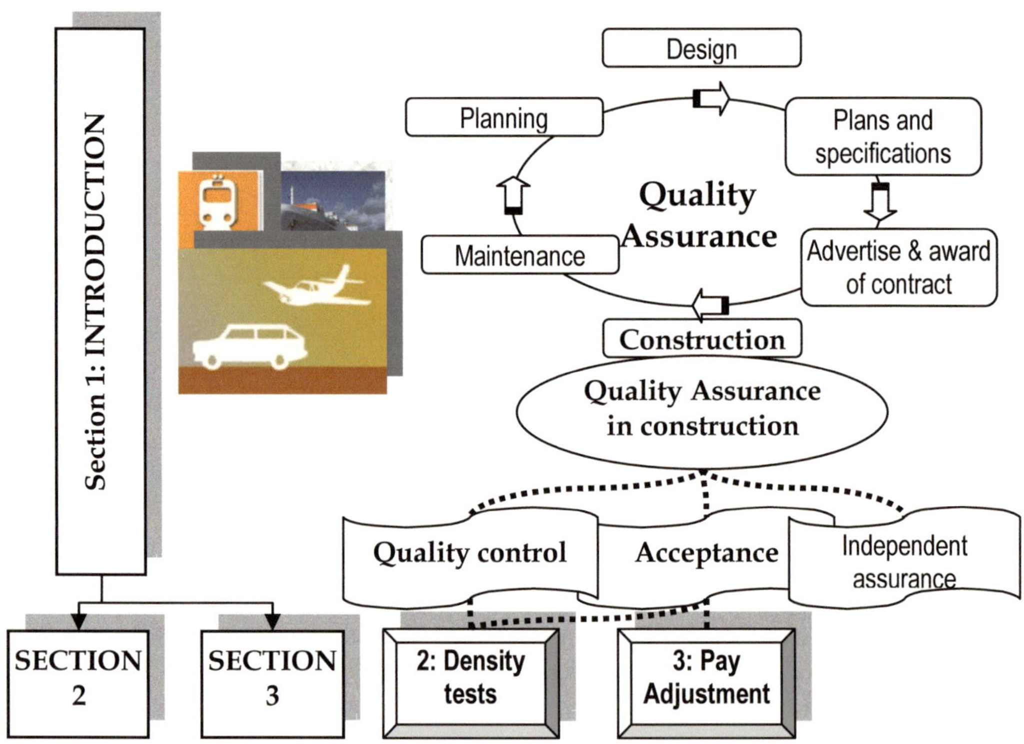

Fig. 2. Conceptual framework of the chapter.

The highway agency's responsibility is to either accept or reject the final product or to apply a price adjustment (PA) commensurate with the degree of compliance with the specifications. In practice, current specifications are neither solely "materials and methods" nor "end result." (Burati et al., 2003). **Quality assurance specifications** (a.k.a. QA/QC specifications or QC/QA specifications) are a combination of end result specifications and materials and methods specifications. **QA specification** consists of two separate functions, i.e. **quality control** or process control, and **acceptance**. As for the development of QC procedures and requirements, the QC procedures and requirements are made up of two parts: the QC requirements (tests to be performed, minimum frequency, qualified laboratories) and the quality characteristics to be measured. For HMAC (hot mix asphalt concrete, or HMA), typical quality characteristics that may be tested for QC include aggregate quality, density (see next section), gradation of critical sieve sizes, plant and discharge temperatures, degree of aggregate coating, moisture content of fine aggregate and/or of finished mix. For PCC (Portland cement concrete) , typical quality characteristics that are tested for QC include aggregate quality, gradation of critical sieve sizes, air content, water–cement ratio, mix temperature, slump. Note that three different subjects can carry out the acceptance testing: contractor, agency, third part. The agency: i) may decide **to do (itself)** the acceptance testing; ii) may assign the testing to the **contractor**; iii) may have a **combination of agency and contractor acceptance testing**; iv) or may require a **third part** to do the testing. In Italy, acceptance testing is usually carried out by a third part, following "general" and "special" Contract specifications (Capitolato special CIRS, Capitolato speciale prestazionale ANAS, etc.). Figure 2 illustrates the conceptual framework of the chapter.

2. Density

By referring to density (more technically: bulk specific gravity and air voids content), this is a very crucial factor for QC/QA procedures used to evaluate contract specifications. Indeed, the life cycle of hot mix asphalt (HMA) depends on the material density. Bulk specific gravity, Gmb, measures the specific gravity of a compacted hot mix asphalt sample (core or in-lab compacted). Life cycle costs, contract requirements, and QC/QA procedures are all modeled as functions of the effective Gmb properties resulting from suitable design and construction techniques and by application of appropriate boundary conditions. A variety of methods for determining (in laboratory or on site) Gmb are available (dimensional, AASHTO T 269-EN 12697-6:2003; parafilm, ASTM D 1188; vacuum sealing principle, ASTM D 6752; paraffin coated, BU N40-1973, AASHTO T 275-A, EN 12697-6:2003; saturated Surface Dry, AASHTO T 166, ASTM D 2726, EN 12697-6:2003; non-nuclear portable device, ASTM D 7113, AASHTO TP68). In the vacuum sealing method (VSD), specimen volume is determined by a vacuum chamber that shrink-wraps the specimen in a high quality plastic bag. Surface texture effects can be appreciable, and are accounted for in by the shrink-wrap process. Archimedes' principle is then applied. The dimensional method uses height, diameter, and width measurements to estimate the volume. Surface irregularities (i.e., the rough surface texture of a typical specimen) can introduce inaccuracies, because, in practice, an "osculatory" volume is computed. The parameter P is a density estimate derived from measurements of a non-nuclear portable device, collected at the tested surface (Williams, 2008; Megali et al., 2010; Rao et al., 2007; Kvasnak et al., 2007; Romero, 2002; Sargand et al., 2005; TransTech, 2003; Gamache, 2004, 2005; Praticò et al, 2009; Praticò et al, 2009a; Praticò and Moro, 2011; Alvarez et al, 2010). In non-nuclear portable device measurements, an electrical current is transmitted from a transmitter through the asphalt concrete pavement at a given location and is detected by the receiver. Note that the current cannot flow through the isolation ring. The impedance (ohm) is measured, and the dielectric constant (dimensionless) is derived. The dielectric constant of the HMA is used to estimate the density. The dielectric constant depends on the HMA composition in that it is derived from the dielectric constants of air (~1) and bitumen &

aggregates (5-6). Water and moisture, if present, strongly influence the dielectric constant estimate because the dielectric constant of water approximately 80. Densities measured in the laboratory (dimensional, parafilm, vacuum sealing principle, paraffin coated, saturated surface dry), when contract specifications are well-defined, are quite reliable and accurate but are obtained very slowly. On-site measurements (non-nuclear portable device) are made using non-nuclear portable devices and result often biased and unreliable. Researchers have attempted to find correlations between the results obtained using different procedures (Brown et al., 2004; Cooley et al., 2002; Crouch et al., 2003; Mohammad et al., 2005; Montepara & Virgili, 1996; Spellerberg & Savage, 2004; Williams et al., 2005; Megali et al., 2009). Studies have been carried out in order to propose a theoretical framework for interpreting in-lab and on-site measurements and in the aim of proposing strategies for using non-nuclear portable devices in QC/QA (Megali et al., 2010). A decrease in porosity (or effective porosity) yielded in-lab specific gravities that converged. In contrast, due to the high dielectric constant of water the divergence of in-site measures was observed. The ranking among procedures was the following: GmbSSD > GmbFIN > GmbCOR > GmbFILM > GmbDIM. Furthermore the ranking did not depend on core diameter (Praticò and Moro, 2011b). From a practical standpoint, two strategies were proposed for estimating the density of an asphalt sample: i) consideration of both W (water content) and P (density measured through non-nuclear portable devices) in an estimate for the effective porosity, which is the most crucial parameter for quality control evaluation; ii) consideration of both W and P in an estimate of density.
In (Praticò and Moro, 2011), two equations were proposed for practical applications:

$$G_{mbCOR} = aP + bW + c \tag{1}$$

$$n_{eff} = dP + eW + f \tag{2}$$

where a = 0.914, b = -0.007, c = 0.303, d = -56.673, e = 0.420, f = 128.698 (coefficients are case-specific).

3. Acceptance procedures

This section deals with acceptance procedures. Premises (§3.1) illustrate how acceptance procedures can be carried out, the main indicators which are used and what are the main issues. Afterwards, section 3.2 (model) describes a model for the derivation of the price adjustment (PA) based on life cycle cost analysis. Due to the need of considering surface properties, section 3.3 addresses life expectancy of surface properties (skid resistance, texture, surface wear, drainability, acoustic performance). Bulk properties still remain the key-factor in determining the expected life of the as-constructed pavement. To this end, section 3.4 deals with several tools for the determination of the expected life of a pavement through empirical to rational models. Section 3.5 presents an example of application and provides other references. Main findings are summarized in section 3.6.

3.1 Premises and symbols

In the aim of providing a useful indicator for acceptance procedures, from each quality characteristic the corresponding value of quality measure is derived (Burati et al., 2003; Leahy et al., 2009). Quality measure is any one of several means that have been established to quantify quality. Some examples of quality measures are the mean, the standard deviation, the percent defective, the percent within limits, the average absolute deviation, and the quality index. Note that it is required to determine specification limits (USL, upper specification limit, and LSL, lower specification limit), and to decide on AQL (Acceptable quality level) and RQL (Rejectable quality level). USL and LSL are the limiting values placed on a quality characteristic, while AQL and RQL refer to the quality measure and are respectively the minimum level of actual quality at which the material or construction can be considered fully acceptable (for that quality characteristic) and that maximum level of actual quality at which the material or construction can be considered unacceptable (rejectable). Specification limits and quality levels are basic inputs to decide pay relationships. Performance-related pay, incentive/disincentive, minimum pay provisions, remove/replace provisions, retest provisions are examples of pay relationships. At the present several classes of models for acceptance procedures depending on input parameters and underlying principle can be listed (Praticò, 2007; Praticò et al, 2008; Praticò et al, 2010a; 2010b; Praticò et al, 2011a): 1) IRI-type models (where IRI stands for International Roughness Index) and average-type models; 2) PD-type models (where PD stands for Percent Defective); 3) LCCA based models (where LCCA stands for Life Cycle Cost Analysis). IRI-type models are based on roughness indicators, as synthetically expressive of the quality of the work and of the related costs (for example the Profile Index (PI), the International Roughness Index (IRI), etc.. In this case the Pay Adjustment (PA) is often empirically determined on the basis of the Pay Factor (PF). PF (and the payment to the contractor) decreases as IRI increases. In the average-based models (see for example typical Italian contracts such as CIRS and ANAS 2008), the pay adjustment (or/and its specific value, PA/C=PA*) usually depends on the difference (DI), for the given j-th quality characteristic, between the average (x_{AC}) and the design value (x_{AD}). Sometimes the percent difference is used (%, PDI=(x_{AC}-x_{AD})/ x_{AD})or its difference from a threshold. It results PA=abs(DI)·k (where abs(DI) indicates the absolute value of DI, while k usually ranges from 0.3 to 3 and the overall PA is the sum of the PAs of the single quality characteristics. For example, for a SFC (side force coefficient) of 36 instead of 40, it results PDI=10% and PA=10·0.5=5%. Note that these models are cumulative and only penalties are involved (no bonuses). Thickness, HMA moduli, pavement bearing, surface properties (such as SFC and sand height, SH) are usually the main quality characteristics which are considered. In PD-type models asphalt content, grading, mix in-place density, air voids, Marshall resistance, and/or thickness are often used (Deacon et al., 1997; Epps et al., 1999; Khaled, 2003; Burati, 2005; Hughes et al., 2005). In this case the PA is often (but not always) computed by taking into account the percentage within the limits (PWL), with particular probabilistic hypotheses in order to calibrate agency and contractor risks; for the i-th variable, for a given layer, from the PWLi a percent defective (PDi) and then a pay factor (PFi) are computed; by combining all the PFi a combined pay factor (CPF) is often derived. Given that, by the means of CPF and cost, the Pay Adjustment is computed. LCCA based models are models in which the concept of life cycle cost analysis (as a methodology useful to compare the total user and agency costs of competing project implementation alternatives) is explicit. Therefore, pay adjustment depends on how long the pavement will last (Weed, 2001; Weed & Tabrizi, 2005; Whiteley et al., 2005). The Pay Factor may be often referred (more or less implicitly) to an expected life. For this reason, the boundary between these different models may be difficult to define. Though many algorithms do exist for assessing pay adjustment, many classes of issues still call for research. Furthermore, note that the real OC curves (Operating Characteristic Curves) and risks will depend on sample size (number of test results used to judge the quality of a lot), lot size and sublot size (sublots are needed in order to ensure that the specimens for the sample are obtained from throughout the lot, and are not concentrated in one portion or section of the lot). The operating characteristic (OC) curve is a graphic representation of an acceptance plan that shows the relationship between the actual quality of a lot and either the probability of its acceptance or the probability of its acceptance at various payment levels (for acceptance plans that include pay adjustment provisions). OCs aid in the selection of plans that are effective in reducing risks, because they provide buyer's and seller's risk.

In the abovementioned processes (state-of-the art in the field) the following critical issues can be listed (Praticò, 2008). Problem 1. As is well known (Di Benedetto et al., 1996; Domenichini et al., 1999), all the properties and characteristics influence the real and perceived economic value of the as-constructed pavement at a given time. So, when one characteristic fails, this constitutes a quality assurance problem for the state agency (Burati et al., 2003; Muench & Mahoney, 2001), and an acceptance plan, with particular acceptance procedures, is needed, in order to estimate a pay adjustment, PA. The use of **road surface condition** measurements for the acceptance of roadwork is becoming more and more relevant (Boscaino and Praticò, 2001; Boscaino et al, 2005) and calls for a synergistic approach. Problem 2. Can a friction course be treated as the remaining layers in estimating pay adjustment? In other terms, how can pay adjustment be estimated when both surface and mechanical defects are involved? Problem 3. Is it possible to apply life cycle cost analysis when **both surface and mechanical performance are involved**? Problem 4. Attention is often entirely focused on the quality of single asphalt layers, without any systemic consideration of the quality of the whole **multilayer** during the life cycle. However, it must be observed that the influence of the bottom layers on the performance of a multilayer can strongly modify both failure typology and pavement expected life. Problem 5. **Percentage of defects or average values**? Is there a connection between models based on percent defective and model/procedures based on the consideration of average values? Is there the potential for a synergetic consideration of both position and dispersion? Problem 6. Is there a **relationship between the position of the mean respect to the limits and the value of the percent of defects**? Probably yes, but it depends on the single type of set data (standard deviation, asymmetry, kurtosis, or in the simplest case on standard deviation). Many uncertainties in this field call for further research (Uddin et al, 2011). Problem 7. Given that performance are usually related to averages and not to percent defective, is it possible to relate **pavement performance and PDs** (percentage of defects)? Problem 8. The logic of percent defective is usually linked to pay factors and a **composite pay factor** must be considered. A possibility is to have a correlation between the different factors and the expected life of the pavement. But how much logical and exhaustive can be such procedures? Problem 9. Is it possible a synergetic consideration of **defects and delays** in pay adjustment models? Problem 10. A drawback of the method of percent defects is the impossibility for taking into account the extended service life due to the surplus in some of the quality indicators (for example thickness), even if other defects are detected. In other terms, **PDs are defined positive**. Is there any possibility to correct this shortcoming within the framework of the PD-models? Problem 11. Another problem does occur when one tries to apply the concept of percent defective to Open Graded Friction Courses or Porous European Mixes. In fact, in these cases, the quality indicators used in PD evaluation (thickness and air voids) seem not to provide a logical estimate of expected life (included the surface performance). This fact is due also to the uncertainties in the field of the

τ, $\tau\tau$	τ refers to each periodical effect on friction (F); $\tau\tau$=maxτ
α, β, χ	real numbers, coefficients
φ_i	i-th phase of the i-th periodical effect on friction (F)
a, a_i, a_{1B}	structural layer coefficients; i refers to i-th layer and B refers to the Bearing component
B	"bearing" component of the pavement
BPN	British Pendulum Number
C_B, C_S C_{DP} C_i, C_{1B} %C_B	Present Costs, referred to "B" characteristics, to "S"ones, to DP, to the i-th layer, to the 1st layer of the "B" component, to "B" characteristics as a percent of C_{DP}, respectively.
CBR	California Bearing Ratio
CP	as-Constructed Pavement, actual pavement constructed by the contractor
CPF	Combined Pay Factor
CT	equivalent Cumulative Traffic
D	Design life of the as-Designed Pavement; also called initial design life, it is the amount of time for which the chosen pavement design is expected to carry the traffic loads without the application of a global rehabilitation.
DFC	Dense-graded Friction Course
DP	as-Designed Pavement; desired pavement, as defined by the agency (buyer)
$\delta_{AC, I}$, $\delta_{AD, j}$	Parameters which take into account for successive resurfacings.
E, EB, ES	Expected life of the CP, general, for only B component, for only S component, respectively
ESALs	Equivalent single Axes Loads
EXPLIF	Expected life
f, f_3	real numbers, ratio between C_2 and C_1 (f), and ratio between C_3 and C_1 (f3), respectively
F_1, F_2, F_3, F	Coefficients F_i in the formula of the friction F
fre_i	i-th frequency (formula for skid resistance dependence on time)
G_0, .., G_3	Real coefficients in the recipe of the expected life (Burati et al, 2003)
h	real number, ratio between t_{1B} and t_2
I	Indicator, for example percolation speed or drainability
INF, INT	Inflation rate and interest rate, respectively
IRI	International Roughness Index
M1B, M1	Moduli, respectively, of the first layer of DP– B component and of the first layer of the DP
MS	Marshall Stability
N, n	Number of layers (above the subgrade) total (N) and to resurface/construct (n)
NDT	Non Destructive Test
O, O_{DP}, O_{CP}	Expected life of successive resurfacing/reconstruction, general, of DP, of CP, respectively
OGFC	Open Graded Friction Course
P	Pavement
PA, PA_B, PA_S, PA%, PA%(1L),..	Pay Adjustments; PA is the total one; B and D mean referred to B or S, respectively; %:expressed in percentage, i.e. referred to CDP; (1L): referred to one layer (1L) or more (2L, etc.)
PD_V, PD_T	Percent Defective (PD) referred to air voids and to thickness respectively.
PEM	Porous European Mixes
PF	Pay Factor
PI	Profile Index (a roughness indicator)
PM	Preventive Maintenance

PWL	Percentage Within Limits
PWL	Percentage Within Limits
REH	Rehabilitation
S	supplementary component of the pavement. It hasn't bearing properties.
SFC	Side Force Coefficient
t_i t_{1B}	Thickness of the i-th layer of DP and of the first layer of the bearing component of DP
T_j	expected life of the j-th supplementary characteristic
UCS	Unconfined Compressive Strength – 7 day break
Y, Y*	Years (real number) and time in years to reach a quasi-constant friction, respectively

Table 1. List of symbols and acronyms.

correlations among air voids, moduli and expected life. Problem 12. Can base and subgrade be part of the considered pavement system in such algorithms? Problem 13. The density and volumetrics of as-built pavements is a vital part of QC/QA procedures.
Expected life, infrastructure management and pay adjustment strongly depend on air voids content, especially when bituminous mixes are involved. Despite this measurement process is affected by several classes of uncertainties and many issues still call for further research: influence of core diameter, reliability of non-destructive testing, etc. Table 1 lists the main symbols used in this section.

3.2 Model

This section deals with model development. Equations 3-13, Fig.3, Table 1 summarize the derivation of the algorithms. Note that the algorithm here presented overcomes and refines the previous formula as stated in (Praticò, 2007) and was successively updated (Praticò et al, 2010a; Praticò et al, 2010b; Praticò et al, 2011). In summary, the model allows to estimate the pay adjustment on the basis of how long the pavement (considered in all its "qualities") will perform adequately. In order to introduce the model, let DP be the as-designed pavement and CP be the as-constructed pavement (symbols are listed in table 1). The main inputs of the model are costs and expected lives, while the output is the Pay Adjustment, PA (negative if a penalty is provided). It is well known that the friction course has supplementary characteristics (friction, fire resistance, etc., Praticò et al, 2010c). Therefore, every pavement P (both DP and CP, for example) can be divided into two main "components": Bearing characteristics, B, and Supplementary characteristics, S. By comparing, separately, the Bearing characteristics (B, addressing substantially moduli, Poisson coefficients and thicknesses) of the as-Designed Pavement (DP) and of the as-Constructed Pavement (CP), the Pay Adjustment PA_B is estimated (where the subscript B means that PA is referred to the bearing characteristics). Similarly, by comparing the Supplementary characteristics (S) of as-designed (DP) and of the as-constructed pavement (CP), the pay Adjustment PA_S is estimated:

$$P=B+S \tag{3}$$

$$PA= PA_B+PA_S \tag{4}$$

where PA_S refers to S, PA_B to B, and PA to all the non-conformities. To estimate PA_S and PA_B, it is necessary to analyze the costs of the pavement during its life. Let N be the total layers of the as-Designed Pavement (above the subgrade) and let n be the layers to be resurfaced (or reconstructed, if base layers are involved): thus n≤N. Each of the n layers has a contract cost (i.e. present cost of the as-Designed Pavement, in €/m²) equal to C_i (i=1, 2, .. n, from the surface towards the subgrade), then the relative cost C_{DP} of the as-Designed Pavement can be expressed as follows:

$$C_{DP}=C_1+C_2+\ldots+C_n. \quad (5)$$

$$C_{DP}=C_S+C_B \quad (6)$$

where C_S and C_B are respectively the cost of the S component (supplementary) and of the B component of the as-designed pavement. The first layer of B will have a present cost C_{1B} (where C stands for cost, 1 for 1st layer, B for bearing component of the pavement) and a thickness t_{1B} (in order to permit to B to last for D years). Thus:

$$C_B= C_{1B}+C_2+\ldots+C_n. \quad (7)$$

$$C_S=C_{DP}-C_B=C_1-C_{1B} \quad (8)$$

where C_{1B}, referred to the first layer of the B component (which is intrinsically designed only to have mechanical properties), is generally lower than C_1. Let INT and INF be the long-term annual interest rate and inflation rate respectively, given in decimal form (typically 0.08 and 0.04). The Rate R is so defined:

$$R=(1+INF)\cdot(1+INT)^{-1}. \quad (9)$$

For the as-Designed Pavement, let Design life (D) be the expected life in years of the B component and D_S of the S component. For the as-Constructed Pavement, let E_B be the expected life of the B component and let E_S be the expected life of the S component. In practice, E_S can be interpreted as the minimum expected life (years) for supplementary characteristics (where j represents a given supplementary characteristic):

$$ES=min\ [ESi],\ i=1,\ 2,\ \ldots,\ k \quad (10)$$

where ES_i is the expected life of the i-th supplementary characteristic. Note that, in order to consider the right number of resurfacing processes the parameters δ (δ_{AC} and δ_{AD}) can be introduced, where, for example, a=1, m=1, n=20000, τ=0.05, t=EB-ES-DS:

$$\delta=a\cdot(1+m\cdot e^{-t/\tau})\cdot(1+n\cdot e^{-t/\tau})^{-1}. \quad (11)$$

In practice, for t=EB-ES-DS=0 or negative, δ approaches 0, while for EB-ES-DS>1 it approaches 1. Note that in a first analysis of the problem δ can be negleted. Let us introduce the concept of expected life (years) of successive resurfacing or reconstruction (typically 10 years). For the as-Designed Pavement (DP) let ODP be this "successive" expected life, both for B (bearing) and S (supplementary). For the as-Constructed Pavement (CP) let O_{CP} be this "successive" expected life, both for B (bearing) and S (supplementary). The reason for separating the concept of O into two different concepts O_{DP} and O_{CP} is that the actual expected life of a resurfacing / reconstruction depends also on the part of the pavement not-resurfaced. For example, it is possible that, after reconstruction, a subgrade weaker than that set out in contract causes a lower expected life of the surfacing, or that a base stronger then that set out in contract causes a greater one. Given the above facts, it is possible to demonstrate the following (see table 1):

$$\mathbf{PA} \cong C_S \cdot (R^{DS} - R^{ES}) + \delta_{AC} \cdot C_S \cdot [-(R^{ES+DS})] + \delta_{AD} \cdot C_S \cdot [R^{2DS}] + (C_S) \cdot \{[(R^{D}))/(1-R^{ODP})] - [(R^{EB}))/(1-R^{OCP})]\} + (C_B) \cdot \{[(R^{D}))/(1-R^{ODP})] - [(R^{EB}))/(1-R^{OCP})]\} \cong$$
$$\cong C_S \cdot (R^{DS} - R^{ES}) + (C_S) \cdot \{[(R^{D}))/(1-R^{ODP})] - [(R^{EB}))/(1-R^{OCP})]\} + (C_B) \cdot \{[(R^{D}))/(1-R^{ODP})] - [(R^{EB}))/(1-R^{OCP})]\} \quad (12)$$

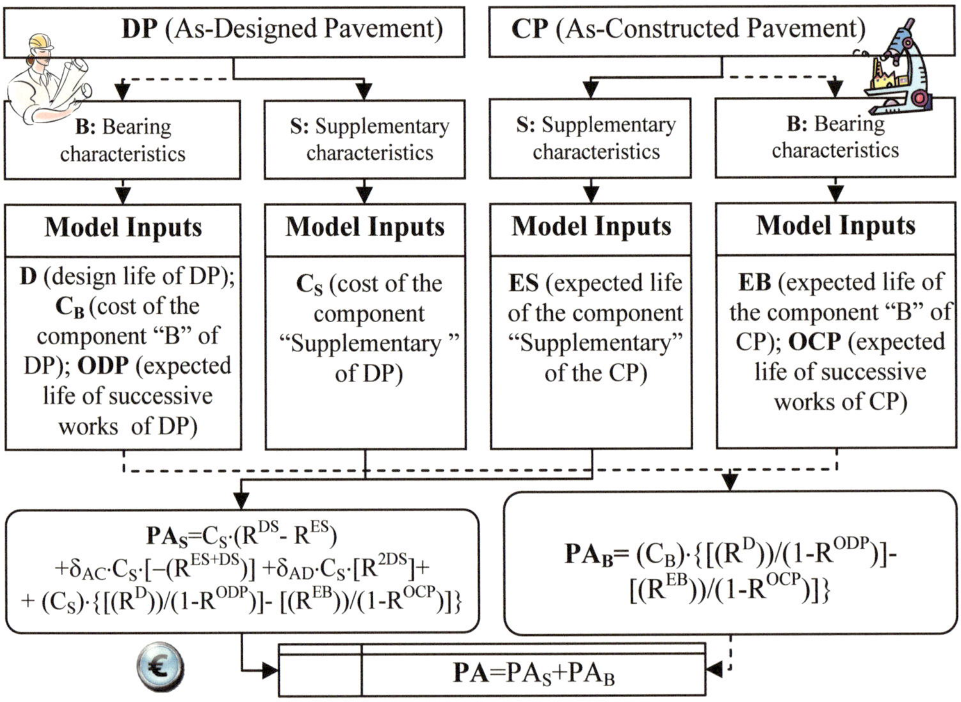

Fig. 3. A synthesis of the model (symbols are listed above).

The term containing C_B is the PA_B, while PA_S is given by the sum of the three terms containing C_S. Note that, when $D_S=E_S$, $E_B=E$, $C_S+C_B=C$, $O_{DP}=O_{CP}=O$ ($\delta_{AD}=\delta_{AD}=0$), it follows (as in Burati et al, 2003; Weeds, 2001):

$$PA = (C) \cdot [(R^{D} - R^{E}))/(1-R^{O})] \quad (13)$$

Table 2 illustrates expenditure flows. In order to test the validity of the hypotheses and the effectiveness of the method the following focal points can be focused:

i) In the case of an as-constructed pavement capable of withstanding the design loading, there is only a pay adjustment for non-conformities of supplementary characteristics:

$$\text{if } D=E_B \text{ and } O_{DP}=O_{CP}, \text{ it is } PA_B=0 \quad (14)$$

or if

$$E_S=0,\ E_B=D \rightarrow \infty,\ O_{DP}=O_{CP},\ PA \rightarrow -C_S; \quad (15)$$

ii) The model is conceived for an expected life of the supplementary component less than that of the bearing component (as-constructed pavement); this limitation is quite obvious and agrees with the state of the art of pavement constructions. The equations system *must* contain also this equation:

$$E_S \leq E_B; \quad (16)$$

iii) As a consequence of the previous point, when the expected life of resurfacing/reconstruction is equal for both the as-Designed and the as-Constructed Pavement, if the expected life of the S component is greater than that of the B component, then the pay adjustment can be computed by the above-mentioned Ref.Eq. after (Weed, 2001):

$$\text{if } O_{DP}=O_{CP} \text{ and } E_S>E_B, \text{ it is } PA=C_{DP}\cdot(R^D-R^E)\cdot(1-R^O)^{-1} \quad (17)$$

iv) The pay adjustment must be compatible with the cost of the layers to resurface; from this another equation to be included in the equation system is derived. This equation originates from an intrinsic limitation of the model (in common with the previous model (Weed, 2001]); for example for DB=20, E_B=0, O_{DP}=O_{CP}=10, ES=DS, it is PA$\cong$-1.7$\cdot C_{DP}$ (which is inconsistent); the new model agrees with common sense if, when E_B=0, one puts also DB$\rightarrow\infty$ and O$\rightarrow\infty$, or/and with D=O. This supplementary equation is:

$$PA \geq -C_{DP} \quad (18)$$

vi) PA is substantially dependent on D-E (with E=EB=ES and O=OCP=ODP); it may be approximated, for example, by linear or quadratic relations:

$$PA\cdot(1-R^O)\cdot C_{DP}^{-1}= -0.\ 02414284\cdot (D-E) \quad (19)$$

$$PA\cdot(1-R^O)\cdot C_{DP}^{-1}= -0.00049374\cdot(D-E)^2 - 0.0164080\cdot(D-E) - 0.00202513 \quad (20)$$

vii) PA is substantially dependent on OCP-ODP; for D=20, with E=EB=ES=15, it may be approximated, for example, by this linear relation:

$$PA\cdot C_{DP}^{-1}= -0.\ 1601\cdot (OCP-ODP)-0.3753 \quad (21)$$

viii) PA% is affected by the difference (R^D-R^{ES}); this relationship depends on the thickness of resurfacing/reconstruction; if $f_3=C_3/C_1$, with $f=C_2/C_1$, OCP=ODP=O, D=EB, and n is the number of layers resurfaced:

$$PA\%= (1-fh)\cdot(1+f+f_3)^{-1}\cdot(R^D-R^{ES}), \text{ for } n=3 \quad (22)$$

$$PA\%= (1-fh)\cdot(1+f)^{-1}\cdot(R^D-R^{ES}), \text{ for } n=2 \quad (23)$$

$$PA\%= (1-fh)\cdot(R^D-R^{ES}), \text{ for } n=1, \text{ where } (1-fh)>(1-fh)\cdot(1+f)^{-1}>(1-fh)\cdot(1+f+f_3)^{-1} \quad (24)$$

3.3 Life expectancy of surface properties

This section deals with the estimate of the life expectancy for the different surface properties (see Figs 4-8, Tabs 2 and 3, equations 25-36). The author is aware that, if inadequately used, even the new model could cause misevaluations(Praticò, 2007). Some tools to optimize the estimate process are proposed below. The estimate of ES (as a minimum life expectancy for the various supplementary characteristics) can be obtained from quality control/assurance tests, if the time-dependence of the supplementary characteristics, for given traffic, is predictable. There are many effects and related indicators that can be considered eligible as supplementary characteristics (AA.VV., 2005): drainagebility, friction (polishing), noise, texture, splash & spray, raveling, reflectivity, chemical spill tolerance, faulting (difference in

AD, As-designed pavement		AC, As-constructed pavement		AD vs. AC
Times	Expenditure	Times	Expenditure	Pay Adjustment
DS	$C_S\cdot(R^{DS})$	ES	$C_S\cdot(R^{ES})$	$C_S\cdot(R^{DS})$- $C_S\cdot(R^{ES})$
$2DS$	$\delta_{AD}\cdot C_S\cdot R^{2DS}$	$ES+DS$	$\delta_{AC}\cdot C_S\cdot[(R^{ES+DS})]$	$\delta\cdot C_S\cdot[(R^{ES+DS})-R^{2DS}]$
DB	$(C_B+C_S)\cdot(R^{DB})$	EB	$(C_B+C_S)\cdot(R^{EB})$	$(C_B+C_S)\cdot(R^{DB}-R^{EB})$
$DB+O_{DP}$	$(C_B+C_S)\cdot(R^{DB+ODP})$	$EB+O_{CP}$	$(C_B+C_S)\cdot(R^{EB+OCP})$	$(C_B+C_S)\cdot(R^{DB+ODP}-R^{EB+OCP})$
$DB+2\cdot O_{DP}$	$(C_B+C_S)\cdot(R^{DB+2.ODP})$	$EB+2\cdot O_{CP}$	$(C_B+C_S)\cdot(R^{EB+2OCP})$	$(C_B+C_S)\cdot(R^{DB+2.ODP}-R^{EB+2OCP})$
$DB+n\cdot O_{DP}$	$(C_B+C_S)\cdot(R^{DB+nODP})$	$EB+n\cdot O_{CP}$	$(C_B+C_S)\cdot(R^{EB+nOCP})$	$(C_B+C_S)\cdot(R^{DB+nODP}-R^{EB+nOCP})$
Sum	$C_S\cdot(R^{DS})+(C_B+C_S)\cdot(R^{DB})/(1-R^{ODP})$		$C_S\cdot(R^{ES})+(C_B+C_S)\cdot(R^{EB})/(1-R^{OCP})$	$C_S\cdot(R^{DS}-R^{ES})+\delta_{AC}\cdot C_S\cdot[-(R^{ES+DS})]+\delta_{AD}\cdot C_S\cdot[R^{2DS}]+(C_B+C_S)\cdot\{[(R^{DB}))/(1-R^{ODP})]-[(R^{EB}))/(1-R^{OCP})]\}$

Table 1. A synopsis of times and expenditures flows for AD and AC.

elevation across a joint), pitting, resistance to wearing, etc. Let Esi be the expected life in relation to the i-th characteristic. Under these hypotheses, Es will be the minimum value among the Esi. In the light of the above facts, if i=1 (for example friction) and the as-designed target is q_a, while the as-constructed value of friction is $q_b<q_a$, in the case of linear law over the time it results: a) expected life of the as-designed friction course: $ES_a=(T-q_a)/m$, where T stands for minimum level and m is the gradient (negative); b) expected life of the as-constructed friction course: $ES_b=(T-q_b)/m$, where T stands for minimum level; c) loss of expected life: $ES_a-ES_b=(q_b-q_a)/m$. For example, if m=-2, T=35, q_a=55, q_b=45, it results ES_a-ES_b =5 years.

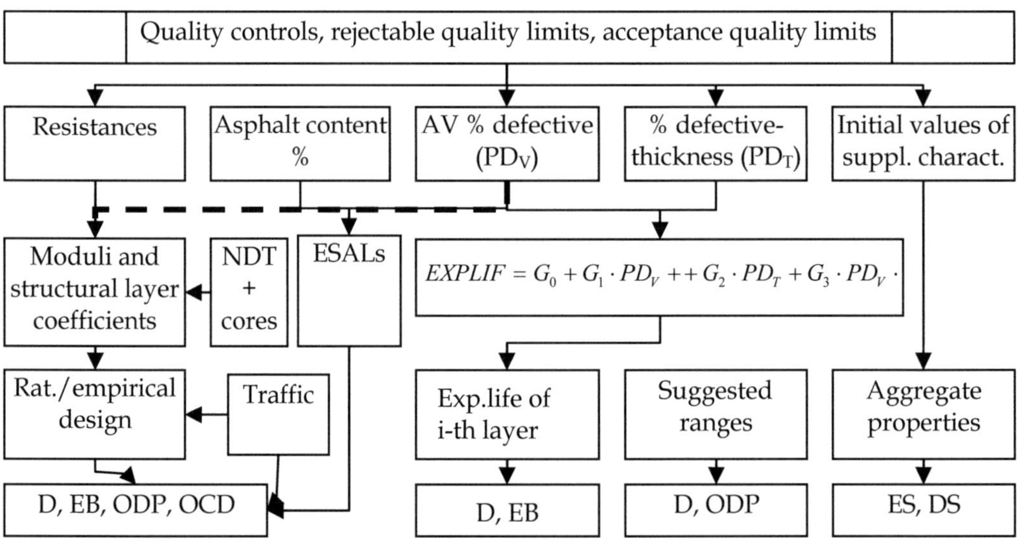

Fig. 4. Expected life: tools to optimize the estimate (symbols are listed in table above).

As is well known, skid resistance changes over time. Typically it increases in the first two years following construction as the roadway is worn away by traffic and rough aggregate surfaces become exposed, then decreases over the remaining pavement life as aggregates become more polished. Skid resistance is also typically higher in the fall and winter and lower in the spring and summer. This seasonal variation is quite significant and can severely skew skid resistance data if not compensated for. Skid resistance deterioration rate depends on the combination of many factors (aggregate properties and gradation, alignment, detritus build-up, rainfall and traffic characteristics, etc) and this can affect substantially any attempt at prediction and modelling. On the basis of the international literature on this topic the following equation can be here derived for skid resistance (*F*) drop due to pavement wear:

$$F = F_1 + F_2 \cdot e^{\left[\frac{Y}{Y^*} \cdot Ln(0.001)\right]} + F_3 \cdot \prod_{\tau=1}^{\tau\tau} sen(2 \cdot \pi \cdot fre_i - \phi_i) \quad (25)$$

where i) Seasonal, monthly, daily effects and variations are considered by the different frequencies (*fre_i*) (Diringer and Barros, 1990; Wilson and Kirk, 2005); initial increase is not explicitly considered here; ii) The terminal value F_1 depends on aggregate shore hardness (both average value and coefficient of variation); this value seems to depend on Polishing Stone Value (PSV, British Standard BS 812, Italian standard CNR BU 140/92) and on Los Angeles (or micro-Deval) (Dupont and Turenq, 1993); both the Los Angeles abrasion number and the PSV, alone, do not correlate well with field performance (Dupont and Turenq, 1993); iii) time in years to reach a quasi-constant friction (Y*) can correspond to 2 million cumulative vehicle passes (Diringer and Barros, 1990) and is usually two to four years, depending on traffic and aggregate properties; iv) $F_2 \cdot F_1^{-1}$ can be estimated in 0.8~1.1 for sedimentary rocks and 0.5~1.4 for igneous rocks (data referred to BPN); v) $F_3 \cdot F_1^{-1}$ can be estimated equal to 0.1~0.3 (Diringer and Barros, 1990). Figure 5 (left) provides a simplified depiction of these hypotheses on friction time-dependence, where Y*=4, F_1=47; F_2=38, fre_1=1year^{-1}, $\varphi_1=\varphi_2=0$; fre_2=4 year^{-1}, $F_3 \cdot F_1^{-1}$=0.22; dotted curve refers to fre_2=0. Figure 5 (right) shows time-dependence for two friction indicators (initial increase is not represented (Brosseaud and Roche, 1997; Kokkalis and Panagouli, 1998). A possible relationship between SFC (Side Friction Coefficient, range 0-1) and SN (Skid Number, range 0-1) (Ullidtz, 1987) is as follows (see figure 5):

$$SFC=-0.014+1.516\ SN \quad (26)$$

The minimum value of SFC over the time can be also estimated through the following equation (see figure 6):

$$Min\ SFC=0.024-0.663*10\text{^}-4\ QCV+0.01PSV \quad (27)$$

where QCV is the number of commercial vehicles/lane/day, PSV is the Polishing Stone Value (Ullidtz, 1987). For limestone, the decrease of SFC over the time as a function of N (number of heavy vehicle equivalents in millions) and SFCi (initial value of SFC) can be expressed as (see figure7):

$$\Delta SFC=-0.48N^{0.373}(SFCi-3) \quad (28)$$

while for basalt as (figure 7):

$$\Delta SFC=-0.30N^{0.503}(SFC_i-3) \tag{29}$$

(Flintsch et al, 2001) proposed the following model:

$$SN(64)_S=26.865+2.079\cdot Binder+1.601\cdot PP200+1.03\cdot VTM \tag{30}$$

$$SN(64)_R=104.211-4.356\cdot NMS+0.1833\cdot VTM \tag{31}$$

where $SN(64)_S$ stands for Skid Number measured at 64Km/h for smooth tires, Binder stands for binder code (-1 for PG 64-22, 0 for PG 70-22, 1 for PG 76-22), PP200 is the percentage of material passing the #200 sieve, VTM represents the total voids in the mix, $SN(64)_R$ stands for Skid Number measured at 64Km/h for ribbed (R) tires, NMS is the Nominal Maximum Size. Note that the model after Flintsch et alia refers to the starting point of diagrams (as-constructed value). In particular, for Binder=-1, NMS=12.5, PP200=5, VTM=20 it results $SN64_S$= $SN64_R$=53 and for Binder=-1, NMS=19, PP200=5, VTM=6, it results $SN64_S$= 39 and $SN64_R$=23. As for texture, (Flintsch et al, 2001) proposed the following model:

$$MPD=-2.896+0.2993NMS+0.0698VMA \tag{32}$$

where MPD stands for Mean Profile Depth, NMS for Nominal Maximum Size and VMA for Voids in Mineral Aggregates. Note that in this case the law doesn't provide the variation over the time. Another model for texture depth was developed (Arnold et al, 2005) according to the following algorithm:

$$\text{MPD} = \text{k1} - \text{k2} \,.\, \log(\text{N}), \tag{33}$$

where $k1$ and $k2$ are constants and N is number of wheel passes. The constants k_1 and k_2 in the equation (the Patrick equation) have been calculated for two different cases (see figure 7). The Surface wear due to the combined action of salt and traffic can be estimated through the following model (Ullidtz, 1987, Praticò et al, 2010):

$$RDW= 2.48\cdot 10^{-5}\cdot PASS^{1.02}\cdot CW^{-0.46}\cdot S^{1.22}\cdot SALT^{0.32} \tag{34}$$

Where RDW is the rut depth due to studded tires in mm, PASS is the number of vehicles with studded tires in one direction expressed in thousand, CW is the carriageway width in m, S is the vehicle speed in Km/h, and SALT is a variable for salting (2 salted, 1 unsalted, see figure). Based on other authors (Smith, 1979), with studded tires in the range 7-23%, concrete pavement wear can be considered as follows:

$$PW\approx 0,5*Y \tag{35}$$

Where PW is the wear in mm, while Y stands for number of years. Also drainability values depend on the chosen indicator (Praticò and Moro, 2007a, 2008a). On the basis of the international literature on this topic, a typical curve for drainability is as follows (where I is drainability indicator, Y stands for years, α and χ are positive and β is negative; PEM indicates porous European mixes; EM: emergency lane):

$$I=\alpha\cdot e^{\beta\cdot Y}+\chi \tag{36}$$

It is important to note that, although some of these estimates may be considered approximate, all the inputs may be "conditioned" by the same methodology and the PA formula is strongly dependent on the differences D-E and O_{CP}-O_{DP}. This can help to reduce the possible conflicts between contractor and buyer without using simple but empirical formulas and models not well-grounded in logic.

Pavement	α	β	χ	Indicator I	I Range	Y Range (years)
PEM	0.469	-0.778	0.862	Percolation speed (cm/s)	0.79~1.5	0~3
PEM	0.449	-2.435	0.795	Percolation speed (cm/s)	0.57~1.6	0~3
PEM	1.049	-0.778	0.248	Permeability (cm/s)	0.3~1.3	0~3.75
PEM - EL	1.119	-1.312	0.168	Permeability (cm/s)	0.15~1.3	0~3.75

Table 2. Time-dependence of drainability (example).

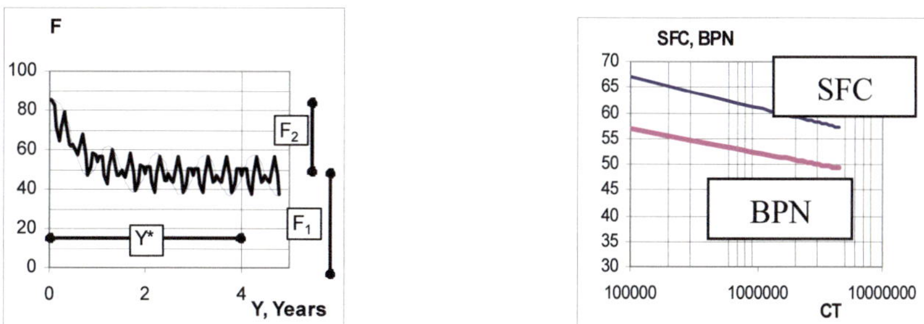

Note: F: Friction indicator; SFC: Side Friction Coefficient; BPN: British Pendulum Number

Fig. 5. Skid resistance (F=BPN) vs. time (Years) and skid resistance (SFC, BPN) versus equivalent Cumulative Traffic (CT).

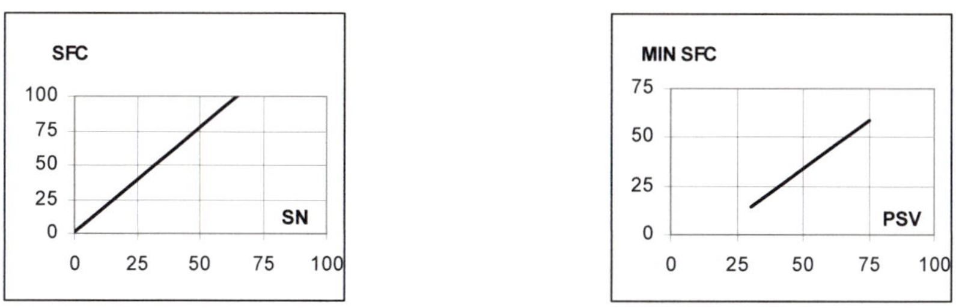

Fig. 6. Example of relationship SN vs. SFC and PSV vs. MINSFC.

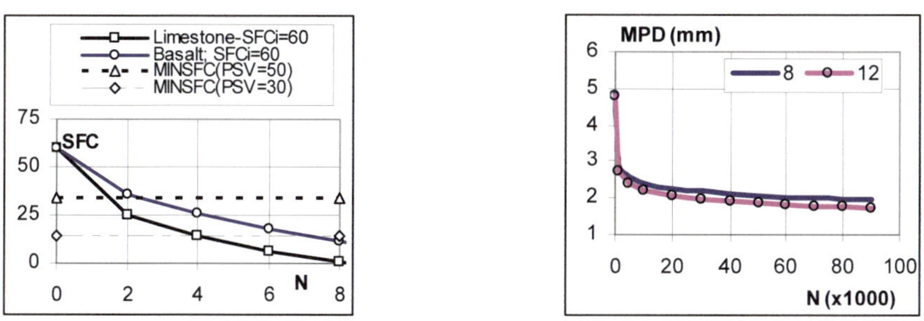

Fig. 7.Example of relationships N vs. SFC and MPD vs. number of axes.

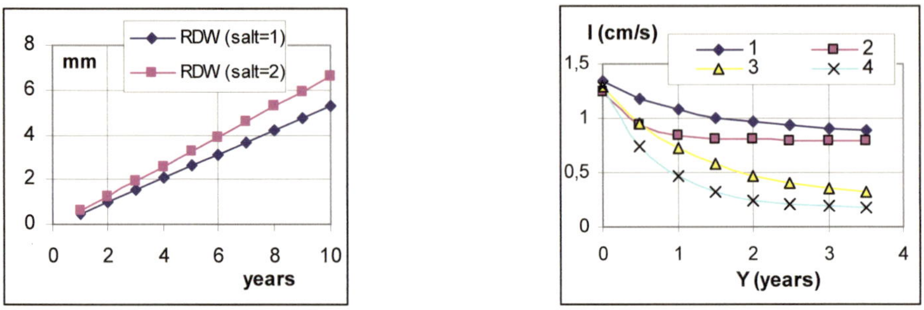

Fig. 8. Example of relationships years vs. wear (mm) or permeability.

As for noise mitigation, based on the international literature the following information is available:

Solution	E	IN	FINAL
DAC (dense asphalt concrete)	Variable	0	-2
PA (porous asphalt)	10-12	4	<3 (?)
TPA (two-layer porous asphalt)	9	6	4
SMA-like thin layers	9.5	4.7	3
Porous-type thin layers	8.5	5	3
E: Expected lifetime (years); IN: Initial noise reduction (db(A)); FINAL: Final/minimum noise reduction (db(A))			

Table 3. Duration of noise mitigation (SPB method).

3.4 Life expectancy of bulk properties

Both for unbound and HMA/PCC layers, volumetrics, mechanistic performance and related indicators are often correlated and vary over the time. Equations 37-38, Figs 9-13, Tables 4 and 5 summarize several practical relationships. In the M-E PDG, the HMA layer modulus is characterized using the dynamic modulus (or backcalulated modulus from FWD data). In more detail, the dynamic modulus at a given loading time and temperature is assumed to be the elastic modulus in the response computation. PCC materials need a static modulus of elasticity adjusted with time. For chemically stabilized materials the elastic modulus or the resilient modulus is needed (lime-stabilized typical value: 45000psi). For unbound materials the resilient modulus is needed (39000 psi: very good; 10000psi: very poor). In the AASHTO guide 1993 the structural layer coefficients and the resilient modulus (subgrade) are used. Design life (D) of the as-Designed Pavement (DP), can be estimated from contract specifications. D can also be estimated on the basis of the design report (in which it is usually specified). Empirical or rational design can be used after having estimated structural layer coefficients or moduli from resistance thresholds set out in the contract (Marshall Stability, MS, for surface course, base course and bituminous treated bases, California Bearing Ratio (CBR), for untreated bases or subbases, Unconfined Compressive Strength - 7 day break (UCS) for cement treated bases, (Huang, 2003), see figures 9 to 13); by knowing thickness and traffic loading it is therefore possible to estimate D. Importantly, on the basis of the international literature (Van Til et al, 1972; Gaspard, 2000; Sebesta, 2005) and current practice, the modulus of cement treated bases which corresponds to a given laboratory resistance (Unconstrained Compressive Strength - 7-day break) can vary to a great extent in relation to many factors (time from base construction, traffic, subgrade drainage, shrinkage cracks, etc). Figure 9 shows the relationship between resistances and structural layer coefficients according to (Van Til et al, 1972). Note that in the following figures MS indicates Marshall stability, M the modulus, a the structural layer coefficient, UCS is the unconstrained compressive strength- 7 day break, CBR is the california bearing ratio; ESAL indicates equivalent single axle load, AC stands for as-constructed. In principle, E_B, expected life of the bearing component of the as-Constructed Pavement, can be estimated as D (except where D is derived from thresholds in contracts or from design report). Moreover, it must be noted that for the asphalt layers there are empirical formulas that can be used to estimate D and E_B in function of Percent Defective of air voids and thickness (PD_V, PD_T, (Burati et al, 2003)), or in function of the air voids and asphalt content of the as-Constructed Pavement. Both for E_B and D, reliable information can be obtained from Non-Destructive Tests, NDT (e.g. Falling Weight Deflectometer), or /and laboratory tests on cores (resilient moduli - test methods AASHTO TP9-94-1B, ASTM D 4123, LTPP P07) (Giannattasio and Pignataro, 1983; Ullidtz, 1987). The estimate of ODP and OCP can be approached by the same methodologies above-mentioned for D and E_B respectively (see also tables 4 and 5). All these values can be modified (and costs upgraded) if extended service life gains for preventive maintenance treatment are considered (overband crack filling, crack sealing, single or double chip seal, slurry seal, microsurfacing, ultrathin, hot-mix asphalt overlay, hot-mix asphalt mill and overlay, etc.) and in relation to particular design philosophies.

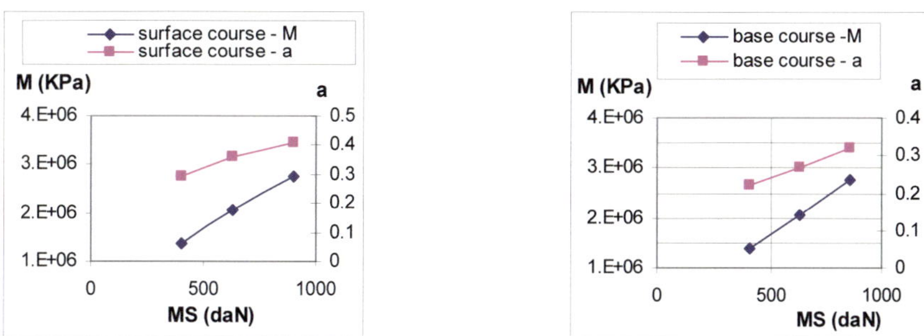

Fig. 9. Example of MS-M-a relationships, for surface courses and base courses.

Mr (psi)	a	a	a	UCS, Mpa	UCS,psi	MD, Mpa	CBR	MD, Mpa	CBR	MD, Mpa	CBR	MD, Mpa
all	CTB (1)	UB (2)	SB (3)	CTB (4)	CTB	CTB (5)	UB (6)	UB (7)	SB	SB (8)	SUB (9)	SUB (10)
9,128			**0.06**						6	31	6	29
12,628			0.09						17	**80**	8	40
15,000			**0.11**						30	144	10	48
15,750		**0.07**	0.11				10	47	40	192	11	**50**
18,300		0.08	0.13				17	**80**	64	309	12	59
20,548		0.10	**0.14**				25	122	100	479	14	66
25,000		0.12					52	248			**17**	80
30,000		**0.14**					100	480			20	96
33,582		**0.15**					150	722			22	107
39,000											**26**	125
45,000											**30**	144
599,081	**0.15**			**2.5**	362	**150**						
752,581	**0.20**			4.7	676	381						
864,318	**0.23**			6.0	867	521						

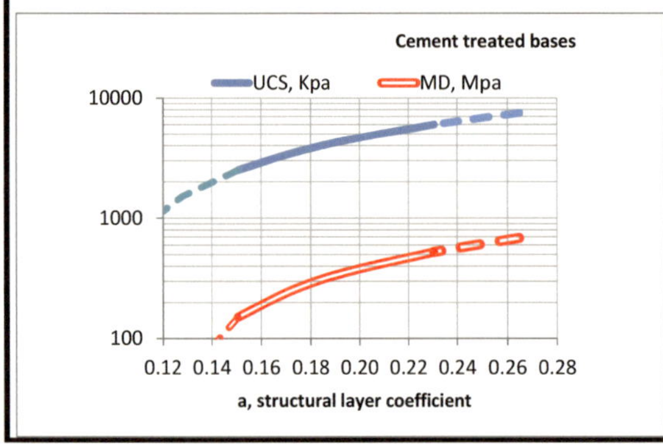

Equations:
(1): a=0.499*LOG10(Mr)-2.7324;
(2): a=0.249*LOG10(Mr)-0.977;
(3):a=0.227*LOG10(Mr)-0.839;
(4): UCS=(a-0.0935)/0.02282609;
(5): MD= 4667.3*a - 552.6;
(6): CBR=10^(a/0.0689);
(7, 8): MD=CBR*4.8;
(9): CBR=Mr/1500.
Symbols: CTB: Cement treated base;
UB: Untreated base; SB: Subbase;
MD: Modulus of deformation;
SUB: Subgrade (bolded characters refer to a very good subgrade support);
Mr: resilient modulus;
a: structural layer coefficient;
UCS: unconfined compressive strength;
MD: Modulus of deformation;
CBR: California bearing ratio.

Fig. 10. Cement treated base courses, untreated bases, subbases, subgrades.

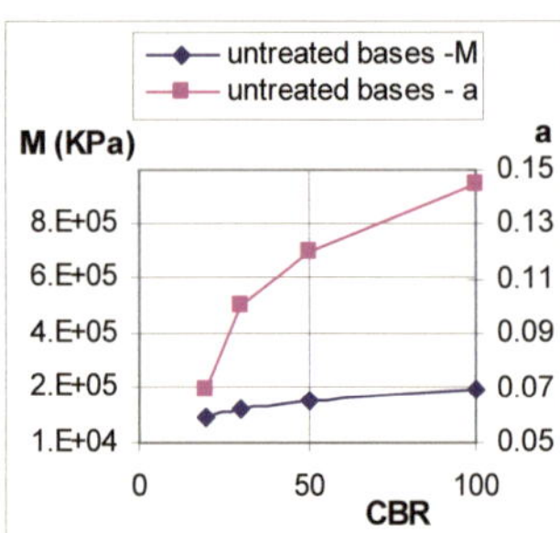

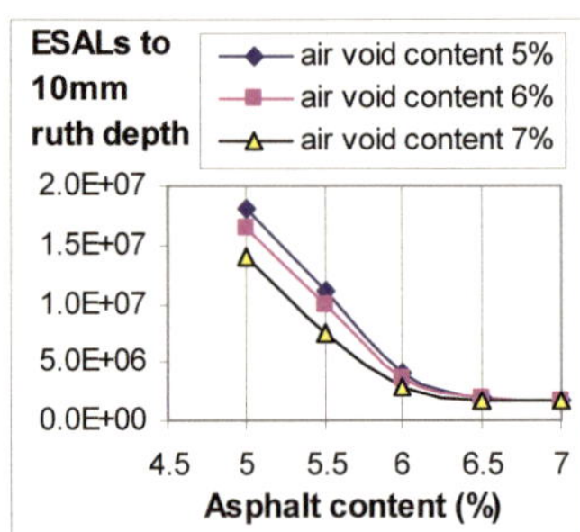

Fig. 11. Left (untreated bases): CBR vs. M and a. Right (surface course - fine gradation): ESALs to 10mm Ruth Depth vs. asphalt content (Epps et al. , 1999).

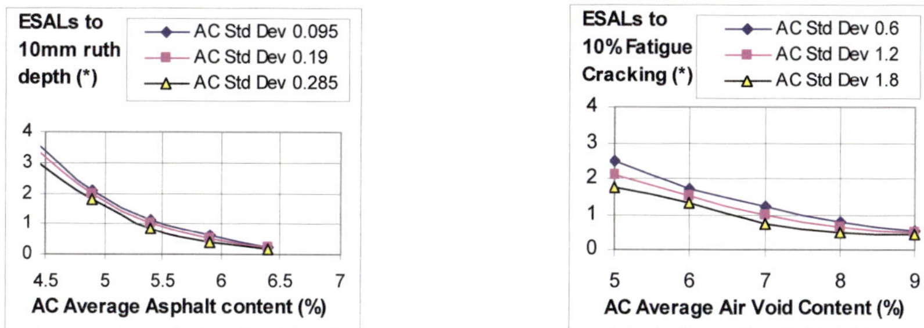

Fig. 12. Left: ESALs to 10mm vs. asphalt content. Right: ESALs to 10% vs. air void content (Surface courses (Epps et al. , 1999)) (*) expressed as multiple of target ESALs.

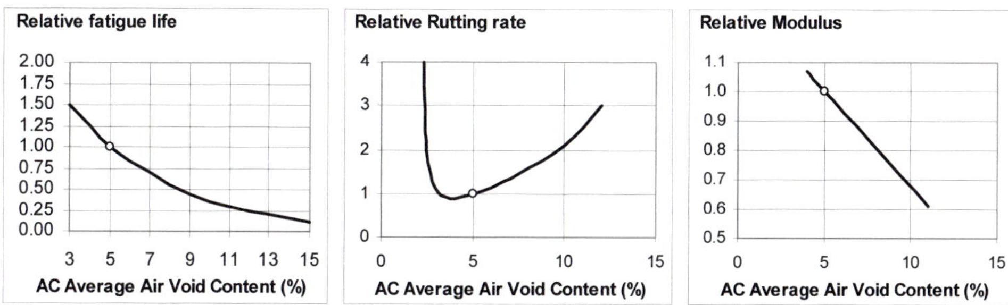

Fig. 13. Air Voids vs. Relative Fatigue Life (left) or Relative Rutting Rate (center) or . Relative Modulus (right) - Surface courses (Austroroads, 1999).

years to 1st rehabilitation (≅D)	years to 2nd rehabilitation (≅ODP)	years to 3rd rehabilitation (≅ODP)	years to 4th rehabilitation (≅ODP)
12~16 (base layer undrained) 15~20 (base layer drained)	10~12 (mill and overlay)	12~16 (full depth mill and overlay) 10~12 (mill and overlay)	10~12 (mill and overlay)

Table 4. Estimates of years to the n-th rehabilitation.

Note that the application of the PA formula depends on the ability to split the surface course into two parts. A tentative method to estimate t_{1B}, C_{1B}, C_B and C_S is to identify the component B of DP in a pavement with a design life D, but with a different friction course (this time with negligible surface properties, for example just the binder course). In particular, two tentative values of t_{1B} may be obtained by the following expressions:

$$t_{1B}\cdot M_{1B}^{1/3} \cong t_1 \cdot M_1^{1/3}\ ;\quad t_{1B}\cdot a_{1B} \cong t_1 \cdot a_1 \qquad (37)$$

where the modulus M_{1B} of the first layer of the B component of the as-Designed Pavement can be tentatively identified in M_2, the structural layer coefficient a_{1B} can be considered equal to a_2, and the thicknesses t_1 and t_2 are known. As above-mentioned, M_1, M_2 (moduli of the 1st and 2nd layer of the DP) and a_1, a_2 (structural layer coefficients of the 1st and 2nd layer of the DP) may be estimated by using correlation charts and algorithms in literature

references (Van Til et al, 1972; Huang, 2003) in function of the traditional resistance tests. Similarly, C_{1B}, C_B and C_S (€/m²), for a given t_{1B}, can be easily estimated by comparing the costs of the 1st and 2nd layer of the DP, given that the cost of the unit of volume of the layers 1B and 2 of the DP are the same. For example, if C_2 and t_2 are the cost and thickness of the second layer of the as-Designed Pavement respectively, then for h=0.5, f=0.6:
if

$$t_{1B}=h\cdot t_2\,,\ C_2=f\cdot C_1,\ C_{1B}\cdot t_{1B}^{-1}=C_2\cdot t_2^{-1}\ (€/m^3) \Rightarrow C_{1B}=f\cdot h\cdot C_1,\ C_S=C_1\cdot(1-f\cdot h)$$

$$C_S=0.7\cdot C_1,\quad C_{1B}=0.3\ C_1 \qquad C_B=0.3\cdot C_1+C_2+\ldots+C_n \tag{38}$$

Treatment	expected Life (Years, Y)		Cost (€/ m²)	Cost for year €/(m²·Y)
	min	max		
Crack seals (PM)	2	3	1.7	0.66
Fog Seals (PM)	3	4	0.8	0.24
Slurry seal (PM)	4	9	1.5	0.23
Microsurfacing (PM)	5	14	2.3	0.24
Chip seals (PM)	4	6	1.3	0.26
Thin hot mix overlay (PM/REH)	2	10	2.8	0.47
HMA- Dense Graded (5cm) (PM/REH)- (ODP→)	5	15	5.3	0.53
Asph. Rubber Hot Mix - Gap Graded (4-5cm) (PM/REH)-(ODP→)	10	20	5.9	0.40
37.5mm mill+37.5mm overlay (PM/REH) - (ODP→)	8	12	7.2	0.72
Milling (37.5mm) +100mm overlay (REH) - (ODP→)	18	22	9.5	0.48
Milling+thick overlay (75-180mm) (REH) - (ODP→)	18	25	10.9	0.51
Note. PM: Preventive Maintenance treatment; REH: REHabilitation				
After (Moulthrop et al, 1998; Hicks and Epps, 2005; Shober and Friedrichs, 2002)				

Table 5. Summary of a variety of maintenance and rehabilitation life expectancy and costs.

3.5 Experimental application

An experimental application was performed on a motorway in Southern Italy. Traditional quality characteristics of the mixes, for single lane and given layer, have been organized in function of the progressive abscissa. Many experimental devices were used in order to measure the actual characteristics of the as-constructed pavement. From the comparison between the requirements of the as-designed pavement (DP) and the actual characteristics of the as-constructed pavement (CP) the moduli for DP(@0) and CP(@0) were derived (see table 7). The ratio t_{1B}/t_2 was about 0.5; asphalt concrete thickness was 27 cm for DP and about 27-5+3.5=25.5cm for the B component of the pavement. Results were obtained by Kenpave™ [Huang, 2003] and successfully compared with the AASHTO Guide 1993 equation. By analyzing the drainability and friction data, E_S=3 was estimated, caused by insufficient drainability. Pay adjustments (in absolute and in percentage, as referred to the cost C_{DP}) are summarized in table 8.

	DP (@0)	DP, B(@0)	CP(@0)	DP(@E_B)	CP(@E_B)
Asphalt (27 or 25.5* cm)	$2.29 \cdot 10^6$	$2.65 \cdot 10^6$ (*)	$1.00 \cdot 10^6$	$2.29 \cdot 10^6$	$2.29 \cdot 10^6$
Cement treated (20 cm)	$1.00 \cdot 10^6$	$1.00 \cdot 10^6$	$1.30 \cdot 10^6$	$0.50 \cdot 10^6$	$0.70 \cdot 10^6$
Subbase + sub-grade	$0.10 \cdot 10^6$	$0.10 \cdot 10^6$	$0.13 \cdot 10^6$	$0.10 \cdot 10^6$	$0.11 \cdot 10^6$
Life (years)	D=22	D=22	E_B=18	O_{DP}=12	O_{CP}=14
Output	D	t_{1B}/t_2=0.50	E_B	O_{DP}	O_{CP}

Table 7. Case-history (years are rounded to the nearest integer; moduli in KPa).

Input	INT	0.08	C_2, asphalt, €/ m²	4,25
	INF	0.04	C_3, asphalt, €/ m²	7,27
	R	0.963	Cost of the cement treated, €/ m²	6,92
	D (years)	22	Cost of the granular sub-base, €/ m²	3,83
	E_B (years)	18	Overall Cost €/m²	29,09
	E_S (years)	3	Asphalt cost €/ m²	18,34
	O_{DP} (years)	12	C_B, €/ m²	13.65
	O_{CP} (years)	14	C_S, €/ m²	4.69
	t_{1B}/t_2	0,5	C_B%	74.42
	C_1, asphalt, €/m²	6,82	C_S%	25.58
Output	PA (€/m²)		-2.51	
	PA_B (€/m²)		-0.52	
	PA_S (€/m²)		-1.99	
	PA%		-13.70	
	PA_B %		-2.85	
	PA_S %		-10.85	

Table 8. PA determination –Inputs and outputs *(the percentages are referred to C_{DP}).*

4. Main findings

In the light of above facts, the main findings of the study can be summarized in terms of drawbacks and point of strengths. Air void content has a vital role in QC/QA. A decrease in porosity (or effective porosity) yields in-lab specific gravities that converge toward the maximum theoretical specific gravity G_{mm} (convergence of in-lab measurements). In contrast, gravities determined through non-nuclear portable devices often increase due to the high dielectric constant of water (divergence of in-site measures).

Uncertainties in the determination of the expected life of the different characteristics and complexity are the main drawbacks of the models of PA based on LCCA. On the contrary, though more research is needed, analyses proved that the LCCA-based models are able to give a solution well-grounded in logic, even in cases where supplementary characteristics, such as drainability or friction, may have a premature failure. In particular, when only the bearing characteristics of asphalt layers are considered, analyses proved that the complex model (in which also supplementary characteristics are considered) gives the same solution as an existing model (Weed, 2001). The effectiveness and importance of the complex models increases when thin resurfacing/reconstruction is considered. The conceptual framework of the complex models, finally, has the potential for considering other supplementary characteristics such as for example the albedo. Analyses and experimental applications proved that the complex models quantify the dependence of the pay adjustment on all the layers. Another point of interest is that life expectancy differences govern PA estimates: this can play a key role in reducing possible conflicts between buyer and contractor. Future research will aim to consider the consequences of the new model here set out in terms of relationship between probability of acceptance and expected lives.

5. References

Alvarez AE, Epps Martin A, Estakhri C. Internal structure of compacted permeable friction course mixtures. Construction and Building Materials, 24 (2010) 1027–1035.

Arnold G., Steven B., Alabaster D. & Fussell A. (2005). Effect on Pavement Wear of Increased Mass Limits for Heavy Vehicles - Stage 4 - Land Transport New Zealand Research Report 280.

Austroroads and Australian Asphalt pavement Association, Air voids in asphalt, 1999.

Boscaino, G., Pratico, F.G. , A classification of surface texture indices of pavement surfaces | [Classification et inventaire des indicateurs de la texture superficielle des revêtements des chaussées], 2001, Bulletin des Laboratoires des Ponts et Chaussees (234), pp. 17-34+123+125+127.

Brosseaud Y., Roche J-P (1997). Chantier experimental d'enrobés drainants, Bilan après huit annèes de traffic sur l'A63, Revue Générale des routes et des aerodromes, n.755, pp.47-52.

Brown ER. Hainin MR, Cooley A, Hurley G. Relationship of Air Voids, Lift Thickness, and Permeability in Hot Mix Asphalt Pavements. NCHRP Report 531, National Center for Asphalt Technology—Auburn University, Auburn, AL, Transportation Research Board, Washington, D.C. 2004.

Burati J., L., Weed R. M., Hughes C. S., Hill H. S.(2003). Optimal procedures for quality assurance specifications, Final Report n. FHWA-RD-02-95, Office of research, development, and technology, Federal Highway administration (FHWA - COTR: Peter A. Kopac) - 6300 Georgetown Pike, McLean, VA 22101-2296.

Burati, J., L., Jr (2005). Risks with Multiple Pay Factor Acceptance Plans, TRB 2005 Annual Meeting, January.

Cooley LAJr, Prowell BD, Hainin MR, Buchanan MS, Harrington, J. Bulk specific gravity round-robin using the corelok vacuum sealing device. National Center for Asphalt Technology Report 02-11, FHWA-IF-02-044, November 2002.

Crouch LK, Badoe DA, Cates M, Borden TA, Copeland AR, Walker CT, Dunn T, Maxwell RA, Goodwin WA. Bulk specific gravity of compacted Bituminous mixtures:

finding a more widely applicable method. Final Report, State of Tennessee, Department of Transportation, July 2003.

Deacon John A., Monismith Carl L., and Harvey John T. (1997). California department of transportation, "Pay factors for asphalt-concrete construction: effect of construction quality on agency costs".

Di Benedetto H. et alia (1996). Mesures in situ et en laboratoire des écoulement dans les BBDr, modélisation et validation, Bulletin des Laboratoires des Ponts et Chaussées, 204.

Diringer K. T., Barros R. T. (1990). Predicting the skid resistance of bituminous pavements through accelerated laboratory testing of aggregates, Surface characteristics of roadways: International research and technologies, ASTM STP 1031, W.E. Meyer and J. Reichert, Eds, ASTM, Philadelphia.

Domenichini L., Fracassa A., La Torre F., Loprencipe G., Ranzo A., Scalamandrè A. (1999). Relationship between Road Surface Characteristics and Noise Emission, 1° Int.Coll. on Vehicle Tyre Road Interaction, Roma.

Dupont P., Tourenq C. (1993). Granulats et microrugosité, Bull. Liaison Labo. P. et Ch., Ref. 3719, mai-juin.

Epps J. A. et al. (1999). "The road to performance-related specifications", International Conference on Accelerated Pavement Testing, Reno, NV, October 18-20.

Flintsch G, Al-Qadi I.L., Davis R., McGhee K.K. (2001). Effect of HMA properties on pavement surface characteristics, 2002 FWD Users Group Presentations, Nevada.

Friedrichs D. A., Pavement design and programmino options: european design to maintain only, Wisconsin Department of Transportation Library, USA, 2002.

Gamache RW. FEA guides the way for sensor redesign. Machine Design, April 15, 2004. p. 25.

Gaspard K.J., Evaluation of Cement Treated Base Courses, Technical assistance report Number 001TA, Louisiana transportation research center, Louisiana department of transportation and development, December, 2000.

Giannattasio P., Pignataro P., La caratterizzazione dei conglomerati bituminosi, dati di ingresso per il calcolo delle pavimentazioni, Scuola di specializzazione in infrastrutture aeronautiche, Napoli, 1983.

Huang Y.H. (2003). Pavement Analysis and Design, Pearson Prentice Hall, Upper Saddle River, NJ, USA.

Hughes C.S. et alia (2005). NCHRP Synthesis 346, State Construction Quality Assurance programs, Transportation Research Board, Washington D.C..

Khaled Ksaibati, Ph.D., P.E., Nathan E. Butts (2003). Asphalt pavement quality control/quality assurance programs in the United States, TRB 2003 Annual Meeting .

Kokkalis A. G., Panagouli O. K.(1998). Fractal evaluation of pavement skid resistance variations. II: surface wear, Chaos, solitons & fractals, Vol.9, 11, pp.1891,1889, Pergamon Press.

Kvasnak AN, Williams C, Ceylan H, Gopalakrishnan K. Investigation of electromagnetic gauges for determining in-place HMA density. IHRB Project TR-547, CTRE Project

05-233, Final Report, Iowa State University, Ames; Iowa Department of Transportation, 2007.

Leahy et al. Committee on Management of Quality Assurance (A2F03) (2002). Glossary of Highway Quality Assurance terms, Transportation Research Circular number E-C037, TRB National research Board, 2101 Constitution Avenue, NW, Washington, DC 20418, April.

Main italian laws and standards concerning pay adjustments (1998-2000). a) L.11.02.94, n. 109; b) L. 18 novembre 1998, n. 415 - c.d. L. Merloni ter; c) Capitolato Speciale d'appalto ANAS 1998 e succ.- artt.12, 13, 19; d)D.P.R. 21.12.99 n. 554 (Reg. di attuazione); e) Norme tecniche di tipo prestazionale per capitolati speciali d'appalto (CIRS, Centro sperimentale Interuniversitario di Ricerca Stradale), 2000; Capitolato prestazionale ANAS 2008; Capitolato Autostrade.

Megali, G., Cacciola, M., Ammendola, R., Moro, A., Praticò, F.G., Morabito, F.C., Assessing reliability and potentiality of nonnuclear portable devices for asphalt mixture density measurement, 2010, Journal of Materials in Civil Engineering 22 (9), pp. 874-886.

Mohammad LN, Herath A, Wu Zhong, Cooper SA. Comparative study of factors influencing the permeability of hot-mix asphalt mixtures. Volume 74E Electronic Journal 2005, AAPT Publications, 2005.

Montepara A, Virgili A. La determinazione della massa volumica e dei vuoti nei conglomerati bituminosi aperti. Atti Convegno S.I.I.V. Ancona, 1996.

Moulthrop J.S., Hicks G.R., Delaiden J.L., Selecting a flexible pavement preventive maintenance, 1998 Western Pavement maintenance Forum, Sacramento, CA, USA, January, 1998.

Muench, S.T. and Mahoney, J.P.(2001). A Quantification and Evaluation of WSDOT's Hot Mix Asphalt Concrete Statistical Acceptance Specification. WA-RD 517.1. Washington State Department of Transportation, Transportation Center (TRAC). Seattle, WA.

Niagara Peninsula, "Special provisions - contract items", 2004.

Praticò F.G., How should payment adjustments be assessed when both surface and mechanical defects are involved? a synergetic study on theory and experiments, SURF 08 - 6th Symposium on surface characteristics, Portoroz, Slovenia, 20/22 October -2008.

Praticò F.G., Quality and timeliness in highway construction contracts: a new acceptance model based on both mechanical and surface performance of flexible pavements, Construction Management and Economics, Volume 25, Issue 3, pages 305 - 313, Routledge - Taylor and Francis 2007, March.

Praticò FG, Moro A, Ammendola R. Modeling HMA Bulk Specific Gravities: A Theoretical and Experimental Investigation. International Journal of Pavement Research and Technology 2/3 (2009) 115-122.

Praticò, F.G., Moro, A , Measurement of air void content in hot mix asphalts: method and core diameter dependence, 2011b, Construction and Building Materials, doi: 10.1016/j.conbuildmat.2011.06.032.

Praticò FG, Moro A. Permeability and volumetrics of porous asphalt concrete: a theoretical and experimental investigation. International Journal: Road Materials and Pavements Design, Vol. 8 N 4/2007a.

Praticò, F.G., Ammendola, R., Moro, A., Factors affecting the environmental impact of pavement wear, 2010, Transportation Research Part D: Transport and Environment 15 (3), pp. 127-133

Praticò, F.G., Casciano, A., Tramontana, D., Pavement life-cycle cost and asphalt binder quality: Theoretical and experimental investigation, 2011a, Journal of Construction Engineering and Management 137 (2), pp. 99-107

Praticó, F.G., Moro, A., Ammendola, R., Factors affecting variance and bias of non-nuclear density gauges for porous european mixes and dense-graded friction courses, 2009a, Baltic Journal of Road and Bridge Engineering 4 (3), pp. 99-107+Ia+IIA+IIIA.

Praticò, F.G., Moro, A., Ammendola, R., Potential of fire extinguisher powder as a filler in bituminous mixes, 2010c, Journal of Hazardous Materials 173 (1-3), pp. 605-613

Praticò, F.G., Moro, A., Flow of water in rigid solids: Development and experimental validation of models for tests on asphalts, 2008a, Computers and Mathematics with Applications 55 (2), pp. 235-244.

Praticò, F.G., Moro, A., In-lab and on-site measurements of hot mix asphalt density: Convergence and divergence hypotheses, 2011, Construction and Building Materials 25 (2), pp. 1065-1071.

Praticò, F.G., Moro, A., Permeability and volumetrics of porous asphalt concrete: A theoretical and experimental investigation, 2007b, Road Materials and Pavement Design 8 (4), pp. 799-817

Praticò, F.G., Tramontana, D., Casciano, A., Attinà, A. , Geosynthetics and pavement life cycle: An analysis through the M-E PDG, 2010a, Geotechnical Special Publication (207 GSP), pp. 302-309.

Praticò, F.G., Tramontana, D., Casciano, A., Bitumen quality, pavement LCCA and contractor's expectations, 2010b, Geotechnical Special Publication (203 GSP), pp. 523-529

Rao C, Von Quintus HL, Schmitt R. Calibration of Non-nuclear Density Gauge Data for Accurate In-Place Density Prediction. Session 596. 86th TRB annual meeting, Washington D.C., January 2007.

Romero P. Evaluation of Non-Nuclear Gauges to Measure Density of Hot-Mix Asphalt Pavements. Pooled Fund Study, Final Report, The University of Utah Department of Civil and Environmental Engineering July 18, 2002.

Salt (1997). Research on skid resistance at the transport and road research laboratory (1972 to 1977), TRR 622, TRB, Washington, DC.

Sargand SM, Kim Sang-Soo, Farrington SP. Non-Nuclear Density Gauge Comparative Study. Draft Final Report. Ohio Research Institute for Transportation and the Environment, 114 Stocker Center, Ohio University, Athens, OH, 45701-2979, August 2005.

Sebesta S., Use of microcracking to reduce shrinkage cracking in cement treated bases, Transportation Research Board 2005 Annual Meeting, 2005.

Shober S.F., Friedrichs D.A. (2002). A Pavement Preservation Strategy, Wisconsin Department of ransportation Library, USA.

Siegwart, R. (2001). Indirect Manipulation of a Sphere on a Flat Disk Using Force Information. *International Journal of Advanced Robotic Systems,* Vol.6, No.4, (December 2009), pp. 12-16, ISSN 1729-8806

Smith R.D. (1979) Pavement wear and studded tire use in Iowa, Final Report, Iowa highway research Board, Project HR-148.

Spellerberg P, Savage D. An investigation of the cause of variation in HMA Bulk Specific Gravity test results using non-absorptive aggregates. National Cooperative Highway Research Program Web Document 66 (Project 9-26 (Phase 2), July 2004.

State of Florida Department of Transportation (2004). Asphalt concrete friction courses, December.

TransTech Systems Inc. Effect of Water and Temperature on Hot Mix Asphalt Density Measurement using Electromagnetic Sensing. TransTech Technical Note 0301, Schenectady, January 15, 2003. Available online: http://www.transtechsys.com/products/pro_lib_pqi.htm.

Uddin, M., Mohboub, K.C., Goodrum, P.M. (2011). "Effects of Nonnormal Distributions on Highway Construction acceptance Pay Factor Calculations", Journal of Construction Engineering and Management, volume 137, n° 2, pp. 108-118.

Ullidtz, P. (1987). Pavement Analysis. Elsevier, Amsterdam.

Van Til, C.J., McCullough B.F., Vallerga B.A., Hicks R.G., Evaluation of AASHO Interim guides for of pavement structures, NCHRP 128, Highway Research Board, 1972.

Weed Richard M. (2001). "Derivation of equation for cost of premature pavement failure", 80th TRB annual meeting.

Weed Richard M., Kaz Tabrizi , Conceptual framework for pavement smoothness specification, TRB 2005 Annual Meeting, January, 2005.

Western Federal Lands Highway Division FP Specification Change (2004). "Pavement Smoothness/Roughness".

Whiteley Leanne, Susan Tighe, Zhanmin Zhang, Incorporating Variability into Pavement Performance, Life Cycle Cost Analysis and Specification Pay Factors, 84th Annual Meeting - January 9-13, 2005.

Williams SG. Non-Nuclear Methods for HMA Density Measurements. MBTC 2075, Final Report, University of Arkansas, June 2008.

Part 5

Quality Control for Medical Research and Process

11

Procedures for Validation of Diagnostic Methods in Clinical Laboratory Accredited by ISO 15189

Silvia Izquierdo Álvarez and Francisco A. Bernabeu Andreu
Servicio de Bioquímica Clínica, Hospital Universitario Miguel Servet, Zaragoza Hospital Universitario Príncipe de Asturias, Alcalá de Henares, Madrid, Spain

1. Introduction

Actually, each clinical and/or biochemical laboratory has responsibility for demonstrating its competence and therefore must obtain results of good quality. Medical laboratories provide vital medical services to different clients: clinicians requesting a test, patients from whom the sample was collected, public health and medical-legal instances, referral laboratories and authoritative bodies. All expect results that are accurate and obtained in an effective manner, within a suitable time frame and at acceptable cost. There are different ways of achieving the end results, but compliance with International Organization for Standardization (ISO) 15189, the international standard for the accreditation of medical laboratories, is becoming progressively accepted as the optimal approach to assuring quality in medical testing. As result, the accreditation of clinical laboratories is shifting from being a "recommendation" to becoming a "requirement" in many countries throughout Europe and in the other countries around the world (Berwouts, 2010). *Accreditation* is defined by ISO as the "Procedure by which an authoritative body gives formal recognition that a body or person is competent to carry out specific tasks". Although accreditation also considers the quality management system (QMS), it has additional formal requirements of technical competence, including initial and continuous training of personnel, *validation of methods and instruments*, and internal and external quality control.

A good QMS in the laboratory has a lot of advantages such as increased transparency, traceability, uniformity, work satisfaction and better focus on critical points. On the contrary, it will require extra time on aspects such as document control and there is a danger of losing critical attitude and curbing innovation and changes. Therefore, a formal accreditation and the linked periodical audits are stimulant for keeping the quality system (QS) alive. Without accreditation, there is a danger of giving less attention to quality improvement. In addition, accreditation is a good way to demonstrate and attest competence and a worldwide tool to recognize laboratories. Finally, all parties (patients, families, the laboratory and clinicians) are benefited through better processes and quality of results (Berwouts, 2010).

All essential elements of QS are covered by the ISO 15189 accreditation standard in two distinct chapters: management requirements and technical requirements. Technical elements enclose personnel and training, accommodation, equipment, *validation* and assuring quality of examination procedures by internal quality control (IQC), external quality control (EQA), maintenance and calibration.

ISO 15189 standard emphasizes so in the quality of contributions to patient care as in laboratory and management procedures and specifies the quality management system requirements, in particular to medical laboratories and stages:

<<The laboratory shall use only validated procedures for confirming that the examination procedures are suitable for intended use>>, <<The validation shall be as extensive as are necessary to meet the needs in the given application or field of application>>, and <<Procedures need to be periodically revalidated in view if changing conditions and technical advances>>.

IQC is an internal verification that the test yields consistent results day after day; in the other words, the identification measure of *precision*, but not necessarily of *accuracy*. ISO 15189 requires that "the laboratory shall design IQC systems that verify the attainment of the intended quality of results". On the hand, the laboratory should avoid mistakes (ISO 15189, 5.6.1.) in the process of handling samples, requests, examinations, reports and so on; on the other, the laboratory should determine *uncertainty* (ISO 15189, 5.6.2) where relevant and possible. For each test, the laboratory should identify and define potential errors, risks and challenges (typically, during the validation phase); subsequently, specific IQC should be defined to assure each risk and potential problem.

EQA is an important complement to IQC in which a large number of laboratories are provided with the same material and required to return results to a coordinating centre. The results are compared to determine the *accuracy* of the individual laboratory. In addition, EQA provides continuous education and training for laboratories as well. Accredited laboratories are required to "participate in interlaboratory comparisons such as those organized by EQA schemes" (ISO 15189, 5.6.4). EQA should, as far as possible, cover the entire range of tests, and the entire examination process, from sample reception, preparation and analysis to interpretation and reporting (ISO 15189 5.6.5). For some specific tests, no EQA scheme exists. ISO 15189 (5.6.5) states "whenever a formal laboratory comparison programme is not available, the laboratory shall develop a mechanism for determining the acceptability of procedures nor otherwise evaluated"; examples include reference materials or interlaboratory exchange. Interlaboratory comparisons should cover the scope of services offered and there should be a formal mechanism of review and comparison of results.

Used together, IQC and EQA provide a method of ensuring accuracy and consistency of results and are vital tools in the laboratory. The relation between precision and accuracy may be illustrated by the familiar example of shooting arrows at a target (Berwouts, 2010; Burnett, 2006) (figure 1).

The results provided by the clinical/medical laboratory must be accurate to allow a correct clinical interpretation and to be comparable with earlier or later and between laboratories. So the purpose of this chapter is to establish a set of guidelines and recommendations to help personnel carry out their work in clinical/medical laboratories that are accredited or under accreditation by ISO 15189. It is necessary to establish and define the different procedures validation, the fundamental guidelines for the proper design of the validation, the recommendations to validate an established method in the laboratory, and the different parameters to be assessed.

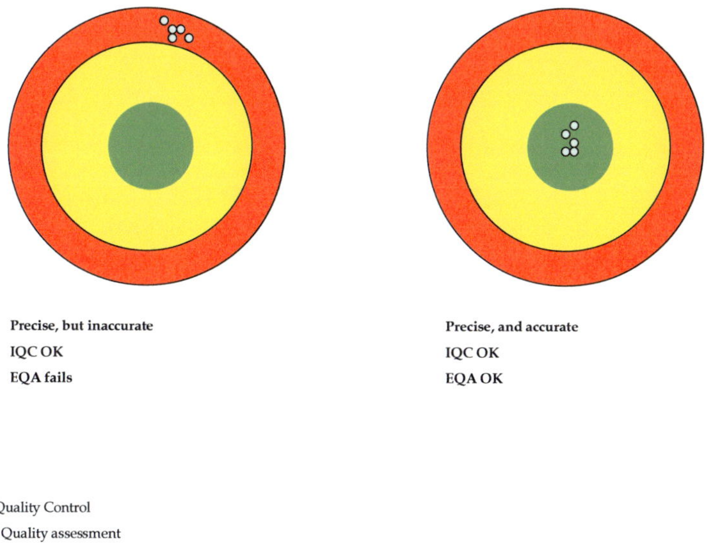

Fig. 1. Accuracy and precision.

2. Validation design of a method

Diagnostic validation is a formal requirement of accreditation standards, including ISO 17025 and ISO 15189, those tests/methods and instruments must be validated before diagnostic use to ensure reliable results for patients, clinicians or referring laboratories and their quality must be maintained throughout use. In other words, the laboratory must demonstrate that their tests/methods are fit for the intended use before application to patient samples. Figure 2 shows a summary of what ISO 15189 states with regard to validation (Berwouts, 2010; Burnett & C. Blair, 2001, Burnett et al., 2002; Burnett, 2006).

Although the concept of validation makes explicit reference to the purely analytical aspects may also include preanalytical and sampling procedures, handling and transport. At a minimum, the techniques used to determine the performance of a method should be one or more of the following:

- Calibration using reference standards or reference materials or traceable to these
- Comparison of results obtained with other methods
- Interlaboratory comparisons
- Systematic evaluation of the factors that influence outcomes
- Estimate of the uncertainty of the results based on scientific knowledge of theoretical principles of the method and practical experience.

In any case, analytical methods must be those that meet customer requirements, that is, those that provide clinically useful information. Thus, an analytical method for determination of aluminium in serum based on the complexation of this element with 8-hydroxyquinoline and quantification by fluorimetry can have high reliability but its detection limit is at least an order of magnitude above the upper limit of the reference element, which makes this method, does not meet customer requirements.

There are publications that provide general methods of evaluation of analytical methods including the following:

1. Definition of the evaluation protocol that registers the results of the measurements.
2. Determining the range of application and dilution mechanisms, if any.
3. Identify the components of precision in the day, every day.
4. To determine the accuracy through recovery studies before a definitive method or reference, if any.
5. To determine the sensitivity.
6. Estimate the limit of detection and quantification (and others if applicable).
7. To study the specificity of the method, checking for interference.
8. Establishing the reference range.
9. Document the validity of the method.

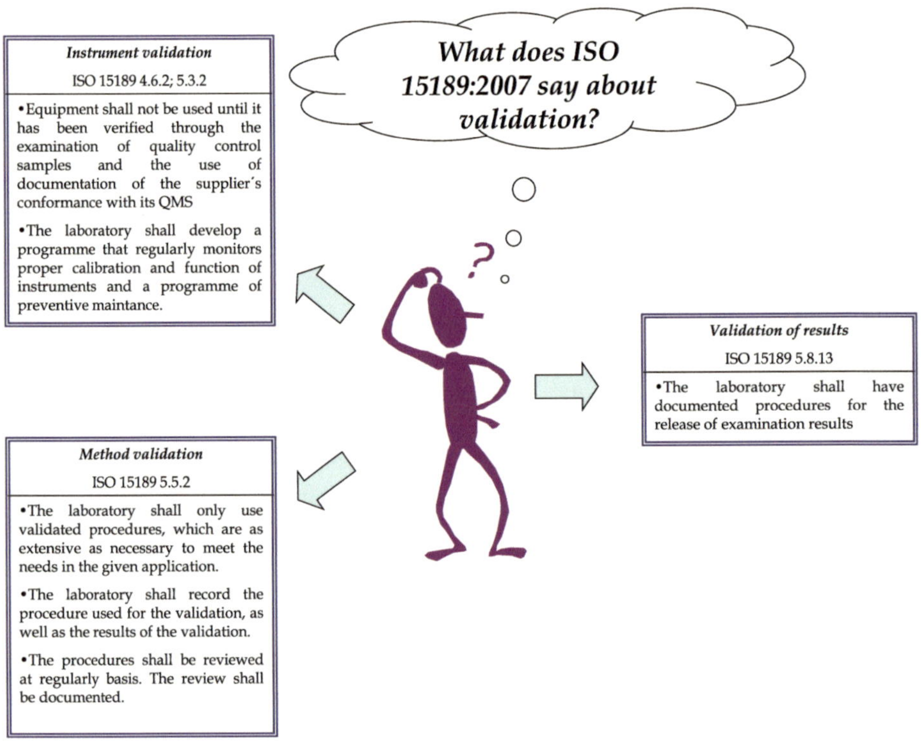

Fig. 2. Validation requirements according ISO 15189:2007.

Regarding the validation and control of analytical procedures, paragraph 5.5 of ISO 15189: 2007 specifies to be used those procedures "that have been published by experts or international guidelines, national or regional". Own procedures "must be properly validated for its intended use and fully documented".

So, ISO 15189: 2007 says "as appropriate, the documentation should include the following: technical specifications (e.g., linearity, precision, accuracy, expressed as a measurement uncertainty limit of detection, measurement, sensitivity and specificity, and interference)" (ISO 15189, 5.5.3).

When a unit or section of a clinical/medical laboratory chooses to engage in the accreditation ISO 15189, must be aware that although the analytical methods which usually works have been validated in its implementation must be validated in time.

A validation process like any other requires a series of planning, execution and control to ensure that the results come to fruition.

a. Planning:
   - Definition responsible for performing the validation process.
   - Definition of objectives and internal requirements applied the method to validate (purpose, parameters to measure in the matrix or matrices to be determined).
   - Definition and documentation of the method (procedure for validation).
b. Implementation: implementation of activities, results obtained and recorded (date, operators and results).
c. Control:
   - Verification of compliance with targets.
   - Final Declaration of the appropriateness of the procedure defined.

a. *Planning* consists of the following phases:
1) Assign responsibility. In this phase it will be defined a person responsible for carrying out the validation process and deciding the outcome. This person can count on help from others, but he is responsible for making decisions so, he must have a proper qualification; 2) Definition of the characteristics and requirements applied to the method: The definition of requirements has to do with the intended use of the method (i.e. as property or analyte, the matrix or matrices in which they will determine the use that will make the test results and legal requirements or economic policy to be applied to test results), from the specified requirements and based on a literature search using other standards, etc. There is a design and optimization phase of the procedure that is performed by laboratory. This is the stage where, for example for an instrumental method, you must establish a priori the linearity of the method, the working range, the limit of detection and quantification is desired, the accuracy and precision fit. In short what features the laboratory can apply the method to the intended use; 3) Description documented procedure: It should be sufficiently detailed to ensure its proper performance and repeatability. This ensures that all laboratory personnel that are qualified can do just as the method with comparable results.
To accomplish this phase can be helpful in establishing a suitable index of the case as the reference standard.
b. *Implementation*: Outcome is based on the realization of a series of tests and experiments that occur as a result values for the parameters defined in the requirements. These parameters can be variable depending on the type of method applied and the requirements and can include accuracy, precision, limit of detection, limit of quantization, selectivity, etc.
c. *Control*: The control is the verification of compliance and the final declaration. 1) Verification of compliance: As a result of the implementation of activities will be decided whether the values meet the specified requirements, in which case proceed to establish which checks should be made to the method as regular monitoring to confirm that remain requirements requested at the time of validation, e.g. using a control pattern periodically check the parameters of the regression line, etc., proceeding to their inclusion in the proceedings and preparing a final edition of the same. Otherwise you may be assessed if an amendment to the previously established requirements. 2) Final Statement: All the validation process should conclude with a formal statement of the adequacy of the procedure defined as stated is suitable for their intended use, according to specified requirements (Burnett & C. Blair, 2001; Burnett et al., 2001; Burnett, 2006).

2.1 Types and methods of validation

The laboratory shall validate examination procedures from non standard methods, laboratory designed or development methods, standard methods used outside their intended scope and modified validated methods.

When examination procedures have been validated by the method developer (i.e., the manufacturer or author of a published procedure), the laboratory shall obtain information from the method developer to confirm that the performance characteristics of the method are appropriate for its intended use. When changes are made to a validated examination procedure, the influence of such changes shall be documented and, if appropriate, a new validation shall be carried out.

Examination procedures from method developers that used without modification shall be subject to verification before being introduced into routine use. The verification shall confirm, through provision of objective evidence (performance characteristics), that the performance claims for the examination method have been met. The performance claims for the examination method confirmed during the verification process shall be those relevant to the intended use of the examination results.

Verification and validation are two slightly different procedures (figure 3). By default, all new laboratory procedures must be validated before application to clinical testing. In addition, a validation is necessary when major technical modifications to existing methods are carried out or when the performance of existing methods has been shown to be unsatisfactory (Berwouts, 2010; Hauck et al., 2008).

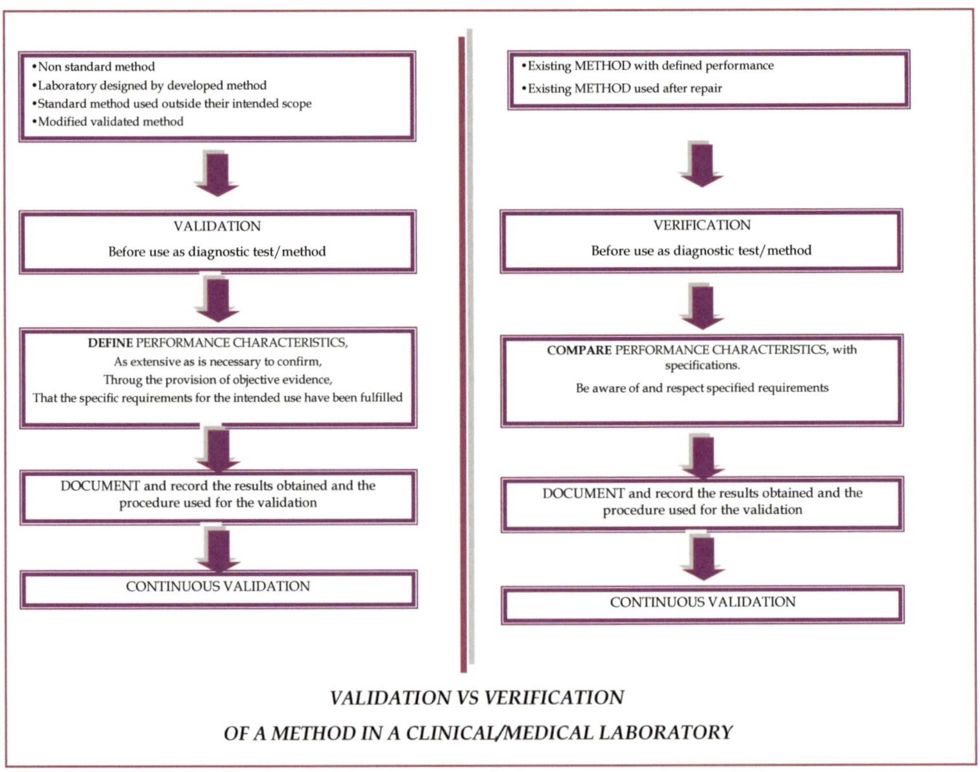

Fig. 3. Validation vs verification in diagnostic methods.

The validation or verification of methods, as defined in figure 4, is a normal requirement for the accreditation of laboratories according to the two major international standards applicable to clinical/medical laboratories, ISO 15189 and ISO 17025. Although the general requirements are clearly stablished (figure 4), the standards provide very little guidance about the detailed requirements or procedures.

Before a test/method can be validated, it is necessary to establish (a) that the particular measurements are diagnostically useful and (b) that the correct analyte(s), and only the correct analyte(s), are measured.

Full validation is required when is no suitable performance specification available, for example, with novel tests/methods or technologies. This process involves assessing the performance of the test/method in comparison with a "gold standard" or reference test/method that is capable of assigning the sample status without error. In simple terms, validation can be seen as a process to determine whether the laboratory is "performing the correct test/method". Validation data can be used to assess the accuracy of either the technology or the specific test/method. Generally speaking, the generic validation of a novel technology should be performed on a larger scale, ideally in multiple laboratories (interlaboratory validation), and should include a much more comprehensive investigation of the critical parameters relevant to the specific technology to provide the highest chance of detecting sources of variation and interference (Berwouts, 2010; Burnett, 2006).

Principle requirements of ISO 157025:2005	*Principle requirements od ISO 15189:2007*
5.4.2 "Laboratory-developed methods or methods adopted by the laboratory may also be used if they are appropiate for the intended use and if they are validated". 5.4.5.2 "The laboratory shall validate non-standard methods, laboratory-designed/developed methods, standard methods used outside their intended scope, and amplifications and modifications of standard methods to confirm that methods are fit for the intended use. The validation shall be as extensive as is necessary to meet the needs of the given application or field or application.The laboratory shall record the results obtained, the procedure used for validation, and a statement as to wether the method is fit for the intended use". 5.4.5.3 "NOTE 1 validation includes specification of the requirements, determination of the characteristics of the methods, a check that the requirements can be fulfilled by using the method, and a statement on the validity. NOTE 3 Validation is always a balance between costs, risks and technical possibilities, There are many cases in which the range and uncertainty of the values (eg accuracy, detection limit, selectivity, linearity, repeatability, robutness and cross-sensivity) can only be given in a simplified way due to lack of information".	5.5.1 "[....]If in-house procedures are used, they shall be appropriately validated for their intended use and fully documented". 5.5.2 "The methods and procedures selected for use shall be evaluated and found to give satisfactory results before used for medical examinations. A review of procedures by the laboratory director or designated person shall be understaken initially and at defined intervals". 5.6.2 "The laboratory shall determine the uncertainty of results, where relevant and possible".

Fig. 4. Principle requirements of ISO 15189: 2007 and ISO 17025:2005 about validation and verification.

2.2 Recommendations to validate a method developed in the laboratory

2.2.1 Quantitative methods

Two components of analytical accuracy are required to characterize a quantitative method: trueness and precision. Trueness expresses how close the methods result is to the reference value. Typically, multiple measurements are made for each point and the rest method result is taken to be the mean of the replicate results (excluding outliers if necessary). As quantitative assays measure a continuous variable, mean results are often represented by a regression of data (a regression line is a linear average). Any deviation of this regression from the reference indicates a systematic error, which expressed as a bias (i.e., a number indicating the size and direction of the deviation from the true result). There are two general forms of bias. With constant bias, method results deviate from the reference value by the same amount, regardless of that value. With proportional bias, the deviation is proportional to the reference value. Both forms of bias can exist simultaneously. Although measurement of bias is useful, it is only one component of the measurement uncertainty and gives no indication of how dispersed the replicate results are. This dispersal is called precision and can be measured by imprecision, that provides an indication of how well a single method results is representative of a number of replicates or repetitions. Imprecision is commonly expressed as the standard deviation of the replicate results, but is often more informative to describe a confidence interval (CI) around the mean result. Precision is subdivided according to how replicate analyses are handled an evaluated.

Repeatability refers to the closeness of agreement between results of test performed on the same method items, by the same analyst, on the same instrument, under the same conditions in the same location and repeated over a short period of time. Repeatability represents "within-run precision".

Intermediate precision refers to closeness of agreement between results of methods performed on the same method items in a single laboratory but over an extended period of time, taking account of normal variation in laboratory conditions such as different operators, different equipment and different days. Intermediate precision therefore represents "within-laboratory, between-run precision" and is therefore a useful measure for inclusion in ongoing validation.

Reproducibility refers to closeness of agreement between results of methods carried out on the same method items, taking into account the broadest range of variables encountered in real laboratory conditions, including different laboratories. Reproducibility therefore represents "inter-laboratory precision".

In practical terms, internal laboratory validation will only be concerned with repeatability and intermediate precision and in many cases both can be investigated in a single series of well-designed experiments. Reduced precision indicates the presence of random error. The relationship between the components of analytical accuracy, types of error and the metrics used to describe them is illustrated in figure 5.

Any validation should also consider robustness, which, in the context of a quantitative method, could be considered as a measure of precision. However, robustness expresses how well a method maintains precision when faced by a specific designed "challenge", in the form of precision does not represent random error. Typical variables in the laboratory include sample type, sample handling, sample quality, instrument make and model, reagent lots and environmental conditions (e.g., humidity, temperature). Appropriate variables should be considered and tested for each specific method. The principle of purposefully challenging methods is also applicable to both categorical and qualitative methods and should be considered in the validation as well. Robustness can be considered as a useful prediction of expected intermediate precision (Berwouts, 2010).

As trueness and precision represent two different forms of error, they need to be treated in different ways. In practice, systematic error or bias can often be resolved by using a correction factor; constant bias requires an additive correction factor, whereas proportional bias requires a multiplicative correction factor.

For quantitative methods, particularly those requiring absolute quantification, it is most effective to estimate analytical accuracy on an ongoing basis by running a set of calibration standards (standard curve) with each batch or run. In this case, it is important that linearity be evaluated and that the lower and upper standards are respectively below and above the expected range of the results as precision cannot be assessed on extrapolated results. Where possible, calibration standards should be traceable to absolute numbers or to recognized international units.

Other factors that may need to be evaluated include the limit of detection defined as the lowest quantity of analyte that can be reliably detected above background noise levels and the limits of quantification that define the extremities at which the measurement response to changes in the analyte remains linear (Berwouts, 2010).

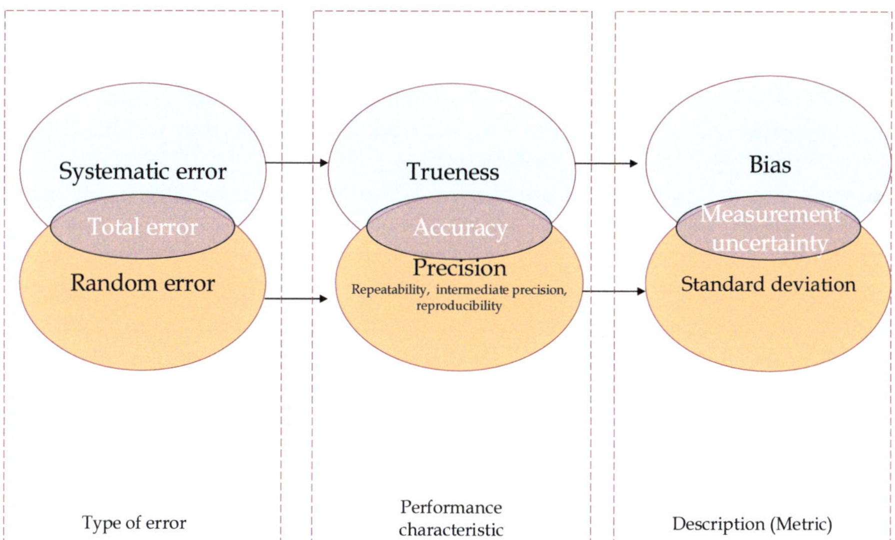

Fig. 5. Performance characteristics, error types and measurement metrics uses for qualitative methods (adapted from Menditto et al., 2007).

2.2.2 Qualitative methods

This is an extreme form of a categorical test/method, in which there are only two result categories, positive and negative. This binary categorization can be based either on a cut-off to a quantitative result. The diagnostic accuracy of a qualitative method can be characterized by two components, both of which can be calculated: sensitivity (the proportion of positive results correctly identified by the method), and specificity (the proportion of negative results correctly identified by the method).

3. Parameters required for considering in the validation or revalidation of a method

There are several measurable parameters that should be taken into account during validation or verification. The estimation of *accuracy* is a key parameter. Accuracy consists of both precision and trueness for quantitative and semiquantitative test/method. *Precision* or "closeness of agreement between results of replicate measurements" includes the following:
Repeatability: within-run variation (same sample, same conditions).
Intermediate precision: between-run variation within a single laboratory (different samples, operators, equipments).
Reproducibility: between-run variation in different laboratories (different samples, operator).
Robustness: variation when confronted with relevant challenges (e.g., sample type, environmental conditions and so on).
Trueness is the "closeness of agreement with a reference value". Appropriate reference materials are, therefore, essential and could include positive and negative/normal controls, certified reference materials, EQA materials, synthetic samples or material characterized by another technique.
The components of accuracy for quality tests/methods are *sensitivity* and *specificity*. *Sensitivity* is a measure of how well the test/method detects positive results, whereas *specificity* describes how well negatives are detected.
Thorough documentation during a validation process is essential, especially in the context of accreditation process pragmatic approaches, reconciling the formal requirements of accreditation standards while respecting the aim that "validation must be practical", such as the design of IQC based on validation results, making full use of data that laboratories are already collecting, for example from IQC or EQA, for continuous validation. There are no detailed practical guides for validation of diagnostic methods in medical laboratories; moreover, the accreditation standards have no specific details about how to fulfil their requirements. The laboratory must decide on this, on the basis of their experience and performance requirements; it is duty of the laboratory to provide evidence that tests results provided are reliable, and that the performance claims are correct (Burnett, 2006; Hauck et al., 2008).
In the end, validation is never finished. The implementation of quality indicators for systematically monitoring and evaluating the laboratory's contribution to patient care is good way to continuously validate diagnostic tests/methods, apart from IQC, EQA and other data (Eurachem, 2010; Hauck et al., 2008).
But when the method has already implemented several years and with performance and optimal quality specifications, we recommend periodic revalidation through the information provided by the treatment program reports or international interlaboratory comparison or testing fitness, programs or external quality control, in which the laboratory has been involved for years. The advantage of this validation methodology described above, is to estimate the imprecision, bias and uncertainty, without much effort and without having to make a lot of trials and experimental trials that would mean stopping the routine work in

the laboratory. You can validate the micro-range or specific concentration range, applicable as diagnostic daily reality for clinical laboratories.

3.1 Estimation of the accuracy using reference materials

ISO 15189 accreditation for clinical laboratories require a verification of the accuracy of the measurement procedures. The study of the accuracy, by estimating the systematic error, should be in the validation of the measurement. To study the accuracy of a measurement procedure is necessary to compare average values obtained with a conventional true value. In the clinical laboratory can be used as true values considered, mainly 3 types: value assigned to a reference material, the consensus value obtained in a program of external quality assessment, the value obtained with a reference measurement procedure.

The accuracy is expressed numerically by systematic error, which is the difference between the average measurement results obtained and a conventional true value. The values assigned to some reference materials may be considered conventionally true values.

The reference materials used in the study must have a value assigned to the magnitude that is measured and the corresponding value of uncertainty. You should also know the traceability of the assigned value. It is preferable that the material has a matrix similar to human samples. The main types of materials: certified reference materials, prepared by metrology institutes or other organizations related to metrology and reference materials business (controls to the truth).

The reference material manufacturer must provide the traceability and uncertainty of assigned values; the latter expressed as standard uncertainty or expanded uncertainty. Along with the uncertainty value must also specify the coverage factor used.

The results of accuracy studies should be used to validate the accuracy of the measurement procedures, ensures the absence of relevant and introduce systematic errors in the calculation of uncertainty of measurement uncertainty components associated with any correction factors.

3.2 Estimation of the accuracy from participation in external quality assessment programs

The external quality control includes different activities aimed at assessing the accuracy of the results through the intervention of an organization outside the laboratory. The most common form of external quality control is comparisons between laboratories or programs of external quality assessment.

These programs are organized by professional associations, government agencies or manufacturers of control materials that have a similar function. Participating laboratories measured once a magnitude of a control material of unknown value. Organization of the program collects the results of laboratory and a study of the data then forwarded to each participating laboratory, informing about the error of its outcome. The duration of the program, the number of measurements that are performed and the number of different materials are used, depending on the different programs. To study the accuracy is recommended that the program in which you participate fulfill the following conditions: high number of participants, the laboratory has a minimum of 12 results for participation and that you know the standard deviation characterizes the dispersion of results among participating laboratories.

3.3 Estimation of measurement uncertainty

The results provided by the clinical laboratory must be accurate (true and precise) to allow a correct clinical interpretation and to be comparable with earlier or later and between laboratories.

The error of measurement of clinical laboratory results is almost always unknown. Instead, it is possible to ascribe a measurement uncertainty and metrological traceability of each result. The uncertainty is a numerical expression of the degree of doubt of the result. Traceability relates the result with reference values established allowing reproducibility over time and between laboratories (Eurachem, 2010).

In the estimation of measurement uncertainty is assumed that any systematic error is eliminated, corrected or ignored, random effects are assessed on the outcome of an action and establishing a range within which lies the true value of the measured magnitude a certain level of confidence. The standard for laboratory accreditation ISO 15189 requires an estimate of the uncertainty of the results. The appropriate methodology for estimating the uncertainty described in the Guide to the Expression of Uncertainty in Measurement (GUM). The GUM was developed jointly by several international organizations for standardization and metrology for use in calibration and testing laboratories and measures applied to physical or chemical analysis. Currently, the GUM is difficult to apply to measures that are performed in clinical laboratories, although they maintained their principles. Moreover, the complexity and cost of obtaining an estimate of the uncertainty of measurement must be commensurate with the quality requirements applicable to the clinical use of the results.

Sources are contributing to the uncertainty of a result as follows: sample collection, sample preparation, calibrators or reference materials, input quantities (e.g., absorbance), computer equipment used, environmental conditions, sample stability and changes in workers.

The uncertainty associated with the collection and sample preparation is difficult to estimate and should be reduced through rigorous standardization of procedures. In this paper only consider the sources of uncertainty in the analytical phase, which begins when the sample interacts with the first technical step of the measurement (for example, placing the sample into an analyzer) and ends with obtaining a value numerical measurement result.

The main components of the uncertainty of the analytical phase correspond to the uncertainty of the measured, the stability of the sample in the measurement system calibration, the volume dispensed, the batch of reagents, instrumentation equipment, operators and environmental conditions. In the following paragraphs, are discussed in more detail the main components.

Measurement uncertainty is a parameter that is specifically associated with each outcome. In clinical laboratories, it is impossible to estimate particular measurement uncertainty for each measurand of each sample, so it makes a rough estimate of the uncertainty of measurement for a measurand defined and values of the same close to decision clinic. Measurement uncertainty does not apply to qualitative tests, in which the result is a numeric value.

3.3.1 Definition of measurand

The measurand is defined by the following parameters:

a. Analyte to be measured. For example, protein, sodium ion, cholesterol, ASO, hemoglobin, white blood cell counts, etc.
b. System. For example, serum, urine, venous blood, pleural fluid, etc.

c. Type size and unity. For example, substance concentration (mmol/L), mass concentration (g/L), catalyst concentration (nkat/L), etc.
d. Measurement procedure.

The existence of different molecular forms of the analyte can introduce uncertainty in the results. This source of uncertainty can be reduced or eliminated by careful definition of the measurement, so they may react differently to some or other molecular forms.

Another source of uncertainty regarding the definition of the measurand are possible cross-reactions and interference that can occur with some samples and must be identified and documented to prevent, where possible, their influence.

In short, uncertainty caused by the uncertainty of the measurand can not be quantified, but may be reduced or eliminated by detailed specification of the measurand.

3.3.2 Imprecision

Most of the components of measurement uncertainty of the analytical phase are contained in the estimation of imprecision (CV_{id}). It is usually obtained using control materials.

This assessment should be a sufficient number of data to collect the different sources of uncertainty apply, i.e., a minimum of six months of data and new estimation every year. In the period of data collection should include several calibrations to collect the uncertainty generated by the calibration process. Moreover it is necessary to use different batches of calibrator if you have the uncertainty of the assigned value.

The estimate of CV_{id} is made for a measurement value close to the values of clinical decision.

3.3.3 Value assigned to the calibrator

The clinical laboratory must know the uncertainty and metrological traceability of values assigned to calibration materials used. As usual it is commercial material the manufacturer must provide such data (Directive 98/79/EC). Along with the uncertainty value must also specify the coverage factor used. Typically, uncertainty is expressed as expanded uncertainty (U) for a confidence level of 95% (coverage factor = 2).

The standard uncertainty (u) is calculated by dividing U by the coverage factor. U on (%) of the value assigned to the gauge should not vary excessively batch to batch and should generally be lower than CV_{id}.

3.3.4 Systematic error (bias)

The estimation of measurement uncertainty is assumed that any significant systematic error of the measurement procedure has been deleted, corrected or ignored. The identification of a possible systematic error should be done during the validation of the measurement procedure.

When systematic error is corrected by a factor, the correction has an associated uncertainty (u_{cf}) that should be considered in calculating the combined measurement uncertainty.

Systematic errors caused in the routine use of the measurement by the inevitable differences between different calibrations behave randomly in the long term, so this component of uncertainty is reflected in CV_{id}.

3.3.5 Uncertainty calculation

The uncertainty is calculated by combining various sources. For this reason, clinical laboratories should identify each measurand, specifying the measurement procedure, and calculate for each of them calculate the combined uncertainty from the data of internal quality control and other data, using the following equation:

$$u_c = \sqrt{CV_{id}^2 + u_{cal}^2 + u_{cf}^2}$$

Where:
u_c: relative combined standard uncertainty (%); CV_{id}: imprecision (coefficient of variation) interday; U_{cal}: relative standard uncertainty (%) of the value assigned to the calibrator; u_{cf}: relative standard uncertainty (%) of the factor used to correct a systematic error.
It is recommended to express the combined uncertainty for a confidence level of 95% (expanded uncertainty, U_c). To do this, multiply the value of u_c for k = 2.

$$U_c = 2 \times \sqrt{CV_{id}^2 + u_{cal}^2 + u_{fc}^2}$$

The relative expanded uncertainty should be expressed to two significant figures, for example: 4.2%, 16%.

3.3.6 Interpretation

The estimation of measurement uncertainty provides a quantitative indication of the level of doubt that the laboratory has in each result and is therefore a key element in the system of analytical quality in clinical laboratories.
The relative expanded uncertainty of a measurand should be less than one third of the Maximum Permitted Error (MPE). If it was superior, should be studied in greater detail the different sources of uncertainty, identify the most significant and perform the appropriate actions to reduce them.

3.3.7 Applications

The uncertainty of measurement should be used primarily for:

- Selection of measurement procedures that fulfill the specifications of accuracy.
- Strict interpretation of the significance of a change between two consecutive values of magnitude biochemistry.
- Strict interpretation of the significance of a result compared with a value of clinical decision.

3.3.8 Limitations

The value of the measurement uncertainty varies with the concentration of the measurand and may be substantially different for very low or very high analyte. For this reason it is recommended that the estimate for a concentration closes to clinical decision values.

3.4 Estimation of precision

Precision is one of the most important metrological characteristics to be considered for selection and implementation of a measurement procedure in the clinical laboratory. In addition, the quantitative understanding of this feature is essential for establishing tolerance intervals of internal control materials for the objective interpretation of the significance of a change between two consecutive values of magnitude biochemistry, and the calculation of uncertainty.

The accuracy can be studied under conditions of repeatability, reproducibility and intermediate. The study conditions that are more interested in the clinical laboratory are the kind of repeatability and intermediate terms, which vary from day to day.
Before starting the study a period of familiarization with the measurement procedure and the operator's experience are recommended.
Also for this study is recommended that samples for this study were commercial materials or control samples. We recommend using at least two samples with different concentrations of the magnitude under study, with a value within the physiological range or close to a discriminate value, and one with a pathological value. When it sees fit, they can not be tested with values close to the limits of the measuring range of the procedure. Samples should be stable during the duration of the study. If using commercial control materials, whenever possible, they should be interchangeable with samples of human origin.
The run imprecision were obtained under the same conditions of repeatability, ie the same samples, the same operator, the same components of the measurement system for a short time and without calibrations between measurements. A minimum of 30 measurements are required for fulfilling statistical criteria.
If the results of the run imprecision are not consistent with previous results, either supplied by the manufacturer or obtained in the literature, the study should be stopped to find and correct the cause of the discrepancy.
Imprecision is obtained under certain conditions. Each laboratory should perform the estimation with standard calibration frequency (daily weekly, etc), and changes of operator, calibrator lot, reagent lot, etc, which are common in everyday work. Following the statistical criteria are recommended to estimate a minimum of 30 days.
Before calculating the mean, standard deviation and coefficient of variation of the results must be detected the presence of possible outliers. An abnormal result will be removed provided that it is related to a documented error or has demonstrated statistically that is an outlier. After the removal of outliers, if any, imprecision is calculated by the coefficient of variation.

4. Documentation: procedures and instructions needed for the validation of a method in the clinical laboratory

International Standard ISO 15189:2007 clearly identifies the documentation requirements necessary to determine compliance with the requirements referred to for quality and competence of clinical laboratories.
Standard clinical laboratory means (paragraph 3.8) that "laboratory devoted to biological, microbiological, immunological, chemical, immunohematological, biophysical, cytological, pathological or other material derived from the human body in order to provide diagnostic information, prevention and treatment of diseases or the assessment of human health and can provide a consultant advisory service covering all aspects of laboratory analysis, including interpretation of the findings and recommendations on any proper analysis additional".
The implications of documentary that suggests the validation of a method, it follows that it must develop a set of documents or records.
Registration means that documentary evidence of a fact that has occurred and is understood by documentary evidence to document that describes how the activities should be conducted.

According to these definitions it is able to state that the laboratory should have a defined overall validation procedure (document) that describes: What activities will be performed; responsibility to perform; records to retain; how to be performed.
To confirm the verification of compliance, apply the method to real matrices, records to keep are: 1. Requirements applied to the method (Must be defined prior to conducting the tests, indicating preserved based on what have been defined); 2.Records of previous tests. (Straight calibration standards used, results obtained from different computers, etc.); 3. Written procedure (approved by qualified personnel); 4. Results of tests for checking compliance with requirements (The laboratory must clearly indicate the results of the parameters and the comparison with the specified requirements); 5. Statement by the head of the validation of the procedure is suitable for their intended use based on the evidence (All these records should include dates, personnel and equipment used in ways that can be reconstructed).
It must have an overall validation procedure describing the activities undertaken; those responsible for conducting, records to keep (the method established requirements, records of tests: calibration lines, patterns, etc.).

<table>
<tr><td rowspan="5">Lab logo</td><td rowspan="2">Registration</td><td>"Name of the Laboratory"</td></tr>
<tr><td>Date:</td></tr>
<tr><td rowspan="3">Design/planning of the method validation</td><td>CODE:</td></tr>
<tr><td>Review:</td></tr>
<tr><td>Page X of Y</td></tr>
<tr><td colspan="3">Method:</td></tr>
<tr><td colspan="3">Responsible for validation:</td></tr>
<tr><td colspan="3">Used:</td></tr>
<tr><td colspan="3">References:</td></tr>
<tr><td colspan="3">Procedure/codification:</td></tr>
<tr><td colspan="3">Objectives:</td></tr>
<tr><td>Parameter</td><td>Value</td><td>Observations</td></tr>
<tr><td></td><td></td><td></td></tr>
</table>

Fig. 6. Template record: Design / planning of the method validation.

It must be developed and used a generic template so that it is not necessary to have to develop a validation process for each method, but simply change the data in the template. Thus, for any method in the laboratory which will continue to want to validate one of the two ways described: the classical or from the results of inter-comparison programs. It is used for the models listed in the Annexes to this case: report validation, design / planning of the method validation, quality control plan method. They detail steps for each procedure to be followed in the validation.

Here are a few examples of formats and / or templates of records that are considered necessary for the validation of a method:

- Template record: Design / planning of the method validation (figure 6).
- Template record: Report of validation (figure 7).
- Registration: Plan quality control method (figure 8).
- Title page of a validation procedure in clinical laboratory methods (figure 9).
- Technical Registration. Spreadsheets (Excel) (figure 10).

<table>
<tr><td rowspan="5">LAB LOGO</td><td colspan="2" rowspan="2"></td><td>"Name of the laboratory"</td></tr>
<tr><td>Date:</td></tr>
<tr><td colspan="2" rowspan="3">VALIDATION REPORT</td><td>Code:</td></tr>
<tr><td>Review</td></tr>
<tr><td>Page X of Y</td></tr>
<tr><td>Date of procedure of validation</td><td>Register</td><td>Observations</td><td>Personnel</td></tr>
<tr><td></td><td></td><td></td><td></td></tr>
<tr><td colspan="4">Additional data</td></tr>
<tr><td colspan="4"></td></tr>
<tr><td>Parameter</td><td>Result /value</td><td>Observations</td><td></td></tr>
<tr><td></td><td></td><td></td><td></td></tr>
<tr><td colspan="4">Declaration:
❑Validated method
❑No validated method
❑Validated method with restrictions</td></tr>
<tr><td colspan="4">Validation criteria:</td></tr>
<tr><td colspan="2">Date of validation of the method:</td><td colspan="2">Signature: (Responsible of validation)</td></tr>
</table>

Fig. 7. Template record: Report of validation.

Lab logo	Registration	"Name of the Laboratory"	
		Date:	
	Quality control plan method	CODE:	
		Review:	
		Page X of Y	
Method for validation (Description)	**Quality Control**	***Periodicity***	***Acceptance criteria***
	Accuracy		
	Trueness		
	Precision		
	Repeateability		
	Reproducibility		
	Uncertainty		
	Robutness		
	Specificity		
	LOQ		

Fig. 8. Registration: Plan quality control method.

LAB LOGO	MANUAL OF PROCEDURES	"Name of the laboratory"
		Date:
	PROCEDURE FOR VALIDATION METHODS	Code:
		Review
		Page X of Y

PROCEDURE FOR VALIDATION METHODS

Developed	*Revised*	*Approved*
Date		
Technical Manager	*Quality Manager*	*Lab Director*

Fig. 9. Title page of a validation procedure in clinical laboratory methods

LAB LOGO	MANUAL OF PROCEDURES	"Name of the laboratory"
		Date:
	PROCEDURE FOR VALIDATION METHODS	Code:
		Review
		Page X of Y

CONTENTS

1. OBJECTIVE AND SCOPE
2. DEFINITIONS
3. REFERENCE DOCUMENTS
4. PROCEDURE
    - 4.1. Accuracy
    - 4.2.Precision
    - 4.3.Trueness
    - 4.4. Repeateability
    - 4.5.Reproducibility
    - 4.6. Uncertainty
    - 4.7. Another validation parameters

5. ANNEXES

Fig. 10. Contents in a validation procedure in clinical laboratory.

5. Conclusion

It is important to have documented procedures for the validation of different diagnostic methods available within a clinical laboratory. There is a need to develop practical guidelines for method validation procedures in clinical laboratories through the various tools available to the laboratory. There is no single way to validate a diagnostic and clinical laboratory validate and verify the validation of their methods over time to meet the requirements of the existing accreditation standards and to demonstrate the laboratory's technical competence to offer quality results.

6. References

Berwouts, S.; Morris, M.A. & Dequeker, E. (2010).Approaches to quality management and accreditation in a genetic testing laboratory. *European Journal of Human Genetics.* No.18, pp. 1-19.

Burnett, D. (2006). ISO 15189:2003-quality management, evaluation and continual improvement. *Clin Chem Lab Med.* No. 44, pp. 133-739.

Burnett, D. & Blair, C. (2001). Standards for the medical laboratory-harmonization and subsidiarity. *Clin Chem Acta.* No. 309, pp. 137-145.

Burnett, D.; Blair C.; Haeney, M.R.; Jeffcoate, S.L.; Scott, K.W. & Williams, D.L. (2002). Clinical pathology accreditation: standards for the medical laboratory. *J Clin Pathol.* No. 55, pp. 729-733.

Eurachem. (May 2010). The firness for purpose of analytical methods a laboratory guide to method validation and related topics, Avaliable from: http://www.eurachem.org/guides/valid.pdf

Hauck, W.W.; Kock, W.; Abernethy, D. &Williams, R.L. (2008). Making sense of trueness, precision, accuracy and uncertainty. *Pharmacopeial Forum*. No.34, pp.838-842.

International Organization for Standardization. Medical laboratories-Particular requirements for quality and competence. ISO 15189:2007.

International Organization for Standardization: General requirements for the competence of testing and calibration laboratories.ISO/IEC 17025:2005.

Menditto, A.; Patriarca, M. & Magnusson, B. (2007). Understanding the meaning of accuracy, trueness and precision. *Accred Qual Assur*. No. 12, pp.45-47.

12

Quality Control and Quality Assurance in Human Experimentation

Stahl, Edmundo
LatAmScience, LLC
U.S.A.

1. Introduction

During the 20th, century the awareness of the need for the ethical treatment of human subjects participating in experimentation has evolved. Various incidents over the years have sparked the creation of government entities dedicated to the regulation of human experimentation. This has brought about the creation of regulations whose objective is the protection human subjects throughout the experimentation process. These regulations call for many checks and balances with the objective of protecting the individual under experimentation through quality control procedures in the monitoring process of the experiment. The quality is assured through auditing the process by independent professionals. This chapter will describe the history of the development of Good Clinical Practices (GCP) and an analysis of some applicable documents and practices developed by the Food and Drug Administration of the USA, and the International Conference on Harmonization of Technical Requirements for Registration of Pharmaceuticals for Human Use (ICH). FDA (USA), EMA (EU) and Pharmaceutics and Medicines Safety Bureau (Japan) as well as pharmaceutical industry representatives of the USA, EU and Japan form the ICH. ICH guidelines provide a unified standard for designing, conducting, recording and reporting clinical trials involving the participation of human subjects and other necessary activities related to human experimentation. ICH is especially concerned with harmonizing the regulatory requirements of its sponsor countries; USA, EU and Japan. It describes the necessary activities and documentation that would allow the evaluation of the ethical conduct of a clinical trial and assure the quality of the information derived from such a study. Many countries all over the world are now including these guidelines in their regulations and are effectively adhering to them.
A significant part of human experimentation is conducted in the development of new drugs for the treatment of human disease as well as devices and instruments used in medical practice. This chapter will also describe the development process, the logic behind it, the non-clinical testing that is necessary for the drug/device development process, the clinical phases of drug development, the role of the ethics committees and Institutional Review Boards in the approval process to conduct human experimentation as well as the role of the government agencies which regulate human experimentation.

2. Evolution of ethical conduct in human experimentation

Since the 5th century B.C. the most prevalent code of ethical conduct for the medical profession has been and still is the "Hippocratic Oath"[1]. It is widely believed to have been written by Hippocrates, often regarded as the father of western medicine. The original text of the Hippocratic Oath is usually interpreted as one of the first statements of a moral of conduct to be used by physicians, assuming the respect for all human life. It has been modified over time in many occasions but the spirit of the concept has been preserved.

It was not until after the end of World War II that the United States authorities conducted in their occupied zone several trials for war crimes committed by the Nazis in Nüremberg[2]. The trials were formally named the "Trials of War Criminals before the Nuremberg Military Tribunals". They were held before US military courts, not before the International Military Tribunal. The defendants were accused of unethical human experimentation and other atrocities. On August 19th, 1947 the tribunal delivered its verdict including their opinion on human experimentation. The Nüremberg Code that emerged from these trials consists of 10 points that represent a set of ethical research principles for human experimentation. The Nüremberg Code includes concepts like: voluntary consent of the research subject; experimentation with clear fruitful objectives; experimentation in humans should be preceded by animal experimentation; the conduct of research in humans should not produce physical or mental injury nor results in death of the study subjects; the experimentation should be conducted with a view of introducing the minimal possible risk to the individual during the experimentation and conducted by a qualified person. It also includes the concept that the subject should be at liberty to stop the experiment at any time for any reason. Likewise, the experimenter should be prepared to terminate the experiment if in their judgment there is any reasonable chance that it may harm the research subject. Subsequently in 1948 the World Medical Association introduced the Declaration of Geneva [3] as a modernization of the Hippocratic Oath. It was designed as a formulation of that oath's moral truths that could be comprehended and acknowledged modernly. The Declaration of Geneva has been amended in 1968, 1984, 1994, 2005 and 2006.

Another important historical document addressing human experimentation is the Declaration of Helsinki[4] adopted in 1964 by the World Medical Association in Helsinki, Finland. It is a set of ethical principles for the medical community specifically related to human experimentation and is widely regarded as a cornerstone document for human research. It has been revised six times since its adoption, the last revision being in 2008. The Declaration of Helsinki adopted the ten principles first stated in the Nüremberg Code and tied them to the Declaration of Geneva. It addresses clinical research reflecting the changes in medical practice. Its various revisions introduced the concept of independent review committees, now known as Institutional Review Board or Independent Ethics Committees; the management of the inclusion of minors in clinical research and the recognition of vulnerable groups; addressed the use of placebos; and the inclusion of human volunteers in clinical trials. This document was not meant to be legally binding but has influenced national and regional regulation and legislation around the world. It introduced the concept that ethical considerations must take precedence over laws and regulations.

In the USA, the Belmont Report[5] was created by the now named Department of Health and Human Services with the title "Ethical Principles and Guidelines for the Protection of Human Subjects of Research". The report was issued in April 1979 prompted in part by problems arising from the Tuskegee Syphilis Study (1932-1972). The Tuskegee Syphilis Study was designed to observe the clinical evolution of syphilis. The patients, 399 impoverished Black individuals from Macon county, Alabama, who thought they were receiving free health care from the government were never told they had syphilis nor were they treated for it. The Belmont Report incorporates the principles of the Nuremberg Code, the Declaration of Geneva and the Declaration of Helsinki. These documents influenced significantly the legislation and creation of regulations for the ethical conduct of human experimentation in the USA. Sections 45 (government sponsored studies, 45 CFR) and 21 (private and industry sponsored studies, 21 CFR) of the Code of Federal Regulations (CFR) base many of their regulations on these important ethical documents and have influenced in many important ways human experimentation in the US and around the world.

3. Regulatory environment

The U.S. Food and Drug Administration (FDA) [6] was created in 1906 by the Federal Food, Drug, and Cosmetic Act, the Wiley Act. The purpose was to prevent the manufacturing, sale, or transportation of adulterated or misbranded or poisonous or deleterious foods, drugs, medicines, and liquors. The FDA evolved over the years to require manufacturers to submit a New Drug Application (NDA) for each newly introduced drug and provide data that demonstrates the safety of the product (1938), and later (1962) to establish efficacy, in order to show that the products were effective for their claimed indication. Several amendments to the law have followed to reflect the evolution and emerging issues in the drug development and approval process, which remain today in the crossroads where science, medicine, politics and business intersect. Because a new drug approval is based largely on clinical data obtained by experiments in humans, the FDA has vested significant effort in ensuring the quality of the clinical data and the conditions under which they are obtained. The set of regulations and guidelines the FDA publishes constitute what is collectively known as good clinical practices or GCP. Through these FDA sets the minimum standards for the conduct of clinical trials, the collection of data and data management and reporting of clinical studies.

The European Medicine Agency, EMA[7] (formerly known as EMEA, European Agency for the Evaluation of Medicinal Products), was founded in 1995 and is a decentralized agency of the European Union responsible for the scientific evaluation of medicines developed by pharmaceutical companies for use in the European Union. Its main function is the promotion and protection of public human and animal health, through the evaluation and supervision of medicines for human and animal use. They are responsible for the evaluation of European Marketing Authorizations for human and veterinary use medicines. The agency monitors the safety of marketed products and provides scientific advice to companies on the development of new medicines. The agency constantly works to forge close ties with partner organizations around the world, including the World Health Organization, the FDA and the other regulatory authorities.

In Japan the Pharmaceuticals and Medical Devices Agency (PMDA) [8] working with the Ministry of Health implements measures for securing the efficacy and safety of drugs, cosmetics and medical devices. The PMDA also has forged close ties with other regulatory agencies, namely the EMA and FDA as partners in the formation of the International Conference on Harmonization (ICH).

The International Conference on Harmonization of Technical Requirements for Registration of Pharmaceuticals for Human Use (ICH) [9] was created in 1990. This organization brings together the regulatory authorities and pharmaceutical industry of Europe, Japan and the

US. The purpose is to discuss scientific and technical aspects of drug registration with the goal of harmonizing drug development and registration across the world. The Global Cooperation Group of this organization has been working to harmonize the increasingly global approach to drug development, so that the benefits of international harmonization for better global health can be realized worldwide. ICH's mission is to achieve greater harmonization to ensure that safe, effective, and high quality medicines are developed and registered in the most resource-efficient manner. ICH has developed a series of guidelines to help regulate the clinical drug development process. The instruments developed include a standardized medical terminology system (MedDRA) to use in the capturing, registering, documenting and reporting adverse events during human experimentation. ICH maintains, develops and distributes MedDRA. ICH also developed a standardized package for the submission of new drug applications, the Common Technical Document (CTD) which has been adopted around the world as the gold standard for new medical products submissions. The CTD assembles all the quality, safety and efficacy information required for regulatory submissions in a common format to facilitate the regulatory review process. The CTD has simplified the assembly of the regulatory packages since reformatting for submissions to different regulatory agencies is not necessary anymore. Another area where ICH has worked to improve is international electronic communication by evaluating and recommending Electronic Standards for the Transfer of Regulatory Information (ESTRI). ESTRI has developed recommendations for electronic individual case safety reports and electronic CTDs. Furthermore, ICH has developed guidelines to standardize and harmonize the areas involved in the drug development process. These include guidelines on quality of the product being developed; on safety focusing on nonclinical studies to uncover potential risks for humans; on efficacy, which is concerned with the design, conduct and safety clinical trials; and other guideless like the ones developed by ESTRI. The clinical guidelines include clinical safety (E1-2), clinical study reports (E3), dose response studies (E4), ethnic factors (E5), Good Clinical Practice (E6), clinical trials (E7-11), clinical evaluation by therapeutic category (12), clinical evaluation (E14), and pharmacogenomics (E15-16). The GCP (E6) document describes the responsibilities and expectations of all participants in the conduct of clinical trials, including investigators, monitors, sponsors and IRBs. It is one of several GCP guidelines published by various organizations, like the World Health Organization (WHO), Pan American Health Organization (PAHO), FDA and EMA. ICH's E6 is the GCP guideline most commonly accepted worldwide. It has been adopted by many countries in their regulations and accepted as the gold standard of GCP for clinical drug development. We will discuss E6 in more detail.

4. Nonclinical drug testing

Animal testing is an imperfect predictor of drug activity in humans. It constitutes the best practical experimental models for identifying and measuring the pharmacologic activity of the drug and predicting its effects in humans. *In vivo* and *in vitro* animal testing is the first major activity in the drug development process. The purpose is to characterize the toxicology, pharmacokinetic activity, and pharmacological activity of the candidate compounds prior to administration to human beings. FDA (21CFR58) as well as ICH (S guidelines) have developed standards for such testing. Initially, short-term effects are evaluated to decide if the drug is sufficiently safe for administration to humans and at what dose should the human testing start. As the drug development in human beings progresses, additional animal studies are conducted. These animal studies include long-term drug administration, and specialized animal tests are conducted to support longer administration in humans. These experiments allow the observation of drug effects that would be impractical or unethical to study in humans. Researchers can observe the effects of the compound over the lifespan of an animal, test dose responses and maximum doses; assess the effects on reproduction, pregnancy and the embryos; effects on genes; assess potential for carcinogenicity; evaluate mechanisms of action of the drug; and characterize the site, degree and duration of action of the compound. Regulatory agencies are involved in determining the amount and type of animal testing required to initiate drug development in humans as well as the requirements to support the whole clinical development program.

The regulatory agencies, specifically FDA, set the minimum standards for laboratories conducting these nonclinical tests through the publication and enforcement of Good Laboratory Practice (GLP)[10]. To ensure the quality and integrity of the data derived, nonclinical laboratories are required to implement quality systems to conduct their experiments and to abide by the animal welfare laws of the country. GLPs establish basic standards for the conduct and reporting of non clinical safety testing, including the organization of the laboratory, personnel qualifications, physical structure of the facility, equipment, maintenance procedures, and operating procedures. It requires the use of a written protocol and its structure, including its purpose, who is sponsoring the study, procedure for identification and evaluation of the test animals or test system. GLP details how to report nonclinical studies, the storage and retrieval of records and data, and the retention of records. FDA conducts inspections to monitor compliance with GLP requirements. Nonclinical laboratories may be disqualified if the laboratory facility fails to comply with the regulations, and the noncompliance affects adversely the results of the study.

In addition, FDA may provide advice to sponsors on the adequacy of the nonclinical testing plans before animal testing has begun, and evaluate independently the results and conclusions of the nonclinical testing. FDA has developed guidances for nonclinical testing also. Other regulatory authorities, namely European Community and Japan, have also developed their testing standards. ICH has stepped in in an effort to harmonize these standards with the Safety guidelines (S).

The basic toxicology studies undertaken to identify and measure a drug's adverse effects in the short- and long-term may include any or all of the studies shown in table 1 depending on the drug, intended use and duration of exposure in clinical trials (Table 1).

The responsibility of the conduction of these animal experiments falls on the sponsor, the animal laboratory and the regulatory authorities. Quality systems are required to guarantee the quality of the data generated. The Sponsor monitors the study and conducts audits, the laboratory needs to have proper standard operating procedures and guidelines in accordance with the regulations and prevailing laws as well as a quality group to ensure compliance with said regulations and laws, and the regulatory authorities perform inspections to make sure the regulations and laws are being complied with.

5. The clinical phases of drug development

In the FDA regulations and regulations by health authorities around the world accept 3 phases of drug development[11]. A fourth phase is frequently included during the post

Acute toxicity studies	Measure the short-term adverse effects of one of more doses administered over no more than 24 hours. Provide information on appropriate dosage for multiple dose studies, potential target organs, timeline of drug induced effects, species specific toxicity, potential acute toxicity in humans and estimate the safe acute dose for humans.
Subacute or subchronic toxicity studies	Evaluate toxic potential over 14 to 90 days depending on the proposed clinical indication and duration of exposure. They are designed to assess the progression and regression of drug induced lesions.
Chronic toxicity studies	Determine the risk in relation to the anticipated dose and duration of treatment, potential target organs, reversibility of observed toxicity and the no observed effect level. These studies last 180 days to 1 year of exposure.
Carcinogenicity studies	To observe the generation of malignant tumors in animals. Generally they are required for drugs which are intended to be used for chronic conditions for 6 months or more, or to be intermittently used over the years for chronic or intermittent conditions. These studies are usually in rodents and last for 2 years.
Special toxicity studies	These are studies appropriate for specific formulations, route of administration, or conducted in particular animal models relevant to a human condition, disease or age. They include immunotoxicity studies.
Reproductive toxicity studies	For drugs to be used in women of childbearing potential. They include fertility and general reproductive performance, teratology and perinatal/postnatal development.
Genotoxicity studies	Mutagenicity studies. Are used to assess the likelihood of the drug causing genomic damage that could induce cancer development.
Toxicokinetic studies	Used to describe the systemic exposure achieved in animals and its relationship to the drug concentration, dose and time course of the toxic effect. The purpose is to contribute in the assessment of the relevance of these findings to clinical safety, and support the choice of species and dose regimen in other nonclinical studies as well as the design of subsequent nonclinical studies.

Table 1. Basic Nonclinical Toxicology Studies.

approval period. These are not mandates that determine the specific structure or design of clinical trials. Although these phases are in general to be conducted sequentially, frequently they overlap. Clinical development programs commonly proceed in the following stages:

5.1 Phase I

This is the phase where initial introduction of an investigational product to humans. The drug is administered cautiously to a few patients or normal human volunteers (usually less than 80), to gain an understanding of the pharmacology, and basic safety of the drug, including tolerability, activity, pharmacodynamics, pharmacokinetics, mechanism of action in humans and optimal route of administration. Drug metabolism, structure-activity relationships and studies in which investigational drugs are used as research tools to explore biological phenomena or disease processes are also included in this phase. The first evidence of the drug efficacy in humans may also be observed in these phase. Subjects are monitored very closely. The studies in Phase II are designed based on the results obtained during this phase.

5.2 Phase II

In this phase a small group of patients are tested, usually 100 to 200, who suffer from the condition the drug is intended to treat or diagnose. The studies include well controlled, closely monitored trials. The investigational product is administered with the objective of increasing the understanding of the safety profile and the initial observations on the efficacy of the drug in the proposed disease. In this phase the aim is to establish a foundation for the phase III trials. The information gathered includes dose, dose regimen and fine tuning of the target population.

5.3 Phase III

The drug in this phase is used in much larger groups of patients, several hundred or thousands, who suffer from the condition that the compound is supposed to treat. This phase includes controlled and uncontrolled studies. The idea is to gather additional safety and efficacy information to determine the benefit-risk ratio of the drug. In this phase the trials follow more rigorous standards since they will serve as the primary basis for the approval of the drug to be marketed.

5.4 Phase IV

In addition to these 3 phases, regulatory authorities may require additional studies after approval to clarify some finding observed during the development program or to produce additional safety data, or treat special populations (e.g. the elderly, patients with renal function impairment, children, etc.). In a general sense the clinical development process continues long after the drug has been approved for marketing. Collection and evaluation of adverse experiences and other information collected while the drug is in the market provides the sponsor and regulatory authorities of a continuous flow of data that allows ongoing review and reassessment of safety and efficacy of the drug. The concept of risk minimization action plans have been introduced recently. Risk minimization action plans are strategic plans to minimize a drug's known risks and for the regulatory agency to monitor the sponsor's implementation of the plan. These postmarketing commitments range from comprehensive literature reviews to large controlled trials. These post marketing studies are usually called post approval trials or phase IV trials. Phase IV trials can be undertaken at the request of the regulatory agency as part of a postapproval commitment, as a specific regulatory agency requirement, or at a company's own decision to learn more about their product.

6. Quality systems in clinical research[12]

Many aspects of Good Manufacturing Practice (GMP) apply to the drug development process. Quality is a measure of the ability of a product, process, or service to satisfy stated or implied needs. A high quality product is one that meets these needs. For human experimentation, quality may apply to data generation and management, or, the processes involved in the implementation of the trials. Quality systems for human experimentation are the formalized practices, e.g. monitoring programs, auditing programs, complaint handling systems, etc., for periodically reviewing the adequacy of the activities and practices during human experimentation, and for revising such activities and practices so that data and process quality are maintained. For human experimentation GCPs are the basis for implementation of quality systems through quality management. This is done through the coordination of activities by the sponsors of the experiments, the clinical investigators and their staff, the institutional review boards and independent ethics committees, and by regulators to direct and control the operations with respect to quality. Quality management consists of three components: quality control, quality assurance, and quality improvement.

In the case of human experimentation, **Quality Control** is the steps taken during the implementation of the clinical trial to ensure the quality of the data generated and the processes involved. These include investigator supervision, sponsor monitoring, and any review by the regulatory authorities, to ensure that the trial meets the protocol and procedural requirements and is reproducible. **Quality Assurance** is the systematic process to determine whether the quality control system is working and effective. In clinical trials this is usually done by the sponsor through independent auditing of quality control activities, and also by the regulatory authorities through inspection of the quality systems and activities.

With the knowledge obtained from the quality assurance, audits and activities changes are made to the systems and activities with the purpose of increasing the ability to fulfill the quality requirements for the moment and in the future. This process can be called **Quality Improvement**.

Another activity central to maintaining and improving quality in clinical trial is the process of monitoring. Monitoring is a quality control activity conducted by the sponsor or a representative of the sponsor. The purpose is to ensure that the research data are accurate, complete, and verifiable from source documents. GCP guidelines (ICH E6)[9] defines monitoring as "the act of overseeing the progress of a clinical trial, and of ensuring that it is conducted, recorded, and reported in accordance with the protocol, standard operating procedures, good clinical practices, and the applicable regulatory requirements." Monitors usually compare the data in the case report forms designed for the study and the source documents, i.e., with the medical chart of the patient, physician notes, laboratory results, etc. Monitors also make sure that the activities related to protecting the rights and welfare of the study subjects were carried out appropriately. On the other hand, auditing is an independent quality assurance activity used by the sponsor to evaluate the effectiveness of the monitoring program. Auditing procedures are similar to the monitoring activities. The difference is that monitoring occurs only during the execution of the clinical study, auditing occurs at any time during or after the clinical study is completed. In addition to quality audits there are inspections conducted by the regulatory authority(ies). An inspection is the act of conducting an official review of documents, facilities, records, and any other resources the authorities deem related to the clinical study. The inspection may be at the clinical trial site, at the sponsor's facilities, and/or at the Contract Research Organization (CRO) facilities, or at any other establishment the authorities deem appropriate. CROs are organizations which are normally contracted by sponsors to monitor their clinical studies. CROs may also conduct a complete development program for a sponsor on occasions, or deliver part of the activities related to the development of the investigational product. The purpose of monitoring is to determine if the research was conducted in compliance with national and local laws and regulations for the conduct of research and the protection of human subjects.

All parties involved in human experimentation (sponsors, clinical investigators, Institutional Review Boards/Independent Ethics Committees, and regulatory authorities) need to adopt and implement quality systems for the processes and activities they are responsible for.

This includes clinical research facilities. Clinical research should include **Quality Systems** to measure the quality of clinical research through the use of standard operating procedures (SOPs), study protocol compliance, internal monitoring and the sponsor's monitoring activities. This is accomplished through training of the personnel involved in clinical trial activities, internal and external audits, and accountability of the personnel.

A Typical quality system would include production and process control, equipment and facilities control; records, documents and change controls; material controls, design controls and corrective and preventive action. (Figure 1). This system can easily be adapted for the development of medical devices.

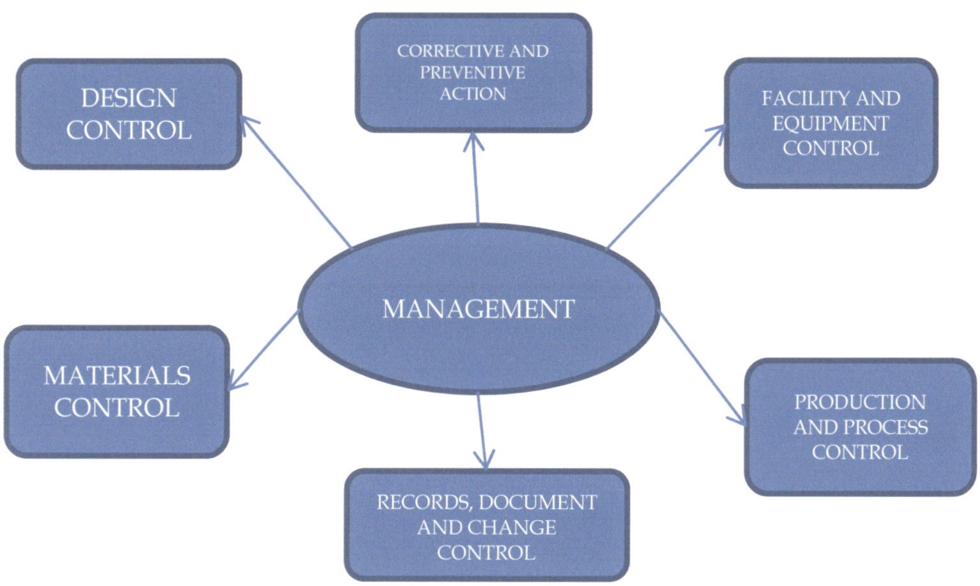

Fig. 1. Typical quality system. This system can easily be adapted to a medical device development facility.

A **Quality System** for an investigational clinical center may be also adapted from this diagram to include the following areas under the control of the clinical investigator (Figure 2):

- Facility and Equipment Evaluation and Documentation
- Source Documentation Generation, Integrity and Retention
- Consent Process and Documentation
- Safety Management and Reporting Processes and Documentation
- Investigational Product Accountability and Integrity and Documentation
- Site Staff Qualifications, Training, and Documentation
- Corrective and Preventive Action Development and Implementation Facility

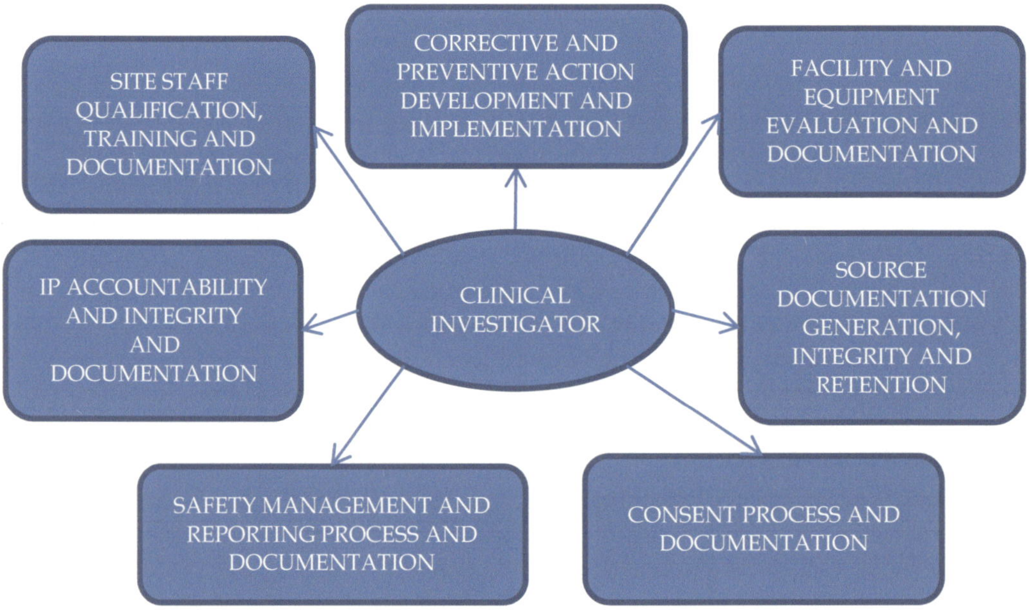

Fig. 2. The organization of a clinical investigational site.

These represent the activities required in a well run clinical investigational site. The investigator is responsible for all activities. The site should have guidelines and/or standard operating procedures for each these areas and activities. In addition, the investigator should have sufficient personnel who are properly trained and qualified to conduct these activities. It is also important that the facility are appropriate in size and configuration to accommodate all these areas.

7. FDA Bioresearch monitoring [12]

The Food and Drug Administration's (FDA) bioresearch monitoring program (BIMO) was established in 1977 with input from the drug, biologics, medical device, veterinary medicine, and food areas. Chapter 48 of the FDA's Compliance Program Guidance Manual is dedicated to Bioresearch Monitoring and delineates the inspection and reporting procedures for studies under FDA jurisdiction. The stated objectives of the bioresearch monitoring program are: to protect the rights, safety, and welfare of subjects involved in FDA-regulated clinical trials; verify the accuracy and reliability of clinical trial data submitted to FDA in support of research or marketing applications; and assess compliance with FDA's regulations governing the conduct of clinical trials. The purpose of the program is to provide instructions for FDA's field personnel for conducting such inspections.

BIMO developed compliance programs to provide uniform guidance and specific instructions for inspections of clinical investigators, sponsors and monitors, in-vivo bioequivalence facilities, Institutional Review Boards, and nonclinical laboratories involved in the testing of investigational products. The purpose of these programs is to adapt a Quality System framework for the oversight and management of clinical studies.

The most useful elements of a quality system that applies to clinical studies are: corrective and preventive action (CAPA) and management controls. CAPA procedures can be adapted to ensure effective and efficient clinical study management.

The application of CAPA to clinical research activities involve:

- Identification of non-conformances, e.g. protocol deviations, errors of omission or transcription.
- Investigation of the cause of the problem identified
- Identification of the actions needed to correct and prevent recurrence of the problem
- Verification that the corrective action is effective
- Making sure that the information is appropriately disseminated
- Submission of the information for management review on problems identified and actions taken
- Documentation of the process

Management controls involve the appointment of a management representative responsible for the research, in this case the investigator or sub Investigator, and to conduct management reviews.

Figure 3 shows the relationship between CAPA, management reviews and audits, external (sponsor monitoring, third party or FDA) and internal through monitoring internal activities.

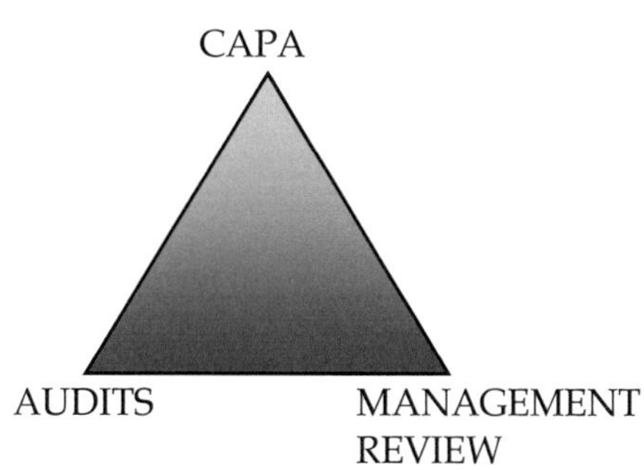

Fig. 3 Relationship between CAPA, Management Reviews and Audits.

Although FDA inspections are focused on clinical investigators, they are of great importance to sponsors. The inspections are designed to determine how well sponsors performed their responsibilities for the conduct of the study; should the inspections uncover serious problems it may result in rejection of the data essential for drug approval. As a result the sponsor may face inspections and compliance actions if it is found to have worked with noncompliant investigators and did not take corrective action.

8. Good clinical practice (E6)[9]

Good Clinical Practices (GCP) is not a set of instructions on how to develop a product or how to design human experiments. GCP is a series of general principles that must be observed during the conduct of human experimentation. This GCP guideline provides a unified standard for designing, conducting, recording, and reporting clinical trials that involve human subjects. Compliance with this guideline provides public assurance that the rights, well-being, and confidentiality of the trial subjects are protected and that the results of the study are credible. GCP are part of the quality systems to cover testing of medicinal products and devices, and conducting clinical studies in human beings. Their objective is to provide a unified standard for the European Union, Japan, and the United States, with consideration to existing GCPs of Australia, Canada, the Nordic countries as well as the World Health Organization, and to facilitate the mutual acceptance of clinical data by the regulatory authorities. It includes also the minimum information that should be included in the information to the investigator, which are the documents considered essential, their purpose, and how to file them. Many countries around the world have adopted these guidelines as their own. The ICH guideline on GCP (E6) outlines the 13 principles of good clinical practices. These principles are in line with the Nüremberg Code, the Declaration of Helsinki and the Belmont Report. These guidelines should be adopted by IRBs/IECs, sponsors, and clinical investigators as well as regulatory authorities who oversee or conduct clinical trials.

8.1 Principles of GCP

1. Clinical studies should be conducted according to ethical principles
2. Foreseeable risks and inconveniences should be weighed against anticipated benefit to the subjects
3. The rights, safety and well-being of the trial subjects should be the most important consideration and prevail over scientific or societal interests.
4. Available preclinical and clinical information on a product should be adequate to support the proposed trial
5. A clinical trial should be scientifically sound and described in a clear detailed protocol.
6. A clinical trial should be conducted in compliance with a protocol previously approved by an IRB/IEC.
7. Medical care given to, and decisions made on behalf of, trial subjects should be always the responsibility of a qualified physician or qualified dentist.
8. Each individual involved in conducting the trial should be qualified by education, training, and experience to perform his/her respective tasks.
9. Freely given informed consent should be obtained from every study subject prior to clinical trial participation.
10. All clinical trial information should be recorded, handled, and stored in a way that allows accurate reporting, interpretation, and verification.
11. The confidentiality of the records should be protected
12. The investigational products should be manufactured, handled, and stored in accordance with applicable good manufacturing practices (GMP), and used in accordance with the approved protocol.
13. Systems with procedures that assure the quality of every aspect of the trial should be implemented.

The GCP outline the duties of the IRBs/IECs, sponsors, and the clinical investigators.

8.2 Independent Ethics Committees (IECs) and Institutional Review Boards (IRBs)

Their main responsibility is to safeguard the rights, safety, and wellbeing of all trial subjects, with special attention to the inclusion of vulnerable subjects to the trial. The IRB/IEC is required to have standard operating procedures and maintain written records of their meeting and decisions. The composition and authority under which the IRB was established should be documented in writing. All meetings, notification to members, and schedules should be disseminated in writing. In summary all information and documentation of activities should be documented and transparent.

The IRB/IEC should consider the qualifications of the investigator, ensure that all subjects have freely provided their informed consent to be included in the study, ensure that payments to the subject for participation in the trial are not coercive or exercise undue influence, and continuously review the progress of the experimentation at intervals appropriate to the degree of risk to human subjects, but at least once per year. The IRB can request additional information if in the judgment of the IRB members the additional information would add meaningfully to the protection of rights, safety and/or well-being of the trial subjects. The IRB should always determine that a protocol or the information provided adequately addresses relevant ethical concerns and meets applicable regulatory requirements.

The IRB is usually composed of at least 5 members, at least one member whose primary area of interest is nonscientific, and at least one member who is independent of the trial site. The investigator may provide information on any aspect of the trial but should not participate in deliberations or in the vote, or opinion of the IRB.

8.3 The informed consent process

For the implementation of the informed consent the investigator should comply with all regulatory requirements and adhere to GCP and the ethical principles originating in the Declaration of Helsinki. The subject should be thoroughly informed of the experiment to be conducted, the risks and the potential benefits. Ample time should be given to the subject to make his/her decision to participate in the study. It should be very clear what the experimentation is all about. No coercion should be applied on the potential study subject.

The informed consent document should include explanations of the following:

- The trial involves research, and some parts of it are experimental.
- The purpose of the trial and the voluntary nature of the subject's participation.
- The treatment and probability for random assignment to treatment. That is in a double blinded study neither the patient nor the investigator may know which treatment is being administered.
- The trial procedures. Including potential for risky procedures and their potential consequences.
- The subject's responsibility to follow the indications from the investigator.
- Reasonable foreseeable risks or inconveniences to the subject, and embryo, fetus or nursing infant, if applicable.
- Reasonable expected benefits.
- Alternative forms of treatment for the condition under investigation and their potential risks and benefits.
- Compensation for trial related injury.
- Payment to the subject, if any.

- Expenses for the subject, if any.
- That the subject's original medical records may be accessed by regulatory authorities, IRB/IEC, the monitor and auditors for verification of the information.
- That confidentiality will be maintained and not be made public.
- All new information related to the trial that becomes available that may be relevant to the subject's willingness to continue to participate in the trial will be forwarded to the subject.
- Who to contact with questions or in the event of a trial related injury
- Foreseeable circumstances or reasons under wich the subject's participation in the trial may be terminated.
- The duration of the trial and approximate number of subjects involved.

8.4 The investigator

The investigators supervise the study staff to ensure they follow established procedures for the conduct of the study. They should be qualified by training, education and experience to conduct clinical trials. The investigators should be thoroughly familiar with GCP, the product under investigation and the study protocol. Investigators are responsible for all medical decisions. In their role they obtain approval to conduct the study from the IRB/IEC; ensure that informed consent is obtained freely and without coercion before the study starts; establish and maintain the subjects' case histories; transcribe the subjects' medical data from the medical files to a case report form for the sponsor; ensure the accuracy, completeness, legibility, and timeliness of the data reported; promptly report all adverse events and other problems; document and explain any deviations from the study protocol; be responsible for the accountability and proper storage as well as the use according to the protocol of the investigational product; and provide all required reports at the end of the study to the sponsor.

Investigators should be in contact with the IRB/IEC and the sponsor frequently. Communications involve,

- Before initiating the trial, obtain a written and dated approval/favorable opinion from the IRB/IEC to start the study
- Provide the IRB/IEC with a copy of the information on the product under investigation (the Investigator's Brochure, IB) and any amendments to the IB during the study.
- Report promptly any serious adverse event or laboratory abnormality immediately to the sponsor, the regulatory authorities and the IRB/IEC, and follow up with a detailed written report with any additional information requested.
- For patients who die during the study the investigator should supply the sponsors, regulatory authorities and the IRB/IEC with all pertinent information on the event.

Upon completion of the trial the investigator should inform the sponsor, the IRB/IEC and the regulatory authorities with a summary of the trial outcome, and any other report required by applicable regulation.

8.5 The sponsor

Sponsors are responsible implementing and maintaining quality assurance and quality control systems with written SOPs to ensure that the trials are conducted, and data are generated, documented and reported in compliance with the protocol, GCP, and applicable regulatory requirements. Sponsors are also responsible for securing agreement from all involved parties to ensure direct access to clinical trial related sites, source documents, and reports for the purpose of monitoring and auditing by the sponsor, CRO and regulatory authorities. Agreement with the investigators or any other party involved in the study should be in writing.

Quality control should be applied to each stage of data handling to ensure that all data are reliable and have been properly processed, for securing the services of monitors to ensure compliance of clinical investigators and verify that the study is carried out according to the approved study protocol. Sponsors also audit the monitor's performance, other quality control activities and systems to ensure performance. The monitors hired by the sponsor to review the records at the clinical centers, and report their finding to the sponsor in written reports of all visits and trial related communications.

Sponsors may transfer in writing any or all their obligations to a contract research organization (CRO), but the ultimate responsibility for the quality and integrity of the data always resides with the sponsor. CROs have the same obligations as the sponsor.

The sponsor is responsible for the medical expertise. Qualified medical personnel should be readily available to advise on trial related matters. An external consultant may be appointed for this function.

Sponsors are responsible for the trial design, trial management, trial data handling, and retention of documents for the specified period required by law and regulations. They are also responsible for the selection of qualified investigators and to apply with the regulatory authorities to conduct the trial.

Finally, the sponsor is responsible to provide insurance or indemnification to the investigator against claims arising from the trials, except for claims arising from malpractice and/or negligence.

8.6 Regulators

The regulators may inspect all parties who conduct or oversee clinical research and verify the information submitted to the regulatory authorities. Regulatory agencies inspect specifically clinical investigators, pharmaceutical companies, device companies, CROs, IRBs/IECs, as well as nonclinical laboratories, to ensure the accuracy and validity of the data generated, and to ensure that the rights and welfare of the research subjects are protected. The regulatory inspectors evaluate how well sponsors, monitors, clinical and non clinical investigators, CROs, and IRBs/IECs comply with the regulations. They may require certain conditions for a study to proceed. They develop policies and procedures for reviewing product applications and for the conduct of GCP inspections as exemplified by the FDA's BIMO compliance programs.

9. Conclusion

Over the last century the scientific community has developed a better understanding of how to protect and respect the rights, safety and wellbeing of research subjects. For centuries the Hippocratic Oath was the only ethical guidance for physicians and scientists on how to treat subjects, and specifically research subjects. The development of Good Clinical Practice was the result of various incidents that resulted in the Nuremberg Code, the Declaration of Geneva, the Declaration of Helsinki and the Belmont Report. ICH is an attempt to harmonize GCP in the most advanced democracies. Today, many regulatory agencies around the globe use these principles to regulate human experimentation in their countries.

The responsibility of GCP is shared by all parties involved in human experimentation, investigators, sponsors, ethics commitees, regulatory authorities, and research subjects. To guarantee the quality and accuracy of the data generated during human experimentation Quality Systems have been developed and are applied around the world.

10. References

[1] Markel, H. (2004) Perspective. Becoming a Physician. "I Swear by Apollo"- On Taking the Hippocratic Oath. N Engl J Med 2004; 350:2026-2029

[2] Shuster, E. (1997) Fifty Years Later: The Significance of the Nuremberg Code. N Engl J Med 1997; 337: 1436-1440, Nov 13, 1997

[3] Jones, D.A. (2006) The Hippocratic Oath II. The Declaration of Geneva and other modern adaptation of the classical doctor's oath. Catholic Medical Quarterly, February 2006

[4] WMA Declaration of Helsinki - Ethical Principles for Medical Research Involving Human Subjects. October 2008 available from http://www.wma.net/en/30publications/10policies/b3/index.html

[5] The Belmont Report. (1979) Ethical Principles and Guidelines for the Protection of Human Subjects of Research. Department of Health, Education, and Welfare available in http://ohrs.od.nih.gov/guidelines/belmont.html

[6] U.S. Food and Drug Administration available in http://www.FDA.gov

[7] European Medicines Agency available in http://www.EMA.europa.eu

[8] Japanese Regulatory Agency: Pharmaceutical and Medical Devices Agency available in http://www.pmda.go.jp/english/index.html

[9] International Conference on Harmonization and Technical Requirements for the Registration of Pharmaceuticals for Human Use (ICH) available in http://www.ICH.org/home.html

[10] Good Laboratory Practice (GLP), 21CFR58 available in http:// www.fda.gov

[11] Phases of Drug Development, 21CFR312.21 available in http:// www.fda.gov

[12] Bioresearch Monitoring in Compliance Program Guidance Manual, Chapter 48 available in: http://www.fda.gov

13

Quality and Quality Indicators in Peritoneal Dialysis

Javier Arrieta
Hospital Universitario de Basurto – Bilbao
Spain

1. Introduction

Physicians have historically shared an intuitive concept of Quality, concerning the care we provide to our patients. Our academic education and practice have been focused on Quality as a technical concept, assessable only by technicians and with no strong correlation with outcomes. The concept of Medicine as an Art is related to the values of vocation, dedication and good practice, recognizing that results can after all be negative

In the XXI Century, we all now accept the scientific nature of Medicine and, therefore, its dependence on the objective assessment of outcomes. In contrast, the patient's perception of Quality strongly depends on the culture and the environment. The current definition of disease given by the World Health Organization (WHO) focuses on self-perceived health and wellbeing. In this context, quality-based medicine should also be oriented towards the health and welfare as perceived by the patient.

Quality is one of the strategic elements on which the transformation and improvement of modern health systems is based. The effort made in recent years towards quality assurance in this field -including in the particular case of nephrology-, entails recognition of the need for objective and standardized measurement tools for health activities: Quality is not just good intentions.

2. Definitions of quality

There are many definitions of Quality, which in itself suggests that none of them are comprehensive. Definitions focused on Quality in Health, mainly date from the 1980's, when Palmer, Donabedian (Donabedian, 1980), the American Medical Association and many other authors tried to develop an adequate definition. As early as 1990, the Institute of Medicine adapted the definition given by the ISO (International Organization for Standardization), which does not specifically refer to Health: "Quality is the degree to which the characteristics of a product or service meet the objectives for which it was created", defining Quality in Medicine as "the degree to which health services for individuals and populations increase the likelihood of desired health outcomes and are consistent with current professional knowledge"(Lohr, 1990). This is the current philosophy of Medical Quality, which assesses the results, both objective and perceived by the patient (Committee on Quality of Health Care in America, 2001), assuming a degree of uncertainty with respect to processes and the final outcomes.

Quality in patient care depends on a large number of factors, but doctors tend to consider only a few, such as efficacy and effectiveness, more recently including accessibility, efficiency, privacy, and safety, among others as respect for the environment. Some factors are of great interest to Society as a whole - like those just listed - while others may be interesting primarily for patients, such as timeliness, convenience, patient participation, etc. The restrictive view of Quality used by doctors explains the differences we find between the technical quality and the quality perceived by the patients (JCAHO: Agenda for change, 1989).

2.1 Quality models

Although all models of quality are based on common ideas, such as reducing the variability in medical practice through standardization -using standards and indicators-, two types of quality models can be distinguished with respect to their underlying purpose. On the one hand, some models pursue standardization. This involves assessment by a qualified and independent entity that will accredit or certify us for providing high quality medical care. On the other hand, there are models that aim for continuous quality improvement based on self-monitoring. These produce continuous feedback that should help eliminate errors and lead to improvements in outcomes.

These two types of models are by no means exclusive. In Europe, public hospitals commonly use the European Foundation for Quality Management (EFQM) model, which not health specific:
(http://www.efqm.org/en/Home/aboutEFQM/tabid/108/Default.aspx); while, providers in the United Stated and private centers in Europe have chosen accreditations based on the standards of the Joint Commission on Accreditation of Healthcare Organizations (JCAHO, 1989; JCAHO update, 1990; JCAHO, 2011). Meanwhile, in Latin America the EFQM model is adopted more widely through the Latin American Foundation for Quality Management (FUNDIBEQ; www.fundibeq.org).

The Joint Commission uses a wide range of indicators and standards from the National Institute of Standards and Technology (NIST) -355 in the international version, of which 171 are mandatory for accreditation-, divided into medical and organizational indicators. They can be accessed from www.jointcommission.org or www.quality.nist/gov/.

The EFQM model allows centers to choose their indicators and standards -as long are they are logical and supported by scientific evidence- and pays greater attention to the evolution of the indicator towards "Excellence", than to the achievement of a standard at a given time. In other words, centers are not valued for their good work, but for their year-on-year improvement.

Another important difference from an operational point of view is that the Joint Commission certifies Centers -although it may also test Units- and assesses both clinical and other organizational, structural, plant safety and accessibility indicators. On the other hand, the European model can readily be applied to processes. For example, it is possible to apply the EFQM to a chronic hemodialysis process, peritoneal dialysis, nephrology hospitalization ward or kidney transplantation unit. Therefore, we can first apply it to one of the processes in our Service or Hospital, and within a few years extend it to other processes. It should be noted that processes are only one part of the EFQM (Figure 1), a useful aspect of the model is that we can start by applying it into individual processes, based on the priorities of clients and employees. Later on, the analysis can evolve to address common issues for the Hospital, such as leadership and strategy.

Finally, the European model can theoretically be implemented with no additional economic cost (the only requirement being training in use of the model), while those who pursue international accreditation need to pay the evaluators. There is a cost calculator on the Joint Commission's website, and it should be noted that the center's accreditations have to be renewed every three years.
In this context, we note that dialysis is a high cost therapy that can rarely be paid for by the patient. Funders have the authority -and obligation- to monitor the quality of the Healthcare for which they pay. Therefore, they increasingly demand the accreditation of Dialysis Units. Evaluators are usually independent from the payer, and they act as intermediaries between the payer and the health provider. Nevertheless, even in the accreditation systems, evaluation is considered as an element to guide these units in making improvements.
In this chapter, we will consider a quality system focused on a continuous improvement (rather than quality accreditation) that every dialysis unit could adopt if so desired.

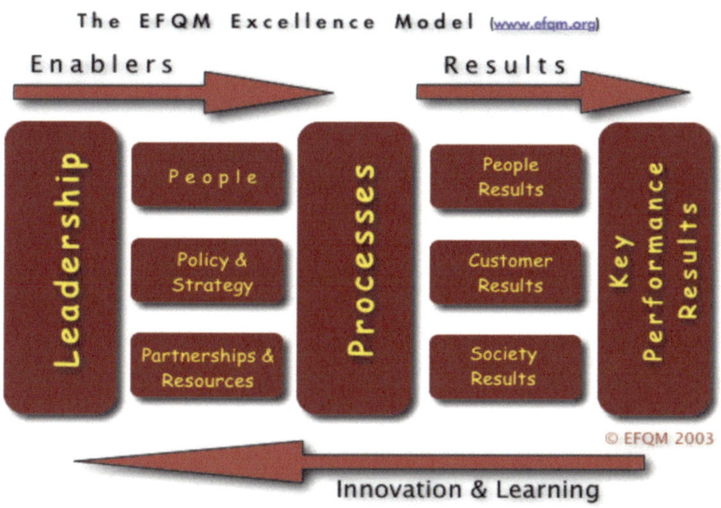

Fig. 1. European Foundation for Quality Management. Model of Excellence.

2.1.1 Quality systems in dialysis

Initially, quality systems have been used in acute care processes (mainly surgical), as well as general services such as laboratories, radiology units, etc. Quality indicators in these cases are derived from different patients who undergo a procedure at different points in time. However in dialysis, patients continue treatment over periods of months and years, and this implies several conceptual changes. It is clear that dialysis is not a curative procedure, but rather a life support technique. Its purpose is then to prolong life and improve its quality. Accordingly, indicators that seek to measure the quality of a certain dialysis therapy should be correlated with those two endpoints: survival and quality of life.

2.1.2 Quality systems in peritoneal dialysis

Quality systems in hemodialysis have been implemented for two decades, fundamentally, due to accreditation requirements. However, peritoneal dialysis (PD) is performed at the patient's home under clinical guidance depending on the general hospital, itself already

under global assessment and accreditations. That dependence explains why quality systems in PD have not been prioritized. The EFQM model can be applied to isolated processes, so it can be used in Peritoneal Dialysis Units.
The full EFQM model (Total Quality Management) includes the assessment of multiple criteria, grouped into facilitators (5 types) and results (4 types).
In this chapter, we will only describe the PD process (as a part of the Dialysis Process) and the most appropriate indicators and standards for its evaluation..

3. The peritoneal dialysis process

The process includes information concerning the alternative techniques of dialysis offered to patients from which they can choose, and withdrawal from the PD program due to death, transplantation, changing to hemodialysis or recovery of renal function. As hemodialysis and PD have a similar start and end, and the same therapeutic purpose, we have grouped them under a single process of chronic dialysis, with its two main variants (Figure 2). Logically, the dialysis process is part of a series of support processes including those of the laboratory, pharmacy, maintenance, etc. The description of each activity in the process (Table 1) should not be exhaustive but rather refer to specific protocols that need to be written, accessible to all staff and regularly updated. However, it is important that there is a designated person in charge of each activity in the process and a record of the activity that could be consulted if necessary (Lopez-Revuelta et al., 2002; Arenas, 2006).
The process of peritoneal dialysis is a part of a more complex dialysis process that includes all the renal replacement therapies (Figure 2). Patients' opinions and medical contraindications determine the decision between the three main alternatives for dialysis, whether as a definitive therapy or as life support waiting for kidney transplantation. In this chapter we consider only the indicators of quality for the home peritoneal dialysis option.

3.1 Characteristics of clinical quality indicators

A clinical indicator is a quantitative measure that can help us monitor and evaluate quality in care activities and support services. It is not a direct marker of quality, but rather can serve to alert to areas which require specific action within a healthcare organization
Indicators express information as numbers of events or ratios. In the latter case, the denominator is the number of patients among whom the event could potentially occur. Although the event selected is undesirable, in general it should occur commonly enough to be used as an index. There is, however, a special kind of indicator that cannot be expressed as an index or a ratio: the Sentinel Event Indicator, that measures events which are undesirable, preventable, rare and have serious outcomes. When detected, such indicators warrant a thorough investigation and urgent action (even if there is just one case).
Indicators can measure either processes or results. The best process indicators are those closely linked to patient outcomes, and for which there is scientific evidence that indicates that the care provided will lead to a specific result. In the event that the result of a process cannot be measured, or there is an excessive delay for corrective action, process indicators are the only type that can be used.
Further, indicators can measure desirable or undesirable results. In the former case, the objective is that the vast majority of patients meet the criteria; while in the latter case, the aim is just the opposite. Ideally, the quality systems used by a given Unit should select only those indicators that represent desirable objectives, in order to avoid confusion. For instance,

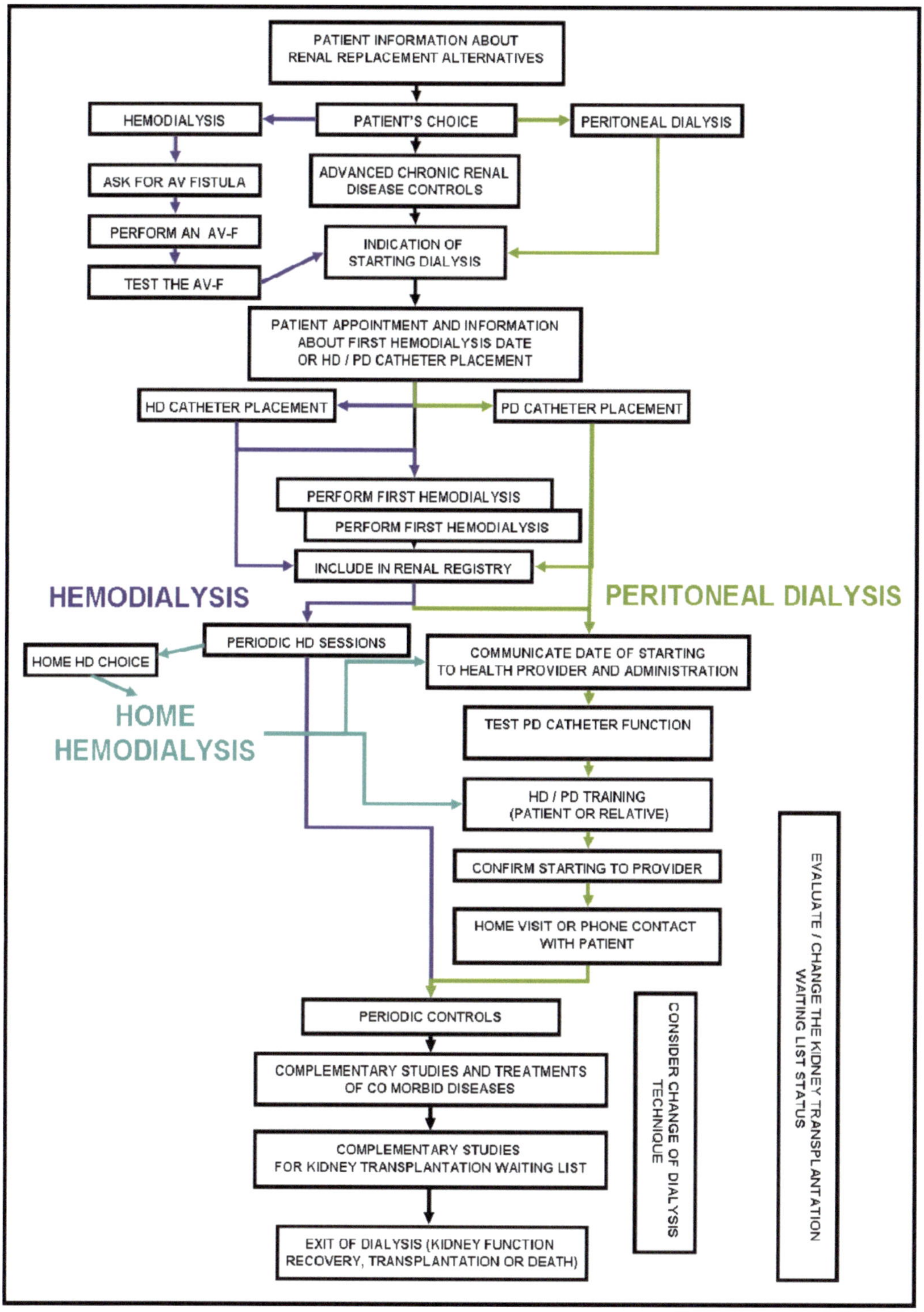

Fig. 2. Dialysis Process.

ACTIVITY	DESCRIPTION	RESPONSIBLE	REGISTRY
Patient information about RRT techniques	DIALYSIS GENERAL PROTOCOL	Nephrologist PD nurse	Clinical Record Nursing record Consent Form
Indication for starting PD	DIALYSIS GENERAL PROTOCOL	Nephrologist PD physician	Clinical Record
Appointment for PD catheter insertion	Written appointment, with date, time and premedication	PD nurse	Nursing record
Deliver to patient Information about appointment for catheter insertion	PERITONEAL DIALYSIS PROTOCOL	PD nurse	Nursing record
Catheter insertion	PERITONEAL DIALYSIS PROTOCOL	PD nephrologist or surgeon	Lab Reports Clinical Record
Incorporate in the RRT Registry and waiting list for renal transplant	On line data communication to RRT Registry Database, including Identification, Clinical and Serological Data, and required data for Kidney Waiting List	Nephrologist	RRT Registry. Waiting List for kidney transplant
Convey to Administration and PD material provider about the patient	To be done by email about patient information and chosen PD technique (CAPD or APD)	Nephrologist	E-mail and letter of approval
Check PD catheter permeability	PERITONEAL DIALYSIS PROTOCOL	PD nurse	Nursing record
Patient PD training	PERITONEAL DIALYSIS PROTOCOL	PD nurse	Nursing record
Call the PD provider	To finalize the supply of PD equipment at patient's home from specific date	PD nurse	Nursing record
Home visit or phone call to patient's home	After starting PD at home, some contacts asking for possible problems or doubts.	PD nurse	Nursing record
Regular controls at hospital, or by phone, mail, web-cam, etc.	PERITONEAL DIALYSIS PROTOCOL	PD nephrologist PD nurse	PD Graphics Nursing record Clinical Record
To consider change in dialysis technique	In case of patient's decision or unsolvable problems. PD PROTOCOL	PD nephrologist	PD Graphics Nursing record Clinical Record
Regular controls about studies and treatments of associated illnesses	PERITONEAL DIALYSIS PROTOCOL	PD nephrologist	PD Graphics Clinical Record
Regular control's studies about Waiting List for kidney transplant	PRE-TRASPLANT STUDIES PROTOCOL	Nephrologist	PD Graphics Clinical Record
Reconsider situation in Waiting List for kidney transplant	PRE-TRASPLANT STUDIES PROTOCOL	Nephrologist	PD Graphics Clinical Record RRT Registry

ACTIVITY	DESCRIPTION	RESPONSIBLE	REGISTRY
Discharge from PD due to partial improvement of renal function, change to hemodialysis, transplant or patient's death.	PERITONEAL DIALYSIS PROTOCOL	Nephrologist	Lab Reports Clinical Record

Table 1. Peritoneal Dialysis: Activities in the Process.

"peritonitis rate with a negative culture" is an indicator of low quality, a high rate suggesting poor quality in sample collection, transport or laboratory processing of peritoneal effluent. As this may not be intuitive, it is preferable to use the "peritonitis rate with a positive culture", aiming for this indicator to exceed 80% of cases.

In addition, indicators must be valid. This means they should identify situations in which quality in the healthcare provided can be improved (reflected in final outcomes). Validity is often only apparent and the indicator has to be "validated" afterwards. Lastly, indicators must be sensitive, able to identify real problems with care, avoiding "false positives", and they must also be specific, so that they detect only these real problems.

The selection of a set of indicators is a complicated task. It is preferable to select only a few, avoiding an increase in workload related to maintaining the database that would have no direct translation into improvements. From the selected recommendations, the quality indicators were drawn up according to a format which includes: their definition, criterion, equation, units, frequency of the assessment, standard, bibliographical references and comments. The methodologies proposed by the Joint Commission and the Standing Committee of the Hospitals of the European Union were followed by systems for monitoring healthcare processes. These have been complemented by the specific HD methodology that is followed by the American ESRD Special Project and implemented by the Centres for Medicare and Medicaid Services (CMS) -such as in the ESRD Clinical Performance Measures (CPM) Project-. Initially, quality criteria were selected from each recommendation for measurement of performance. The indicator is a quantitative measurement to evaluate a criterion. A "standard" was set for each indicator (namely, the required degree of performance to ensure an acceptable level of quality) based on scientific evidence or, in its absence, by consensus. On many occasions, sufficient scientific evidence was not available, but experience derived from the follow-up of indicators will help us adjust and redefine them in the future. Those interested in understanding the subject more deeply, should consult the 1989 and 1990 JCAHO references.

3.2 Quality indicators in peritoneal dialysis

Traditionally (Donabedian 1980; JCAHO, 1989), we distinguish between structure, process and outcome indicators. Variations in the quality of the structure or the process tend to influence the outcomes. Structure Indicators are highly valued for accreditation, as adverse results caused by structural defects imply a greater responsibility if patients file lawsuits. However, we assume process indicators are a more accurate reflection of quality than those directly related to outcomes, as they detect systematic errors and their correction more commonly produces improved outcomes (Williams et al, 2006; Ballard, 2003).

Indicators must monitor quality. Therefore, they should be correlated to survival or quality of life of the patients, and be based on scientific evidence. In our case, we based them on the Clinical Practice Guidelines in PD, recently published by the Spanish Society of Nephrology (Arrieta, 2006). Following the publication of these guidelines, a panel of peritoneal dialysis, with the support of the Quality Management in Nephrology Group (a working group of Spanish Society of Nephrology), designed a definition for quality indicators and standards that can be used by all the nephrology community -especially those dedicated to PD-. The new definitions would also serve as a framework or terms of reference for future areas of improvement, filling the gap between the development of guidelines and subsequent monitoring.

Often, we found that there was not sufficient scientific evidence to define a standard. In these cases, we proposed a provisional framework that should be evaluated later. Earlier in this chapter, we have explained that continuous improvement objectives should be set by each unit, based on local outcomes.

Whatever the result of applying an indicator in a given unit, what is important is that they guide improvement activities, and there will then be ongoing monitoring of whether such measures are effective. In fact, indicators are basically an internal tool that permits comparisons with our own previous performance and helps us assess our own improvement. In the future, the pooling of results from different institutions would determine the appropriate quality standards in peritoneal dialysis for the Spanish population.

Having similar quality criteria in all centers is a medium-term objective, as we are all interested in comparing our results and assessing whether variations in clinical practice lead to different final outcomes (Jha et al., 2005).

On the other hand, it has been shown that regular measurement of quality indicators –and the fact of having set up targets and standards- encourages monitoring and improves the outcomes of the process (Williams, 2005; Fink, 2002).

The initial list of indicators, standards and objectives selected includes a large number of indicators that have been already established for hemodialysis. As the most prevalent renal replacement technique, many Quality Systems have already been developed in that field (Lopez-Revuelta et al., 2007). Nevertheless, we should always consider those indicators or standards that have not been specifically validated for PD patients as provisional, and focus on the survival and quality of life outcomes instead.

There are usually too many indicators. Each unit should select those that seem most relevant to its daily routine. In addition, data management technologies become a priority. A wide range of computer software (Renalsoft®, Nefrolink®, Nefrosoft®, Versia® etc.) is used in peritoneal dialysis and hemodialysis units in Spain. In some cases, more advanced programs are being developed and adopted than enable quality indicators to be estimated automatically and rapidly.

In the following sections, we will describe the initial selection of Quality Indicators used by the Spanish Society of Nephrology (currently, at the evaluation stage). They are Clinical Indicators, so they have to be supplemented with Structure Indicators, Satisfaction Surveys and Quality of Life Questionnaires for patients. From a business point of view, and in order to obtain Accreditations of our units, it is also a good idea to carry out Satisfaction Surveys of our staff and suppliers.

3.3 Classification of peritoneal dialysis indicators

We use Global Indicators and Comorbidity Indicators to describe patients (Table 2). Most of these are not quality indicators but Registry data, local practice frameworks or terms of reference which enable us to identify certain patient and PD unit characteristics that may influence outcomes and modify other indicators. It is interesting to see how their evolution pans out over time. In some cases, they do indicate aspects of the quality of medical attention before starting PD, but our intention is to use them to adjust the evaluation of Outcome Indicators. The modified Charlson Index (Bedhu et al., 2000) extends the item "Documented History of Myocardial Infarction" to include another one namely "Ischemic Heart Disease (CHD)", which includes all forms of coronary heart disease (angina, myocardial infarction, angiographic evidence of coronary artery disease history of angioplasty or bypass surgery). For this reason, we consider it more appropriate for the usual profile of PD patients. Global and Comorbidity Indicators are collected annually, as they are not used to make improvements. The Charlson Index is measured at the start of PD and, as it can only increase, it is reassessed once or twice a year.

Outcome Indicators (Table 3) (Arrieta et al., 2009; Bajo et al., 2010) include more up-to-date data, such as the rate of infections associated with the technique, the adequacy of the dialysis dose, test results and medications taken. These Indicators can alert us to deficiencies in the initial stages of treatment, and early correction can rapidly improve outcomes. Usually, they are compiled twice a year, but with a good IT system they can be calculated and consulted as often as is agreed to be appropriate in each unit, though clearly this involves additional work.

Other indicators such as rates of hospitalization or withdrawals from DP should be explored more carefully, as they are influenced by local characteristics, the socio-cultural context and the availability of replacement therapy.

GLOBAL INDICATORS
PD Incidence
Period Prevalence (prevalents at begin of period + incidents)
Mean age of incidents
Mean age of prevalents
Sex rate of incidents y prevalents
Mean time in PD treatment of prevalents
Percent of diabetics among incidents
" of incidents not dialyzed before
" " coming from HD
" " coming from TX
" of incidents with a signed Informed Consent about all RRT techniques
" of prevalents on CAPD (vs total in PD techniques)
COMORBIDITY INDICATORS
Median of Modified Charlson Index in incidents
Median of Modified Charlson Index in prevalents

Table 2. Quality Indicators at starting PD.

OUTCOME INDICATORS (1) (ANNUAL INDICATORS)	
Hospitalization	
admissions	
average days by admission	
Exits from PD	
totals	
change to HD	
deaths	
transplants	
Transplants	
percent of patients in Kidney Tx Waiting List (WL) (among prevalents in PD)	
mean time in PD before inclusion in WL	
annual rate of transplants in PD patients (among patients in WL)	
mean time in PD before kidney Tx	
mean time between Tx and PD catheter extraction	¿< 1-3 months?
OUTCOME INDICATORS (2) (SEMESTER INDICATORS)	
Infections (limited to PD technique)	
rate of peritonitis	¿< 0.5 / pte / yr?
partial rates in APD and CAPD	
percent of peritonitis with a positive culture (identified germ)	
" of peritonitis by Gram +	
by Gram –	
by fungus	
" of peritonitis "catheter dependent"	
rate of infections of catheter exit site	
rate of patients with nasal cultures (positive or not)	>80%
Adequacy and membrane function	
percent of patients with an urea KT/V measured in the semester	>80%
" of patients with urea KT/V > 1.7	>80%
" of patients not anurics with Renal Residual Function measured	>80%
" of patients with a daily UF rate > 1000ml/ day	>80%
" of patients using daily one or more hypertonic bags (3.86 / 4.25%)	<20%
" of patients with a PET performed in the 3 months alter starting PD	>80%
" of patients with a PET performed annually	>80%
" of patients resulting High Absorbers in PET. (D/P Cr 4h > 0.81)	
Analysis and medication	
percent of patients within Hb objective (11 to 13)	>80%
" of patients with serum ferritin > 100	>80%
" of patients with Index of Resistance to EPO < 9	>80%
" of patients with I.R. to darboepoetin < 0.045	>80%
" of patients with serum cholesterol LDL < 100	>80%
" of patients with serum albumin > 3.5	>80%
" of patients with serum phosphate < 5.5	>80%
" of patients with serum corrected calcium > 8.4 and < 9.5	>80%
" of patients with Ca x P < 55 (in mg/dL)	>80%
" of patients with PTH < 300	>80%

Table 3. Quality Indicators of Outcomes.

Calculation of the rate of occurrence of a certain outcome may present problems in units with few patients. We recommend estimating the prevalence of "at-risk" patients per month, to determine the "real" total number of patients to be used in the denominator of the ratios (Jager et al., 2007).

3.4 Standards and objectives of quality indicators

Every indicator should have a clear definition, a target or objective (threshold or range), and a standard for assessing compliance. We have defined objectives when there is a reasonable amount of scientific evidence to support them. However, such evidence is often not sufficiently tested in PD (though it may have been tested in HD patients or in the general population, as is the case of LDL cholesterol levels). The original standard is commonly set at the percentage of patients who meet the target. For clarity, we prefer to express the degree of compliance than the rate of "non-compliance".

It is important that targets are always to be established based on scientific evidence. For instance, the hemoglobin target is set at 11 mg/dL or above because the Guidelines for Good Clinical Practice (based on hemodialysis) agree on this level; nevertheless; PD patients may have Hb higher than 13 in the absence of EPO. Accordingly, we will not set a maximum target as we do in HD. The standard is a given rate of compliance with objectives -usually 80% to 85%-, and is later adapted to the real results obtained and the real possibilities of achieving the Standard in our healthcare context.

When we initially apply an indicator in our units, we may find that our compliance rates are very low. This could mean that the target was too high, the indicator was not appropriate or, even, that the sample of patients on which the assessment if based are really ill. The objective must be based on high-grade evidence. If it is well established, we must strive to achieve it over time and accept a low compliance rate, re-evaluating the rate once or twice a year.

I insist that a good progress is more important than a good result. Evidence is often drawn from clinical trials involving highly selected patients, with a high rates of adherence to prescribed medication (which is often free during the trial) and under close medical supervision. These results would be very difficult to achieve in routine practice. In any case, it is absolutely not permissible for the threshold for compliance with an objective to be lowered as a means of achieving a better rate of compliance, unless on reconsideration it is judged that the target is not supported by current evidence, or that the effort required to achieve the target is not justified by real improvements in the final outcome measures (namely, survival and quality of life).

Finally, we must remember that just measuring outcomes tends to produce an improvement in clinical practice (Williams et al., 2005; Fink et al., 2002). It has also been proven that, in hemodialysis, the level of compliance with quality standards is directly related to mortality and morbidity, although most of the standards applied have not yet been validated (Rocco et al., 2006; Plantiga et al., 2007). From a theoretical standpoint, this introduces a bias towards the validation of an Indicator or a Standard, but it should also encourage doctors to use the quality control systems as tools for continuous improvement of our daily practice, rather than consider them as management tools with little relevance to medical practice.

4. Conclusion

It has already been demonstrated that the regular measurement of quality indicators –as well as having standards and establishing objectives-, helps to improve the monitoring and results of the dialysis process, and contributes to improving outcomes in terms of patient morbidity and mortality. Access to management software becomes a priority. A Quality System should be focused on achieving Continuous Improvement of Quality expressed in terms of Survival and Quality of Life. Patients' opinion about self-perceived health and wellbeing and about quality of health care must be considered. Accreditation of the Unit should not be a final objective.

5. Acknowledgment

Groups of Quality in Hemodialysis and Peritoneal Dialysis of Spanish Society of Nephrology have played an essential role in the process of selecting indicators and testing the suitability of proposed standards of Quality in PD.

6. References

Arenas, MD.; Lorenzo, S.; Alvarez-Ude, F.; Angoso, M.; López- Revuelta, K. &Aranaz, J. (2006). Quality control systems implementation in the Spanish Dialysis Units. *Nefrología,* Vol 26, No.2, pp. 234-245. Online ISSN: 2013-2514. Print ISSN: 0211-6995.

Arrieta, J.; Bajo, MA.; Caravaca, F.; Coronel, F.; García-Perez, H.; Gonzalez-Parra, E.; et al. (2006). Guidelines of the Spanish Society of Nephrology. Clinical practice guidelines for peritoneal dialysis. *Nefrología.* Vol 26. Suppl 4, pp 1-184. Online ISSN: 2013-2514. Print ISSN: 0211-6995.

Arrieta, J. (2009). Calidad en Diálisis Peritoneal. In: *Tratado de Diálisis Peritoneal.* (Chapter 31). Montenegro, J.; Correa-Rotter, R & Riella, MC., pp 573-582. Elsevier. ISBN: 978-84-8086-394-0, Madrid.

Bajo, MA.; Selgas, R.; Remón, C.; Arrieta, J.; Alvarez-Ude, F.; Arenas, MD.; Borrás, M.; Coronel, F.; García-Ramón, R.; Minguela, I.; Pérez-Bañasco, V.; Pérez-Contreras, J.; Fontán, MP.; Teixidó, J.; Tornero, F. & Vega N. (2010). Scientific-technical quality and ongoing quality improvement plan in peritoneal dialysis. *Nefrologia.* Vol 30, No. 1, pp. 28-45. : 2013-2514. Print ISSN: 0211-6995.

Ballard, DJ. (2003). Indicators to improve clinical quality across an integrated health care system. International Journal of Quality in Health Care. Vol. 15, Suppl 1, pp i13-i23, Online ISSN 1464-3677 - Print ISSN 1353-4505.

Bedhu, S.; Bruns, FJ.; Saul, M.; Seddon, P. & Zeidel, ML. (2000). A simple comorbidity scale predicts clinical outcomes and costs in dialysis patients. *American Journal of Medicine,* Vol 108, pp 609-613, ISSN 0002-9343.

Committee on Quality of Health Care in America. (2001). Crossing the quality chasm: a new health system for the 21st Century. Washington, DC: National Academy Press. ISBN: 0-309-07280-8.

Donabedian, A. (1980). Explorations in quality assessment and monitoring. Ann Arbor, MI: Health Administration Press. ISBN: 0914904477, 0914904485

Fink, JC.; Zhan, M.; Blahut, SA.; Soucie, M. & McClellan, WM. (2002). Measuring the Efficacy of a Quality Improvement Program in Dialysis Adequacy with Changes in Center Effects. Journal of American Society of Nephrology, Vol 13, pp. 2338-2344. Online ISSN: 1533-3450, Print ISSN: 1046-6673.

Jager, KJ.; Zoccali, C.; Kramar, R. & Dekker, FW. (2007). ABC of Epidemiology (1): Measuring disease occurrence. Kidney International, Vol 72, No 4, pp. 412-415. Online ISSN: 1523-1755, Print ISSN: 0085-2538

Jha, AK.; Li, Z.; Orav, EJ. & Epstein, AM. (2005). Care in U.S. hospitals--the Hospital Quality Alliance program. *New England Journal of Medicine,* Vol 353, pp. 265-274. On line ISSN 1533-4406 Print ISSN 0028-4793.

JCAHO, 1989: Joint Commission on Accreditation of Healthcare Organizations: Agenda for change. (1989). Characteristics of clinical indicators. *Quality Review Bulletin.* Vol 15, No 11, pp 330-339. ISSN: 0097-5990.

JCAHO, 1990: Update: clinical indicators. (1990). *Hospital Food & Nutrition Focus.* Vol 6, No 11, pp. 6-7. ISSN: 0747-7376.

JCAHO, 2011: Joint Commission on Accreditation of Healthcare Organizations: Agenda for change (2011). Advanced Chronic Kidney Disease Certification. http://www.jointcommission.org/certification/chronic_kidney_disease.aspx

Lohr, KN. (editor). Committee to Design a Strategy for Quality Review and Assurance in Medicare. Institute of Medicine. (1990). A strategy for quality assurance. Washington, DC: National Academy Press. ISBN: 0-309-04230-5

López-Revuelta, K.; Lorenzo, S.; Gruss, E.; Garrido, MV. & Moreno-Barbas, JA. (2002). Application of process management in nephrology. Hemodialysis process management. *Nefrología,* Vol 22, No 4, pp. 329-339. Online ISSN: 2013-2514. Print ISSN: 0211-6995.

López-Revuelta, K.; Barril, G.; Caramelo, C.; Delgado, R.; García-López, F.; García-Valdecasas, J. et al. (2007). Developing a Clinical Performance Measures System for hemodialysis. Quality Group, Spanish Society of Nephrology. *Nefrología,* Vol 27, No 5, pp. 542-559. Online ISSN: 2013-2514. Print ISSN: 0211-6995.

Platinga, LC.; Fink, NE.; Jaar, BG.; Sadler, JH.; Levin, NW.; Coresh, JK.; et al. (2007). Attainment of clinical performance targets and improvement in clinical outcomes and resource use in hemodialysis care: a prospective cohort study. *BMC Health Services Research,* Vol 7, pp. 5-18. On line ISSN: 1472-6963

Rocco, MV.; Frankenfield, MV.; Hopson, SD. & McClellan, VM. (2006). Relationship between Clinical Performance Measures and Outcomes among Patients Receiving long-term Hemodialysis. *Annals of Internal Medicine,* Vol 145, pp. 512-519. On line ISSN: 1539-3704. Print ISSN: 0003-4819.

Williams, SC.; Schmaltz, SP.; Morton, DJ.; Koss, RG. & Loeb, JM. (2005). Quality of care in U.S. hospitals as reflected by standardized measures, 2002-2004. *New England Journal of Medicine,* Vol 353, pp. 255-264. On line ISSN 1533-4406 Print ISSN 0028-4793.

Williams, SC.; Watt, A.; Schmaltz, SP.; Koss, RG. & Loeb, JM. (2006). Assessing the reliability of standardized performance indicators. *International Journal of Quality in Health Care,* Vol 18, pp. 246-255. Online ISSN 1464-3677 - Print ISSN 1353-4505.

14

From a Quality Assurance and Control System for Medical Processes, Through Epidemiological Trends of Medical Conditions, to a Nationwide Health Project

Yossy Machluf[1,2], Amir Navon[1], Avi Yona[1], Avinoam Pirogovsky[1,3], Elio Palma[1,4], Orna Tal[5], Nachman Ash[6], Avi Cohen[1] and Yoram Chaiter[1]

[1]*Quality Assurance and Control Committee, Medical Corps, IDF,*
[2]*Weizmann Institute of Science, Rehovot,*
[3]*Head of Standards and Regulation Department in the Division of Community Medicine, Ministry of Health, Tel Aviv,*
[4]*Head of Department of Occupational Medicine, Clalit Health Services, Afula,*
[5]*Israeli Center for Technology Assessment in Health Care; The Gertner Institute for Epidemiology and Health Policy Research, Head of Emerging Technologies Unit, Tel Aviv,*
[6]*Chief Medical Officer, Medical Corps, IDF,*
Israel

1. Introduction

A health policy is an integral part of the general welfare policy in every state. According to Dye (1987), interior public policy constitutes the same action as a government that chooses to "do or not do" (Dye, 1987), and thus it is assumed that any governmental action is derived from its will to preserve and provide quality services to the public. Quality of patient health care is defined by two principal dimensions: access and effectiveness (Campbell *et al.*, 2000).
Over the last few decades, quality control has been gaining a central place in public organizations in order to improve the quality of services and treatment (Blumenthal, 1996; Landon, *et al.*, 2003; Mandel, *et al.*, 2003; 2004). Quality assurance is a key component in the processes aimed at improving the quality of service and medical care.

1.1 Quality control and the health system

The quality index in the health system is a criterion that shows measurable values in morbidity and service levels (Campbell *et al.*, 2000). Since the 1960s, the quality indices in health systems were divided into three main levels: infrastructure and structure, process, and outcome. The last two are usually included as preferred measures (Donabedian, 2005).
Infrastructure and structure indices are related to organizational issues of the health services, the attending person's nature, and the procedures and medical policy that are being implemented by the organization on both the private and public levels. The assumption is that any health organization should be capable of providing quality health services according to its resources, which are made up of human, economical, and

infrastructure components. Process indices are aimed at examining the extent of the medical actions that are taken to achieve the desired target, assuming that well performed operations increase the chance of accomplishing the desired effects. By contrast, outcome indices focus on the individual level rather than on the organization level. Outcome measurements provide an indirect measure of the overall quality assessment and may provide a benchmark for tracking progress. In general, the more the first two indices (the infrastructure and structure and the process indices) are involved in measuring quality, the greater the reliability of the outcome measures (Donabedian, 2005).

Many approaches were developed to assess the quality of health care and to improve both medical and patient processes, such as the Six Sigma, which utilizes the DMAIC model (Define, Measure, Analyze, Improve, and Control); ISO (International Organization for Standardization); BOS (Business Operating System); CI (Continuous Improvement); TQM (Total Quality Management); etc. (Donabedian, 2003; Munro, 2009; Ovretveit, 1992; Ransom *et al.*, 2008). Many of these approaches are derived and adapted from quality assurance systems in industry, where processes are straightforward and the implementation of such methods is easier. Yet, the medical process represents an intricately interwoven and dynamic process (Donabedian, 2003, Ovretveit, 1992) where many variables are interconnected. For instance, in medical committees, managing authorities, health care medical professionals, technical and administrative personnel, and patients, as well as medical policy, regulations, and goals are all part of medical processes. Therefore, difficulties may arise while applying these methods to medical procedures.

Although there are many models of health care quality control and assurance, most focus on specific issues and are a tool for managerial decisions and not for the day-to-day surveillance of the processes on a clinical level. They may address a specific issue at a local medical facility and may try to improve the specific circumstances, such as the high rates of certain infections at a specific ward or at a specific medical center. Most of the quality assurance processes measure outcomes rather than the completeness and intactness of the continuous process and use a set of tracers or specific indicators, such as Hemoglobin A1C levels in diabetics, cholesterol levels, blood pressure, etc. There are also many economical issues addressed by such quality assurance systems and only rarely do they achieve an optimal resolution of the processes that they aim to control. Many decisions are made in an effort to save expenses and by that to achieve control, while the clinical issues of the processes are not addressed in detail. Most quality systems deal with a limited aspect of the health care process, rather than dealing with the whole process and its various components, including both personnel (policy makers, managers, medical professionals, technical staff, administrative shell, patients) and non-personnel (policy, regulations, infrastructure, economy). In this chapter, the design, application, and outcomes of a unique quality control and assurance program within the framework of medical committees will be described.

1.2 Overview of medical processes at recruitment centers

In Israel, adolescents aged 16-19 are obligated by law to enlist for military service and are examined by medical committees at conscription centers in order to determine their medical status. The medical process at the recruitment centers necessitates the coordinated action of medical professionals, technical and administrative personnel, and managing authorities, on both the local and national levels. The medical process is based mainly on a medical interview and an examination by a medical committee and is supplemented and supported by information from the family physician, medical consultants, and experts (Fig. 1).

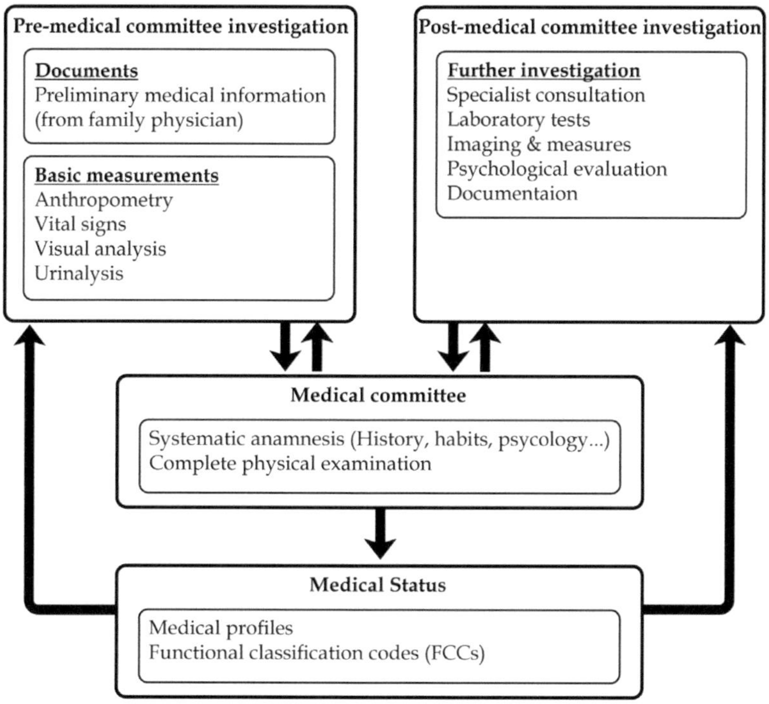

Fig. 1. Schematic representation of the medical process at the recruitment centers.

The administrative medical department employees at the recruitment centers constitute the administrative shell of the medical process, are thought of as medical facility workers, and serve as an extension of the medical process. They are exposed to medical information and follow strict laws of keeping medical confidentiality. They are in charge of the appointments of the recruits for various medical procedures at the recruitment centers and at medical facilities outside the recruitment centers; queue regulations; exchanges of letters and documents, and submitting requests from clinics, hospitals, and government facilities.

The technical medical assistants are part of the medical committee. Prior to medical committee examinations, they measure vital signs and anthropometric values (such as height, weight, blood pressure, and pulse) and check visual acuity and color vision according to Ishihara color tables. Urinalysis testing is done by a laboratory technician. They also collect all previously sent documents, add all required forms, and provide all of this accumulated medical documentation and forms to the medical committee.

The physician's work on the medical committee consists of a thorough medical anamnesis, including family history, habits, and a psychological evaluation. A systematic and comprehensive physical examination is performed. According to the findings, the committee chairman decides whether additional tests, such as a specialist consultation, laboratory tests, imaging, or other measures, are required or if the information is sufficient to apply functional classification codes (FCCs) and complete a recruit's medical profile. The medical profile reflects the recruit's current health status and is used by the Personnel Directorate to assign a military position that is consistent with that health status. Similar to coding systems like the International Classification of Diseases (ICD), Medical FCCs describe disorders, their severity, and determine the profile (Chaiter *et al.*, 2010).

For a complete and successful medical process, a close collaboration between the medical, administrative and technical personnel and harmony in their working relationship are needed.

1.3 Complex and dynamic medical processes at recruitment centers

Based on our annual analysis of the last decade (Conscription Administration Data, 2001-9), each year about 100,000 new recruits start the medical processes, and more than 400,000 medical encounters are performed at the recruitment centers. About 50% and approximately 13% of the recruits have ≥2 and ≥4 encounters, respectively, mostly with specialists or in hospitals. Most recruits (≥75%) have at least one FCC, while many (~20%) have 3 or more co-existent medical conditions, in spite of their age. Furthermore, it is not just the prevalence of medical conditions but also their severity and morbidity. For instance, ~8% of recruits suffer from chronic asthma, while ~1.5% and ~10% suffer from various cardiac anomalies and mental disorders, respectively (Machluf *et al.*, 2011). In about 10% of the recruits, the discovery of new conditions (following accidents, operations, or a change in the severity of the medical conditions from the first check-up) leads to the modification, addition, or cancellation of FCCs. This may result in the determination of a new medical profile. The thorough examination at a recruitment center might reveal a new, previously unknown, medical problem that warrants further follow-up or treatments at a primary clinic by the primary care physician. Among these conditions, we have revealed even cases of severe disorders, such cardiac anomalies, nephropathies, cancer, etc. In these cases, the diagnostic-treatment loops need to be closed.

1.4 Databases

The information acquired during the medical process in stored, organized, and archived in a database. Computer-based tools allow analysis and visualization of data. With regard to the administrative aspects of the medical process, the computerized system consists of three main components (Machluf *et al.*, 2011):

A status system: a specific status code is assigned to each file/recruit using code numbers reflecting the specific specialist, test, or documentation needed. A status code and its beginning and end dates allow the administrative medical department to actively and dynamically follow up and manage the medical and administrative processes on both the individual and collective levels.

An appointment system: this system is used to assign a specific date for an appointment to a certain specialist or medical procedure, to generate invitations, and to document the appointment outcome. Its design principles allow better control and management of human and medical resources according to the capacities and limitations of the medical system.

A directing, monitoring, and controlling system: a local smart card-based system for real-time follow-up and regulation of waiting lists of patients at each station. It automatically directs recruits either to an available station or according to priorities pre-set by the medical personnel that the specific recruit is required to pass. It provides the medical administrative personnel and the medical committee members with relevant information on both the individual (recruit) and collective (queues) levels.

Each component system is directed toward answering a particular need and, although each is independent, they are all compatible with each other and provide the user with a comprehensive view of the medical process and information. The combined computerized system improves the control and management of the medical processes and informatics from the point of view of both the patients and the system operators. Implementation and automation of medical regulations and procedures within the computerized system also make this system play a key role and serve as a control tool during the decision-making process (Machluf *et al.*, 2011).

2. Quality assurance and control system

During 1997, a quality assurance and control system was set up (Chaiter *et al.*, 2008; 2010; 2011; The State Comptroller and Ombudsman Office, 2002). This unique system, unlike most policies or systems for quality assurance of medical care, originated from within the medical profession, rather than from industry or academia, and its regulation and modes of action reflect the real daily health care activity in medical departments.

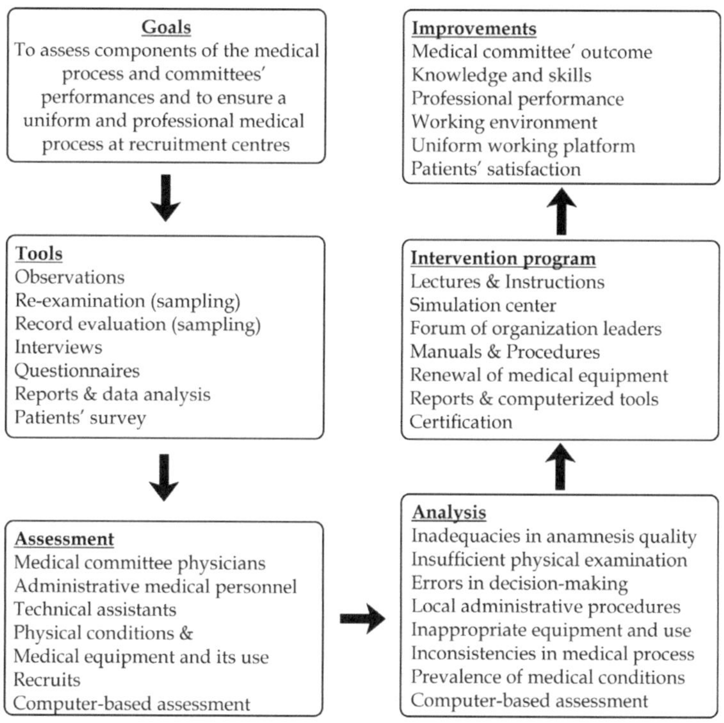

Fig. 2. Scheme of the quality assurance and control project of medical processes at the recruitment centers.

The goals of the quality assurance and control committee are not solely to assess and improve the medical committee outcomes. They are also to identify the limitations and needs of whole medical process regarding the medical, technical, administrative, procedural, and physical aspects and to evaluate knowledge, skills, judgement and working patterns, which count for quality care (Holmboe *et al.*, 2008), among all medical personnel; to develop and implement intervention programs to answer the needs of all of the parties participating in the medical process; and to achieve a higher quality of patient care and patient satisfaction (Fig. 2). Accordingly, the quality assurance and control system evaluates and analyzes in detail the various facets of the activities of the medical policy makers, managers, administrative staff, medical professionals (physicians, experts, consultants), and technical assistants, utilizing different complementary methods (Fig. 2). Experts from all these fields (and others), with their high skills and experience, are incorporated into the quality assurance and control system. Of note, performance measures are evidence-based and valid, feasible to collect, applicable to a large enough population of patients, attributed to the performance of individuals, adjustable to the patient level, and representative of the activities of a specialty (Landon *et al.*, 2003).

2.1 Approach

Two main approaches are deployed by the quality assurance and control committee: (i) physical visits to recruitment centers, during which the procedures, work, decision-making processes, and outcomes are directly assessed, and (ii) data mining and processing from the computerized databases. The different means used in each approach to assess the work and results of the different medical process-related staff are summarized in Table 1. Incorporation of such complementary assessment methodologies provides both quantitative and qualitative analyses of daily activities and practices in the medical departments.

Methodology		Components of the medical process			
Approach	Tools	Medical physicians	Technical assistants	Administrative staff	Recruits
Direct assessment	Observation	✓	✓	✓	-
	Re-examination	✓*	-	-	-
	Record evaluation	✓	✓	✓	-
	Interviews	-	✓	✓	✓
	Questionnaires	✓*	✓#	✓#	✓
Data mining	Reports-QC	✓	✓	✓	-
	Epidemiology-like	✓	✓	-	✓

*-distributed to recruits; #-distributed to soldiers

Table 1. The various methodologies used to assess components of the medical process.

2.2 Direct assessment

Direct assessment during the physical visits to the recruitment centers provides an opportunity to evaluate the work of the medical process as a whole. It also enables direct interaction and brainstorming with local medical department members, from administrative staff though technicians, physicians and experts to managers.

2.2.1 Direct assessment of medical committee performance

The medical committee is central to the medical process. Direct assessment of the medical committee's performance consists of direct observation and clinical evaluation of the physicians' work, random samplings, re-examination of recruits that were examined by the medical committee, distribution of questionnaires to recruits following their examination by the medical committee, and analysis of random samples of completed medical files.

Observation and clinical assessment (clinical audit): Upon receiving a recruit's consent, a physician from the quality and assurance system joins the medical committee as an observer and evaluates the completeness and adequacy of the anamnesis, physical examination, and decision-making process (using a pre-designed form), as well as the quality of the communication and service to the recruits. For each of these components, quantitative (a numerical scale following specific criteria) and qualitative (description and comments) assessments are used. For example, for each issue in the anamnesis or physical examination, codes are used to determine whether and how the issue was addressed (according to regulations and clinical merits as determined in the medical literature). The decision-making process is also assessed for the quality of referrals to further investigation and the quality of decision making according to clinical criteria and regulations.

For each medical committee, 5-10 cases are observed. Findings are shared and discussed with the observed physicians, and proper instructions and guidance are provided, if necessary.

Sampling for recruits' re-examination and record evaluation: With a recruits' consent, physicians from the quality and assurance system re-examine a random sampling (8-10 on average) of recruits immediately following their examination by the medical committee. The findings of the complete re-examination, including the "assigning" of a medical profile and FCCs, are compared to those of the local medical committee, and discrepancies are recorded and discussed for each case. In addition, a random sampling of files (30-50 files from each recruitment center), where the profiling process was completed, are re-checked by a physician from the quality and assurance system to assess the anamnesis, the medical findings and documents, the decision-making process, and the assignment of profiles and FCCs.

Questionnaire (patients' survey): In each recruitment center, questionnaires are randomly distributed to recruits (15-20 on average) following their examination by the medical committee to gain more insight regarding the medical processes (basic measurements, anamnesis, and physical examination), preserving medical confidentiality and right of privacy, as well as the recruits' rates of satisfaction with the service, during the medical process. Recruits are also asked to express their general impression of the medical process.

2.2.2 Direct assessment of technical and administrative medical staff

There are many similarities that are shared by the work of administrative staff and the technical assistants and also by their quality control and assessment. Direct assessment of each person's performance consists of direct observation and evaluation of their work, distribution of questionnaires to professional personnel, interviews, and analysis of random samples of completed medical files.

Observation (audit): The work of all of the technical medical staff is assessed prior to, during, and after the medical committee examinations. First, medical equipment (scales, altimeters, chart tables for the visual acuity examination, sphyngomanometers, and Ishihara color tables) and its use are examined during the process of taking basic measurements. Also, documentation handling, data recording, and proper directing to further processes are checked. When required, correct instructions are provided to prevent future mistakes. Similarly, observation of the administrative staff involves the same parameters, except for an evaluation of basic measurement techniques.

Questionnaire: A detailed questionnaire is used in order to evaluate the skills of all technical assistants in measurement techniques (weight, height, etc.), knowledge of the normal range and abnormal findings regarding different measurements and the corresponding FCCs (blood pressure, pulse, urinalysis, etc.), and administrative issues (recording of measurements and medical history, signatures, etc.). By contrast, questionnaires are given to the administrative staff to evaluate their knowledge of regulations (status, appointments, and the smart card-based system) and administrative issues (special populations, reports, etc.). Correct instructions are given and even demonstrations are done to promote an increase in knowledge and skills.

Interviews: The technical assistants and administrative personnel are interviewed about their work, the findings of the control system are discussed individually, and ideas of how to improve the infrastructure, process, and procedures are shared.

Sampling for records evaluation: a random sampling of files (10 on average of each kind, at each recruitment center), where the profiling process was completed, are re-checked by members of the quality and assurance system to assess the documentation and administrative processes.

2.3 Data mining and processing

The demographic information and medical-administrative data are stored, organized, and archived in a database. This data can be visualized and retrieved. Computerized tools, such as reports and regulation-based automated procedures, allow in-depth data analysis. Computerized databases and tools, when integrated into the medical process, allow efficient follow-up and management of medical processes and informatics (Machluf *et al.*, 2011). These databases and reports can also serve as quality control and assessment means (Chaiter *et al.*, 2008; 2010; Machluf *et al.*, 2011; Navon *et al.*, 2011). Using the reports, one can assess the work of medical professionals (physicians and experts, for example), technical assistants, and administrative staff members. For example, reports are aimed at identifying discrepancies between the medical information (such as anthropometric and basic measurement data) and FCCs or medical profiles, inconsistencies in the medical-administrative information between the status and appointment systems, inadequate or incomplete medical processes (deviations from defined regulations), etc. (Chaiter *et al.*, 2008; Machluf *et al.*, 2011; Navon *et al.*, 2011). Such populations, at the individual level, are monitored, and reports are distributed monthly to the relevant personnel at each recruitment center and to the managing authorities.

Reports also support the design, planning, monitoring, and use of human and medical resources and are a component of the decision making that is made by the medical-administrative managers. For instance, various aspects of the availability and need for medical services, such as specialists, consultants, medical procedures, waiting queues, and various causes of congestion that need to be taken care of, are all assessed regularly by reports. The findings are distributed to the relevant personnel at the recruitment centers and to the managing authorities.

Reports also allow a comparison between the performance of the medical personnel within and between the recruitment centers and across a longitudinal time axis. Furthermore, since medical profiles and FCCs are indicators of medical conditions among recruits, an epidemiological investigation of the profile distribution and prevalence of FCCs can be performed, providing inter- and intra-recruitment center analysis (Chaiter *et al.*, 2010). This valuable information can be related to or crossed with gender, geographical area, country of origin, ethnicity, socio-economic background, education, and morbidity trends in the general population in Israel and other countries.

3. Findings

Using these tools, we analyzed key parameters related to the performance, integrity, and completeness of the medical processes and procedures. Furthermore, we were able to identify difficulties concerning physical conditions, administrative or medical procedural deficiencies, and insufficient knowledge or skills of medical/technical/administrative personnel within the medical departments.

3.1 Medical committee performance

In general, during the years 1997-2010, more than sixty visits to the different recruitment centers were carried out. Regarding the performance of the medical committees, the work of more than 110 chairmen (of medical committees) was assessed by different means. Six hundred and fifty-five recruits were re-examined, 452 filled out questionnaires, and approximately 1700 records (medical files) were examined and evaluated, providing a 3-pronged approach to observational findings. For the sake of clarity, we will present the findings from the first years and from the period after the intervention programs, focusing on recruitment centers with bigger populations.

During the first years of the program, we found that at most of the recruitment centers there was an inadequate amount or a lack of medical equipment, and often it was found that the equipment was used inappropriately (Chaiter *et al.*, 2008). The following were among the more common medical equipment-related faults that were found: broken altimeters, inadequately balanced scales, unsuitable Ishihara books for color vision examination, incomplete regulation manuals, and inadequate equipment for visual acuity examinations. Following our visits, a major effort was made to improve the physical conditions and equipment at the medical committee examinations. Examination rooms were redesigned according to the needs of the physician and the patients. Inadequate and old medical equipment were replaced by new machines, complete manuals and regulations were provided, and attention was paid to the physical surroundings and environment. After a gradual improvement in this area throughout the recruitment centers, such faults, even minor ones, are now only rarely detected.

Among the physicians, we observed inadequacies in anamnesis quality, insufficient physical examinations, and errors in decision making (Chaiter *et al.*, 2008). Furthermore, significant differences were observed between recruitment centers.

Anamnesis includes both specific tracer questions and general systematic queries. The tracer questions are all obligatory and refer to night vision, refractive eye surgery, dyslexia, drug use, gynaecological issues among female recruits, prolonged staying abroad at specific areas that are endemic to HIV, and psychological/psychiatric treatments. The main findings show a lack of anamnesis about most of these tracers, except the last two. Yet, differences were observed between recruitment centers. For instance (Chaiter *et al.*, 2008), in 2002, complete anamnesis for refractive surgery was found to range between 12.5% of recruits at center 2 to 90% of recruits at center 5 ($p<0.001$). Anamnesis of drug use was found to be only 25% of the cases at center 2 as compared to 100% of the recruits at center 5 ($p=0.04$). Similarly, at center 3, 10% of recruits there were asked about dyslexia, while at center 2 the rate was 100% ($p<0.001$). During and especially following the intervention program, a clear improvement in the anamnesis process was evident by the completion of anamneses and specific tracer questions at all of the recruitment centers. For example (Chaiter *et al.*, 2008), at center 1, the rates of recruits who were asked about the night vision trace and about prolonged stays abroad at specific areas that are endemic to HIV significantly increased (from 4.35% to 50% ($p=0.003$) and from 8.7% to 50% ($p=0.01$) during the years 2006 and 2007).

Noted findings concerning physical examinations included a lack of examination of the lateral motion range of the spine and an incomplete examination of heart auscultation (at one point only - left sternal border), lung auscultation (one or two points on each side), the abdomen (partial palpation of one or two quadrants), the lower extremities (especially the feet), and male genitalia (lack of examination of the inguinal canal for possible inguinal hernia or incomplete examination of testicles) (Chaiter *et al.*, 2008). Moreover, similar to anamnesis, clear differences were observed between recruitment centers. In 2002, the rate of complete examination of the abdomen ranged from 25% (center 2) to 90% (center 5) of recruits ($p=0.048$), while complete foot examinations were carried out on 38% (center 4) to 70% (center 5) of recruits. During the years following the intervention program, noticeable progress was observed in the physical examinations. For example, at center 1, a significant improvement was noted from 2006 to 2007 in abdominal examinations (from 8.7% to 75% ($p<0.001$)), foot examinations (from 47.8% to 100% ($p=0.01$)), and in male genitalia examinations (from 4.35% to 75% ($p<0.01$)).

Decision-making processes are integral and central to the medical process. Over the years, no clear trends were observed in the rates of correct decision-making procedures (Chaiter *et al.*, 2008). This may be attributed to the turnover of human resources (chairmen, physicians), and the persistence of local medical procedures at recruitment centers which are not in line with the general guidance. Nevertheless, an improvement in the decision-making processes at specific time points, or over several years, at specific centers was associated with cooperation with the quality control and assurance system.

Together, better medical history recording, physical examinations, and decision-making processes by the physicians of the medical committees were noted.

Data mining and processing by computerized reports revealed a major decrease in the rate of discrepancies between the medical information (such as anthropometric and basic measurements data) and FCCs or medical profiles (Chaiter *et al.*, 2008).

Analysis of medical profile distribution and the prevalence of FCCs, the indicators of medical conditions among recruits, uncovered significant differences between recruitment centers (Chaiter *et al.*, 2010). Analyzing all of the FCCs revealed the 26 most common FCCs, which comprised approximately 90% of all assigned FCCs. Almost 90% of these common FCCs (23 out of 26) were found to vary significantly between different recruitment centers. Data stratification according to ethnic origin did not affect the results (Conscription Administration Data, 2001-9). These 26 FCCs include overweight, underweight, anemia, asthma, cardiac anomalies (either valvular or non-valvular), hypertension, varicocele/hydrocele, epilepsy, mental illness conditions (personality disorders, neurosis, psychosis, depression, mental retardation, and autism), hernia, visual acuity, allergic rhinitis/sinusitis, and flat feet. Some of these FCCs (such as bee sting allergy; anemia; valvular and non-valvular cardiac anomalies, including mitral valve prolapse; hypertension; hydrocele/varicocele; flat feet; hernia; hearing loss; visual acuity problems; and color blindness) were found not to significantly affect the final profile outcome. Yet, a set of only 8 FCCs (those indicating recruits who were underweight or suffered from asthma, chronic headache/migraine, mental illness, scoliosis/kyphosis, chronic back pain, knee joint disorders, or rhinitis/sinusitis) accounted for 90% of the medical profiling differences between recruitment centers (Conscription Administration Data, 2001-9). Of these key profile-affecting FCCs, the prevalence of all of them except scoliosis/kyphosis and mental conditions was found to be 1.5 to 2.5 times higher at recruitment center 1 as compared to the other centers (Chaiter *et al.*, 2010).

Over the years, significant trends were observed: the prevalence of chronic headaches (increased), symptomatic scoliosis/kyphosis (increased), and active asthma (decreased) in recruitment center 1 (Table 2). Interestingly, a trend toward a significant increase in prevalence until the year 2005 and then a significant decrease in prevalence until the year 2009 was common to all other FCCs, except for that of underweight (Table 2).

Medical condition	Year									Correlation
	2001	2002	2003	2004	2005	2006	2007	2008	2009	p value
Chronic headache	4.31	4.82	4.88	5.77	6.66	6.21	6.06	6.24	6.17	0.820, <0.01
Symptomatic scoliosis/kyphosis	1.90	3.41	3.93	5.35	5.52	4.88	4.91	5.16	4.72	0.709, <0.05
Active asthma	11.29	9.29	8.66	9.62	9.94	9.54	8.52	8.94	8.12	0.684, <0.05
Under-weight	4.44	4.56	4.91	5.49	4.79	3.57	3.44	3.85	4.14	0.558, n.s.
Chronic back pain	8.32	7.99	8.27	8.52	9.03	8.50	7.28	7.63	7.50	0.515, n.s.
Allergic rhinitis/ sinusitis	11.89	11.75	11.19	12.39	12.86	12.31	11.18	11.19	10.30	0.471, n.s.
Knee joint disorders	3.62	3.19	3.79	4.37	4.94	4.16	4.24	4.02	3.09	0.082, n.s.
Mental conditions	9.83	8.17	9.14	8.94	11.13	10.20	10.03	9.75	8.10	0.040, n.s.

n.s. - not significant

Table 2. Percentages of main medical conditions contributing to a difference in profiling at recruitment center 1 from 2001 to 2009 (Modified with permission from Chaiter *et al.*, 2010).

Some of these 26 FCCs were found to vary significantly between sub-districts in all of the recruitment centers (data not shown and Chaiter *et al.*, 2010). This may be attributed to both demographic-environmental parameters and professional-human causes. Across recruitment centers, and over the years, clear and significant disparities were found in the prevalence of how a majority of these FCCs were assigned by different medical committee chairmen in the year 2006 (data not shown and Chaiter *et al.*, 2010). This further strengthens the supposition that there are differences among recruitment centers and among the chairmen's reporting of medical histories, performing physical examinations, and interpreting various medical conditions, all of which lead to differential assignments of FCCs. Differences in FCC assignment by different chairmen at the same recruitment center indicate decision-making disparities or lack of experience and expertise in specific fields. The intervention program led to an improvement in certain aspects of the chairmen's knowledge and skills and, as a result, to an improvement in the performance of the committees; while variations between chairmen and between recruitment centers still exist, it is to a slightly lesser extent. The impact of professional development and understanding of the whole medical process (see below), as well as the establishment of a uniform working platform, resulted in decreased variability of medical conditions in the various conscription centers and among physicians. However, it could have been greater if the turnover of professional medical human resources would have been lower.

It is noteworthy that a subset of the 26 most commonly assigned FCCs were found to vary significantly among all three stratification criteria (recruitment centers, sub-districts, and chairmen assigning the profiles) (data not shown and Chaiter *et al.*, 2010). The common characteristic of these FCCs (such as underweight, asthma, chronic headache, symptomatic scoliosis/hypnosis, chronic back pain, knee joint disorders, and allergic rhinitis/sinusitis) is that their assignment procedure is prone to a relatively high degree of variation in anamnesis, examination, chairman discretion, and interpretation.

3.2 Technical assistant staff performance

The work of more than 110 technician assistants and laboratory staff members was assessed by questionnaires, interviews, and direct observation of their performance before, during, and after the medical committee examinations (Conscription Administration Data, 2006-9; Chaiter *et al.*, 2010; 2011). Prior to the intervention program, insufficient knowledge was revealed mainly with regard to the normal range of values for blood pressure and pulse, urinalysis, visual acuity, and color vision and to the interpretation of abnormal values of these measurements. Furthermore, some technician soldiers suffered from inadequate execution of the techniques, such as incorrect weight and height measurements. Inadequate knowledge regarding the relationship of all of the above mentioned measurements to specific medical FCCs was found. Some of these findings (insufficient knowledge, technical skill, and their relationships and meanings) were common to the technical staff at all of the recruitment centers, while other aspects were evident at specific recruitment centers.

After the intervention program, a higher level of expertise, increased skills in measuring medical parameters, and a more accurate interpretation of these values were observed among the technical staff. Improvements were found in the measurement techniques of weight, height, color vision tests, and determining of visual acuity, as well as an increased understanding of the normal parameters of these and other measurements, such as pulse and urinalysis, and their interpretation and relationship to medical FCCs (Chaiter *et al.*, 2011). For example, in recruitment center 2, a significant and sustained improvement was observed in the interpretation of low systolic or diastolic blood pressure and in the determination of color blindness and other issues, and in recruitment center 1 there was a higher rate of correct measurements for weight, height, and visual acuity. The number of medical inconsistencies was progressively and dramatically reduced.

3.3 Administrative staff

During the years 2007 to 2009, 23 visits were carried out to assess the work of the administrative medical departments at the recruitment centers. During these visits, the work of almost 200 medical administrative personnel and managers was analyzed. The main findings (Navon *et al.*, 2011) include incomplete knowledge of medical-administrative processes (such as appointments and statuses), a lack of professional collaboration between medical departments at different recruitment centers that was inevitably caused by differences in working patterns and operational procedures at all of the centers, a partial management of diaries with abnormal laboratory results, and local procedures which deviated and were not in line with regulations and instructions (such as those related to recruit identification or the management of medical questionnaires that were received from primary care physicians). Furthermore, in some cases, the managers of the medical administrative departments were only partially or inappropriately trained for their duty, and therefore their performance was far from optimal during the first period with regard to the professional-administrative-medical aspects and the management of human resources.

As a result of the intervention program (Navon *et al.*, 2011), improvements were found in all of these aspects of knowledge, skills, working procedures, professional collaboration, and management. This led to a considerable decrease in the rates of errors in records, such as inconsistencies in medical-administrative information between the status and appointment systems, and a significant increase in the rate of proper and complete administrative medical processes and profiling processes. Nevertheless, a significant increase was observed in the rate of recruits with medical profile that did not take into account the information written in the medical questionnaires that were filled out by their primary care physicians.

3.4 Computer-based tools – not just for quality control and assurance

The quality control and assurance system and the local medical departments at the recruitment centers continuously analyze data from the computerized system (mainly by means of reports) in order to assess and evaluate the performance of the medical and administrative processes, as well as to identify errors and discrepancies in individual medical files. In light of the findings, the medical committees and departments take action to correct mistakes, follow up specific populations, etc. Moreover, reports can uncover difficulties and vulnerabilities in global processes, which in turn lead to further improvements or modifications of either the medical procedure or the computerized medical database and system.

The combined computerized system (including the appointment system, the status system, and the directing, monitoring, and controlling system) improves the control and management of the medical processes and informatics from the point of view of both the patients and the system operators (Machluf *et al.*, 2011). Different parameters of quality control regarding the medical and administrative processes are assessed (such as efficiency), and solutions are sought. Computerized system-based design and re-allocation of human and medical resources are implemented according to the capacities and limitations of the medical system. For instance, at recruitment center 1, reports during late '90s revealed a significant number of recruits waiting for specific medical processes (ophthalmologist, cardiologist, pulmonologist, orthopede, neurologist, etc.), and the intervention led to a reduction of at least 50% in the number of recruits holding a specific status over a period of 16 months; this reduction was achieved for most (12) of these statuses (15) (Machluf *et al.*, 2011). In addition, analysis revealed the specialists in each recruitment center for which recruits wait long periods of time for successful completion of the process. In all of the recruitment centers, a higher number of recruits who were waiting for more than 3 months was found with recruits who were in the process of medical documentation (centers 1 and 5) or who were waiting to see a neurologist (centers 2-4). Also in this case, computer-aided planning and re-allocation of human and medical resources played a key role in the intervention and in the solutions found for the specific specialists and recruits at each center (Machluf *et al.*, 2011). Another parameter regarding both quality control of the medical process and quality service to the recruits is the number of attendances at a recruitment center for each recruit until the final profile is assigned. Among the recruits who received their medical profile during 2010, approximately two-thirds of the recruits were required to report to the medical departments up to two times (Fig. 3). This rate was better in recruitment centers 2, 4, and 5 (>70% of recruits). On the other hand, a considerable fraction of recruits were obligated to attend six times or more, especially in recruitment centers 1 and 5 (2.5% and 2.3% of recruits, respectively) (Fig. 3). In some cases, the profile was assigned but no arrival to a recruitment center was recorded. These findings, and others, are the basis to analyze the medical and administrative aspects of the medical process, such as coordinated invitations, unnecessary or insufficient investigations, and the proper recording of arrivals.

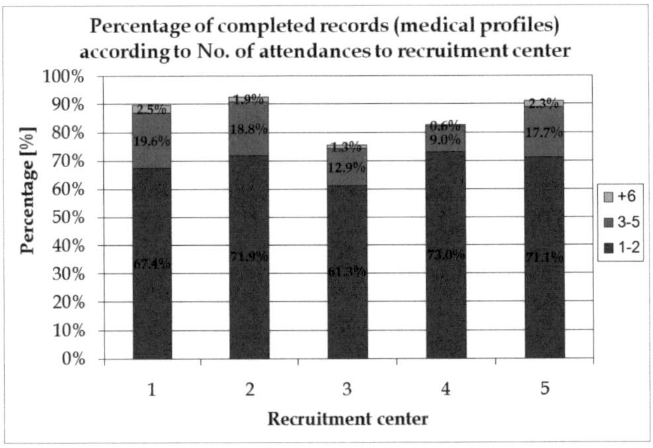

Fig. 3. Analysis of the number of attendances to a recruitment center in completed records during 2010.

The combined computerized system together with reports and automated tools allows the management of populations with special needs, such as mentally retarded and cerebral palsy patients. According to the regulations, these special populations are exempt from reporting to the recruitment centers, and their files are managed with maximum discretion and sensitivity to the individual, thus respecting them and their families' wishes.

Based on data mining findings, modifications were also introduced in working procedures. A reduction in the daily number of invited recruits improved the quality of the medical encounters. Specific combined status codes were introduced for the efficient planning of the medical encounters. Implementation and automation of medical regulations and procedures within the computerized system cause this system to play a key role and serve as a control tool during the decision-making process.

4. Intervention

The quality control and assurance system operates via an analysis→ design→ implementation→ evaluation→ modification loop.

First, all medical processes were mapped and analyzed, and particular infrastructure, medical, technical, management, and administrative needs were characterized by experts. The design principles of the intervention program with regard to the components, mode of action, time lines, etc. were determined to address those needs, in line with the general medical processes and goals. Then, a small-scale pilot program was launched in one recruitment center and was systematically analyzed. Throughout these steps, the quality control and assurance system personnel collaborated with the experts and experienced staff members from the medical departments and with policy makers. This collaboration allowed for better intervention planning, an efficient feedback process, contributed to the participants' sense of ownership and commitment to the process, increased their confidence, and made them receptive to the intervention program. After the required modifications were made, the same procedure was carried out as a large-scale intervention program that was gradually implemented at other recruitment centers, until full implementation of the system was achieved throughout the whole organization. The gradual implementation process allowed for both the full support of each recruitment center and an assimilation of adjustments to specific local needs.

It is noteworthy that acceptance of and cooperation with the quality control and assurance system were not trivial. During the first years of the program, physicians and policy makers, and to a lesser extent technical assistants and administrative staff members, were reluctant to comply with the quality control and assurance activities, during all of the stages and particularly during the intervention. The main reasons for that were professional autonomy and time/procedural requirements. This phenomenon and its causes were found also in France (Giraud-Roufast & Chabot, 2008). To build trust and cooperation, except for the fact that it was an obligatory process, few key actions were taken. First, the main goals of performance assessment were to improve professional competence, rather than to take sanctions. Medical personnel were considered partners, taking part in activities and brainstorming (mainly during the design and implementation stages). During intensive educational activity, the benefits (professional development, money, time, resources, quality care, etc.) to the individual service-providing staff (physicians, managers, etc.) and to the patients (quality of service and care, safety, etc.), to the medical departments, and to the organization were highlighted. Gradually, quality control and assessment became an integral component of the medical department's work, both routinely by the medical personnel and occasionally by the quality control and assessment system.

Except for the physical condition issues, three main needs were acknowledged in every aspect of the medical processes, in the medical department: uniform medical and administrative processes at all recruitment centers; proper and comprehensive acquisition of medical, technical, and administrative knowledge, expertise, and skills; and sharing between all conscription centers. In light of these needs, an intervention program was designed. It utilizes various means:

Forums of organization leaders: A forum was established, which includes leading physicians, medical administrative personnel, and the managers of the medical departments. Meetings are held on a regular basis in order to update all involved personnel on the novelties introduced into the medical system, to exchange ideas, and to formulate recommendations (and thereafter their implementation) for ongoing quality assessment and improved working procedures. This led to the development and production of manuals for administrative and technician soldiers, the printing of a catalogue of medical equipment, and the initiation of theoretical and practical training programs.

Training and simulation center for physicians: A training program was implemented for physicians, which includes lectures, clinical training at a simulation center, and continuous observation and feedback on their work. Each chairman physician is invited once a year to participate in a workshop that provides simulated scenarios of patient-physician encounters. Each scenario is played by specially trained actors performing the role of the recruits. A detailed anamnesis, a recording of the findings, and further investigation is performed by the physician. Each encounter is recorded on video and is then discussed in detail with the special training team to point out mistakes and inconsistencies in the process. Each physician receives a personal feedback summary and accreditation is given for participation in the course.

Instruction of physicians: After being observed, the physicians are instructed and trained in all issues assessed by the quality control and assurance system. Proper physical examination, partial or complete, is demonstrated upon request or, if necessary, is based on the findings.
Manual for technical medical staff performance: A comprehensive manual was written describing measurement techniques that are carried out by the technical assistants (and laboratory staff) at the medical committee examinations. It was distributed to relevant personnel and is used on a daily basis. The manual also contains information about the normal range of systolic and diastolic blood pressure and pulse measurements and correct interpretation of visual acuity and color vision tests (Ishihara and D15), as well as about the interpretation of abnormal values with instructions of how to act if an abnormal value is encountered during measurement or is written in the medical committee protocol. The manual contains information that involves the technician soldiers in the coordination and quality assurance processes of the medical committee examinations. For instance, the technician soldiers are instructed to return a file to the physician if they find an abnormal blood pressure value recorded in the protocol of the medical committee without any instructions from the examining physician on how to proceed or if the physician determines the profile of the recruit without assigning a FCC that indicates hypertension.
Frontal lectures: All technical medical personnel and administrative staff at all of the recruitment centers were given lectures at each recruitment center and in special meetings arranged at the medical assessment branch of the IDF Medical Corps. The issues discussed in the lectures further stressed what was described in the written manuals and also emphasized cases of risk management in order to strengthen the notion about the importance of the technical medical staff's work and the administrative processes as part of the quality assurance of the medical committee examinations and as assistants of the physicians in the process of medical profiling. During some of the lectures at the recruitment centers, training in measurement techniques, such as blood pressure measurements, was performed. In addition, the relationships and links between the medical, technical, and administrative processes were highlighted.
Instruction of technical medical staff: After the observations, questionnaires, and interviews were conducted, the technician soldiers were given detailed feedback and were instructed and trained in all of the issues that were assessed by the quality control and assurance system. In addition, all problematic areas that were identified by the system were discussed with the medical department managers.
Instruction of administrative staff: After the observations, questionnaires, and interviews were completed, the administrative staff were given detailed feedback and were instructed and trained in all of the issues that were assessed by the quality control and assurance system. In addition, all problematic areas that were identified by the system were discussed with the administrative staff's managers.
Written reports: Written reports summarized the findings with an emphasis on the recommendations required to make improvements and to correct the mistakes found in each recruitment center. The reports were distributed to the relevant medical and management authorities at the local recruitment centers and at headquarters. These reports also allowed a comparative overview between medical departments at different recruitment centers and between different time periods or specific assessments.
Computerized tools: The quality control and assurance system plays a key role in the design and planning of computerized systems or in their modifications so that the medical department's needs, mainly with regard to procedures and regulations, will be answered.

The quality control and assurance system also gives complementary support to the medical department with implementation, adjustments, and instruction in the proper use of the computerized tools in the medical processes for the purpose of higher quality of control, management, and service. The computerized system, through its implementation and automation of medical regulations and procedures, plays a central role and serves as a control tool during the decision-making process and as a way to prevent inconsistencies in the medical information. In addition, electronic medical files were incorporated into the computerized system.

Certification: Certification is virtually awarded for the completion of requisite training/instruction by accredited authorities, including the quality control and assurance system. It is important to note that, unlike most licensures and board certifications (Landon *et al.*, 2003), the qualification process and examinations are periodic and specific to particular medical fields and ensure professional competency according to pre-determined standards.

5. Effects of intervention

During the years of intervention, benefits were observed in the quality of all of the areas that were examined. As these effects were described before (see the section entitled Findings) or earlier (Chaiter *et al.*, 2008; 2010; 2011; Machluf *et al.*, 2011; Navon *et al.*, 2011), we can summarize them in a few main categories: 1) infrastructure and physical conditions; 2) consistent improvement in the knowledge, skills, and performance of the physicians (anamnesis, examination, decision-making processes, etc.), technicians (measurements, recording, interpretation, etc.), and administrative staff (procedures, regulations, etc.); 3) uniform working platforms and procedures in terms of the medical and administrative processes at all of the recruitment centers; and 4) a launching of the framework and forums for sharing knowledge and skills between all of the conscription centers.

One important contribution of the quality control and assurance system is in terms of the diagnosis, design, and implementation of the intervention program and its analysis. The impact on the medical department's performance is clear and evident. Furthermore, after the intervention program's implementation, the satisfaction rate, sense of belonging, and responsibility were higher among all of the medical department personnel. It led to the increased perception of the administrative and technical medical staff to feel that their work was part of the medical profiling process, acting as case managers and part of quality assurance aimed at providing the best medical service for recruits. Therefore, it is only natural that this would consequently lead to a consistent, major increase in the satisfaction level reported by recruits.

As mentioned above, not all of the goals were achieved fully, and in some respects there was a disparity between what was achieved and the desired objectives. Clearly, some changes are time-dependent, and the benefits of current efforts will be evident in the near future. Other challenges may require different solutions. We are always looking for new and better tools and modes of action to achieve these goals.

6. Epidemiological aspects

The medical processes for the adolescent population of Israel present a unique opportunity to assess the health status of the young Israeli population on a nationwide level and to identify risk factors that can affect present and future morbidity.

A pilot study, including over 105,000 adolescents, has been carried out using data from one of the recruitment centers to assess trends in weight, height, and other medical parameters (e.g., blood pressure) and conditions (e.g., congenital heart disease) among 16-19 year olds in Israel (born in 1971-1992). Our preliminary analysis suggests that variability between recruitment centers may affect the consistency and reliability of nationwide analysis, as opposed to information from a single recruitment center. Furthermore, in one selected center, the differences between chairmen in assigning profiles and FCCs were much smaller than in other centers. Clear trends were demonstrated, and their association with demographic variables was examined. The findings are out of the scope of this manuscript, yet we wish to illustrate two possible insights drawn from such analysis.

6.1 Anthropometric values

Among the young (16-19 year olds) Israeli population born between 1971 and 1992, a significant increase in average body weight was demonstrated in females and mainly in males (Fig. 4, upper panel). In both genders, the increase in average weight is more dramatic in teenagers who were born during 1982-1992. On the other hand, no dramatic change was observed in average height in females or in males (Fig. 4, lower panel). This might be an indication of an increase in the body-mass index (BMI).

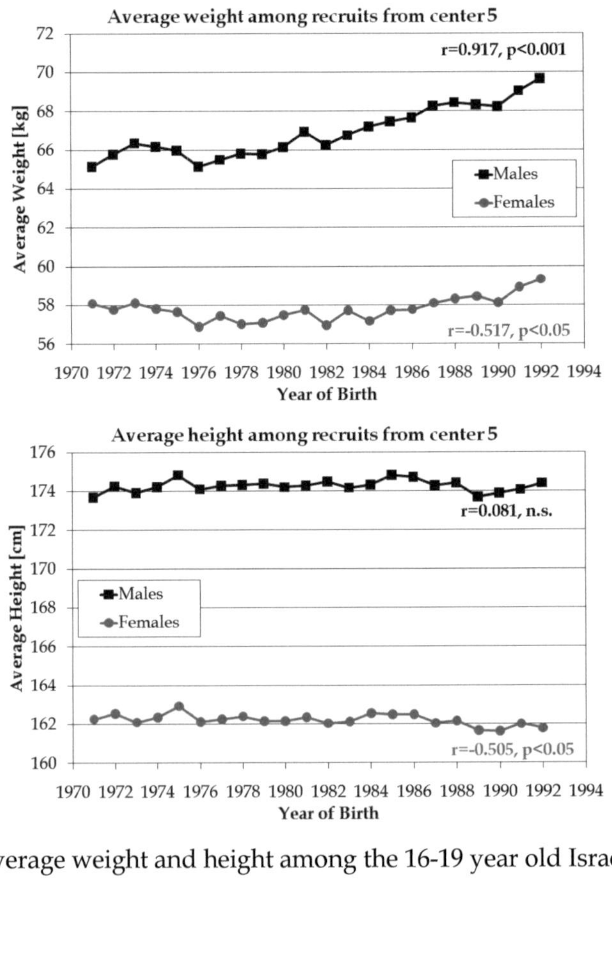

Fig. 4. Trends of average weight and height among the 16-19 year old Israeli population.

General population data and military records suggest that US heights essentially remained stable after World War II (mainly in birth cohorts of 1955-74), which is concurrent with continual rapid increases in height in Western and Northern Europe (Komlos & Lauderdale, 2007). A historical retrospect of German military male recruits found an increase in both the average body weight and height, where the changes in body weight were greater than those of body height (Jaeger *et al.*, 2001).

The implications of these findings with regard to different demographic parameters, socioeconomic status, anthropometric indices, medical conditions, and other risk factors are now under investigation.

6.2 Medical conditions

Analysis of the prevalence of specific FCCs provided the opportunity to gain knowledge about the trends of different medical conditions. Such an analysis in recruits from center 5 uncovered an increase in the prevalence of valvular and non-valvular congenital heart anomalies among male recruits (Fig. 5. upper panel). Referring to cardiac diseases, the prevalence of congenital valvular heart disorders was higher than the prevalence of non-valvular heart disorders. Furthermore, an increase in the prevalence of solid tumors in both males and females was also demonstrated (Fig. 5. lower panel).

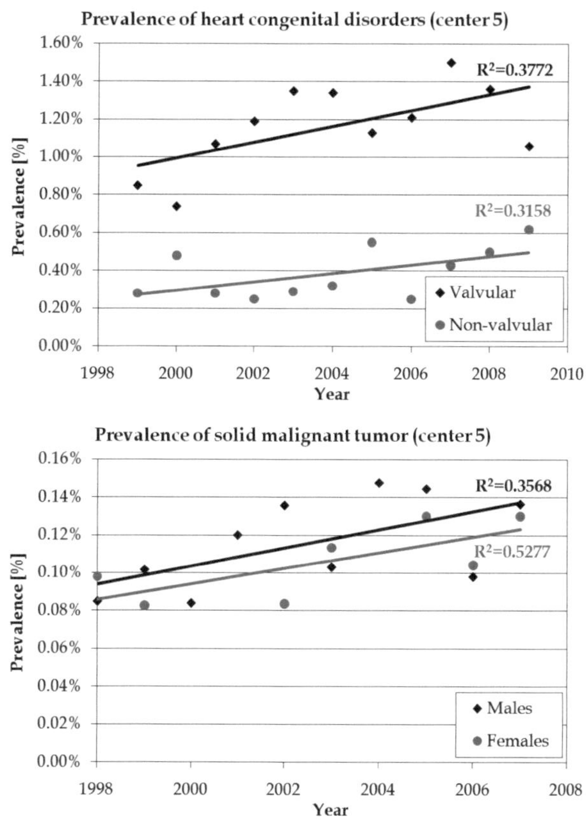

Fig. 5. Prevalence of solid tumors and congenital heart anomalies among the 16-19 year old Israeli population.

Previous analyses of specific morbidity prevalence in Israeli conscripts from all recruitment centers also revealed an increase in the prevalence of heart defects among male recruits (Farfel *et al.*, 2007), as well as a higher prevalence of congenital valvular heart disease compared to non-valvular heart disease (Bar-Dayan *et al.*, 2005).

7. A nationwide program

These findings demonstrate the need for a nationwide intervention program to reduce morbidity, future illness, and even mortality. Furthermore, a project of information sharing and cooperation was established with family physicians at primary clinics on a nationwide basis, the Ministry of Health, the National Insurance Institute, the Israeli National Cancer Registry, and the Ministry of Social Affairs and Social Services. This national project (Fig. 6) is aimed at education, prevention, and early intervention in target populations.

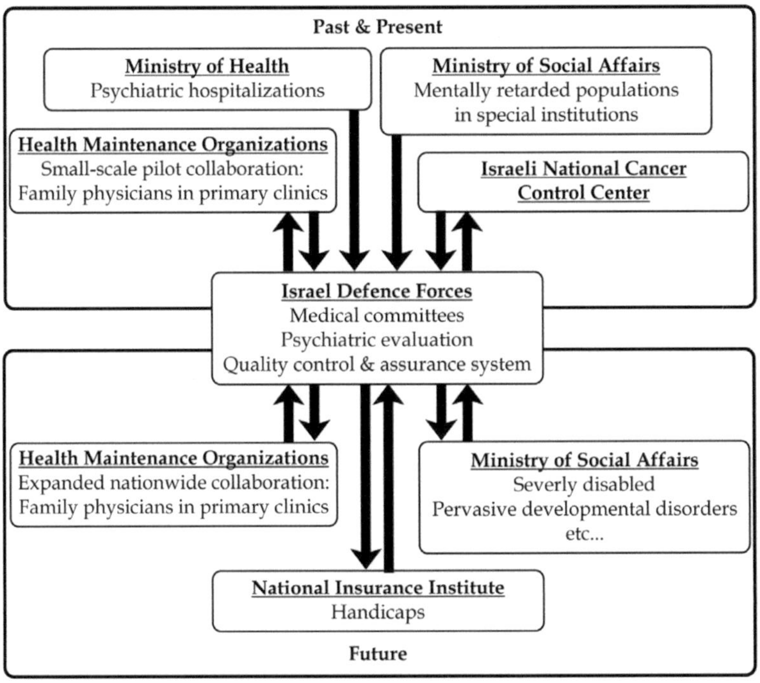

Fig. 6. Scheme of national project for information sharing and collaboration.

So far, although this information-sharing project was only initiated on a small-scale pilot format, covering only a minor fraction of the relevant population, bi-directional benefits are evident. For example, medical investigations in recruitment centers uncover medical conditions that were unknown before to the civilian authorities and vice-versa. These medical conditions include vision and hearing problems, essential hypertension, asthma, cardiac anomalies, tumors, urological conditions (hernia, varicocele, hydrocele, etc.), nephrological disorders (nephropathies, microscopic hematoria, severe proteinuria, etc.), orthopaedic problems, neurological problems, and mental disorders. Such findings are then reported to the primary care physician for further investigation with a referral to specialists and/or for treatment. The medical departments at the recruitment centers are then informed about further findings and about any change in the medical condition and status of the recruit by the primary physicians. Beyond the issue of medical informatics and process, recruits are provided with better medical service, which might contribute to early diagnosis, quick and appropriate treatment, and better and faster recovery. In the future, if this information sharing is combined with an epidemiological study, we believe it may also contribute to the prevention of health conditions, either via the medical education of teenagers or by active medical intervention.

8. Concluding remarks

Quality assessment, control, and improvement in a health system should include the infrastructure and structure, process, and outcome levels. The quality control and assurance system for medical committees at recruitment centers operates via an analysis→ design→ implementation→ evaluation→ modification loop. It first relies upon the identification of the limitations and needs of the whole medical process and of each department with regard to the medical, technical, administrative, procedural, and physical aspects and with regard to official policy, including the systematic evaluation of the knowledge, skills, judgement, and working patterns of all of the medical personnel. To this end, complementary methodologies are utilized to provide both quantitative and qualitative analyses of daily activities and practices. Among the main tools utilized are observations and assessment, a sampling of recruits for re-examination and of records for evaluation, questionnaires, interviews, data mining and analysis by reports, and patients' surveys. Then, intervention programs are designed and implemented to answer the needs of all of the parties participating in the medical process in order to improve their quality and performance; to increase the quality of patient care; and to achieve a higher patient satisfaction rate. Intervention programs include the establishment of a training and simulation center, lectures and instruction to all of the medical department's personnel, a forum of organization leaders, production of manuals, certification, written reports, and the design of computerized tools. These intervention programs and their impact need to be continuously evaluated and modified according to the specific needs and effects in particular recruitment centers. Significant improvements have been observed in various key parameters, such as the knowledge, skills, and judgement of the personnel and their professional performance, the conditions of their working environment, uniform working platforms, and the patients' satisfaction rate. Incessant monitoring and intervention are important to maintain quality in a medical organization.

Quality improvement at the individual (physicians, assistant technicians, administrative staff, managers, etc.) and global levels (performance, outcomes, physical conditions, procedures and regulations, etc.) is a goal in and of itself but is also a means of improving patient care and safety. Furthermore, the collaboration of all the participants in the process, on all levels - medical professionals (physicians, experts, etc.), technical assistants, administrative staff, and the active support and involvement of the managing authorities and policy makers - is a critical determinant for a successful outcome.

The successful application of a quality control and assurance project can lay the foundation for a population-based investigation, namely an assessment of the health status of the young Israeli population, by measuring anthropometric values and the prevalence of various medical conditions among recruits on a nationwide level. They may result in the identification of risk factors that can affect present and future morbidity. A national project

of information sharing and cooperation was set up, aimed at screening, education, prevention, and early intervention in target populations.
There are multi-directional effects and relationships between the quality assurance and control process and its implications on improving the quality of health care and nationwide projects of preventative medicine through collaboration and information sharing (Fig. 7).
Recently, Zalmanovitch and Vashdi (2010) proposed an inherent trilemma, a trade-off between three desirable objectives (Iverson & Wren, 1998), in any debate on health care policy. In this context, the critical broad objectives are quality, funding, and coverage. In this context, quality refers to the efficiency and effectiveness of the health care services provided; funding refers to the public expenditures for health care that are incurred by taxpayers; and coverage refers to the percentage of a country's population eligible for state health care services and the comprehensiveness of these services. A trade-off means that, at most, only two of the three objectives can be satisfied simultaneously, and satisfying any two will always come at the expense of the third (Zalmanovitch & Vashdi, 2010). The cumulative experience of the quality assurance and control system and its effects suggest that it is central to the successful balancing between the three objectives, at least in the context of the medical departments at the recruitment centers. Considering the common characteristics of medical processes at the recruitment centers and at primary medical facilities, implementation of multi-armed quality assurance and control systems at clinics and hospitals holds great promise in finding the best solution to the trilemma, beyond the direct and clear impact on the medical staff's performance, medical process outcomes, service quality, and patient safety. Together with an epidemiologic investigation and preventive action, this system may contribute to the identification of risk factors and a reduction in future morbidity.

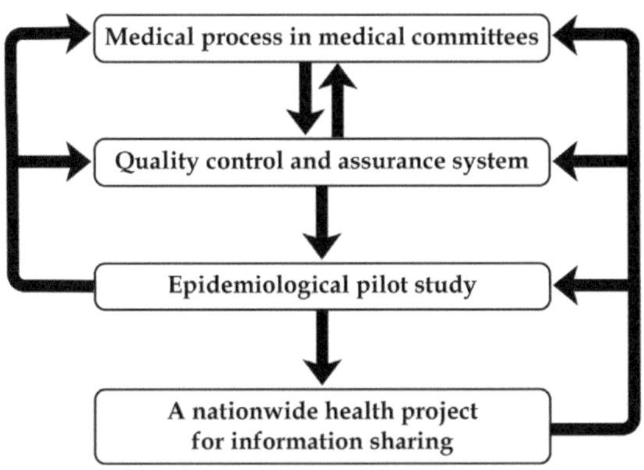

Fig. 7. Establishment of a quality control and assurance system for medical processes in recruitment centers led to an epidemiological study and a nationwide health project.

It remains to be seen in future studies whether the system can efficiently address these issues.

9. References

Blumenthal, D. (1996). Quality of health care part 1: Quality of health care - what is it?. *The New England Journal of Medicine*, Vol.335, No.12, pp. 891-4.

Bar-Dayan, Y.; Elishkevits, K.; Goldstein, L.; Goldberg, A.; Ohana, N.; Onn, E.; Levi, Y. & Bar-Dayan, Y. (2005). The prevalence of common cardiovascular diseases among 17-year-old Israeli conscripts. *Cardiology*, Vol.104, No.1, pp. 6-9.

Campbell, S.M.; Roland, M.O. & Buetow, S.A. (2000). Defining quality of care. *Social Science and Medicine*, Vol.51, No.11, pp. 1611-25.

Chaiter, Y.; Machluf, Y.; Pirogovsky. A.; Palma, E.; Yona, A.; Shohat, T.; Yitzak, A.; Tal, O. & Ash, N. (2010). Quality control and quality assurance of medical committee performance in the Israel Defense Forces. *International Journal of Health Care Quality Assurance*, Vol.23 No.5, pp. 507-15.

Chaiter, Y., Palma, E., Machluf, Y., Yona, A.; Cohen, A.; Pirogovsky, A.; Shohat, T.; Ytzhak, A. & Ash, N. (2011). Quality assuring intervention for technical medical staff at medical committees. *International Journal of Health Care Quality Assurance*, Vol.24, No.1, pp. 19-30.

Chaiter, Y.; Pirogovsky, A.; Palma, E.; Yona, A.; Machluf, Y.; Shohat, T.; Farraj, N.; Tal, O., Campino-Abbebe, G. & Levy, Y. (2008). Medical quality control in conscription centers- ten years of activity. *Journal of Israeli Military Medicine*, Vol.5 No.2, pp. 75-9.

Conscription Administration Data. (2001-9), *Biannual control on medical profiles in recruitment centers*, Personnel Directorate, IDF, Israel (unpublished data analysis).

Donabedian, A. (2003). *An introduction to quality assurance in health care*, Oxford University Press, ISBN 0195158091, New York, USA.

Donabedian, A. (2005). Evaluating the quality of medical care. *Milbank Quarterly*, Vol.83, No.4, pp. 691–729.

Dye, T.R. (1987). *Understanding Public Policy* (6th edition), Prentice-Hall, ISBN 0139369732, Englewood Cliffs, New Jersey, USA.

Farfel, A.; Green, M.S.; Shochat, T.; Noyman.; Levy, Y. & Afek, A. (2007). Trends in specific morbidity prevalence in male adolescents in Israel over a 50 year period and the impact of recent immigration. *The Israel Medical Association Journal*, Vol.9, No.3, pp. 149-52.

Giraud-Roufast, A. & Chabot, J.M. (2008). Medical acceptance of quality assurance in health care, *The Journal of the American Medical Association*, Vol.300, No.22, pp. 2663-5.

Holmboe, E.S.; Lipner, R. & Greiner, A. (2008). Assessing quality of care: knowledge matters, *The Journal of the American Medical Association*, Vol.299, NO.3, pp. 338-40.

Iverson, T. & Wren, A. (1998). Equality, employment, and budgetary restraint: the trilemma of the service economy. *World Politics*, Vol.50, No.4, pp. 507-46.

Jaeger, U.; Zellner, K.; Kromeyer-Hauschild, K.; Lüdde, R.; Eisele, R. & Hebebrand, J. (2001). Body height, body weight and body mass index of German military recruits. Historical retrospect and current status. *Anthropologischer Anzeiger*, Vol.59, No.3, pp. 251-73.

Komlos, J. & Lauderdale, B.E. (2007). The mystery trend in American heights in the 20th century. *Annals of Human Biology*, Vol.34, No.2, pp. 206-15.

Landon, B.E.; Normand, S.L.; Blumenthal, D. & Daley, J. (2003). Physician clinical performance assessment: prospects and barriers, *The Journal of the American Medical Association*, Vol.290, No.9, pp. 1183-9.

Machluf, Y.; Pirogovsky, A.; Palma, E.; Yona, A.; Navon, A.; Shohat, T.; Ytzhak, A.; Tal, O.; Ash, N.; Nachmann, M. & Chaiter, Y. (2011). Coordinated computerized systems aimed at management, control, and quality assurance of medical processes and informatics. *International Journal of Health Care Quality Assurance,* Accepted for publication.

Mandel, D.; Amital, H.; Zimlichman, E.; Wartenfeld, R.; Benyamini, L.; Shochat, T.; Mimouni, F.B. & Kreiss, Y. (2004). Quality assessment program in primary care clinics: a tool for quality improvement. *International Journal for Quality in Health Care,* Vol.16, No.2, pp. 175-80.

Mandel, D.; Zimlichman, E.; Ash, N., Mimouni, F.B.; Ezra, Y. & Kreiss, Y. (2003). Quality assessment of primary health care in a military setting, *Milltary Medicine,* Vol.168, No.11, pp. 890-2.

Munro, R.A. (2009). *Lean six Sigma for the healthcare practice: A pocket guide,* Amer Society for Quality Press, ISBN 0873897609, Milwaukee, Wisconsin, USA.

Navon, A.; Machluf, Y.; Cohen, A.; Pirogovsky, A.; Palma, E.; Tal, O.; Frenkel-Nir, Y.; Ash, N. & Chaiter, Y. (2011). Quality Assurance of Administrative Aspects of Medical Processes within the Framework of Medical Committees. *Journal of Israeli Military Medicine,* Accepted for publication.

Ovretveit, J. (1992). *Health service quality: An introduction to quality methods for health services,* Blackwell Scientific Publications, ISBN 0632032790, Oxford, England.

Ransom, E.R.; Joshi, M.S.; Nash, D.B. & Ransom, S.B. (2008). *The Healthcare Quality Book: Vision, strategy, and tools* (2nd edition), Health Administration Press, ISBN 1567933017, Chicago, USA.

The State Controller and Ombudsman Office. (2002). *Annual Report 53a,* Israel.

Zalmanovitch, Y. & Vashdi, D.R. (2010). Trade-offs are unavailable. *British Medical Journal,* Vol.340, pp. c1259.

15

The Significance of Board-Certified Registered Breast Specialist of the Japanese Breast Cancer Society in Regional Medical Activities

Noriyuki Tohnosu[1,2], Jun Hasegawa[2,3], Yosio Masuda[2,4], Taku Kato[5], Satoru Ishii[6] and Kanae Iwata[7]

[1]*Department of Surgery,Funabashi Municipal Medical Center*
[2]*Breast cancer screening committee of Funabashi Municipal Medical Association,*
[3]*Funabashi Futawa Hospital,*
[4]*Masuda Clinic of breast and thyroid diseases,*
[5]*Laboratory Section of Cytology, Funabashi Municipal Medical Center,*
[6]*Radiotechnical Department, Funabashi Municipal Medical Center*
[7]*Department of Pharmacy, Funabashi Municipal Medical Center*
Funabashi, Chiba,
Japan

1. Introduction

Although the mortality of breast cancer patients in the Western countries has declined due to high screening rate, the number of Japanese breast cancer patients has seen a sharp rise and the most common cause of death of middle-aged women. Since one of every sixteen Japanese women have been diagnosed with breast cancer and more than 10,000 patients die from breast cancer every year, it is a goal to reduce the mortality through detection and treatment in its early stages. Board certified breast specialists of the Japanese Breast Cancer Society have been established in 1997 to contribute for the benefit of welfare of the nation and meet the social needs. In addition, The Ministry of Health, Welfare and Labor authorized the advertisement of specialists via the home page (http://www.jbcs.gr.jp/) in October, 2004 (Sonoo, 2005). As our institution has been designated as the region-based affiliated hospital for cancer treatment since January, 2007, the significance of the breast specialists was surveyed.

2. Breast specialists, board certified institutions and its affiliated institutions

2.1.Breast specialists

It is the minimum requirement for breast specialists to be authorized experts or qualified doctors in any one of the fields of surgery, oncological internal medicine, radiology and gynecology. The standards for qualifying breast specialists are different depending on each field. The standards for surgeons is: (1) It is required to be specialists in surgery and authorized breast doctors whose titles can possibly be acquired 4 years after graduation. (2) It is necessary to deal with breast diseases for over 7 years and experience the treatment and/or diagnosis of more than 100 breast cancer patients. (3) It is mandatory to be engaged in clinical works at the certified institutions. (4) Academic achievements on breast diseases (publications or presentations) have to exceed the compulsory score. (5) Passing the written and oral examinations is needed (Table 1). The qualification has to be renewed every 5 year.

1. Concerning surgeons, it is required to be specialists in surgery and authorized breast doctors whose titles can possibly be acquired 4 years after graduation.
2. It is necessary to deal with breast diseases for over 7 years and experience the treatment and/or diagnosis of more than 100 breast cancer patients.
3. It is mandatory to be engaged in clinical works at the certified institutions.
4. Academic achievements on breast diseases (publications or presentations) have to exceed the compulsory score.
5. Passing the written and oral examinations is needed.

Table 1. Standards of board-certified registered breast specialist.

The breast specialists have been registered from seven regional blocs in Japan: Hokkaido, Tohoku, Kanto, Kinki, Chubu, Chugoku-Shikoku and Kyusyu-Okinawa. The present number of the nationwide breast specialists has been still as small as 837 and 303 in Kanto bloc or eastern part of Japan, 40 in Chiba prefecture of Kanto bloc and 2 in Funabashi city of Chiba prefecture, respectively (Fig.1.)

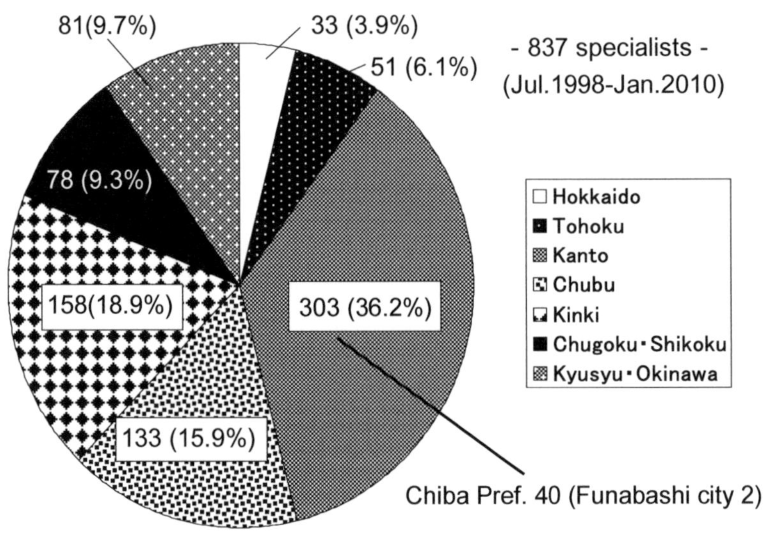

Fig. 1. Distribution of the numbers of board-certified registered breast specialists in Japan.

2.2 Board certified institutions and its affiliated institutions

In1998, the Japanese Breast Cancer Society has designated certified institutions and its affiliated institutions in the seven blocs of Hokkaido, Tohoku, Kanto, Chubu, Kinki, Chugoku-Shikoku and Kyusyu-Okinawa throughout Japan and our hospital has acted as the certified institution since then. Certified institutions have to meet the following standards (Tab. 2).

1. It is required to have beds for surgical treatment or diagnosis and/or non surgical treatment of more than 20 breast cancer patients in a year.
2. Board -certified registered breast specialists have to regularly work and adequately instruct.
3. Laboratories and libraries are well-equipped.
4. Records of anamnesis are well-written and preserved in ample care.
5. Autopsy room is equipped.
6. Instructive events on breast diseases are regularly held.
7. Publications or presentations on breast diseases have to be continued.
8. It is compulsory that board-certified registered breast specialists belonged to certified institutions instruct at the certified affiliated institutions and report its contents.

Table 2. Standards of board certified institutions.

It is impossible for doctors who aim at breast specialists to be qualified although they even practice hard in the non-certified institutions where non-certified but skillful surgeons treat many breast cancer patients. Therefore, considering the present small number of breast specialists in Japan, it is necessary for non-certified institutions to be affiliated with the regional certified institutions so that the doctors could be qualified. If the number of breast specialists is large in the future, the affiliated institutions may possibly be abolished (Sonoo, 2005,2008). The standards of the affiliated institutions are shown in Table3. The number of board-certified institutions and its affiliated institutions are 357 and 410, respectively, Kanto bloc occupying most in the former and Chubu bloc occupying most in the latter (Fig. 2).

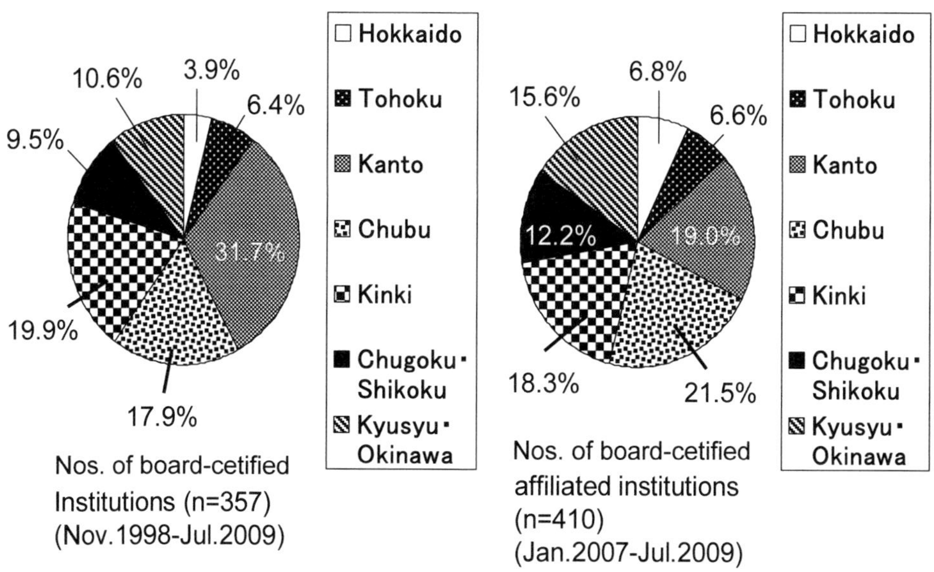

Fig. 2. Distribution of the number of board-certified institutions and its affiliated institutions in Japan.

1. It is required to have beds for surgical treatment or diagnosis and/or non surgical treatment of more than 20 breast cancer patients in a year.
2. Board -certified registered breast specialists have to regularly work and adequately instruct.
3. Laboratories and libraries are well-equipped.
4. Records of anamnesis are well-written and preserved in ample care.
5. Autopsy room is equipped.
6. Instructive events on breast diseases are regularly held.
7. Publications or presentations on breast diseases have to be continued.
8. It is compulsory that board-certified registered breast specialists belonged to certified institutions instruct at the certified affiliated institutions and report its contents.

Table 3. Standards of board-certified affiliated institution.

3. Medical activities as region-based affiliated hospital for cancer treatment

Japanese government has issued an Act of Strategy for Cancer in June, 2006 to treat the Japanese major cancers of lung, stomach, liver, colorectum and breast with equal high medical quality throughout the nation. For that purpose, affiliated hospitals for cancer treatment have been designated in each region and our institution has been the hospital since January, 2007.

The present number of the hospitals is 13 in Chiba Prefecture and the qualification has to be renewed every 5 year. The main works for breast specialists in our region are raising the percentage of examinees, reducing the percentage of detailed examinations, maintaining quality control in breast cancer screening, promoting close cooperation with the board certified affiliated hospital and community hospitals or clinics, optimal team management for breast cancer, an education of trainee doctors and providing citizens with information on breast cancer including extension lectures.

3.1 Breast cancer screening

As for breast cancer screening in Funabashi city, a mammography has been applied to women aged 40 and over since 2004. The percentage of examinees has annually risen from 8.9% in 2004 to 17.0% in 2009, whereas the percentage of detailed examinations has been almost constantly 6−7% exept in 2008 due to the effect of good quality control, the detection rate of breast cancer being 0.28% (Fig.3).

Although there were only four qualified doctors to read mammography when mammographic screening started, there have been 10 qualified doctors at present to form five teams in which a pair of two doctors reads alternately every week. On mammographic technics and knowledge, a qualified technologist at our hospital who plays a leading role in Japan has called a monthly meeting to educate technologists involved in screening mammography in Funabashi city. Breast cancer screening using stereotactic guided Mammotome has begun in our hospital by request of the community hospitals and clinics in March, 2011.

3.2 Coordination with the affiliated hospital

As far as coordination with the affiliated hospital is concerned, a conference has been held monthly to compare mammography with pathology of the postoperative cases of both

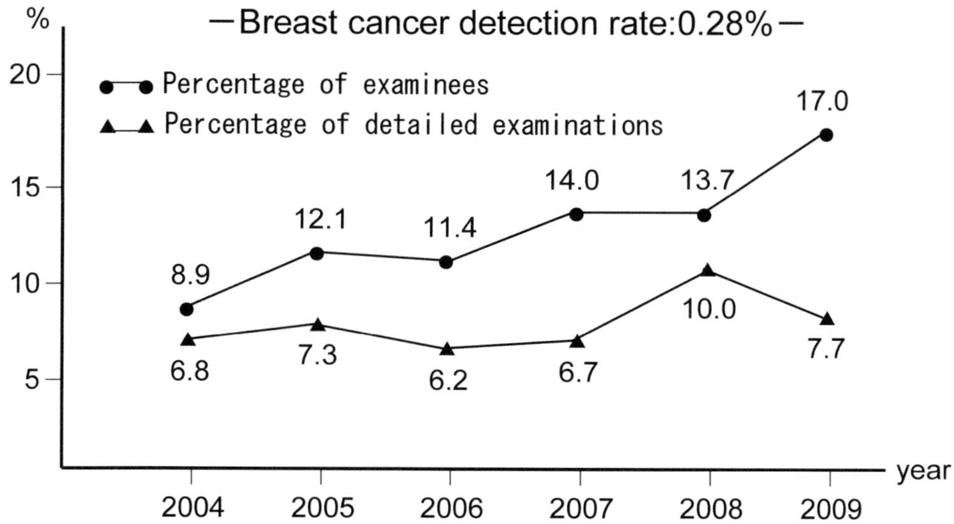

Fig. 3. Change in percentage of breast cancer examinees and detailed examinations in Funabashi city.

hospitals. Doctors including gynecologists, cytologists and technologist attend to share knowledge and information on breast cancer. In addition, breast cancer patients who want to undergo surgery at an earlier date are referred to the affiliated hospital. Those women who are indicated to screening are referred to the hospital also.

3.3 Team management

In our hospital activities on breast cancer patients, patient and the family-centered team management has been carried out and comedical staffs play an important role in each field. Apart from nurses, a pharmacist not only immediately reacts to advise doctors on optimal use of drugs whenever asked but also kindly responds to the patients soon after adverse events particularly of chemotherapy are seen. She is well informed about recent drug news both domestic and abroad on breast cancer and often has presentations even in the Japanese Breast Cancer Society. An experienced cytologist who plays a leading part in Japan is specifically of great help in the diagnosis of breast diseases both on the outpatient and intraoperative basis to contribute much for the benefit of not only our hospital but also the community hospitals or clinics which ask consultations of microscopic specimen.

Education of trainee doctors contain mammography reading, academic presentation and surgery of breast cancer. On mammographic reading ability, a third-year trainee doctor in our hospital was the first successful candidate in Chiba prefecture in the nationwide examination several years ago and it was also quite rare in Japan. Six trainee doctors of 2 general surgeons, 2 lung surgeons and 2 gynecologists have been accredited up until now.

Since present team management basically requires certified nurses (breast care nurse) in particular, it is an urgent task to have the staffs in our team although the nationwide examination is relatively difficult to pass. As for the accredited pharmacist for cancer drug, The Japanese Society of Pharmatheutical Health Care and Sciences has also adopted a board-certified system since 2006 and the standards of certified pharmacist for cancer drugs are as follows (Table 4).

1. It is required to have a career as a pharmacist for over five years and have to be a member of the Society for over five years when applied for the examination.
2. It is necessary to attend the Society or symposium of the Society more than two times.
3. Academic presentation on medical pharmacy has to be carried out as a coauthor over three times in meetings and one of them must be given as an author.
4. Publishing more than three papers on medical pharmacy is needed.

Table 4. Standars of board certified pharmacist for cancer drugs.

Strengthening a supportive system of cancer consultation is so important also that we started the team formed of a nurse who serves exclusively and three social workers in November, 2008. Concerning best supportive care, a specialist of former lung surgeon has gone into action in a newly built ward with 20 beds since April, 2009.
As the region-based affiliated hospital for cancer treatment, we will have to achieve the goal to have common clinical path of breast cancer among the community hospitals or clinics in our city and carry out breast cancer screening using ultrasound especially for the examinees aged 30-40 years.

3.4 Clinical path

A clinical path is useful for low-risk postoperative breast cancer patients in shortening time to visit and wait for receiving the same standard treatment without concentrating only in the region-based affiliated hospital . It is also beneficial for the region-based affiliated hospital to have more ample time on treating more seriously ill patients with metastases or recurrence. When the patients are operated on at the region-based hospitals and found to be at low risk (i.e. node negative, positive hormone receptors for prescription of oral medicines alone and no chemotherapy or no therapy), they are treated there with or without hormonal medicines for 1- 6 months, then are referred to community hospitals or clinics if they consent the use of path. They are checked up only with prescription and blood collection and then return to the region-based affiliated hospitals every 6-12 month for detailed examinations like mammography, ultrasound and CT or bone scan if required. If some emergencies, recurrence and/or metastases happen, the region-based affiliated hospitals have to respond immediately not only for the patients but also the community physicians (Fig. 4).

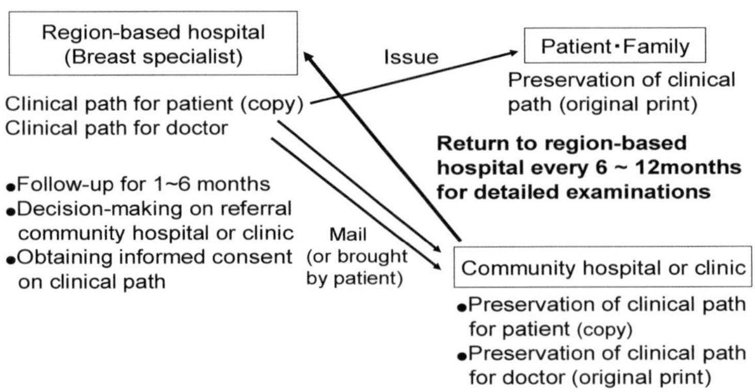

Fig. 4. Flowchart of clinical path for postoperative breast cancer patients.

3.5 Extension lectures

Together with the affiliated hospital, community hospitals and clinics, we have held four-time extension lectures for citizens on breast cancer each with certain theme at every other year since 2006 (Tab. 5.). Specifically, a lecture on liaison clinical path was given for the first time in October 2010, and the majority of the audience including men was in their 40s and 50s, younger than the age group at the previous lectures.

Year	Sponsored organization	Attended institutions
Sep. 2006	A private enterprise	Our hospital and three community hospitals
Mar. 2008	Funabashi city	Our hospital
Jun. 2008	A private enterprise	Our hospital, two ommunity hospitals and a breast clinic
Oct. 2010 *	A private enterprise	〃 〃

Table 5. Extension lectures ever held. ＊A theme 'clinical path ' was included.

4. Discussion

Quite different from Western countries, Japanese specialist system has been privately established to improve and maintain the quality of the members in each society. In 1962, specialist system for anesthetists started for the first time in the Japanese Anestheology Society followed by those for radiologists and brain surgeons in 1964. Since around mid-1970, the Japanese Society of Internal Medicine has taken the iniative not only to let the board-certified doctors and specialists be socially accepted but also to enable them to advertise in April, 2004 (Sakai, 2005). Health, Welfare and Labor Ministry has also authorized the advertisement of breast specialists through the home page in October, 2004. Although the field of breast cancer is related to not only surgery but also radiology, gynecology and oncological internal medicine, surgeons occupy 95% of breast specialists in Japan, quite unlike Western countries. Therefore, Japanese breast specialists are busy working even for rapidly-advancing chemotherapy, however, they can possibly acquire a broad range of knowledge and experience through medical activities in team management. Since there are few emergency patients, even woman doctors including surgeons and gynecologists could work by taking advantage of being female while they rear children. As the number of breast specialists is still small in Japan, they will turn out to be important human resources for Japanese breast cancer patients who have continued to rise in number (Sonoo, 2007, 2008).

Before applying to the examination for breast specialist, the following curriculum has to be finished; 1) to master general knowledge on breast diseases, clinical judgement, the ability to solve problems without regard to each expertise, 2) to master basic knowledge on anatomy, physiology, hormonal environment, epidemiology, pathology, biology of the mammary gland and breast cancer screening and to be able to clinically respond, 3) to master basic treatment technics on diagnosis by imaging modalities, aspiration cytology, core needle biopsy, biopsy using Mammotome, surgical biopsy , sentinel lymphnode biopsy and treatment with surgery, radiation, anticancer drugs, hormonal medicines, best supportive care and postoperative rehabilitation, medical ethics including informed consent, second opinion and clinical trials. 4) to master special treatment technics on breast diseases required for each expertise. 5) to actively attend conference or academic meetings, research or treat via evidence based medicine and give academic presentations on case reports or clinical study. 6) to understand the importance of medical administration involving risk management, medical cost, team management, etc. for carrying out practical medical activities (Sawai, 2006).

The number of board certified institutions and their affiliated institutions has seen a gradual rise and the largest number is centered on Kanto bloc or eastern part of Japan as well as breast specialists. There have been only two breast specialists in Funabashi city with a population of 600,000, therefore it is essential to increase the number of breast specialists for treating the steadily increasing breast cancer patients.

Under a guidance of The Ministry of Health, Welfare and Labor, nationwide mammographic breast cancer screening for women aged 50 and over has been introduced at intervals of two years since April, 2000 followed by the screening for women aged 40-49 since 2003 (Ohuchi, 2007). However, the percentage of examinees aged 40 and over has still been as low as 20.3% in Japan and the percentage of mammographic screening at the age of 50-69 was 23.8% compared to 60-90% in Western countries in 2006 (OECD Health Data 2009). According to the Japanese government statistics in fiscal year 2008, the nationwide average percentage of examinees is as low as 1.5% at mammography screening alone and 7.6% at combined mammography screening with palpation. The Ministry of Health, Welfare and Labor has started to distribute free coupons to raise the percentage up to 50% since October, 2009. The average rate of using coupons has been approximately 30% at the age of 40-69 (Japanese government bulletin, 2009). There have been some reports showing the effect of distributing free coupons to improve the percentage of examinees (Komoto & Ishiyama, 2010). Similarly, the percentage of examinees was raised from 81% to 88% in some age groups after the introduction of free of charge in Finland (Kauhava, 2008). Whereas the percentage was improved from 18.8% to 40.7% in Sendai, 2011's quake and tsunami-hit Miyagi prefecture, northern part of Japan through not using coupons but making various efforts of increase in consultations with universities or institutions to exchange views, women's cancer screening promotion project and distribution of application form to all houses (Satake, 2011). The percentage of detailed examinations and the detection rate of breast cancer screening in 2008 in Japan is 8.6% and 0.32%, respectively (Japanese government report, 2010). Our series have shown almost same as the nationwide data.

In order to maintain a quality control not only on facility but also personnel qualification of mammography reading ability for doctors and mammographic skills for technologists, The Central Committee on Quality Control of Mammographic Screening was launched as a non-profit organization in November, 1997 and has offered nationwide training seminars frequently including examinations since November, 2000 (Tsunoda, 2008 & Endo, 2009).

The examination of reading mammography includes 100 cases half with two views and candidates have to fill out the marksheets for judging categories within a limited time of 100 minutes. The results of examinations are classified into the ranks of A, B, C , D and A+B are certified. Rank A is accepted as the ability of reading and teaching of screening mammography. The possibility of obtaining rank A in a review of rigorous testing is less than 10 percent of the total candidates. The Committee has not only accredited the doctors and technologists involved in breast cancer screening program but also given the members examinations every five year for the renewal of accreditation. Futhermore, the lecturer's and staff member's meeting has also been held once a year for keeping the knowledge and information about mammography screening system (Tsunoda, 2008). Although there has been legally-authorized Mammography Quality Standard ACT (MQSA) of the United States, the Japanese committee is considered the pioneer system in the world to evaluate the ability of individuals who engage in mammographic screening (Ohuchi, 2007).

Although ultrasound screening has been strongly recommended for examinees with dense breast particularly in their 40s or younger (Hashimoto & Ban 2010), there has been no worldwide evidence to prove whether ultrasound screening would have a potential to reduce the mortality of breast cancer for women aged 40-49. Therefore, in order to clarify sensitivity, specificity and detection rate as primary endpoint and cumulative advanced breast cancer incidence rate as secondary endpoint, 5-year Japan Strategic Anti-Cancer Randomized Trial (J-START) (http://jsrtfall.umin.jp/) has been conducted between the two groups of mammography combined with ultrasound and mammography alone in the respective number of 100,000 examinees in their 40s since 2006 with the initiative of The Ministry of Health, Welfare and Labor and the outcome will be shown in 2012. If effectiveness of ultrasound is confirmed, ultrasound screening would start in Japan for the first time in the world and we will have to prepare for the screening to especially secure a certain number of well-trained breast sonographers in Funabashi city also.

As far as clinical path in Japan is concerned, The Ministry of Health, Welfare and Labor authorized the payment for using paths for neck of femur fracture for the first time in 2006. Then, the Ministry issued Stragetic Anti-Cancer Promotion Project in 2007 to oblige the use of clinical paths of the five major cancers of stomach, colorectum, liver, lung and breast. According to the Ministry's survey in 2010, the number of the clinical paths used and patients enrolled has been larger in a comparison between before-2008 and 2009. Specifically, breast cancer patients increased even five-fold, largest in number compared to the other cancer patients. However, questionnaires to 410 hospitals by Health, Welfare and Labor Ministry in 2010 revealed that those patients including their families who understand paths have occupied only 30% even in the hospitals using paths. In other words, hospital staffs seem unlikely to give ample explanation or enlighten on paths considering that there actually have been many other path-using community hospitals throughout Japan. Likewise, the number of Chiba prefecture-based common paths used has been still as small as four, similar to lung cancer (1) and liver cancer (4) from April 2010 to January 2011. According to the other questionnaire by a certain group formed of breast cancer patients and their families, there have been the following anxieties for patients; 1) Whether the physicians of the community hospitals or clinics can treat enough or not, as well as the breast specialists in the region-based hospitals. 2) When clinical path is suddenly mentioned under treatment, some patients feel like losing a relationship of mutual trust. 3) Without close communication between the breast specialists and the community physicians, patients may possibly think to forcedly be sent out.

Region-based affiliated hospitals for cancer treatment have now been required to more closely cooperate to promote coordination with the community hospitals or clinics using clinical paths for the purpose of improving and maintaining treatment quality (Ando, 2004 & Aogi, 2008).

Since The Ministry of Health, Welfare and Labor revised the score of payment for treatment in fiscal year 2010, the region-based affiliated hospitals and community hospitals or clinics can get a certain additional score if the patients requiring a 10-year follw-up are treated using the common clinical path. Apart from the low-risk breast cancer patients, a path for postoperative chemotherapy using trastuzumab has been applied in Kitakyusyu city, Fukuoka prefecture, southern part of Japan (Ohno, 2008). For success of the path, Ohno et al stress the importance and benefit of close communication with the community physicians by holding regular meetings three times a year before starting its use.

We have had regular meetings since 2009 like Ohno et al to start Funabashi city-based clinical path in April, 2011. In addition to the meetings, it is required for success to fully inform the patients both before and after admission for surgery and citizens through regular extension lectures like ours, bulletins or website.

5. Conclusion

It is essential to have a more number of breast specialists in order to treat the steadily increasing Japanese breast cancer patients. To raise the percentage of examinees in breast cancer screening, the efforts must be continued including various campaigns by non-profit organization or consecutive distribution of free coupons by the government. Mammographic reading ability for doctors and mammographic skills for technologists have to be renewed every five year as before to maintain quality control. Evidence-based ultrasound breast cancer screening may possibly start in Japan for women aged 30-40. Clinical path has to be popular for low-risk breast cancer patients via adequate information before and/or after admission for surgery, regular extension lectures for citizens and bulletins or website.

6. Acknowledgement

I wish to thank Dr. Hasegawa for collecting continuous data on breast cancer screening in Funabashi city, Dr. Masuda for referring many early breast cancer patients, cyotologist Mr. Kato for diagnosing quickly on the outpatient and intraoperative basis, pharmacist Ms. Iwata for offering quick assistance on drug information and radiotechnologist Mr. Ishii for maintaining quality control on mammography as well as continuing technical education to other technologists involved in municipal breast cancer screening.

7. References

Sonoo, H. (2005). Certification system of breast diseases of the Japanese Breast Cancer Society (in Japanese with English abstract). *Jpn J Breast Cancer*, vol.20, No.1, 2005, pp. 59-63

Sakai, O. (2005). For the establishment of specialist system (in Japanese). *Treatment*, vol87, No.2, 2005, pp.400-403

Sonoo, H. (2007). Breast surgery; progress, perspective and future directions (in Japanese). *Surgery,* vol.69, No.4, pp. 396-401

Sonoo, H. (2008). Board certified breast specialists of the Japanese breast cancer society (in Japanese). *Surgical Treatment,* vol.98, No.3, 2008, pp. 260-266

Sawai, S. & Nakajima, H. (2006). Breast specialist (in Japanese). *All about easy-to-understand breast cancer,* Nagai Book Co. 2006, pp. 515-519.

Ohuchi, N. (2007). Current status of breast cancer screening (in Japanese). *Nippon Rinsho,* vol 65, Suppl 6, 2007, pp. 213-219

http://www.gankenshin50.go.jp/. OECD Health Data 2009 (in Japanese).

http://www.mhlw.go.jp/stf/shingi. Promotion of the Status of the 2009 women's cancer screening in 1990 article (in Japanese), pp. 5

Komoto, S. et al. (2010). The effect of free shipping coupons and breast cancer screening handbook (in Japanese). *Proceeding of J. Jpn Assoc. Breast Cancer Screen,* vol.19, No.3, 2010, pp. 282

Ishiyama, K. (2010). Current status of breast cancer screening in Akita city in 2009—Status after introduction of free coupons (in Japanese). *Proceeding of J Jpn. Assoc. Breast Cancer Screen,* vol.19, No.3, 2010, pp. 357

Kauhava, L. et al. (2008). Results on mammography screening in Turku, Finland. *Proceeding of J. Jpn.Assoc. Breast Cancer Screen,* vol7, No.3, 2008, pp.341

Satake H. et al (2011) Effort toward examination rate 50% of breast cancer screening with special reference to activity of Sendai city (in Japanese). *J. Jpn. Assoc. Breast Cancer Screen.* Vol.20, No.2, 2011, pp.102-105

http://www.mmjp.or.jp/kawakami-clinic/data. *The Ministry of Health Care Community Health Project Report FY* 2009 (in Japanese), 2010, pp. 1

Tsunoda-Shimizu H. et al. (2008). Quality control of breast cancer screening and future problems (in Japanese with English abstract). *Jpn J Breast Cancer,* vol.23, No.3, 2008, pp.191-196

Endo, T. (2009). Progress of quality control in breast cancer screening and future problems (in Japanese). *J Jpn. Assoc. Breast Cancer Screen,* vol.18, No.2, 2009, pp.107-114

Hashimoto,H. (2010). Present and future of ultrasound breast cancer screening (in Japanese). *Inner Vision* , vol.25,No.8, 2010, pp.27-29

Ban, K. et al. (2010). Consideration of apporopriate scrutiny of concurrent ultrasound screening with mammography screening (in Japanese). *J Jpn. Assoc. Breast Cancer Screen,* vol.25, No.6, 2010, pp.649-656

http://jsrtfall.umin.jp/. Present status and perspective of J-START (in Japanese). *Proceeding of an instructive lecture,* 2010, pp.1-2

Ando, T. et al. (2004). Breast cancer surgery and clinical path (in Japanese). *Surgical Treatment* , vol.90, No.5, pp.937-943

Aogi, K. et al. (2008). Breast cancer surgery—The present status of development of clinical path (in Japanese). *Surgical Treatment* , vol.99, No.1, pp.48-56

Ohno S. et al. (2008) Present status and problems of introduction of regional medicine clinical path of breast cancer (in Japanese). *J New Rem. & Clin,* vol.57, No.12 , pp.12-21

16

Dose Optimization for the Quality Control Tests of X-Ray Equipment

Mana Sezdi
Istanbul University
Turkey

1. Introduction

Radiation is a major risk in diagnostic and therapeutic medical imaging. The problem is caused from incorrect use of radiography equipment and from the radiation exposure to patients much more than required. Exposure of different dose values for the same clinical examination, is an enough reason to draw attention to this issue.

International Commission on Radiation Protection (ICRP), the International Atomic Energy Agency (IAEA) and other various independent institutions have been making publications in relation to ionizing radiation protection for more than fifty years. Report 60 of the ICRP and the Basic Safety Standards that was published in the IAEA report have three basic principles related to the radiation protection (ICRP, 1991; IAEA, 1996).

The most important issue in these principles is the optimization of radiation. In the mentioned policy, the lowest dose is aimed by considering the country's economic and social factors for acceptable applications. Personnel already receive low dose with protection systems in the working areas. However, the patient doses must be taken under control based on the principle of optimization as much as possible.

There are two important points when performing a radiological procedure:

- To obtain the best possible image for a clear diagnosis of the disease,
- To apply the lowest dose for protecting the patient while getting the best image.

The second point indicates that the patient's radiation dose level must be kept at the lowest possible dose. In other words, it indicates dose optimization. The dose optimization meaning "the minimum radiation dose of the optimum image quality", is achieved by applying quality control procedures, calibration and dosimetric measurements.

In the Radiology Quality Control systems, the biggest problem is dose control and dose optimization. Neither patient nor users knows how much dose is exposed because there is no any system in the x-ray device for measuring or showing dose during exposure.

Since there is no dose adjustment on the equipment, the systems are operated by using the usual parameters; kVp and mAs. Because dose can not be adjusted, the patient may receive more dose than the aimed dose.

For dose optimization, all exposures should be kept at the minimum dose level in according to the ALARA principle (ALARA-as low as reasonably achievable). The aim of the optimization is not to download the risks of irradiation to zero. It is to reduce them to an acceptable level. This can be possible only by examining all parameters that affect the X-ray, by investigating the relationship between dose and these parameters, on the basis of this relationship, by performing the necessary regulations.

In all x-ray equipment, the operator can control the quantity and the quality of the radiation with kVp and mAs controls. If the equipment is not properly controlled, it will not be possible to control the radiation output. For this reason, optimization consists of not only improving of image quality and low dose but also establishing quality assurance and quality control programmes to ensure a proper performance of the x-ray equipment.

As frequently documented in the scientific literature, patient dose and image quality are basic aspects of any quality control (QC) tests in diagnostic radiology. Image quality must be adequate for diagnosis and it must be obtained with low doses.

The following QC tests are performed for both patient dose and image quality evaluation;

- kVp Accuracy and Repeatability
- Dose-kVp Linearity Test
- Dose-mAs Linearity Test
- X-ray Tube Output-kVp Relation
- HVL (Half Value Layer)
- Image Quality (Beam alignment, collimation alignment, contrast and resolution)

The quality control tests' methods, as well as the criteria for scoring the results, are in full agreement with those specified in the American Association of Physicists in Medicine (AAPM) Report No.4 and IEC 61223-3-1 (AAPM, 1981; IEC 61223-3-1, 1999).

There are a number of recent studies about dose optimization. Some of them are the surveys about image quality and patient dose in radiographic examinations in the authors' countries (Bouzarjomehri, 2004; Ciraj et al., 2005; Ramanandraibe, 2009; Papadimitriou, 2001; Shahbazi-Gahrouei, 2006). Some investigators focused only patient dose optimization (Brix et al., 2005; Vano & Fernandez, 2007; Seibert, 2004; Williams & Catling, 1998), whereas the others examined both the patient dose and image quality in radiographic devices (Aldrich et al., 2006; Schaefer-Prokop et al., 2008; Geijer, 2002). There are also studies that give reference values for clinical x-ray examinations by measuring phantom dose (Gray et al., 2005). But there is no any study focused to the dose optimization during quality control tests of x-ray devices. Dose optimization is very important because of the quality and quantity of quality control tests of x-ray equipments.

The aim of this study is to provide optimal x-ray parameters that may be used for quality control tests in order to make quality control activities more efficient and can be controlled. The staff know how the quality control tests are performed, but they don't know which parameters' values give which qualified image. They have problems during evaluation of test results, although there are some recommendations in the standards (AAPM, 1981; IEC 61223-3-1, 1999). They need proven parameter values for comparison. In this study, it was examined during quality control tests which parameters give a high quality image and how much dose is measured when these parameters were applied.

This study was performed by investigating the effects of X-ray parameters' changes on dose and by modeling of dose related to these parameters. After the modeling, in according to the related parameters, the dose level can be controlled, and in different x-ray units the dose levels that are obtained by applying the same parameter setting, can be compared.

Thus, in addition to obtain optimal parameters, controlling of the accuracy of the measured dose values may be possible by calculating the dose value during quality control tests.

2. Parameters of x-ray

In radiography, dose and image quality are dependent on radiographic parameters. This study is concerned with the quantification of these parameters and an assessment of their effect on patient dose and image quality. The focus of this study is on the relationship between dose, image quality and other radiographic parameters.

2.1 Absorbed dose

Absorbed dose is the quantity that expresses the radiation concentration delivered to a point, such as the entrance surface of patient's body. Absorbed dose in air is recognized as air kerma and it is a measure of the amount of radiation energy, in the unit of joules (J), actually deposited in or absorbed in a unit mass (kg) of air. Therefore, the quantity, kerma, is expressed in the units of J/kg which is also the radiation unit, the gray (G) (Sprawls, 1987; Hendee et al., 1984).

In this study, the word of "dose" will be used instead of air kerma (absorbed dose in air).

2.2 kVp

The high energy of the x-ray spectrum is determined by the kilovoltage applied to the x-ray tube. The maximum photon energy is numerically equal to the maximum applied potential in kilovolts. The maximum photon energy is determined by the voltage during the exposure time. This value is generally referred as the kilovolt peak (kVp) and is one of the adjustable factors of x-ray equipment (Sprawls, 1987).

2.3 mAs

The x-ray cathode is heated electrically by a current from a separate low voltage power supply. The output of this supply is controlled by the mA selector on the x-ray unit. Additionally, the duration of the x-ray exposure is controlled by the time selector. mAs is described by multiplying of these two values (mA x second) (Hendee et al., 1984).

2.4 Half Value Layer (HVL)

Half value layer describes both the penetrating ability of specific radiations and the penetration through specific objects. HVL is the thickness of material that reduces the intensity of an x-ray beam by half, and is expressed in unit of distance (mm) (Sprawls, 1987).

2.5 Image quality

The purpose of the radiographic image is to provide information about the medical condition of the patient. A quality image is one that provides all the information required for diagnosis of the patient's condition (Hendee et al., 1984).

Image quality is not a single factor but is described with beam alignment, collimation alignment, contrast and resolution. Contrast means differences in the form of gray scales or light intensities, whereas the resolution is a measure of its ability to differentiate between two objects a small distance apart; such that they appear distinct from one another.

An image is acceptable as qualified only if it has high resolution and high contrast.

3. Material and method

The radiographic measurements were performed in ten stationary X-ray units in five hospitals. The X-ray units including: Siemens, Philips, Toshiba, General Electric and Shimadzu were participated in this study. The reason for chosing these x-ray units is that their age is between 5 and 7 years old and the machines have 3 phase generators, thus their HVL value is kept in a narrow range, such as between 3 and 3,2mmAl.

Dosimax Plus A (Wellhöfer, Scanditronix, IBA, Germany) dosimeter was used to measure radiation dose. Dosimax Plus A dosimeter is a universal basic device and is designed according to IEC 61674 for acceptance tests and for quality checks at radiographic X-ray units. In Dosimax Plus A, dose measurements are performed by using solid state detectors (RQA). The dose range is from 200nGy to 9999mGy (Iba Dosimetry, 2008). It was calibrated by the Iba Laboratory of Germany and found to be capable of performing within recommended level of precision and accuracy.

Dose measurement applications has been included in recent recommendations (AAPM, 1981; IEC 61223-3-1, 1999). The measurement procedures that were realized in this study, are explained below step by step.

Before starting dose measurements, kVp accuracy tests were performed for 10 units and it was seen that they have acceptable accuracy in according to the standards (AAPM, 2002).

3.1 Measurement procedure of X-ray dose variation with kVp

The dosimeter was positioned in central beam axis such that the X-ray tube focal spot-dedector distance (FDD) was 100cm for the measurements. The radiation field size was set to cover the dosimeter in order to avoid the possible scatter radiation to the dosimeter.

In order to investigate the effect of kVp to the dose, the unit was set at 20mAs and 50kVp value. An X-ray exposure was made and the dosimeter reading was recorded. This step was repeated at same constant mAs and different kVp settings (50, 70, 80 and 100kVp) and dosimeter reading was determined. Similar X-ray dose measurements were also determined for 40 and 50mAs settings for each kVp value (50, 70, 80 and 100kVp). All measurements were repeated for 60cm (FDD). The measured dose values were plotted against the corresponding kVp for each X-ray unit separately.

3.2 Measurement procedure of X-ray dose variation with mAs

The dosimeter was positioned at 100cm (FDD) from the focal spot of the X-ray tube.

In order to determine the effect of mAs to the dose, the exposures were performed with constant kVp (50kVp), but with gradually increasing mAs (10, 20, 40 and 50mAs). Similar X-ray dose measurements were also determined for 70 and 100kVp settings for each mAs value (10, 20, 40 and 50mAs). All measurements were repeated for distance of 60cm. The measurement results for each X-ray unit were plotted against the corresponding mAs.

3.3 Measurement procedure of X-ray tube output variation with kVp

The X-ray tube output was determined as the ratio of dose reading to the mAs setting. The values of X-ray tube output were plotted against kVp by using dose values obtained from two measurement procedures (Section 3.1 and 3.2).

3.4 Measurement procedure for Half Value Layer (HVL)

For dose measurements, filtration was realized by using aluminum (Al) filters with 1mm and 0,5mm.

During the measurements, mAs and kVp were stable (20mAs, 50kVp) and the distance was determined as 100cm. Initially, the dose measurement without the filter was generalized. After this, the dose measurement was repeated by using filter with different thickness. Each filter thickness was obtained by adding 1mmAl and 0,5mmAl. The dose measurements were taken in the conditions; without filter, 1mmAl, 2mmAl, 3mmAl and 3,5mmAl.

3.5 Observing of image quality

Test tool ETR1 (Iba Dosimetry, 2008) was used for image quality tests. The ETR1 is a multi-purpose test tool. With a single exposure on X-ray film made by using this tool, all criterias (alignments, contrast and resolution) can be checked for quality control of image.
Before exposure, a cassette with x-ray film was placed on the patient table. The distance between the film and the focal spot was set to 100cm. The test tool was placed over the cassette and the collimator was adjusted to ensure that the light beam covers exactly the inner pattern of the test tool. An exposure was performed with 50kVp and 20mAs. The exposure was repeated for each setting value adjusted for dose measurements mentioned in Section 3.1 and 3.2 (50kVp-40mAs, 50mAs; 70kVp-20mAs, 40mAs, 50mAs; etc…).
After developing the film, the image on the film was compared with the real test tool image. Beam alignment, collimation alignment, contrast and resolution factors were determined and recorded.

4. Results

During the quality control of x-ray equipment, it is essential to know the effects of x-ray parameters to the image quality. The x-ray parameters' effects were measured by using quality control test procedures and they were analysed graphically.
In result, the optimized dose in which parameters' value gave the high quality image was determined.

4.1 Assessment of X-ray dose variation with kVp

The measured doses by changing kVp are given in Table 1. During measurements, mAs was firstly kept stable (20mAs) and kVp was changed as 50, 70, 80 and 100kVp to investigate the effects of kVp to the dose at stable mAs.
After this, the same measurement procedure was applied to other mAs values (40 and 50mAs). All measurements were performed at distance of 100 and 60cm.
Graphical representations of the relationship between dose and kVp value for constant mAs (20, 40 and 50mAs) at 100cm and 60cm are given in Fig. 1, 2, 3 and Fig. 4, 5, 6, respectively.

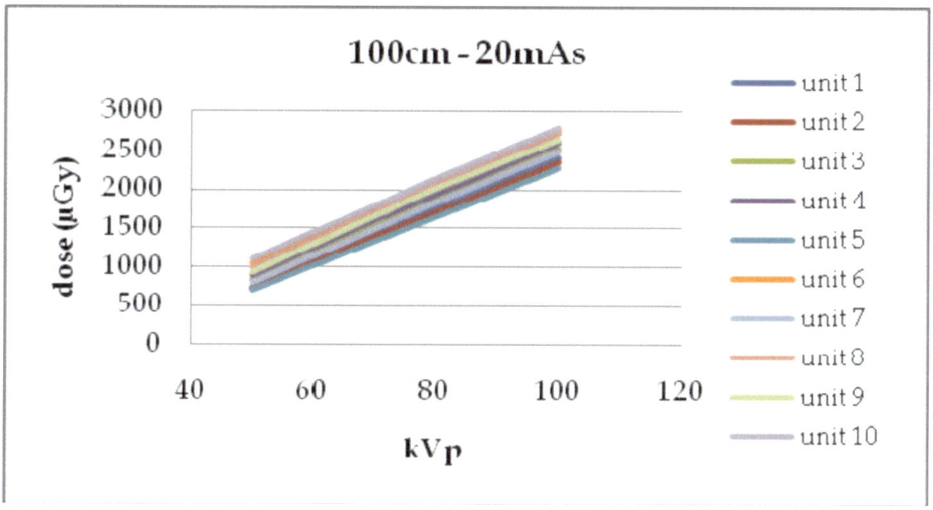

Fig. 1. Measured dose values of 10 x-ray units versus kVp for 20mAs at distance of 100cm.

Dose (μGy)										
	Unit	**50 kVp**	**70 kVp**	**80 kVp**	**100 kVp**	**50 kVp**	**70 kVp**	**80 kVp**	**100 kVp**	
100cm - 20mAs	**1**	736,0	1439	1766	2420	1950	4230	5450	8050	**60cm - 20mAs**
	2	710,7	1388	1707	2357	1883	3955	5243	7588	
	3	824,3	1538	1866	2538	2327	4640	5905	8608	
	4	894,4	1605	1935	2601	2478	4915	6181	8909	
	5	690,4	1324	1642	2282	1554	3717	4981	7268	
	6	1048	1734	2101	2757	3310	5758	7006	9498	
	7	792,1	1481	1826	2483	2077	4353	5627	8277	
	8	988,4	1704	2070	2726	2999	5477	6722	9300	
	9	934,7	1657	2017	2652	2796	5157	6445	9062	
	10	1101	1784	2136	2792	3484	5872	7195	9741	
100cm - 40mAs	**1**	1482	2938	3622	5056	4190	8260	10800	15600	**60cm - 40mAs**
	2	1493	2863	3580	4912	3774	7983	10511	15123	
	3	1587	3129	3797	5276	4511	8588	11383	15515	
	4	1600	3136	3872	5335	4684	8981	11515	16089	
	5	1400	2738	3460	4861	3464	7613	10099	14656	
	6	1923	3461	4207	5716	5468	10097	12415	17247	
	7	1528	3082	3702	5185	4271	8716	10984	15797	
	8	1858	3335	4104	5561	5249	9773	12037	16986	
	9	1692	3290	3946	5405	4870	9408	11788	16487	
	10	2053	3581	4320	5855	5805	10618	12869	17610	
100cm - 50mAs	**1**	1790	3612	4590	6380	5350	12548	15800	22700	**60cm - 50mAs**
	2	1722	3443	4384	6194	4991	12168	15553	22313	
	3	1933	3678	4609	6608	6144	13528	15973	23548	
	4	1985	3697	4481	6652	6588	13249	16348	23704	
	5	1604	3345	4185	6008	4585	11712	15033	21972	
	6	2245	4020	4965	6995	8101	14631	17750	24808	
	7	1853	3665	4575	6518	5574	12999	15799	23255	
	8	2198	3813	4759	6823	7623	14240	17234	24484	
	9	2091	3799	4691	6712	7064	13751	16857	23892	
	10	2391	4213	5167	7194	8724	15027	18203	25283	

Table 1. Measured doses (μGy) for constant mAs but increasing kVp at different distances.

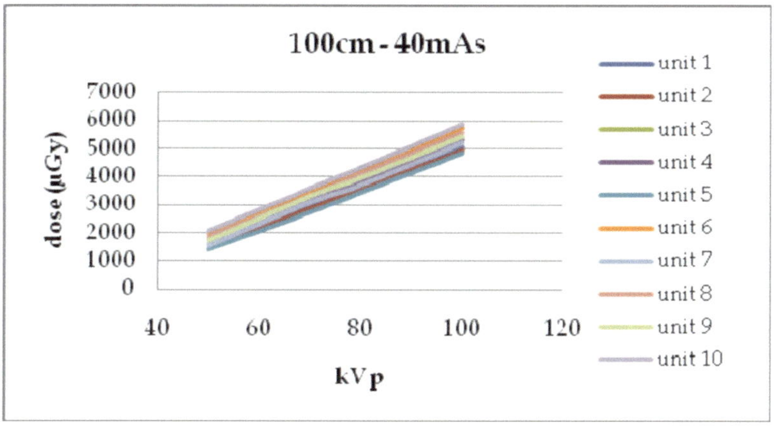

Fig. 2. Measured dose values of 10 x-ray units versus kVp for 40mAs at distance of 100cm.

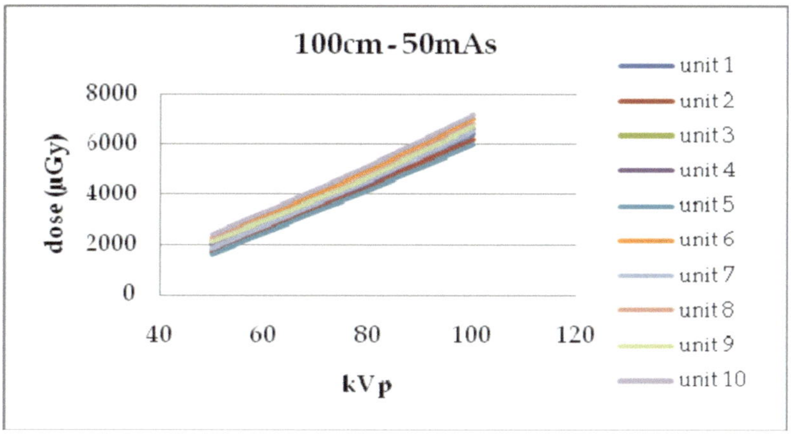

Fig. 3. Measured dose values of 10 x-ray units versus kVp for 50mAs at distance of 100cm.

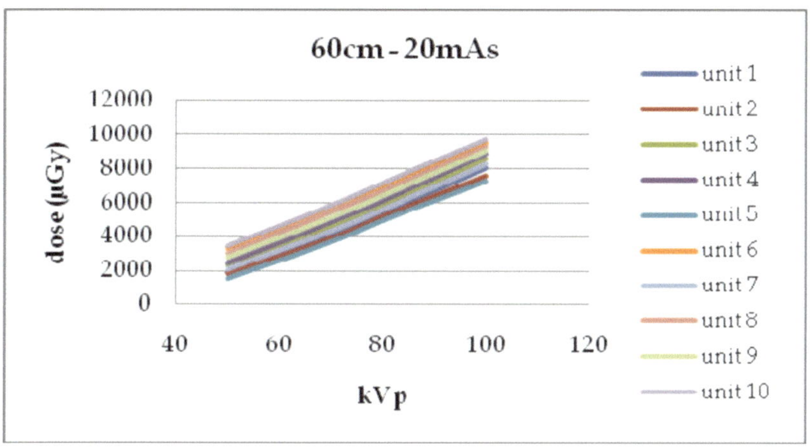

Fig. 4. Measured dose values of 10 x-ray units versus kVp for 20mAs at distance of 60cm.

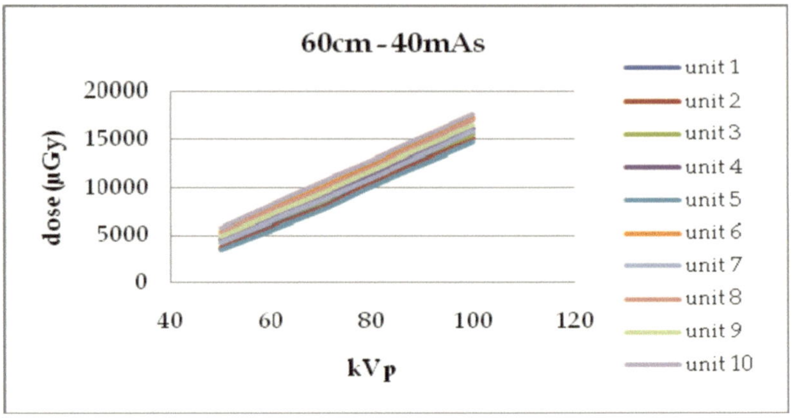

Fig. 5. Measured dose values of 10 x-ray units versus kVp for 40mAs at distance of 60cm.

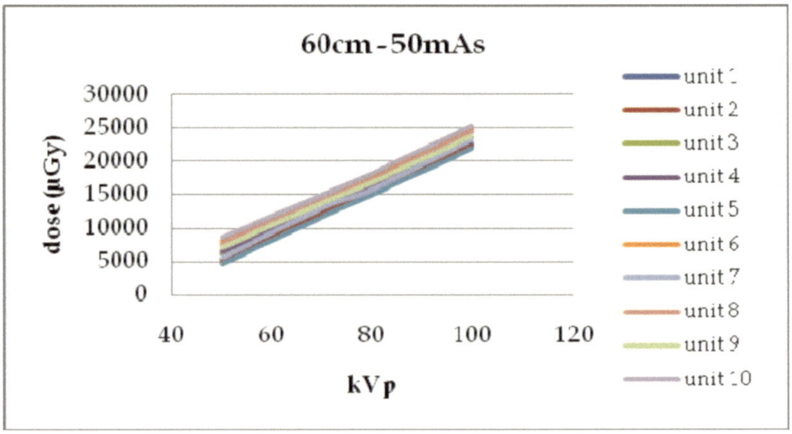

Fig. 6. Measured dose values of 10 x-ray units versus kVp for 50mAs at distance of 60cm.

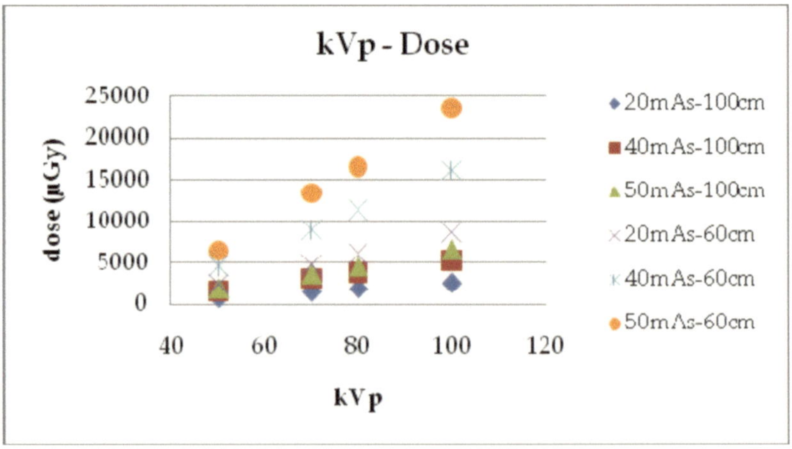

Fig. 7. Mean dose values of 10 x-ray units versus kVp for different mAs and distance setting.

The dose values obtained from 10 x-ray units were analysed statistically and the mean dose values for each setting parameter were defined with standard deviation in Table 2.
For different distance and mAs settings, the mean dose values were plotted against kVp (Fig. 7). Hence, the small differences that are caused from unit changes were eliminated, and the effect of kVp to dose variation was focused.

Dose	100cm - 20mAs				60cm - 20mAs			
(μGy)	50kVp	70kVp	80kVp	100kVp	50kVp	70kVp	80kVp	100kVp
Mean	872,0	1565,4	1906,5	2560,7	2485,8	4807,5	6075,5	8630,2
std	144,3	155,9	172,4	174,8	645,3	753,0	760,0	823,6
Dose	100cm - 40mAs				60cm - 40mAs			
(μGy)	50kVp	70kVp	80kVp	100kVp	50kVp	70kVp	80kVp	100kVp
Mean	1661,6	3155,3	3861,1	5316,0	4628,7	9003,7	11440,1	16110,9
std	215,4	266,5	283,4	329,3	742,1	962,9	866,6	957,6
Dose	100cm - 50mAs				60cm - 50mAs			
(μGy)	50kVp	70kVp	80kVp	100kVp	50kVp	70kVp	80kVp	100kVp
Mean	1981,3	3728,4	4640,7	6608,4	6474,2	13385,3	16455,0	23595,8
std	249,7	254,1	280,0	356,4	1387,9	1069,0	1024,1	1075,5

Table 2. The statistic analysis of dose from 10 units for different kVp at constant mAs.

4.2 Assessment of X-ray dose variation with mAs

The obtained dose values at constant kVp by increasing mAs can be seen in Table 3. The dose measurements were performed at 50, 70 and 100kVp with changing mAs (10, 20, 40 and 50mAs) in distance of 100 and 60cm.
Graphical representations of the relationship between dose and mAs for constant kVp (50, 70 and 100kVp) at 100 and 60cm are shown in Fig. 8, 9, 10 and Fig. 11, 12, 13, respectively.

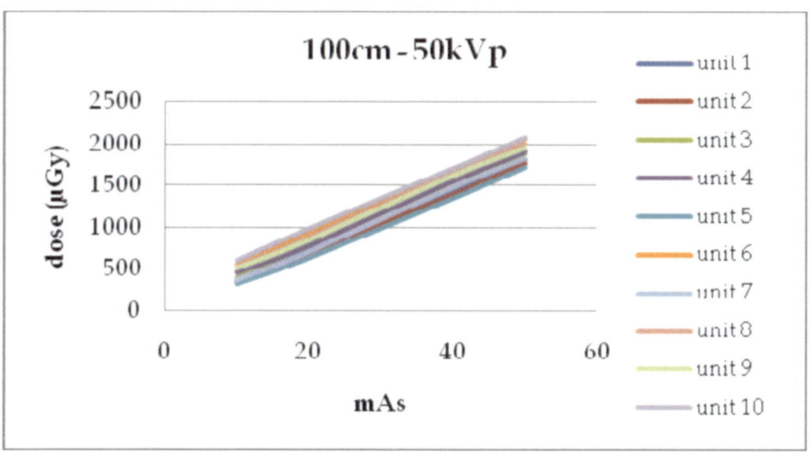

Fig. 8. Measured dose values of 10 x-ray units versus mAs for 50kVp at distance of 100cm.

	Dose (μGy)									
	Unit	10 mAs	20 mAs	40 mAs	50 mAs	10 mAs	20 mAs	40 mAs	50 mAs	
100cm - 50kVp	1	378,0	736,0	1432	1810	1050	2089	4188	5120	60cm - 50kVp
	2	339,8	695,4	1397	1773	932	1940	4056	5075	
	3	402,2	796,4	1527	1886	1202	2231	4276	5245	
	4	464,0	786,5	1554	1911	1247	2333	4329	5394	
	5	321,1	632,5	1344	1717	826	1846	3975	5026	
	6	565,8	934,5	1688	2068	1494	2532	4680	5715	
	7	370,2	712,0	1483	1834	1123	2152	4245	5206	
	8	530,9	879,0	1641	2006	1342	2402	4516	5657	
	9	510,4	844,6	1610	1952	1272	2342	4447	5551	
	10	606,9	987,6	1711	2091	1597	2614	4801	5862	
100cm - 70kVp	1	714,4	1469	2871	3652	2310	4180	8140	10077	60cm - 70kVp
	2	641,8	1372	2787	3528	2187	4040	7966	9846	
	3	801,0	1618	3055	3874	2604	4401	8428	10480	
	4	874,9	1687	3180	3902	2799	4675	8672	10585	
	5	574,9	1292	2683	3414	1993	3828	7692	9613	
	6	1001	1881	3401	4157	3317	5186	9123	10959	
	7	754,2	1558	2955	3773	2580	4309	8359	10228	
	8	987,4	1764	3387	4084	3221	4956	8972	10745	
	9	956,6	1718	3292	3952	2989	4936	8832	10645	
	10	1075	1926	3515	4219	3503	5387	9280	11210	
100cm - 100kVp	1	1404	2808	5616	7020	3250	7288	15435	19426	60cm - 100kVp
	2	1198	2623	5387	6819	3006	6870	14811	18926	
	3	1641	2927	5751	7255	3870	8097	15872	20058	
	4	1690	2922	5964	7285	4278	8296	16080	20415	
	5	1045	2444	5308	6622	2421	6342	14354	18465	
	6	1854	3182	6215	7558	4940	8955	17157	21594	
	7	1582	2885	5703	7131	3537	7601	15343	19500	
	8	1797	3060	6070	7429	4509	8735	17102	21306	
	9	1717	3055	6070	7407	4448	8584	16399	20958	
	10	1889	3343	6266	7638	5290	9310	17697	21944	

Table 3. Measured doses (μGy) for constant kVp but increasing mAs at different distances.

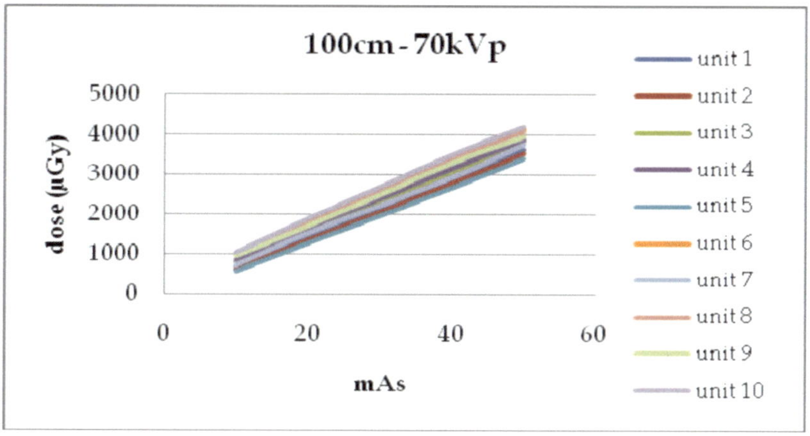

Fig. 9. Measured dose values of 10 x-ray units versus mAs for 70kVp at distance of 100cm.

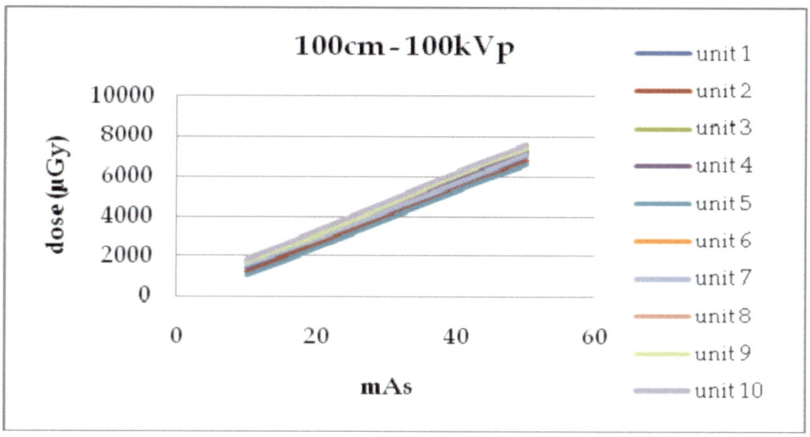

Fig. 10. Measured dose values of 10 x-ray units versus mAs for 100kVp at distance of 100cm.

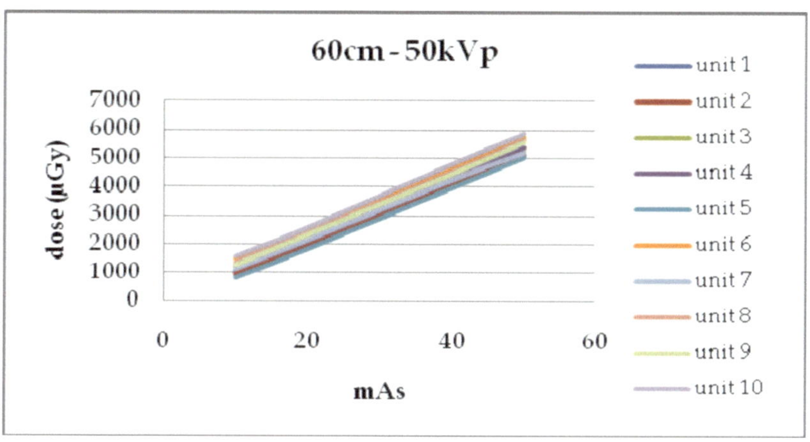

Fig. 11. Measured dose values of 10 x-ray units versus mAs for 50kVp at distance of 60cm.

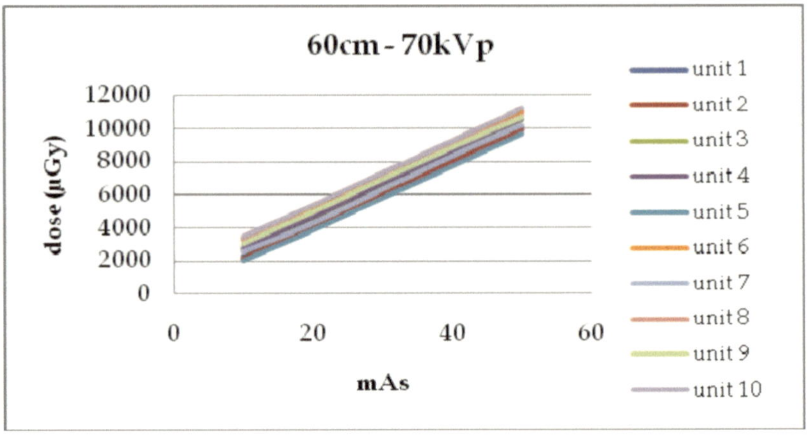

Fig. 12. Measured dose values of 10 x-ray units versus mAs for 70kVp at distance of 60cm.

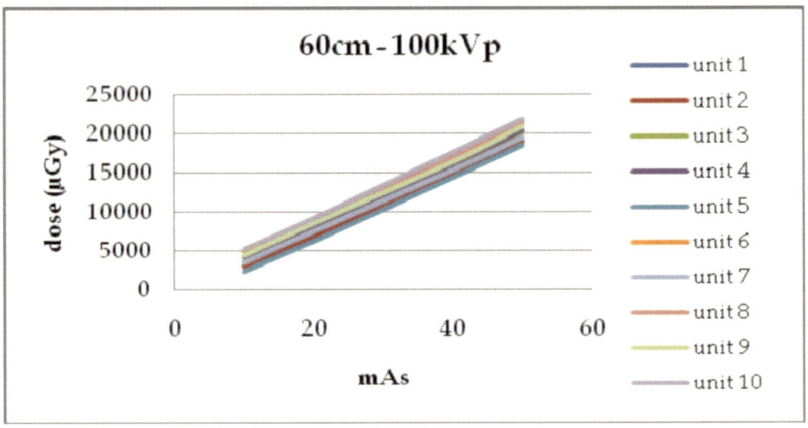

Fig. 13. Measured dose values of 10 x-ray units versus mAs for 100kVp at distance of 60cm.

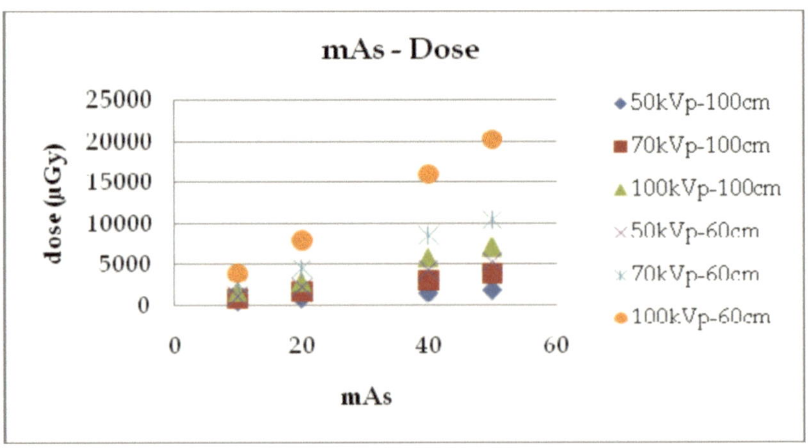

Fig. 14. Mean dose values of 10 x-ray units versus mAs for different kVp and distances.

The dose values obtained from 10 x-ray units were analysed statistically and the mean dose values for each setting parameter were defined with standard deviation in Table 4.
For different distance and kVp settings, the mean dose values were plotted against mAs (Fig. 14). Hence, the small differences that are caused from unit changes were eliminated, and the effect of mAs to dose variation was focused.

Dose	100cm - 50kVp				60cm - 50kVp			
(µGy)	10mAs	20mAs	40mAs	50mAs	10mAs	20mAs	40mAs	50mAs
Mean	448,9	800,5	1538,6	1904,9	1208,4	2248,1	4351,4	5385,1
std	100,5	111,8	125,0	125,0	238,3	246,1	262,3	295,3
Dose	100cm - 70kVp				60cm - 70kVp			
(µGy)	10mAs	20mAs	40mAs	50mAs	10mAs	20mAs	40mAs	50mAs
Mean	838,2	1628,5	3112,6	3855,6	2750,4	4589,7	8546,2	10438,6
std	167,4	208,6	285,7	266,1	526,1	518,1	519,6	498,2
Dose	100cm - 100kVp				60cm - 100kVp			
(µGy)	10mAs	20mAs	40mAs	50mAs	10mAs	20mAs	40mAs	50mAs
Mean	1581,7	2924,9	5835,1	7216,4	3954,8	8008,0	16025,0	20259,2
std	281,9	261,2	334,7	322,8	903,1	960,3	1079,2	1180,2

Table 4. The statistic analysis of dose from 10 units for different mAs at constant kVp.

4.3 Assessment of X-ray tube output variation with kVp

In order to investigate the relationship between the x-ray tube output and kVp, firstly the x-ray tube outputs for 10 x-ray units, were calculated by dividing the measured dose values to the mAs values. It was seen that there is a dose distribution because of the measured different doses of each x-ray unit. Therefore, the mean of the x-ray tube output values for each mAs values were used for plotting of the x-ray tube output against the kVp.
Additionally, the graphics show that there is a different distribution that are caused from different distances although all distributions were similar for each mAs value.
For each different distance, the mean of the calculated tube output for different mAs were plotted with equations.

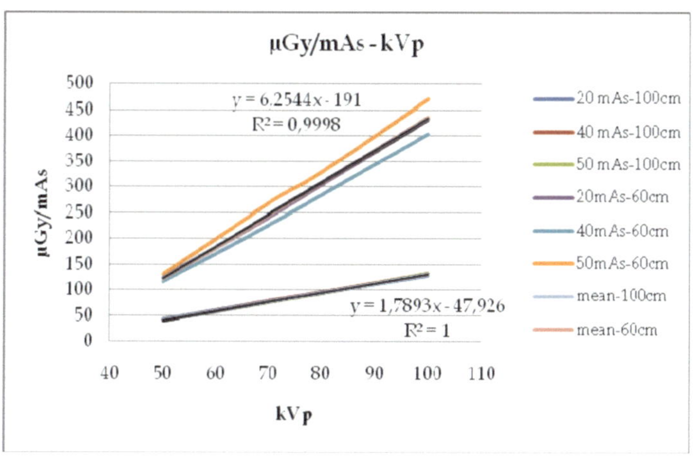

Fig. 15. X-ray tube output changes with kVp (dose values from procedure in Section 3.1).

In Figure 15, the dose values that were obtained by applying the procedure mentioned in Section 3.1 and the procedure settings (mAs, kVp), were plotted, whereas in Figure 16 the values obtained from procedure in section 3.2, were used.

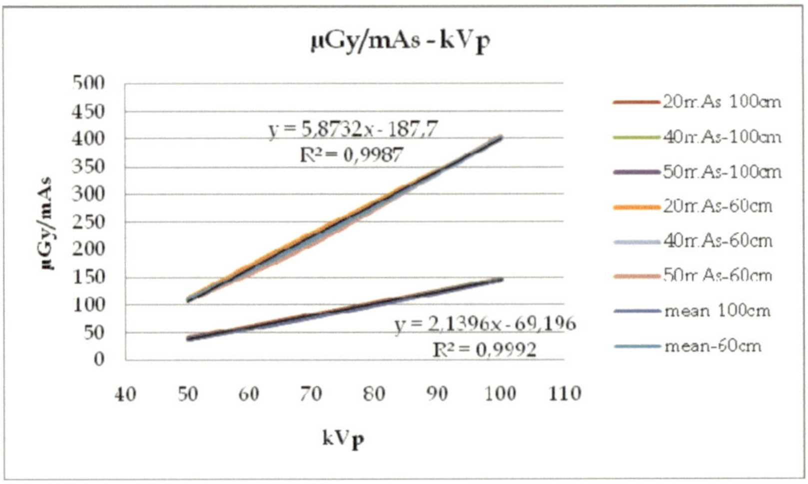

Fig. 16. X-ray tube output changes with kVp (dose values from procedure in Section 3.2).

The tube output which is derived from direct measurement can be expressed in equations obtained from Figure 15 and Figure 16, because kVp is related to tube output directly.
For distance of 100cm, tube output can be written separately in different two equations that were obtained from graphics in Fig. 15 and Fig. 16.

$$\text{Tube output } (\mu Gy/mAs) = 1{,}7893kVp - 47{,}926 \qquad R^2 = 1 \tag{1}$$

$$\text{Tube output } (\mu Gy/mAs) = 2{,}1396kVp - 69{,}196 \qquad R^2 = 0{,}9992 \tag{2}$$

Dose can be determined from tube output, and mAs can be placed in Equations 1 and 2.

$$\text{Dose } (\mu Gy) = (1{,}7893kVp - 47{,}926) \times mAs \tag{3}$$

$$\text{Dose } (\mu Gy) = (2{,}1396kVp - 69{,}196) \times mAs \tag{4}$$

For distance of 60 cm, again tube output can be written separately in different two equations obtained from Figure 15 and Figure 16, respectively.

$$\text{Tube output } (\mu Gy/mAs) = 6{,}2544kVp - 191{,}0 \qquad R^2 = 0{,}9998 \tag{5}$$

$$\text{Tube output } (\mu Gy/mAs) = 5{,}8732kVp - 187{,}7 \qquad R^2 = 0{,}9987 \tag{6}$$

When mAs is placed in Equations 5 and 6, the following Equations 7 and 8 are derived.

$$\text{Dose } (\mu Gy) = (6{,}2544kVp - 191{,}0) \times mAs \tag{7}$$

$$\text{Dose } (\mu Gy) = (5{,}8732kVp - 187{,}7) \times mAs \tag{8}$$

To test the validity of these equations, an external set of dose values obtained from measurements (10mAs-50kVp, 10mAs-70kVp and 10mAs-100kVp for 100cm and 60cm) was selected. kVp and mAs values were plugged into the Equations 3, 4, and Equations 7, 8 to predict the dose values for applied kVp and mAs values of measurement procedures (Section 3.1 and 3.2). The results were then compared with the measured dose values, as shown in Table 5.

It is seen from Table 5, the predicted dose values are within the measured dose value with standard deviations for each measurement procedure (in different distance, firstly constant mAs with increasing kVp, afterly constant kVp with increasing mAs). The predicted dose values obtained from equations that shows the relationship between x-ray tube output and kVp during measurement procedure (mAs is increased with constant kVp), are approximately similar with the predicted dose values that were derived from measurement procedure of constant mAs and increasing kVp. Relatively it can be said that dose measurements are not affected from the application style of parameters (kVp and mAs). Not only keeping mAs as constant and increasing kVp, but also keeping kVp as constant and increasing mAs doesn't affect the measured dose values for the same kVp and mAs. For example, the dose value obtained from measurement of 50kVp-40mAs are approximately similar during application of both constant 50kVp with increasing mAs and constant 40mAs with increasing kVp. This result showed that taking into account the results that were obtained from only one measurement procedure is sufficient. Especially, Equation 3 and Equation 7 can be preferred for dose estimation because of their best R^2.

Distance 100cm	Mean dose from direct measurement (μGy)	Dose calculated from Equation 3 (μGy)	Dose calculated from Equation 4 (μGy)
10mAs-50kVp	448,9 ± 100,5	415,39	377,84
10mAs-70kVp	838,2 ± 167,4	773,25	805,76
10mAs-100kVp	1581,7 ± 281,9	1310,04	1447,64
Distance 60cm	**Mean dose from direct measurement (μGy)**	**Dose calculated from Equation 7 (μGy)**	**Dose calculated from Equation 8 (μGy)**
10mAs-50kVp	1208,4 ± 238,3	1217,2	1059,60
10mAs-70kVp	2750,4 ± 526,1	2468,08	2234,24
10mAs-100kVp	3954,8 ± 903,1	4344,4	3996,20

Table 5. Measured and calculated dose values.

4.4 Assessment of Half Value Layer (HVL)

Testing of half value layer is performed by measuring dose values with different Al thickness and it verifies that half value layer is sufficient to reduce patient exposure to low energy radiation. The obtained dose measurement results of each x-ray unit in this study for stable mAs and kVp are given in Table 6.

The dose measurement results were plotted against the aluminum (Al) thickness (Fig. 17). Dose (μGy) equations were obtained as a function of Al thickness and from these equations, the Al thickness in which the dose decreased to its half value was calculated (Table 7).

kVp = 50, mAs = 20, Distance = 100cm					
Unit	**Dose (μGy)**				
	0mmAl	**1mmAl**	**2mmAl**	**3mmAl**	**4mmAl**
1	736,0	520,9	451,8	374,0	342,8
2	710,7	519,9	446,8	370,1	337,7
3	824,3	647,5	522,0	431,6	388,9
4	894,4	643,1	554,2	462,3	415,7
5	690,4	502,1	431,8	359,5	322,0
6	1047,8	767,3	654,0	524,1	493,8
7	792,1	598,8	485,7	401,8	360,0
8	988,0	702,1	588,9	499,6	454,9
9	934,7	704,2	577,1	474,8	425,2
10	1101,0	782,2	656,0	556,6	506,8

Table 6. Dose measurements for different aluminum thickness (mm).

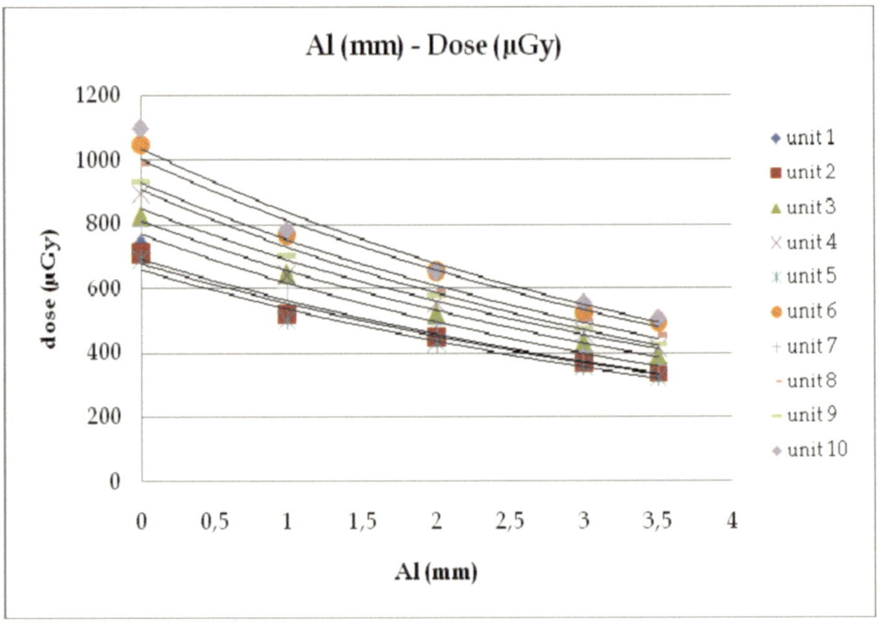

Fig. 17. Al(mm)-Dose(μGy) graphic for each X-ray unit.

Unit	Dose (μGy) = f(Al (mm))	Calculated HVL (mm)
1	$y = 693{,}18\ e^{-0{,}208x}$ $R^2 = 0{,}9714$	3,0
2	$y = 678{,}13\ e^{-0{,}204x}$ $R^2 = 0{,}9819$	3,1
3	$y = 811{,}49\ e^{-0{,}212x}$ $R^2 = 0{,}9979$	3,2
4	$y = 848{,}79\ e^{-0{,}208x}$ $R^2 = 0{,}9775$	3,1
5	$y = 658{,}65\ e^{-0{,}207x}$ $R^2 = 0{,}9812$	3,1
6	$y = 1002{,}5\ e^{-0{,}211x}$ $R^2 = 0{,}9830$	3,1
7	$y = 769{,}50\ e^{-0{,}220x}$ $R^2 = 0{,}9940$	3,0
8	$y = 929{,}33\ e^{-0{,}212x}$ $R^2 = 0{,}9723$	3,0
9	$y = 907{,}66\ e^{-0{,}219x}$ $R^2 = 0{,}9939$	3,0
10	$y = 1035{,}5\ e^{-0{,}212x}$ $R^2 = 0{,}9722$	3,0

Table 7. Dose=*f*(Al) equations and calculated HVL values.

As it is seen from the table, the observed x-ray units' HVL values change from 3,0 to 3,2 mmAl. Because it is required that the HVL of an acceptable x-ray unit with 3 phase generator must exceed 2,9mm, the observed 10 x-ray units were appropriate to the international standards (AAPM, 1981).

4.5 Image quality

Image quality tests were performed by controlling of beam alignment, collimation alignment, contrast and resolution of image.

As a result of the beam aligment and collimation alignment tests, it was seen that beam alignment and collimation alignment are only related to the quality of x-ray tube, are not dependent to the x-ray parameters, such as kVp, mAs and dose. For this reason, the test results that were obtained from only one measurement setting (50kVp, 20mAs, 100cm), are sufficient to obtain information about alignments of each x-ray units (Table 8).

Beam alignment test gives the deviation of the centre from the middle of the exposed film to the middle of the test tool (point "a" in Figure 18). The test's results that were given in Table 8 show that the beam misalignments were less than 10mm for 10 x-ray units.

In the collimation alignment test, the vertical misalignment was defined as the sum of the deviation of the top and bottom edges, horizontal as the sum of the deviation of the right and left edges (point "b" in Figure 18). In according to the international standards, the misalignment must each be less than 25mm (AAPM, 1981). As it is seen from Table 8, all misalignmenst for 10 units are appropriate to the standards.

For the measurement of the resolution, parallel lead strips separated by a distance equal to the width of the strips, that are placed on the test tool (point "c" in Figure 18) were used. The common practice is to describe the line width and separation distance in terms of line pairs (lp) per unit distance (millimeters) (Lp/mm). One line pair consists of one lead strip and adjacent separation space. The number of line pairs per millimeter is actually an expression of spatial frequency. As the lines get smaller and closer together, the spatial frequency increases (Sprawls, 1987). The test pattern contains areas with different spatial frequencies. To evaluate an imaging system, the visible line group is recorded as line pairs per mm. In according to the international standards, resolution below 0,8Lp/mm is not acceptable (AAPM, 1981). The obtained test results in this study, are shown in Table 9.

Evaluating the contrast was performed by looking at the copper step wedge from the test pattern that are placed on the test tool (point "d" in Figure 18). The visible copper step wedges were recorded in order to describe the resolution quality (Table 10). In according to the international standards, all copper steps have to be clearly visible (AAPM, 1981).

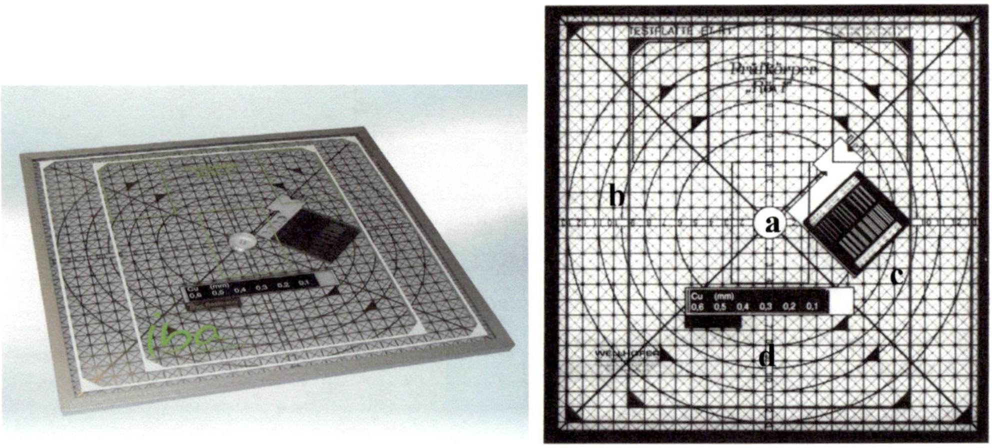

Fig. 18. ETR1 test tool used for image quality test.

Unit	Beam Alignment (mm)		Collimation Alignment (mm)			
1	< 10mm	OK	Top: 3mm Right: 1mm	Bottom: 2mm Left: 1mm	Total: 5mm Total: 2mm	OK
2	< 10mm	OK	Top: 5mm Right: 3mm	Bottom: 3mm Left: 1mm	Total: 8mm Total: 4mm	OK
3	< 10mm	OK	Top: 1mm Right: 2mm	Bottom: 1mm Left: 1mm	Total: 2mm Total: 3mm	OK
4	< 10mm	OK	Top: 3mm Right: 2mm	Bottom: 2mm Left: 3mm	Total: 5mm Total: 5mm	OK
5	< 10mm	OK	Top: 2mm Right: 2mm	Bottom: 1mm Left: 2mm	Total: 3mm Total: 4mm	OK
6	< 10mm	OK	Top: 6mm Right: 3mm	Bottom: 4mm Left: 3mm	Total: 10mm Total: 6mm	OK
7	< 10mm	OK	Top: 2mm Right: 3mm	Bottom: 2mm Left: 1mm	Total: 4mm Total: 4mm	OK
8	< 10mm	OK	Top: 4mm Right: 1mm	Bottom: 3mm Left: 1mm	Total: 7mm Total: 2mm	OK
9	< 10mm	OK	Top: 5mm Right: 2mm	Bottom: 4mm Left: 2mm	Total: 9mm Total: 4mm	OK
10	< 10mm	OK	Top: 7mm Right: 4mm	Bottom: 5mm Left: 3mm	Total: 12mm Total: 7mm	OK

Table 8. Beam alignment and collimation alignment test results.

As it is seen from Table 8, both beam alignments and collimation alignments of 10 x-ray units are appropriate to the international standards and there is no unwanted effect on the image quality.
Although beam alignment and collimation alignment are not dependent to the kVp and mAs value, resolution and contrast are directly related to these parameters. On Table 9, it is seen that resolution increases with increasing parameter setting values, especially with kVp. From Table 10, it can be said that contrast is good on the values of 70kVp, especially on 70kVp-40mAs for 10 x-ray units. At this value of parameters, all copper steps on the test tool can be seen definitely. While the values on the Table 10 decreases from 0,6 to 0,1, the contrast also decreases and the seeable points on the film loss step by step.

5. Discussion

In this study, the x-ray units with ages between 5 and 7 years old were selected to prevent the wide distribution of measured dose because the x-ray tubes don't produce the same exposure and the output decreases with age of x-ray unit.
Again, in this study, three phase generators were preferred because they produces more radiation exposure per unit mAs. This characteristic is essential for modeling of dose.
A difference in tube output among tubes is often caused by variations in the filtration. For this reason, this study were performed on the x-ray units with HVL values that changes approximately from 3,0 to 3,1mmAl. It also prevented the wide distribution of measured dose. The obtained HVL values in this study are acceptable in according to the international standards (AAPM, 1981).
It is known that dose is more sensitive to the kVp changes than mAs changes. Exposure errors can occur if the actual kVp generated by the x-ray generator is different from the adjusting kVp value. Before dose measurements, kVp accuracy testing were performed correctly and it was seen that the kVp during exposure was the close within the acceptable deviation to the selected kVp value.
All dose measurements were performed at different distance of 100cm and 60cm. With this application, the distance effects on dose were investigated and it was used for dose modeling because of the inverse-square effect.
For dose measurements, two different measurement procedures were used. In the first procedure, mAs value was kept constant and kVp values were changed to investigate the dose variation with kVp. In the second procedure, kVp value was kept constant and mAs values were changed to investigate the dose variation with mAs. Thus, the effects of kVp and mAs were examined separately.
Because the x-ray units were selected in according to the criterias mentioned above, the measured dose values didn't show wide distribution for each measurement setup in all 10 x-ray units. In this condition, the mean of the dose values of 10 x-ray units for each measurement setup was used to show the tube output variations with kVp. Plotting of tube output to kVp (Figure 15 and 16) were performed by using dose values obtained from two different measurement procedures. By this way, it was seen that the tube output variations related to kVp were approximately similar at different mAs value. Hence, the mean variations were used for modeling of dose.
Modeling was realized twice for dose values at different distances of 100cm and 60cm, because the different variations were seen between measurement values obtained different distances. By using equations (Equation 3 and Equation 7) in the models related to the

Resolution (Lp/mm)										
	Unit	50kVp	70kVp	80kVp	100kVp	50kVp	70kVp	80kVp	100kVp	
100cm - 20mAs	1	2,2	3,1	3,4	4,0	2,2	3,4	3,4	4,0	60cm - 20mAs
	2	2,2	3,4	3,7	4,3	2,2	3,1	3,4	4,0	
	3	2,5	3,1	3,7	4,0	2,0	2,8	3,0	3,7	
	4	2,5	3,4	3,4	4,0	2,2	3,1	3,4	4,3	
	5	2,5	3,4	3,4	4,3	2,5	3,4	3,4	4,3	
	6	2,0	2,8	3,7	4,3	2,2	3,1	3,4	4,0	
	7	2,5	3,1	3,7	4,3	2,2	3,1	3,4	4,3	
	8	2,0	2,8	3,4	4,0	2,0	2,8	3,1	4,0	
	9	2,0	2,8	3,4	4,0	2,2	3,1	3,4	4,3	
	10	2,2	3,1	3,7	4,3	2,2	3,1	3,7	4,3	
100cm - 40mAs	1	2,2	3,4	3,7	4,3	2,5	3,4	3,7	4,3	60cm - 40mAs
	2	2,2	3,4	3,7	4,3	2,0	3,1	3,4	4,0	
	3	2,5	3,4	4,0	4,3	2,5	3,4	3,7	4,3	
	4	2,5	3,1	3,4	4,0	2,2	3,1	3,4	4,0	
	5	2,2	3,4	3,7	4,3	2,5	3,4	3,7	4,0	
	6	2,8	3,4	3,7	4,3	2,8	3,7	4,0	4,3	
	7	2,8	3,7	4,0	4,3	2,5	3,4	3,7	4,0	
	8	2,2	3,4	3,7	4,0	2,5	3,4	3,7	4,0	
	9	2,5	3,4	3,7	4,3	2,2	3,1	3,7	4,3	
	10	2,2	3,1	4,0	4,3	2,2	3,4	4,0	4,3	
100cm - 50mAs	1	2,5	3,1	4,0	4,3	2,2	3,1	3,7	4,3	60cm - 50mAs
	2	2,8	3,4	4,0	4,6	2,5	3,1	3,7	4,3	
	3	2,8	3,4	4,0	4,6	2,8	3,4	4,0	4,6	
	4	2,5	3,1	3,7	4,3	2,2	3,1	3,7	4,3	
	5	2,5	2,8	3,7	4,3	2,2	3,1	3,7	4,6	
	6	2,8	3,1	4,0	4,6	2,0	2,8	3,7	4,3	
	7	3,1	3,4	4,3	4,6	2,8	3,1	4,0	4,6	
	8	2,8	3,1	4,0	4,3	2,0	2,8	3,7	4,0	
	9	2,8	3,1	4,0	4,6	2,5	3,1	4,0	4,6	
	10	3,1	3,7	4,3	4,6	3,1	3,4	4,3	4,3	

Table 9. Resolution test results for 10 x-ray units

Contrast (mmCu)										
	Unit	50kVp	70kVp	80kVp	100kVp	50kVp	70kVp	80kVp	100kVp	
100cm - 20mAs	1	0,3	0,5	0,4	0,1	0,4	0,5	0,3	0,1	60cm - 20mAs
	2	0,4	0,4	0,3	0,2	0,3	0,5	0,3	0,1	
	3	0,3	0,5	0,4	0,1	0,4	0,4	0,4	0,2	
	4	0,4	0,5	0,3	0,1	0,4	0,4	0,4	0,1	
	5	0,4	0,4	0,3	0,3	0,5	0,5	0,3	0,2	
	6	0,3	0,5	0,3	0,2	0,4	0,5	0,4	0,1	
	7	0,4	0,5	0,4	0,3	0,3	0,6	0,4	0,3	
	8	0,5	0,6	0,4	0,1	0,4	0,5	0,3	0,1	
	9	0,4	0,5	0,3	0,2	0,5	0,6	0,3	0,1	
	10	0,3	0,4	0,4	0,2	0,4	0,4	0,4	0,2	
100cm - 40mAs	1	0,5	0,6	0,5	0,3	0,4	0,6	0,5	0,2	60cm - 40mAs
	2	0,5	0,6	0,4	0,2	0,5	0,6	0,4	0,2	
	3	0,4	0,6	0,4	0,2	0,3	0,6	0,5	0,3	
	4	0,4	0,6	0,5	0,3	0,5	0,6	0,5	0,2	
	5	0,5	0,6	0,5	0,3	0,4	0,6	0,4	0,2	
	6	0,4	0,6	0,5	0,2	0,4	0,6	0,5	0,3	
	7	0,5	0,6	0,4	0,1	0,5	0,6	0,4	0,1	
	8	0,5	0,6	0,4	0,2	0,5	0,6	0,5	0,2	
	9	0,4	0,6	0,5	0,3	0,3	0,6	0,5	0,3	
	10	0,4	0,6	0,4	0,2	0,5	0,6	0,4	0,2	
100cm - 50mAs	1	0,4	0,6	0,5	0,3	0,5	0,5	0,4	0,3	60cm - 50mAs
	2	0,5	0,6	0,5	0,2	0,5	0,6	0,5	0,2	
	3	0,5	0,5	0,4	0,3	0,4	0,5	0,4	0,3	
	4	0,4	0,6	0,5	0,3	0,4	0,6	0,5	0,3	
	5	0,4	0,6	0,5	0,4	0,4	0,6	0,5	0,4	
	6	0,5	0,6	0,5	0,3	0,5	0,6	0,4	0,3	
	7	0,4	0,6	0,5	0,2	0,5	0,6	0,5	0,2	
	8	0,5	0,5	0,4	0,2	0,4	0,6	0,4	0,2	
	9	0,5	0,6	0,5	0,3	0,4	0,6	0,5	0,3	
	10	0,5	0,6	0,5	0,3	0,5	0,6	0,5	0,3	

Table 10. Contrast test results for 10 x-ray units.

distances, the dose was calculated for the parameter settings that are different from the parameter settings used for dose modeling. After estimation, the measured and calculated dose values were compared. And, it was seen that the dose estimation was very successful.

For observing of image quality, a film was exposed during each dose measurement, after this, it was developed. Contrast and resolution tests were performed on these films. From Table 10, it can be said that contrast decreased with increasing kVp. It was seen that the best contrast is possible at the values of 70kVps, especially at 70kVp-40mAs. Although the other mAs values with constant 70kVp show good contrast, the best contrast with low dose is determined at 70kVp-40mAs.

In the resolution tests, from Table 9, it can be said that resolution increased related to increasing kVp. Because of this, the resolution is good in kVp values of 100kVp with different mAs.

But, in this study, because our aim is to obtain high image quality (both good contrast and good resolution), the optimum parameter values were selected as recommendation. The parameter setting values of 70kVp-40mAs can be accepted as the recommended technical parameters to obtain high quality image and low dose. If it is wanted to increase the number of recommended parameters, all mAs changes with constant 70kVp (20, 40, 50mAs) can be used as quality control test parameters.

If a radiographic staff adjusts these recommended parameters in an x-ray device, he/she will know which characteristics will appear on the image and how much dose will be measured. Hence, by this way, the staff can control and evaluate his/her tests' results during quality control tests of x-ray units.

6. Conclusion

The technical x-ray parameters are very important to reduce the dose and to obtain the image with good quality. The dose reduction can be obtained by adequate changes of physical parameters without lose of image quality. The optimal radiation dose for optimal image quality can be achieved by understanding of the parameters that affect radiation dose and image quality. The dose optimization process also consists of quality control programs to test radiographic devices periodically. In this study, it was studied in which parameters' values were appropriate to obtain high quality image and to reduce dose, in other words, dose optimization, during quality control tests of x-ray units.

This study shows that optimization of technical factors may lead to a substantial dose reduction. If the optimized parameters are applied to X-ray equipment during quality control tests, it is possible to determine how much good image quality will be obtained with this optimized parameters and how much dose will be measured when this qualified image is developed.

The results show the importance of radiographic staff training about the recommended parameters that are applied to the x-ray units for a qualified quality control system. It is essential to provide relevant education and training to staff in the radiology departments.

It can be sure that with such a study the questions on many professional staff's mind will be answered, and the dose and the image characteristics will be parameters that are controlled and managed.

7. Acknowledgment

I would like to thank the co-operation of radiographers at all of the radiological departments participating in this study.

8. References

Aldrich, J., Duran, E., Dunlop, P., & Mayo, J. (2006). Optimization of dose and image quality for computed radiography and digital radiography. *Journal of Digital Imaging*, Vol.19, No.2, (June 2006), pp. 126-131

AAPM (1981). Basic quality control in diagnostic radiology. *AAPM Report No. 4*, (1981)

AAPM (2002). Quality control in diagnostic radiology. *AAPM Report No. 74*, (2002)

Bouzarjomehri, F. (2004). Patient dose in routine X-ray examinations in Yazd state. *Iran. J. Radiat. Res.*, Vol.1, No.4, (2004), pp. 199-204

Brix, G., Nekolla, E., & Griebel, J. (2005). Radiation exposure of patients from diagnostic and interventional X-ray procedures. *Radiologe*, Vol.45, No.4, (April 2005), pp. 340-349

Ciraj, O., Markovic, S., & Kosutic, D. (2005). First results on patient dose measurements from conventional diagnostic radiology procedures in Serbia and Montenegro. *Radiation Protection Dosimetry*, Vol.113, No.3, (March 2005), pp. 330-335

Geijer, H. (2002). Radiation dose and image quality in diagnostic radiology. Optimization of the dose-image quality relationship with clinical experience from scoliosis radiography, coronary intervention and a flat-panel digital detector. *Acta Radiol. Suppl.*, Vol.43, (March 2002), pp. 1-43

Gray, J., Archer, B., Butler, P., Hobbs, B., Mettler, F., Pizzutiello, R., Schueler, B., Strauss, K., Suleiman, O., & Yaffe, M. (2005). Reference values for diagnostic radiology: Application and impact. *Radiology*, Vol.235, No.2, (May 2005), pp. 354-358

Hendee, W., Chaney, E., & Rossi, R. (1984). *Radiologic Physics, Equipment and Quality Control*, Year Book Medical Publishers, Chicago, USA

IAEA (1996). *International Basic Safety Standards for Protection against Ionizing Radiation and for the Safety of Radiation Sources*. IAEA Safety Series 15, ISBN 92-0-104295-7, Vienna, Austria

Iba Dosimetry (2008). Dosimax Plus A User Manual, 21 March 2011, Available from: www.iba-dosimetry.com

ICRP (1991). 1990 Recommendations of the international commission on radiological protection. ICRP Publication 60, *Annals of the ICRP*, Vol.21, No.1-3, (1991)

IEC 61223-3-1:1999 (1999). *Evaluation and routine testing in medical imaging departments. Acceptance tests. Imaging performance of X-ray equipment for radiographic and radiscopic systems*, BSI, ISBN 0-580-32753-1, London, England

Papadimitriou, D., Perris, A., Molfetas, M., Panagiotakis, A., Manetou, A., Tsourouflis, G., Vassileva, J., Chronopoulos, P., Karapanagiotou, O., & Kottou, S. (2001). Patient dose, image quality and radiographic techniques for common X-ray examinations in two Greek hospitals and comparison with European guidelines. *Radiation Protection Dosimetry*, Vol.95, No.1, (2001), pp. 43-48

Ramanandraibe, M., Andriambololona, R., Rakotoson, E., Tsapaki, V., & Gfirtner, H. (2009). Survey of image quality and patient dose in simple radiographic examinations in Madagascar: Initial results, *Proceedings of HEP-MAD 09*, Antananarivo, Madagascar, August 21-28, (2009)

Schaefer-Prokop, C., Neitzel, U., Venema, H., Uffmann, M., & Prokop, M. (2008). Digital chest radiography: an update on modern technology, dose containment and control of image quality. *Eur. Radiol.*, Vol.18, (April 2008), pp. 1818-1830

Seibert, A. (2004). Tradeoffs between image quality and dose. *Pediatr. Radiol.*, Vol.34, No.3, (2004), pp. 183-195

Shahbazi-Gahrouei, D. (2006). Entrance surface dose measurements for routine X-ray examinations in Chaharmahal and Bakhtiari hospitals. *Iran. J. Radiat. Res.*, Vol.4, No.1, (2006), pp. 29-33

Sprawls, P. (1987). *Physical Principles of Medical Imaging*, Aspen, ISBN 0-87189-644-3, Maryland, USA

Vano, E., & Fernandez Soto, J. (2007). Patient dose management in digital radiography. *Biomedical Imaging and Intervention Journal*, Vol.3, No.2, (2007)

Williams, J., & Catling, M. (1998). An investigation of X-ray equipment factors influencing patient dose in radiography. *The British Journal of Radiology*, Vol.71, (November 1998), pp. 1192-1198

17

Infectious Aetiology of Cancer: Developing World Perspective

Shahid Pervez
Professor, Section of Histopathology, Department of Pathology & Microbiology, Aga Khan University Hospital Karachi, Pakistan

1. Introduction

Infection attributable cancers contribute over 1/4th of all cancers in the developing countries (26.3%) compared to the developed countries (7.7%), (Parkin, 2006). Overwhelming majority are related to viral infections. In contrast to other carcinogens where it is usually a *'hit and run'* kind of situation, with infectious agents particularly viruses one may precisely demonstrate and prove its presence and integration within host neoplastic cells. Oncogenic DNA viral genome incorporates itself directly into host cells DNA while oncogenic RNA viral genome is transcribed into host cell DNA by reverse transcriptase. Neoplastic transformation usually follows. Oncogenic mechanisms include acting as promoter, transforming protooncogenes into oncogenes. Credit goes to Dr Peyton Rous, a noble laureate pathologist who demonstrated that it was possible to transmit tumours from one animal to other like transmission of an infection.

Human tumours with proven or proposed viral aetiology include 'Human papillomavirus (HPV)', Epstein-Barr Virus (EBV), Hepatitis B and C viruses, RNA retroviruses like 'Human T-lymphotropic virus (HTLV1)', 'Human Herpes Virus-8 (HHV-8). Bacteria with proven carcinogenic potential include 'Helicobacter pylori'. Among fungi aflatoxins produced by 'Aspergillus flavus' are potent carcinogens. Among parasites 'Schistosoma' and 'Clonorchis sinensis' are implicated in the causation of cancer.

2. Human papillomavirus (HPV)

HPV is a small epitheliotropic, non enveloped DNA virus belonging to papovaviridae family. Its genome comprises 7000-8000 base pairs of double-stranded closed-circular DNA. At least 70 genetically distinct types of HPV have been identified in humans. According to their oncogenic potential HPV is classified in a high oncogenic risk group (i.e., HPV16, 18, 31, 33, 35, 39, 45, 51, 52, 56, 58, 65, 66) and low oncogenic risk group (i.e., HPV6, 11, 42, 43, 44). High risk HPV association with cervical and anogenital cancers is established beyond doubt. HPV16 and 18 are declared as human carcinogens by 'international Agency for Research on Cancer (IARC)'. HPV association with other cancers in particular with *'oral cancer'* is also being investigated with evidence of significant association.

2.1 HPV & cervical carcinoma

Cervical carcinoma is one of the most common malignancies in women worldwide. However with effective preventive measures like *'cervical screening'* programs in developed countries more and more cases are picked at an early stage where complete cure is possible. A significant recent breakthrough has come in the form of *'HPV vaccine'* against high risk HPV16 & 18. This is gaining momentum in developed countries with high risk and burden of disease. It is administered at age 11-13, three shots are given intramuscularly. In contrast in the developing countries data is patchy or non-existent. In most countries no cervical screening programs are in place. In developing muslim countries situation is even worse. For instance in Pakistan, a populous muslim country of about 170 million inhabitants there is no cervical screening program and the only source of cervical smears are the sporadic smears obtained at the time of consultation in obstetrics & gynecology clinics. The problem is further compounded by the social taboos on matters of sexual practices and sexually transmitted infections (STI). These socio-cultural prohibitions create a substantial barrier to such investigations.

Recently a study was carried out with the help of IARC in women of Karachi, Pakistan (largest port city of Pakistan with an estimated population of 15 million from diverse ethnic backgrounds), (Raza et al, 2010). A sum of 899 married women aged 15-59 years living in a densely populated suburb of Karachi consented to participate. HPV prevalence was found to be 2.8%. Cervical abnormalities were diagnosed in 2.4% of whom 27.3% were HPV positive. HPV16 was detected as the most common type among women with both normal (0.5%) and abnormal (9.1%) cytology. This study also included 91 invasive cervical carcinomas (ICC) from two major university hospitals of Karachi, Pakistan. HPV16 was also the predominant HPV type (75.8%) in ICC followed by HPV18 (6.6%). This study led to the suggestion of very low burden of HPV infection in general female population, considerably lower than neighboring India (17%, Franceschi et al, 2005), China (15-18%, Dai et al, 2006, Wu et al, 2007) and Nepal (9%, Sherpa et al, 2010).

In another study from Karachi, Pakistan (Khan et al, 2007) women visiting two major tertiary care hospitals in Karachi, diagnosed with ICC, sixty (60) paraffin-embedded biopsies were analysed for HPV subtypes by PCR. Out of the 60 samples only one was negative for HPV, the rest were positive excluding two samples where subtype could not be determined. Fifty six (56) were HPV16 positive and only one was HPV18 positive.

2.1.1 Conclusion

- In most developing countries particularly developing muslim countries data regarding sexually transmitted infections (STI) is either non-existent or sparse.
- A comprehensive well planned study utilizing samples from asymptomatic married women of Karachi, Pakistan revealed very low incidence of HPV (2.8%).
- Samples from both asymptomatic women as well as from invasive cervical cancer (ICC) showed overwhelming predominance of HPV16.
- Results from Karachi, Pakistan underscore the importance of cervical screening programs & HPV vaccination in resource constrained economies.

3. HPV and oral cancer

Oral cancer (OC) / Oral squamous cell carcinoma (OSCC) excluding salivary gland cancers ranks 6th overall in the world in both sexes with much higher incidence in the developing countries. In Karachi it ranked 2nd with an identical risk in both genders (Bhurgri et al, 2003). However if combined with pharynx and larynx cancers which have the same histologic type (squamous cell carcinoma) & risk factors it ranks number 1. Major risk factors in the developed world include *'smoking'* and *'alcohol;* however in the developing world though smoking is a common major risk factor, role of alcohol drinking is possibly a minor risk factor particularly in developing muslim countries. In subcontinent (Pakistan, India & Bangladesh) alternate chewing habits like betel quid and areca nut are major risk factors. Areca nut is now declared by WHO as a bonafide carcinogen. People using paan (betel leaf) are about 8- 9 times more likely to develop oral cancers as compared to non-users (Merchant et al, 2000). Smokeless tobacco, including *'gudka'* and *'niswar'* is an extremely addictive substance with a high rate of use in younger age groups, as well is contributing toward endemic rise of oral cancers in Pakistan (Ali et al, 2009 & Nair et al, 2004). (Figure 1) This habit commonly leads to a pre-malignant condition *'Submucosal fibrosis'* which commonly transforms into OSCC. Poor oral hygiene is another contributory factor in this population.

A significant proportion of OSCC patients however deny exposure to conventional and well known risk factors. This has led to search of other risk factors and associations including microbes (Scully et al, 1985). The striking commonality between oral cavity and cervical cavity paved the way to look for epitheliotropic viruses like HPV. Although these two areas are anatomically different, the squamous epithelium found in both areas has several similarities. For instance the squamous epithelium of ecto-cervix and oral cavity including pharynx and larynx are composed of squamous epithelium with a thin layer of keratin or no keratin. In both areas the epithelium is subject to microtrauma of various types as well as to bacteria and varying chemical irritants. Most common malignancy at both anatomic sites is also SCC with varied differentiation. (Figure 2) These factors may directly expose to HPV infections of cells resulting in malignant transformation. Furthermore the HPV subtypes isolated from lesions of squamous epithelium of cervix are similar to the type found in both normal epithelium and various lesions of the oral cavity, pharynx and larynx. These include HPV subtypes16, 18, 31 & 33.

The reported prevalence of HPV in OSCC varies widely in various studies depending on the population and ethnicity studied and/or sensitivity of the methods used and viral DNA sequence targeted. HPV in particular HPV-16, like in cervix is implicated in the aetiology of OSCC (Gillison, 2004; Miller & Johnstone, 2001) About 40 – 60% of patients with tumours of oropharynx are reported to be positive for HPV infection (Gillison, 2004; Kreimer et al, 2005). HPV-positive tumours are distinct from HPV-negative tumours in their biological characteristics and clinical behaviour. Data from retrospective analyses as well as a prospective clinical trial demonstrated that HPV-positive oropharyngeal tumours are more sensitive to chemotherapy and radiation treatment and have a markedly improved prognosis and favourable clinical outcome compared with HPV-negative tumours (Fakhry et al, 2008; Settle et al, 2009)

Recently, another significant observation has emerged in terms of HPV status of oropharyngeal tumours and racial disparities. Black Americans are known to have a higher incidence of and mortality from head and neck squamous cell carcinoma (HNSCC) than the whites and present with more advanced disease at a younger age (Goodwin et al, 2008; Morse & Kerr 2006; Ryerson et al, 2008; Shiboski et al, 2007). The greatest survival difference between blacks and whites was detected specifically in oropharyngeal cancers, but there was no racial difference between the overall survival rates of patients with non-oropharyngeal tumours (Settle et al, 2009). Most importantly, the recently published prospective analysis demonstrated that a marked difference exists between black and white Americans in terms of HPV infection. HPV positivity was about 9-fold higher in white (34%) than in black (4%) patients, directly correlating HPV infection with significant survival disparities between the two populations (Settle et al, 2009). Clearly, the HPV status of patients with OSCC would be an important determinant for prognosis and treatment options in the future.

Recently, in a retrospective study of 140 patients with primary OSCC and a long-term follow up, Ali et al reported from Karachi, Pakistan, 68% of cases to be positive for HPV (Ali et al, 2008). Approximately 90% of these cases were infected with HPV16, (Figure 3 & 4) the predominant subtype in the US population as well. HPV infection was detected entirely in tumours of the cheek and tongue in the oral cavity; this was consistent with the occurrence frequency in the Karachi population for oral cancers which is as follows: 55.9% for cheek, 28.4% for tongue, 6.8% for palate, 4.4% for gum, 3.1% for lip and 1.4% for floor of the mouth (Bhurgri et al, 2003). Furthermore, though HPV positive patients had comparatively prolonged overall survival when compared with HPV negative patients but the difference was not statistically significant (P=0.97) (Figure 5). Betel quid chewers were comparatively more prone to HPV positivity (OR=2.34; 95 CI= 1.1-4.31). These findings are in contrast with the results from US studies where the ratio of oropharyngeal tumours with respect to other sites was 2:1 and the HPV-positive tumours were consistently associated with a better clinical outcome in terms of both overall and disease-free survival (Fakhry et al, 2008; Settle et al, 2009). The reason(s) for these different findings are not clear.

3.1 Oncogenic HPV pathways

The chief oncoproteins of HPV16 are encoded by the genes E6 and E7. The E6 protein targets the tumour suppressor gene p53 for degradation. In fact, degradation of p53 in HPV positive cells is fully dependent on the presence of E6 (Ali et al, 2010, Figure 6). The E7 oncoprotein is involved in suppression of retinoblastoma protein (pRb) function. Reduced pRb expression is common in HPV-positive tonsillar cancer.

3.2 Mode of transmission

Two questions immediately come to mind, first how HPV gets there and second why patients with HPV association will have better survival. In response to question 1, haematogenous spread from genital tract is proposed besides atypical sexual habits. In response to question 2 one possible explanation is that HPV infection may lead to genome instability, paradoxically making tumour cells more susceptible to radiotherapy.

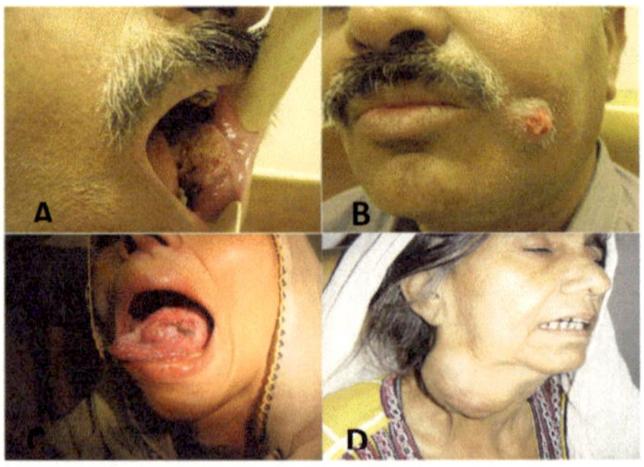

Fig. 1. Clinical presentations of patients with oral squamous cell carcinoma (OSCC) in Pakistani patients (Photographs were taken with patient's consent).

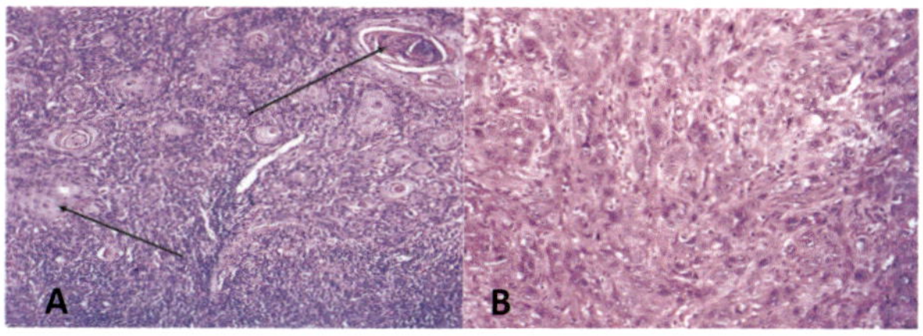

Fig. 2. Photomicrograph of H & E stained (A) well differentiated oral squamous cell carcinoma showing diffuse sheets of squamous cell with prominent keratinization and keratin pearl formations ↖ ↗, Magnification X 10. (B) poorly differentiated oral Squamous cell carcinoma, Magnification X 20.

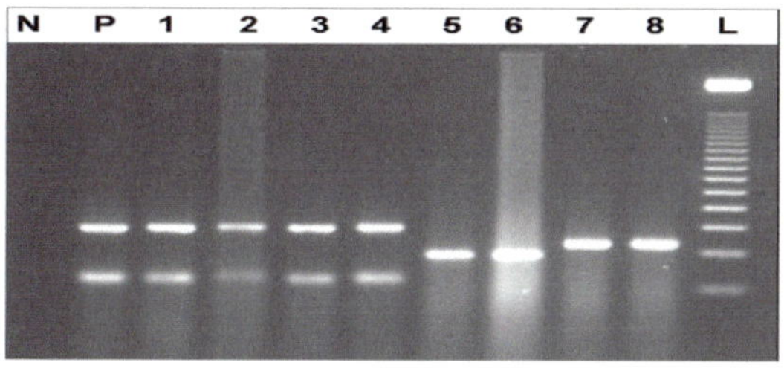

Fig. 3. PCR amplification of HPV general, HPV type 16 and HPV type 18 in OSCC samples. The products were electophoresed on 2% agarose gel and stained with ethidium bromide. Lane N: negative control, lane P: positive control, lanes 1-4 HPV (general primer) positive tumour samples, lanes 5-6 HPV 16 positive tumour samples, lanes 7-8 HPV 18 positive tumour samples, Lane L: molecular size marker (50-bp ladder marker).

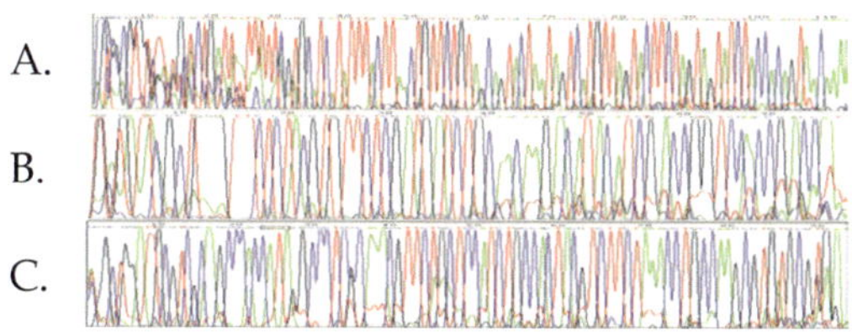

Fig. 4. Result of sequence analysis of PCR products, (A), Gene Sequencing HPV General. (B), Gene Sequencing HPV Type 16. (C), Gene Sequencing HPV Type 18.

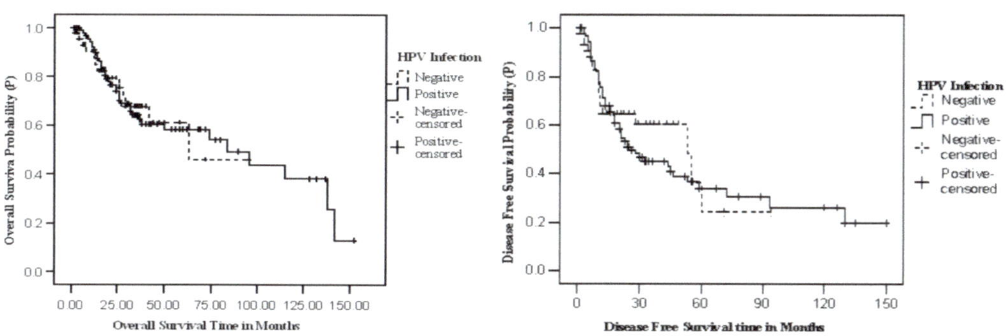

Fig. 5. Kaplan-Meier curves of overall survival (OS) of (A) human papillomavirus (HPV) positive patients as compared with HPV-negative patients. (B) Disease Free Survival of human papillomavirus (HPV) positive patients as compared with HPV-negative patients.

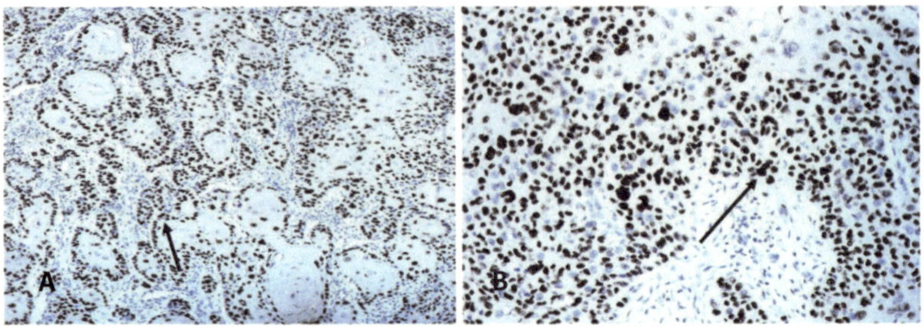

Fig. 6. Photomicrograph of a well-differentiated OSCC demonstrating diffuse strong nuclear TP53 staining. The arrows indicate positive dark brown intranuclear staining (magnification x 10 (A) & 20 (B)).

3.3 Conclusions

- Incidence of Oral Cancer in subcontinent (Pakistan, India & Bangladesh) is one of the highest in the world.
- Alternate chewing habits alongwith cigarette smoking are major risk factors in this part of the world.
- High risk HPV association was seen in 68% of the cases of OC in a high risk population of Karachi (Pakistan) with 90% containing HPV 16.
- Survival advantage was seen in OC patients with HPV association albeit not coming to statistical significance as seen in American whites.

4. Epstein-Barr virus (EBV)

EBV was initially discovered from cell cultures of a high grade B-cell lymphoma *'Endemic (African) Burkitt Lymphoma (BL)'*, which is highly prevalent in paraequatorial Africa and New Guinea. The disease affects children and adolescents and has strong association with malaria. Endemic BL commonly involves extra-nodal sites particularly jaw. In rest of the world *'sporadic form of BL'* is seen having a weaker association with EBV and commonly affecting gastro-intestinal tract (GIT) particularly small intestine.

EBV associated other lymphomas include 'classic Hodgkin lymphoma (cHL)' particularly 'mixed cellularity' type (~60%), 'B-cell lymphoma in immunosuppressed', 'mature T-cell lymphoproliferative disorders' in particular 'Angioimmunoblastic T-cell lymphoma (AILT)', 'Angiocentric (Nasal) T-cell lymphoma'. Non-lymphoid associations include 'Nasopharyngeal carcinoma'.

4.1 EBV & mature T-cell non-Hodgkin lymphoma (T-NHL)

EBV association with certain subsets of T-NHL is now well established. In a study conducted by us in Pakistan (Noorali et al, 2003), mature T-NHL comprised 22.2% of total mature NHLs. These cases were characterised on the basis of morphology, immunohistochemistry and T-cell receptor (TCR) gene rearrangement studies. This study demonstrated frequent presence of EBV in mature T-NHL cases (55.4%) by *'PCR'* (Figure 7) and *'in-situ hybridization (ISH)'*. While analysing various subsets of mature T-NHL 'Peripheral T-cell lymphoma (PTCL) - unspecified' (n=88) showed 51.2% EBV positive cases. EBV can be differentiated according to size polymorphism depending on the number of internal repeats in the Bam HI, E, K, N and Z regions. We also studied the extent of polymorphism in EBV genome by *'single stranded conformation polymorphism (SSCP)'* technique for *'Bam HI E, K, N and Z regions'*. Hypervariability in Bam HI, K and N regions was noticeably higher compared to *E or Z* regions. All in all no association was established between EBV variants differentiated on the basis sequence heterogeneity in *Bam HI, K, N, E and Z regions* in different subsets of T-NHL.

Mode of infection of T-cells by EBV is complex and poorly understood. Nazaruk et al, 1998 proposed that initially virus infects the B-cells and remains in the latent phase but under immunosuppressive conditions IL10 is secreted by EBV specific CD8+ T-cells activating B-cells. Subsequently reactivation of EBV lytic cycle occurs that may contribute to the development of EBV-associated T-cell lymphoproliferative disorders.

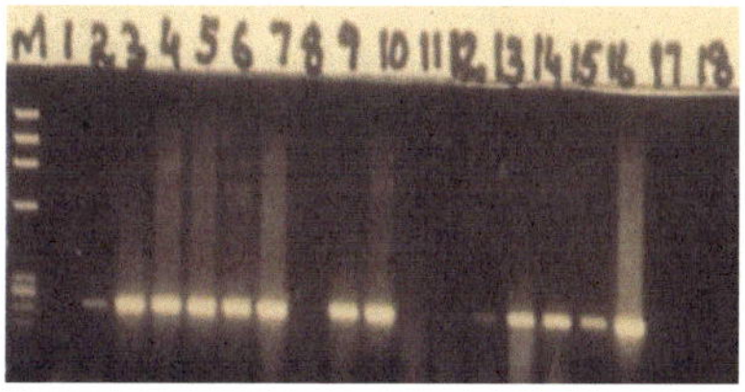

Fig. 7. Ethidium Bromide stained agarose gel showing PCR products of EBV-DNA amplified with primers specific for gp200 region.

4.2 EBV & angioimmunoblastic T-cell lymphoma (AILT)

AILT is an uncommon form of mature T-NHL characterised by systemic disease that occurs predominantly in middle-aged and elderly patients. The clinico-pathologic syndrome is characterised by fever, night sweats, weight loss, generalised lymphadenopathy, hepatomegaly and splenomegaly. Histologic examination of lymph nodes typically shows effacement of lymph nodes architecture, a polymorphous infiltrate including immunoblasts, lymphocytes, plasma cells, eosinophils, epithelioid histiocytes and a prominent arborizing postcapillary vasculature (Figure 8). In a study conducted by us a total of 13 well characterised cases of AILT based on morphology, IHC and TcR gene rearrangement studies were analysed for EBV by PCR and ISH (EBER). Association of EBV was seen in 11 out of 13 cases (84.6%) by PCR. By ISH (EBER), EBV was detected in 8 out of 9 cases (88.8%) cases. (Figure 8) So all in all strongest correlation of EBV was seen in this type of T-NHL. (Noorali et al, 2005).

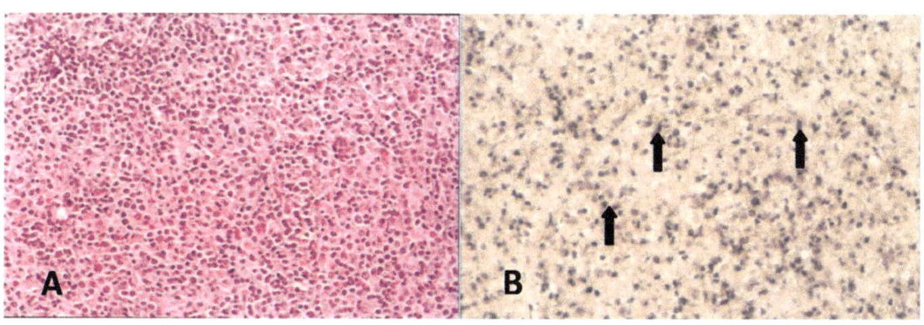

Fig. 8. I*n situ hybridization* photomicrograph of a lymph node showing the localisation of EBV in the nuclei of neoplastic lymphocytes indicated by blackish signal (↑) (B), H& E of the same (A).

4.3 EBV & Mycosis fungoides (MF)

MF is an indolent T-cell lymphoma of skin. In a study conducted by us a total of 14 well characterised cases of MF were analysed for EBV by PCR and ISH (EBER). EBV was identified in 3 out of 6 cases (50%) by PCR but all these were negative on ISH (EBER). This discrepancy is most likely caused by low copy number of infected cells in tissue sections not amplified as in PCR based studies (Noorali et al, 2002).

4.4 EBV & anaplastic large cell lymphoma (ALCL)

ALCL is a peculiar type of T-NHL. In a recent study by us (Syed et al, 2011) ALCL was turned out to be the most common T-NHL in the archives of the largest referral centre of Pakistan. This variant of T-NHL however has the weakest association with EBV (Noorali et al 2004).

5. HTLV-1 & T-NHL

HTLV1 is a RNA oncogenic virus which is associated with *'adult T-cell leukemia /lymphoma'* and is endemic in southern Japan and Caribbean basin. Like HIV which causes AIDS, HTLV1 also shows tropism for CD4+ T cells, hence this subset is the main victim for neoplastic transformation. In our local studies HTLV1 association was absent in mature T-lymphoproliferative disorders. This is in line with relatively low burden of HIV-AIDS in Pakistan so far (Noorali et al, 2004). (Figure 9)

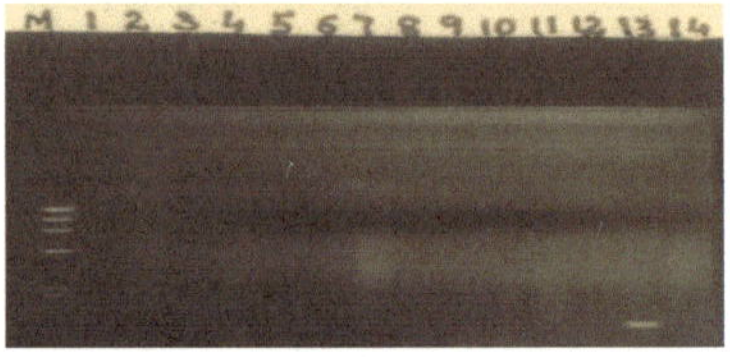

Fig. 9. Agarose gel showing samples of mature T-NHL negative for HTLV-1 DNA by PCR.

6. Role of EBV detection by PCR, ISH & IHC in diagnostic pathology

The ability to amplify specific regions of DNA from paraffin-embedded tissue by PCR has a profound impact on diagnostic pathology. For routine histopathological diagnosis of various lymphoproliferative disorders EBV-ISH (EBV-encoded nuclear RNA -1(EBER-1) and IHC by using an antibody to 'Latent Membrane Protein-1 (LMP-1) are frequently used in diagnostic dilemmas. For instance in the differential diagnosis of cHL and ALCL, EBER or LMP-1 positivity in neoplastic cells will strongly favour cHL as EBV association with ALCL is very *weak.* (Figure 10)

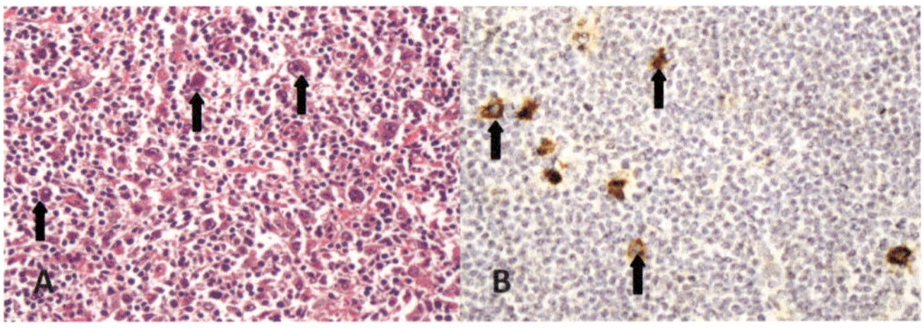

Fig. 10. Photomicrograph of a case of Hodgkin lymphoma (mixed cellularity, H&E, A ↑) stained with an antibody to LMP-1, note cytoplasmic staining of large Hodgkin cells (↑ B).

6.1 Conclusions

- EBV association with *'classic Hodgkin lymphoma'* is well established and is strongest with *'mixed cellularity variant'*
- Endemic (African) *'Burkitt lymphoma (BL)'* involving jaw is strongly associated with EBV while association with *'sporadic BL'* is relatively weak.
- EBV association with mature T-NHL is variable
- Strongest association of EBV is seen in AILT subtype while weakest association with ALCL subtypes.
- EBV immunohistochemistry using an antibody for LMP-1 and *'in situ hybridization (EBER)'* are commonly used in routine diagnostic pathology
- HTLV-1 association with 'adult T-cell leukemia/lymphoma' endemic in Japan & Caribbean is insignificant in our experience in Pakistan.

7. EBV & nasopharyngeal carcinoma

Nasopharyngeal carcinomas are particularly common in some parts of Africa and southern China. In former they constitute most frequent childhood cancer while in the later adults are mostly affected. Association of EBV with nasopharyngeal carcinoma is well established. In fact this association is literally 100%. EBV associated protein LMP-1 is expressed in most cases. (Figure 11)

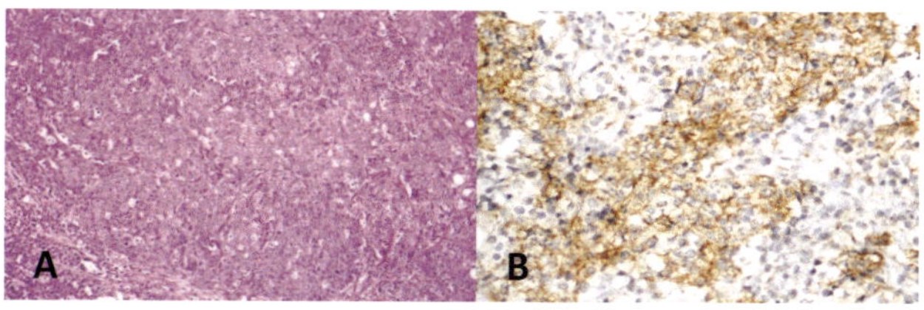

Fig. 11. Photomicrograph of a case of Nasopharyngeal carcinoma (H& E, A) stained with an antibody to LMP-1 (B), note cytoplasmic staining of neoplastic cells.

7.1 Conclusions

- EBV association in *'nasopharyngeal carcinoma'* is literally 100%

8. HHV8 & Kaposi sarcoma

Relatively recently in 1994 *'Human Herpesvirus-8 (HHV-8)* was identified in an AIDS patient with cutaneous *'Kaposi sarcoma (KS)'*. Later it was found that over 95% KS are associated with HHV-8. This virus is largely transmitted sexually. An antibody against HHV-8 shows positive reactivity in about 100% of cases and is a useful tool to confirm the diagnosis. Although *'Kaposi sarcoma'* is uncommon in our practice in Pakistan, it is highly prevalent in developing world with high AIDS incidence. (Figure 12) Four forms are recognized based primarily on population demographics and risk factors. These include a) *'Chronic KS'*, also called European KS b) *'Lymphoadenopathic KS'* also called African or endemic KS c) *'Transplant associated KS'* and d) *'AIDs-associated KS'*.

8.1 Conclusions

- HHV-8 association with KS is near 100%

- KS incidence in countries like Pakistan with relatively low burden of HIV carriers & AIDS is very low

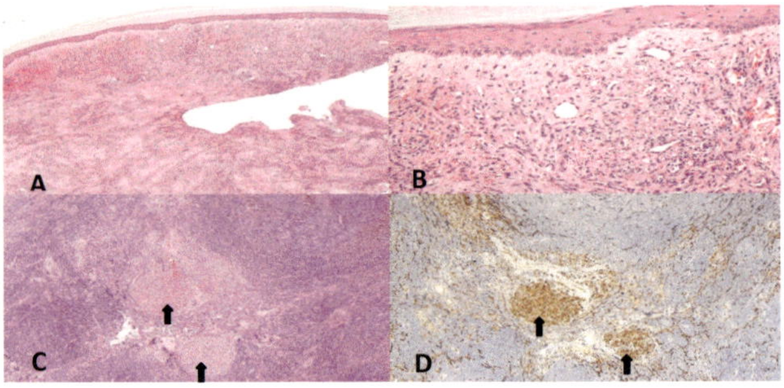

Fig. 12. Photomicrograph of Kaposi sarcoma resected from skin (H&E, A & B). Figure C shows lymph node metastases of the same (↑) highlighted on immunohistochemistry with CD31 (↑D).

9. *Helicobacter pylori* & gastric MALT lymphoma

'MALT lymphomas' were first described in 1983 by Peter Isaacson and Dennis Wright. They noted that primary low grade gastric B cell lymphomas recapitulate the histology of *'Mucosa Associated Lymphoid Tissue (MALT)'* exemplified by the Peyer patches and coined the term *'MALT lymphoma'*. These lymphomas are currently recognized as *'Extranodal marginal zone B cell lymphomas of MALT type'* according to the *'WHO Classification for Tumours of Haematopoietic and Lymphoid Tissues'* (Issaacson et al, 2008) (Figure 13). The stomach is the most reported and best studied site of *'MALT Lymphomas'*. An intimate relationship has been reported between the presence of *'Helicobacter pylori (HP)'* in the stomach and the development of *'MALT Lymphoma'* (Figure 13). In fact the pathogenesis of gastric *'MALT Lymphoma'* is believed to be caused by repeated antigenic stimulation of the immune system in the stomach by HP. The role of HP in the pathogenesis of *'gastric MALTomas'* can be illustrated by the fact that 75% of the patients who have gastric MALToma undergo remission if treated with antibiotics to eradicate HP (Ono et al, 2008). About half the people in the world have HP colonized in their gastrointestinal tract. Of these most remain asymptomatic. Despite the fact that, a high prevalence of HP is reported from Pakistan (Pervez et al, 2011), the prevalence of *'gastric MALTomas'* is very low in our experience. Seroprevalence of HP infection in the Pakistani population has been reported as high as 58%. This correlates with the *'Asian enigma'* described by various authors where less developed Asian countries like Pakistan, India, Bangladesh and Thailand have lower rates of gastric carcinoma compared to well developed countries like Japan and China, despite a higher prevalence of HP infection in the population. HP has been established to have a role in the aetiology of gastric carcinoma and its paradoxical high prevalence in areas with few cases of gastric carcinoma has long puzzled researchers. Available evidences do not support difference in HP strains as the sole explanation for this enigma.

9.1 Conclusions

- Helicobacter Pylori (HP) association with gastric adenocarcinoma & MALT lymphoma is well established.

- In Pakistan though prevalence of HP is very high, associated gastric adenocarcinoma & MALT lymphoma is low.

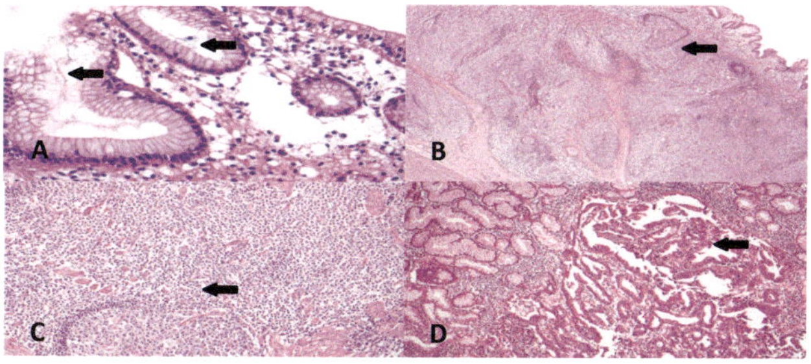

Fig. 13. Photomicrograph of gastric biopsy showing abundant *'Helicobacter pylori'* organisms on epithelial surface (←A), Figure B & C shows gastric MALT lymphoma arising from the marginal zone of lymphoid follicle ←. Figure D shows well differentiated adenocarcinoma of stomach (intestinal type ←).

10. Immunoproliferative small intestinal disease (IPSID) & *Campylobacter jejuni*

Immunoproliferative small intestinal disease (IPSID) is a special variant of, *'Extranodal marginal zone B cell lymphoma'*, which affects the small intestine. In early to mid 1960s it was referred to as *'Mediterranean lymphomas'*, during late 1960s the term *'a-heavy chain disease'* was also used for patients with similar clinico-pathological presentations. Later it was realized that both *'Mediterranean lymphomas'* and *'a-heavy chain disease'* represented a spectrum of the same disease which presents in different stages i.e., benign, intermediate and overtly malignant (stage A, B & C) and the disease was named IPSID (Fine & Stone 2000). IPSID is predominantly found in patients of *'Mediterranean origin'*; however a few cases of IPSID are also diagnosed in the subcontinent (Pervez et al, 2011). IPSID involves the production of truncated alpha heavy chains which may appear in the serum and other body fluids. It can be treated with broad spectrum antibiotics at its early stages.

It is postulated that IPSID occurs in patients with repeated intestinal infections. Recent studies suggest association with *Campylobacter jejuni* (Lecuit et al, 2004). It is postulated that this results in continuous chronic antigenic stimulation of IgA secreting lymphoid tissue common in small intestine with a resultant clonal proliferation of IgA secreting lymphoid cells. Subsequently most cases lose the ability to synthesize light chain. In early stages it may be very difficult to differentiate IPSID, from chronic inflammatory process by the reporting pathologists. In such circumstances it may be impossible to diagnose without the help of clonal studies for IgH chain gene rearrangement (Figure 14). The other close mimicry includes *'Coeliac disease'* as both IPSID and *'Coeliac disease'* are characterised by lympho-plasmacytic infiltrate and villous atrophy. In these cases demographics are important; also gluten free diet will lead to improvement of *'Coeliac disease'* cases. Intra-intraepithelial lymphocytosis with surface epithelial damage shall also favour Coeliac disease. As some cases of IPSID particularly if untreated may transform into aggressive lymphomas like *'Diffuse large B-cell lymphoma' (DLBCL)*, recognition of subtle features and follow-up is of paramount importance, particularly in endemic regions.

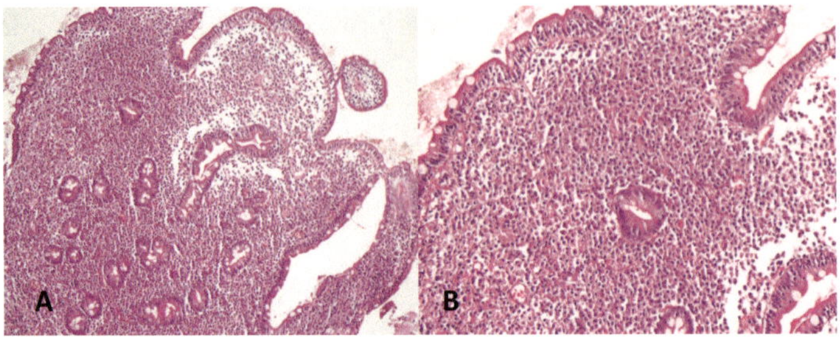

Fig. 14. Photomicrograph of duodenal biopsy from an IPSID patient diagnosed at stage A. Note flattening of mucosa with loss of villous architecture. Lamina propria shows diffuse sheets of plasma cells (H&E A & B).

10.1 Conclusions

- IPSID is a special variant of *'Extranodal marginal zone B-cell lymhoma'* predominantly found in Mediterranean region with sporadic cases in sub-continent.
- Recent studies suggest association with *'Compylobacter jejuni'*.
- In early stages treatment with broad spectrum antibiotics like tetracycline is curative.

11. Hepatitis B virus (HBV), Hepatitis C virus (HCV) & Hepatocellular carcinoma (HCC)

Developing countries bear major burden of 'Hepatitis B & C' for the obvious reasons i.e., insufficient or no screening of transfused blood, multiple use of contaminated needles, drug abuse and overall poor safety standards (Jafri et al, 2006). Pakistan for instance carries a very high burden of hepatitis B & C. There are estimated 7-9 million carriers of hepatitis B with a carrier rate of 3-5% (Ali et al, 2011). Genotype D (63.71%) is the most prevalent genotype in Pakistani population (Ali et al, 2011). The overall anti-HCV prevalence rate is 14-15% in general population of Pakistan (Idrees et al, 2009). Though hepatitis C is a major culprit for the reasons including increased potential to cause *'chronic liver disease'* and *'no vaccination'*; hepatitis B is still highly prevalent as well. A large proportion of population is still not vaccinated for hepatitis B, though now it is included in EPI (Extended Program of Immunization) program by the government and all newborns do get it.

In a recent study from Pakistan out of 161 subjects with HCC, chronic HCV infection was identified as a major risk factor (63.44% of tested HCC patients) for the development of HCC (Idrees et a, 2009). The time from HCV infection to the clinical appearance of cancer ranged from 10-50 years. In this population with HCC among various genotypes of HCV, genotype 3a was predominant (40.96%), followed by 3b in 15.66%, 1a in 9.63% and 1b in 2.40%.

On the face of such a high burden of Hepatitis B & C, hepatocellular carcinoma (HCC) is one of the common malignancies in our practice arising in a background of liver cirrhosis (Figure 15). Besides several other environmental factors are also playing their role in the causation of HCC. In Karachi, a port city of about 15 million inhabitants with hot and humid climate, it is reported that in wholesale markets selling food commodities without proper packing and preservation, a very high content of *'aspergillus flavus'* is isolated which is a known cause of HCC. Unfortunately HCC is a bad cancer and in our experience life expectancy at the time of diagnosis is not more than six months.

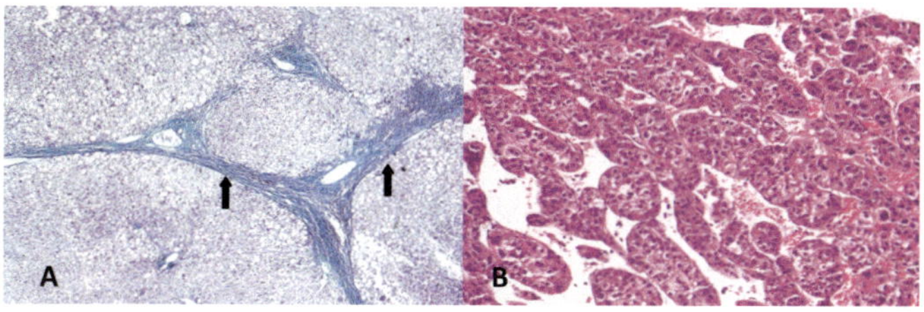

Fig. 15. Photomicrograph of a liver biopsy in a patient infected with Hepatitis C. Note fibrous band dividing the liver parenchyma into varying size nodules (↑A, Trichrome). Figure B shows a well differentiated hepatocellular carcinoma (HCC) arising in this patient (H&E).

11.1 Conclusions

- Hepatitis B & C are highly prevalent in the developing countries like Pakistan
- In Pakistan Hepatitis B carrier rate is 3-5% while anti-HCV prevalence rate is up to 15%.
- HCC is s common cancer in Pakistan mostly arising in a background of liver cirrhosis secondary to Hepatitis B & C.
- Over 60% of HCC are associated with HCV in Pakistan.

12. Acknowledgement

Dr Samina Noorali & Dr Syed Adnan Ali who completed their PhD under my supervision and participated in experimental work & original work on EBV and HPV in T-NHL & Oral cancer respectively included in this chapter are duly acknowledged. Ms. Shamsha Punjwani is acknowledged for her help in formatting this manuscript.

13. References

Ali M, Idress M, Ali L, Hussain A & Rehman I U, Saleem S, Afzal S & Butt S. (2011). Hepatitis B virus in Pakistan: A systematic review of prevalence, risk factors, awareness status and genotypes. *Virology journal*, 8:102-110.

Ali NS, Khuwaja AK, Ali T & Hameed R. (2009). Smokeless tobacco use among adult patients who visited family practice clinics in Karachi, Pakistan. *J Oral Pathol Med.* May;38(5):416-421.

Ali SM., Awan MS., Ghaffar S,., Slahuddin I., Khan S., Mehraj V & Pervez S. (2008). Human Papillomavirus infection in oral squamous cell carcinomas: Correlation with histologic variables and survival outcome in a high risk population, *Oral Surgery*, 1, 96-105

Ali SM, Awan MS, Ghaffar S, Azam SI & Pervez S. (2010). TP53 protein overexpression in oral squamous cell carcinomas (OSCC);correlation with histologic variables and survival oucome in Pakistani patients. *Oral Surgery*, 3, 83-95.

Bhurgri Y, Bhurgri A, Hussainy AS, Usman A, Faridi N, Malik J, Zaidi ZA, Muzaffar S, Kayani N, Pervez S & Hasan SH.(2003). Cancer of oral cavity and pharynx in Karachi-identification of potential risk factors. *Asia Pacific J Cancer Prevention,* Apr-Jun, 4:125-130.

Dai M, Bao YP, Li N, Clifford GM, Vacarrela S, Snijders PJ, Huang RD, Sun LX, Meijer CJ, Qiao YL & Franceschi S (2006). Human papillomavirus infection in Shanxi province, People's republic of China: a population based study, *Br J Cancer, July,* 95:96-101.

Fakhry C, Westra WH, Li S, Cmelak A, Ridge JA, Pinto H, Forastiere A & Gillison ML. (2008). Improved survival of patients with human papillomavirus-positive head and neck squamous cell carcinoma in a prospective clinical trial. *J Natl Cancer Inst,* Feb, 100(4): 261-269.

Fine KD & Stone MJ. (1999). Alpha-heavy chain disease, Mediterranean lymphoma and immunoproliferative small intestinal disease:a review of clinicopathological features, pathogenesis and differential diagnoses. *Am J Gastroenterology,* May, 94:1139-1152.

Franceschi S, Rajkumar R, Snijders PJF, Arslan A, Mahe C, Plummer M, Sankaranarayanan R, Cherian J, Meijer CJ & Weiderpass E. (2005). Papillomavirus infection in rural women in southern India. *Br J Cancer*, Feb, 92:601-606

Gillison ML. Human papillomavirus-associated head and neck cancer is a distinct epidemiologic, clinical, and molecular entity. (2004), *Semin Oncol,* Dec 31(6): 744-754.

Goodwin WJ, Thomas GR, Parker DF, Joseph D, Levis S, Franzmann E, Anello C & Hu JJ. (2008). Unequal burden of head and neck cancer in the United States. *Head & Neck,* March, 30(3): 358-371.

Idrees M, Rafique S, Rehman I, Akbar H, Yousaf MZ, Butt S, Awan Z, Manzoor S, Akram M, Aftab M, Khubaib B & Riazuddin S. (2009). Hepatitis C virus genotype 3a infection and hepatocellular carcinoma: Pakistani experience. *World J Gastroenterology,* Oct,15(40):5080-5085.

Issaacson PG, Chott A, Nakamura S, Hermelink M, Harris Nl & Swerdlow SH, editors. (2008). WHO classification of tumours of haematopoeitic and lymphoid tissues. Extranodal marginal zone lymphoma of mucosa-associated lymphoid tissue (MALT lymphoma), 4th edition, France IARC; 214-219.

Jafri W, Jafri N, Yakoob J, Islam M, Farhan S, Tirmizi A, Jafar T, Akhtar S, Hamid S, Shah HA & Nizami SQ. (2006). Hepatitis B and C: prevalence and risk factors associated with seropositivity among children in Karachi, Pakistan. *BMC infectious diseases,June,* 6:101-110.

Khan S, Jaffer NN, Khan MN, Rai MA, Shafiq M, Ali A, Pervez S, Khan N, Aziz A, Ali SH. (2007). Human papillomavirus subtype 16 is common in Pakistani women with cervical carcinoma. *Int J Infect Diseases*, July, 11(4):313-317.

Kreimer AR, Clifford GM, Boyle P & Franceschi S. (2005). Human papillomavirus types in head and neck squamous cell carcinomas worldwide: a systematic review. Cancer Epidemiol Biomarkers Prev. Feb 14(2): 467-475.

Lecuit M, Abachin E, Martin A, Poyart G, Suarez F, Bengoufa D, Feuillard J, Lavergne A, Gordon JI, Berche P, Guillevin L & Lortholary O. (2004). Immunoproliferative small intestinal disease associated with Campylobacter jejuni. *N Engl J Med,* Jan, 350 (3):239-248.

Merchant A, Husain SS, Hosain M, Fikree F, Pitiphat W, Siddiqui AR, Hayder SJ, Haider SM, Ikram M, Chuang SK & Saeed SA.. (2000). Paan without tobacco: an independent risk factor for oral cancer. *Int J Cancer, April,* 86 (1): 128-131.

Miller CS & Johnstone BM. (2001). Human papillomavirus as a risk factor for oral squamous cell carcinoma: a meta-analysis, 1982-1997. (2001). *Oral Surg Oral Med Oral Pathol Oral Radiol Endod.* June 91 (6): 622-635.

Morse DE & Kerr AR (2006). Disparities in oral and pharyngeal cancer incidence, mortality and survival among black and white Americans. *J Am Dent Assoc,* April, 137(4): 203-212.

Nair U, Bartsch J & Nair J. (2004). Alert for an epidemic of oral cancer due to use of the betel quid substitutes gutkha and pan masala: a review of agents and causative mechanisms. *Mutagenesis,* July, 19 (4) 251-262.

Nazaruk RA, Rochford R, Hobbs MV & Cannon MJ. (1998). Functional diversity of the CD8+ T-cell response to Epstein-Barr virus (EBV): implications for the pathogenesis of EBV-associated lymphoproliferative disorders. *Blood,* May, 91(10): 3875-3883.

Noorali S, Pervez S, Nasir MI & Smith JL. (2005) Characterization of Angioimmunobalstic T-cell lymphomas (AILT): and its association with Epstein-Barr virus (EBV) in Pakistani patients. *JCPSP,* 15(7):404-408.

Noorali S, Pervez S, Yaqoob N, Moatter T, Nasir MI, Hodges E and Smith JL. (2004). Prevalence and Characterization of Anaplastic Large Cell Lymphoma and Its Association with Epstein-BarrVirus in Pakistani Patients. *Pathology Research & Practice;* 200(10): 669-679.

Noorali S., Pervez S., Moatter T., Soomro IN., Kazmi SU., Nasir MI & Smith JL.(2003). Characterization of T-cell Non-Hodgkin's Lymphoma and its Association with Epstein - Barr virus in Pakistani patients. *Leukemia & Lymphoma,* 44(5), 807-813.

Noorali S., Yaqoob N., Nasir MI., Moatter T & Pervez S. (2002). Prevalence of Mycosis Fungoides and its association with EBV and HTLV1 in Pakistanian patients. *Pathol Oncol Res,* 8(3): 194-199.

Ono S, Kato M, Ono Y, Itoh T, Kubota K, Nakagawa M, Shimizu Y, Asaka M.. (2008). Characteristics of magnified endoscopic images of gastric extranodal marginal zone B-cell lymphoma of the mucosa associated lymphoid tissue including changes after treatment. *Gastrointestinal Endos,* Oct, 68(4): 624-631.

Parkin DM. (2006). The global health burden of infection-associated cancers in the year 2002. *Int J Cancer.* Jun 15;118(12):3030-3044,

Pervez S, Ali N, Aaqil H, Mumtaz K Ullah SS & Akhtar N. (2011). Gastric MALT lymphoma; a rarity, *JCPSP,* Mar, 21(3):171-172.

Pervez S, Mumtaz K, Ullah SS, Akhtar N, Ali N & Aaqil H. (2011) Immunoproliferative small intestinal disease. JCPSP, Jan; 21(1):57-58.

Raza SA, Franceschi S, Pallardy S, Malik FR, Avan BI, Zafar A, Ali SH, Pervez S, Serajuddaula S, Snijders PJ, van Kemenade FJ, Meijer CJ, Shershah S & Clifford GM. (2010). Human papillomavirus infection in women with and without cervical cancer in Karachi, Pakistan. *Br J Cancer,* May, 102(11):1657-1660.

Ryerson AB, Peters ES, Coughlin SS,. Chen VW, Gillison ML, Reichman ME, Wu X, Chaturvedi AK, Kawaoka K. (2008). Burden of potentially human papillomavirus-associated cancers of the oropharynx and oral cavity in the US, 1998-2003. *Cancer,* Nov, 113 (10 suppl): 2901-2909.

Scully S, Prime S & Maitland NJ. (1985). Papillomavirus: their possible role in oral disease. *Oral Surg Oral Med Oral Pathol,* Aug, 60(2):166-174.

Settle K, Posner MR, Schumaker LM, Tan M. Suntharalingam M. Goloubeva O, Strome SE, Haddad RI, Patel SS, Cambell EV 3rd, Sarlis N, Lorch J & Cullen KJ. (2009). Racial survival disparity in head and neck cancer results from low prevalence of human papillomavirus infection in black oropharyngeal cancer patients. *Cancer Prev Res* (Philadel) Sept, 2(9) 776-781.

Settle K, Taylor R, Wolf J, Kwok Y, Cullen K, Carter K, Ord R, Zimrin A, Strome S, Suntharalingam M. (2009). Race impacts outcome in stage III/IV squamous cell carcinomas of the head and neck after concurrent chemoradiation therapy. *Cancer,* April, 115 (8) 1744-1752.

Sherpa ATL, Clifford G, Vaccarella S, Shrestha S, Nygard N, Karki BS, Snijders PJ, Meijer CJ, & Franceschi S.. (2010). Human papillomavirus infection in women with and without cervical cancer in Nepal. *Cancer Causes Control, March,* 21 (3):313-330.

Shiboski CH, Schmidt BL & Jordan RC. (2007). Racial disparity in stage at diagnosis and survival among adults with oral cancer in the US. Community. *Dent Oral Epidemiol,* June, 35(3):233-240.

Syed S, Khalil S & Pervez S. (2011). Anaplastic Large Cell Lymphoma: The most common T-cell lymphoma in Pakistan, Asia Pacific J Cancer Prev, 12(3):685-689.

Wu RF, Dai M, Qiao YL, Clifford GM, Liu ZH, Arslan A, Li N, Shi JF, Snijders PJ, Meijer CJ &, Franceschi S. (2007). Human papillomavirus infection in women in Shenzhen city, People's republic of China, a population typical of recent Chinese urbanisation. *Int J Cancer,* Sept, 121(6):1306-1311.

18

Blood Irradiation

Sezer Saglam[1], Aydin Cakir[2] and Seyfettin Kuter[3]
[1]*Istanbul University, Oncology Institute, Department of Medical Oncology, Fatih, Istanbul*
[2]*Istanbul University, Oncology Institute, Department of Medical Pyhsics, Fatih, Istanbul*
[3]*Istanbul University, Oncology Institute, Department of Medical Pyhsics, Fatih, Istanbul, Turkey*

1. Introduction

Transfusion-associated graft-versus-host disease (TA-GVHD) is a possible complication of blood transfusion that occurs when viable donor T-lymphocytes proliferate and engraft in immunodeficient patients after transfusion. Presently, the only method accepted to prevent TA-GVHD is the irradiation of blood and its components before transfusion (Moroff and Luban 1997)). Ionizing irradiation eliminates the functional and proliferative capacities of T-lymphocytes leaving other blood components, especially erythrocytes, granulocytes and platelets, functional and viable. This is possible because T-lymphocytes are more radiosensitive than other blood components (Masterson and Febo, 1992).To carry out the irradiation of blood specially designed commercial irradiators exist, usually localized in blood banks, and dedicated exclusively to this task.

Blood and blood components may be treated with ionizing radiation, such as gamma rays from ^{137}Cs or ^{60}Co sources, and from self-contained X-ray (bremsstrahlung) units and medical linear X-ray (bremsstrahlung) and electron accelerators used primarily for radiotherapy. However, teletherapy machines, such as linear accelerators or ^{60}Co units already available at the hospital, may also be used for the same purpose (Moroff 1997), improving the cost/benefit ratio of the process.

Blood irradiation specifications include a lower limit of absorbed dose, and may include an upper limit or central target dose. For a given application, any of these values may be prescribed by regulations that have been established on the basis of available scientific data. The absorbed dose range for blood irradiation is typically 15 Gy to 50 Gy. In some jurisdictions, the absorbed dose range for blood irradiation is 25 Gy to 50 Gy. The energy range is typically from approximately 40 keV to 5 MeV for photons, and up to 10 MeV for electrons.

For each blood irradiator, an absorbed-dose rate at a reference position within the canister is measured by the manufacturer as part of acceptance testing using a reference-standard dosimetry system. That reference-standard measurement is used to calculate the timer setting required to deliver the specified absorbed dose to the center of the canister with blood and blood components, or other reference position. Either relative or absolute absorbed dose measurements are performed within the blood or blood-equivalent volume for determining the absorbed-dose distribution. Accurate radiation dosimetry at a reference position which

could be the position of the maximum absorbed dose (Dmax) or minimum absorbed dose (Dmin) offers a quantitative, independent method to monitor the radiation process.
Dosimetry is part of a measurement quality assurance program that is applied to ensure that the radiation process meets predetermined specifications.

2. Blood irradiators

The basic operating principles and configurations of a free-standing irradiator with either ^{137}Cs source or a linear accelerator are shown schematically in Figure 1. With a freestanding ^{137}Cs irradiator, the blood components are contained within a metal canister that is a rotating turntable. Continuous rotation allows for the γ rays, originating from one to four closely positioned pencil sources, to penetrate all portions of the blood component. The number of sources and their placement depend on the instrument and model. The speed of rotation of the turntable also depends on the make or model of the instrument. A lead shield encloses the irradiation chamber. Free standing irradiators employing ^{60} Co as the source of γ rays are comparable except that the canister containing the blood component does not rotate during the irradiation process; rather, tubes of ^{60} Co are placed in a circular array around the entire canister within the lead chamber. When free standing irradiators are used, the rays are attenuated as they pass through air and blood but at different rates. The magnitude of attenuation is greater with ^{137}Cs than with ^{60} Co.
Linear accelerators generate a beam of X-rays over a field of given dimension. Routinely, the field is projected on a table-top structure. The blood component is placed (flat) between two sheets of biocompatible plastic several centimeters thick.The plastic on the top of the blood component (ie,nearer to the radiation source) generates electronic equilibrium of the secondary electrons at the point where they pass through the component container.The plastic sheet on the bottom of the blood component provides for irradiation back-scattering that helps to ensure the homogenous delivery of the x-rays. The blood component is usually left stationary when the entire x-ray dose is being delivered. Alternatively it may be flipped over when one half of the dose has been delivered; this process involves turning off and restarting the linear accelerator during the irradiation procedure. Although it seems as if the practice of flipping is not required,further data are needed.

3. Blood components

The risk of GVHD for patients, all components that might contain viable T lymphocytes should be irradiated. These include units of whole blood and cellular components (red cells, platelets, granulocytes),whether prepared from whole blood or by apheresis. All types of red cells should be irradiated, whether they are suspended in citrated plasma or in an additive solution. There are recent data supporting the retention of the quality of irradiated red cells after freezing and thawing.If frozen thawed units are intended for GVHD-susceptible individuals and have not been previously irradiated,they should be irradiated because it is known that such components contain viable T lymphocytes.Filtered red cell products should also be irradiated.Extensive leucoreduction through filtration may decrease the potential for GVHD and serve as an alternative to irradiation in the future when questions about the minimum level of viable T lymphocytes that can lead to GVHD are resolved. There are reports of TA-GVHD in patients who received leucodepleted (filtered) red cells; however,the extent of leucoreduction of the components was not uniformly quantified in such reports. In addition, investigators have suggested that the number of T lymphocytes present in a product that causes GVHD may depend on the extent of patient immunocompetence at the time of transfusion . It is likely that the greater the degree of immunosuppression, the fewer the viable T lymphocytes that will be required to produce GVHD in susceptible patients. In a recent review, it was suggested that cytotoxic T lymphocytes, or interleukin-2-secreting precursors of helper T lymphocytes,may be more predictive of GVHD than the number of proliferating T cells alone. Accordingly,this suggests that until further data are available to confirm adequate removal of these T-cell subtypes by leucoreduction, irradiation should be used for blood products destined for patients at risk for GVHD. Irradiated red cells undergo an enhanced efflux of potassium during storage at 1^0 to 6^0 C.Comparable levels of potassium leakage occur with or without prestorage leucoreduction. Washing units of red cells before transfusion to reduce the supernatant potassium load does not seem to be warranted for most red cell transfusions because posfinfusion dilution prevents the increase in plasma potassium. On the other hand, when irradiated red cells are used for neonatal exchange transfusion or the equivalent of a whole blood exchange is anticipated, red cell washing should be considered to prevent the possible adverse effects caused by hyperkalemia associated with irradiation and storage. Blood components given to recipients, whether immunocompromised or immunocompetent, that contain lymphocytes that are homozygous for an HLA haplotype that is shared with the recipient, pose a specific risk for TA-GVHD. This circumstance occurs when first and second degree relatives serve as directed donors s-ll and when HLA matched platelet components donated by related or unrelated individuals are being transfused. Irradiation of blood components has been recommended in these situations.

Platelet components that have low levels of leucocytes because of the apheresis process and/or leucofiltration should also be irradiated if intended for transfusion to susceptible patients. This is because the minimum number of T lymphocytes that induces TA-GVHD has not yet been delineated.

Fresh frozen plasma does not need to be irradiated routinely because it is generally accepted that the freezing and thawing processes destroy the T lymphocytes that are present in such plasma.

During the past 2 years, there have been two brief articles suggesting that immunocompetent progenitor cells may be present in frozen-thawed plasma; the authors therefore suggested that frozen-thawed plasma may need to be irradiated. Further studies are needed to validate these findings and to assess whether the number of immunocompetent cells, that may be present in thawed fresh frozen plasma, is sufficient to induce GVHD. In rare instances, when nonfrozen plasma (termed *fresh plasma)* is transfused, it should be irradiated because of the presence of a sizable number of viable lymphocytes, approximately 1×10^{7} cells in a component prepared from a unit of whole blood.

4. Quality assurance guidelines

Various dosimetry techniques have been used to measure the dose to blood products. These include thermoluminescent dosimeters (TLD); alanine, ferrous sulphate, red perspex, metaloxide semiconductor field effect transistors (MOSFETs) and chloroform /dithoizone/parafin mixture (Hillyer *et al* 1993). Recently radiochromic film was shown to be an adequate dosimeter for blood irradiation (Butson *et al* 1999). The most prevalent method relies on TLDs and tries to ascertain the causes and variations in delivered *in vitro* dose across an 'active' treatment volume in a dedicated blood box for standard x-ray beams.

Most dosimeters have significant energy dependence at photon and electron energies less than 100 keV, so great care must be exercised when measuring absorbed dose in that energy range.

This practice outlines irradiator installation qualification, operational qualification, performance qualification, and routine product processing dosimetric procedures to be followed in the irradiation of blood and blood components by the blood-banking community. If followed, these procedures will help to ensure that the products processed with ionizing radiation from gamma, X-rays (bremsstrahlung), or electron sources receive absorbed doses within a predetermined range.

One must document that the instrument being used for irradiation is operating appropriately and confirm that blood components had been irradiated.To assure that the irradiation process is being conducted correctly, specific procedures are recommended for free-standing irradiators and linear accelerators, which are summarized in Tables 1 and 2. The procedures to be used with free-standing irradiators are an update to the guidelines provided several years ago by Anderson. Included are current recommendations from the FDA.

Measure	Frequency
Isotop decay factor	Annually for ^{137}Cs ;montly for ^{60}Co
Dose map	Annualy for ^{137}Cs; annually for ^{60}Co
Radiation leakage	daily
Timer accuracy	montly
Turntable	daily

Table 1. Recommended Quality Assurance Measures to be Used with Free-Standing Gamma Irradiators.

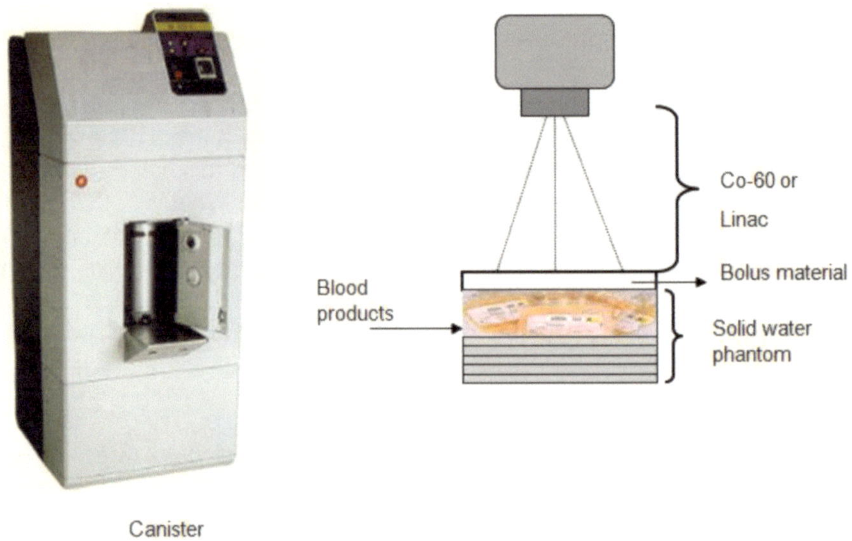

Fig. 1. With a freestanding ^{137}Cs irradiator, the blood components are contained within a metal canister that is positioned on a rotating turntable.

Dose mapping measures the delivery of radiation within a simulated blood component or over an area in which a blood component is placed. This applies to an irradiation field when a linear accelerator is used or to the canister of a free-standing irradiator. Dose mapping is the primary means of ensuring that the irradiation process is being conducted correctly. It documents that the intended dose of irradiation is being delivered at a specific location (such as the central midplane of a canister), and it describes how the delivered irradiation dose varies within a simulated component or over a given area. This allows conclusions to be drawn about the maximum and minimum doses being delivered. Dose mapping should be performed with sensitive dosimetry techniques. A number of commercially available systems have been developed in recent years. Other quality assurance measures that need to be done include the routine confirmation that the turntable is operating correctly (for ^{137}Cs rradiators), measurements to ensure that the timing device is accurate, and the periodic lengthening of the irradiation time to correct for source decay. With linear accelerators, it is necessary to measure the characteristics of the x-ray beam to ensure consistency of delivery. Confirming that a blood component has, in actuality, been irradiated is also an important part of a quality assurance program. At least one commercial firm has developed an indicator label for this purpose.

5. Dose mapping with free-standing irradiators

For free-standing irradiators, a dose-mapping procedure will measure the delivered dose throughout the circular canister in which the blood component is placed. To establish a two-dimensional map, a dosimetry system is placed in a canister that is completely filled with a blood/tissue-compatible phantom composed of water or an appropriate plastic such as polystyrene. The dosimetry material is placed within the phantom in a predetermined way. This approach provides data that describe the minimum levels of irradiation that would be absorbed by a blood component placed in the canister and recognizes that maximum attenuation will occur when the canister is completely filled with a blood-compatible material. Relevantly, it was shown recently that the absorbed dose at the central midplane of a canister (ie, at the center point) decreased by approximately 25% (from 3100 to 2500 cGy) in a ^{137}Cs irradiator (JL Shepherd and Associates, San Francisco, CA) when the loading of the canister was changed from 0% (air) to 100% (with blood components). An irradiationsensitive film dosimetry system (International Speciality Products) that will be described later in this report was used for this purpose. A linear relationship was observed between the amount of fill and the measured central dose. With 1 and 2 units of blood components, the central dose relative to air was 0.98 and 0.93. The minimum and maximum levels were influenced in the same manner as the central dose on decreasing the proportion of the canister that contained air. Other studies have shown that the extent of variability in the dose delivered to the interior of simulated blood units (water or saline in plastic blood storage containers) depended on the model of the ^{137}Cs free-standing irradiator. An immobilized grid of thennoluminescent dosimeters in a plastic sheet were placed within the simulated blood units to measure dose delivery. See the section on dosimetry systems in use. It was also shown that a spacer into the bottom of the canister increases the minimum level of radiation within the simulated blood units as expected from the results of full-canister dose mapping involving a phantom.The extent of variability with ^{137}Cs irradiations is influenced by a number of factors, including the number of sources, turntable speed, and the presence of a spacer at the bottom of the canister.These studies underscore the need for consistency in loading the canister.Attenuation of the irradiation dose delivered is a function of physical density, electronic density, and atomic number with three major processes: photoelectric, Compton, and pair production. In practical terms, attenuation is caused when the irradiation enters a liquid, such as water or blood. The extent of attenuation depends on a number of factors,including the dimensions of the canister. In a fully filled canister, as is used for dose mapping, the attenuation will increase as the irradiation transverses to the center point. The dose map that is generated describes the dose distribution. As depicted in the theoretical dose map shown in Figure 2, the edges of the canister are exposed to a greater dose of irradiation compared with the center line because the attenuation is less in the periphery. The attenuation with ^{60}Co is less than that seen with ^{137}Cs.

When an irradiator is purchased, the distributor will provide a central dose level that is determined in a blood-compatible environment. In the 1970s and 1980s manufacturers provided a central dose that was determined in air, resulting in the use of timer settings that provided for a dose level that was somewhat less than what was expected. Subsequent to the issuing of the FDA guidelines in July 1993 and the use of dose mapping, it has been necessary to readjust irradiation times with some instruments because the attenuation effect had not been considered previously.

A theoretical two-dimensional dose map describing the irradiation dose distribution through a fully filled canister of a free-standing ^{137}Cs irradiator is shown in Figure 2. To obtain this dose map,dosimeters would have been positioned in the central axis and the edge of circular canister from the top to the bottom of the canister. The y dimension of the map depicts the top to bottom axis of the canister, whereas the x dimension depicts the cross-sectional axis. For the theoretical situation described in Figure 2, the central midplane dose is 2560 cGy, slightly above the minimum standard of 2500 cGy, and the minimum dose is 1750 cGy. In this irradiation dose map, the minimum dose is at the central bottom of the canister, a common finding in actual practice.

The dose map can also be used to assess whether the turntable of a ^{137}Cs irradiator is rotating in an appropriate manner. The occurrence of comparable readings at the two edges of the two-dimensional map, as depicted in the theoretical dose map, indicates that the canister is rotating evenly in front of the ^{137}Cs source. If the turntable were not rotating, the dose levels at the edge of the map closest to the source would be much higher than that found on the opposite edge, ie, the side located distant to the source. According to the 1993 recommendations from the FDA, dose mapping should be performed routinely on an annual basis and after a major repair, especially one involving the sample handling apparatus such as the turntable.

6. Dosimetry systems

The delivered irradiation dose can be measured by a variety of dosimetry systems. In recent years, several commercial interests have developed complete systems for use with free-standing irradiators; each system consists of a phantom that fills the canister and a sensitive dosimetry system. Three main types of dosimetry measurement systems are available (Table 3).

These dosimeters are referred to as routine dosimeters. They are calibrated against standard systems, usually at national reference laboratories such as the National Institute of Standards and Technology in the United States. The routine dosimeter measurement systems were initially developed for use with ^{137}Cs irradiators because this is the predominant irradiation source for blood. More recently, they have been developed also for use with ^{60}Co irradiators. Thermolumeniscent dosimeters (TLD chips) are one type of routine dosimeter. TLD chips are small plastic chips with millimeter dimensions having a crystal lattice that absorbs ionizing radiation. Specialized equipment is used to release and measure the energy absorbed by the TLD chip at the time of the test irradiation. In one commercially available system, chips are placed at nine different locations within a polystyrene phantom that fits into the canister of the IBL 437C irradiator (CIS US, Inc,Bedford, MA). The timer setting used routinely for an instrument is used in the test procedure.

Method	Measurement type
Thermoluminescent Dosimetry	Emission of light
Radiochromic(GafChromic) film	Optic density
Mosfet(metal -oxide field effect transistors)	Voltage detection
Alanine/ESR	ESR signal-Magnetic field

Table 3. Dosimetric systems in different clinics.

There are two systems that use radiochromic film. On exposure to irradiation, the film darkens, resulting in an increase in optical density. The optical density, determined at various locations on the film, is linearly proportional to the absorbed irradiation dose. Standard films that are irradiated at a given dose level with a calibrated source at a national reference laboratory provide the means to assess the absolute level of absorbed irradiation.This type of dosimeter is basically an x-ray film comparable with that used in clinical practice. With this device, the map that is developed identifies the absorbed irradiation dose that is measured at a large number of locations. In one system, a film contained in a thin water-tight casement is placed into the canister (International Specialty Products,Wayne, NJ), This approach is being used with a variety of irradiators. The canister is filled completely with water before the irradiation procedure.This system provides a direct readout of the dose that is delivered throughout the canister. The timer setting used routinely is employed for the test procedure. In a second system, a film having different radiation sensitive characteristics is embedded between two halves of a circular-fitting polystyrene plastic phantom (Nordion Internation, Canada, Ontario). Irradiation of specialized films is performed with a number of timer settings, each being larger than that used routinely. The map produced is normalized for a central midplane dose of 2500 cGy. The time to produce the 2500 cGy will have been predetermined with a different dosimeter system, the Fricke system, in which absorbed radiation causes a change in the state of a iron salt that can be assessed spectrophotometrically. Another approach to irradiation dose mapping employs a solid-state electronic dosimeter that is technically referred to as a *metal-oxide silicon field effect transistor* (MOSFET). A board contains a number of small transistors in an arrangement that provides data for a dose map. This board is placed between two halves of a circular polystyrene phantom that fits into the canister. This dosimeter absorbs and stores the radiation dose imparted to it electronically . The radiation causes the formation of holes in the metal-oxide layer that becomes trapped within the transistor. The magnitude of the holes is evaluated by measuring the voltage across the transistor with a voltmeter. The voltages measured are converted to absorbed dose. With each dosimetry system, measurements are used to express the absorbed irradiation dose of cGrays. All dosimetry measurements are associated with a degree of uncertainty or possible error. The magnitude of the uncertainty depends on the kind of dosimeter used. For most dosimeters, the level is 5% of the measured value. For a central absorbed dose level of 2560 cGy (see theoretical dose map in Fig 2) the value could be as high as 2788 cGy or as low as 2432 cGy. Correspondingly, a measured value of 2400 cGy could be as high as 2520 cGy or as low as 2380 cGy. Because the measured value could in actuality meet the 2500 cGy standard, it is appropriate to accept a value of 2400 cGy as meeting the current standard. The same approach should be used when evaluating the minimum value on a dose map. Albeit arbitrary and cautious, the actual minimum on an irradiation dose map should not be below 1500 cGy.

7. Precautions with free-standing irradiators

It is important periodically to lengthen the time of irradiation to correct for decay of the isotopic source that emits the gamma irradiation. Until recently, this was the only major quality assurance measure that was performed routinely. With the half-life for ^{137}Cs being 30 years, annual lengthening of the timer setting is appropriate.On the other hand, with the half-life of ^{60}Co being only 5,27 years, the time of irradiation should be increased on a quarterly basis. The additional seconds of irradiation that are needed can be calculated using formulae that can be found in any physics text. Alternatively, distributors of irradiators provide a chart that specifies the appropriate setting as a function of calendar time.

7.1 Turntable rotation

For ^{137}Cs irradiators, it is essential that the turntable operates at a constant speed in a circular pattern to ensure that each part of a blood component is exposed equally to the source. Daily verification of turntable rotation is an appropriate quality assurance measure. With some free-standing irradiator models rotation of the turntable can be observed before the door of the compartment in which the canister is positioned is closed. In other models, this can be done only indirectly by ensuring that an indicator light is operating appropriately. With some older models, there have been occasional reports that the turntable failed to rotate because of mechanical problems. Such problems should not be encountered with the newer models because of changes in the turntable mechanisms. In any event, daily verification of turntable rotation is a prudent quality assurance measure.

7.2 Radioactivity leakage

Irradiators are constructed so that the isotopic sources are contained in a chamber heavily lined with a protective lead shield to prevent leakage of radioactivity. Accordingly, gamma irradiators are considered to be very safe instruments. Although there have been no reports of source leakage of radioactivity, periodic measurements are warranted to ensure that this is the case. Attaching a film badge to the outside of the irradiator, using a Geiger counter periodically, and performing a wipe test of the inside of the chamber where the canister is positioned at least semiannually are measures that are being used.

8. Dose mapping with linear accelerators

Linear accelerators that are used therapeutically to provide radiation therapy are carefully monitored to ensure appropriateness of dose to an irradiation field. When blood components are treated with x-rays, the instrument settings are very different than those used to treat oncology patients. Hence, additional periodic quality control measures, primarily to assess the dose delivered to blood components, are needed to ensure that linear accelerators are being operated appropriately when used for blood irradiation. Currently, there are no commercially available systems for assessing the dose delivered throughout the area of an irradiation field in which blood components are placed for treatment with x-rays. An ideal dosimeter for this purpose would be made of a tissue-compatible plastic phantom, containing appropriate dosimeter material and a covering that could be placed at the appropriate distance from the source. An alternative approach might involve the use of a blood bag filled with water (simulating a blood unit) containing TLD chips, as described earlier. In comparative studies using such simulated blood units, it was determined that radiation delivery was more uniform with linear accelerators than with ^{137}Cs free-standing irradiators. This reflects the relative homogeneity of x-ray beams. In the absence of an available system modified for the irradiation of blood bags, the dose delivered throughout an irradiation field should be mapped with the dosimetric measuring system known as an *ionization chamber*. The ionization chamber is used to calibrate linear accelerators for patient use. In addition, on a yearly basis, dose mapping should be performed using a tissue-compatible phantom.

In view of the widely divergent conditions that are used during the operation of linear accelerators, other parameters pertaining to the x-ray beam should be evaluated on at least a quarterly basis to provide assurance that the instrument is being used appropriately for the irradiation of blood components. The goal is to ensure that the instrument is being set in a consistent fashion. When setting a linear accelerator for blood component irradiation, the following should be measured: the distance between the x-ray source and the position where the blood components are to be placed; consistency in the strength of the x-ray beam; and (3) the intensity of the x-ray beam. The distance between the source and position on the table where blood components will be placed (referred to as the *target)* can be evaluated easily with a calibrated measuring device. This is a simple task that can be performed on a routine basis. The consistency of beam output can be evaluated by measuring the beam current. Beam intensity can be evaluated by measuring the ionization current in a monitoring ionization chamber array that can be expressed in terms of the number of photons delivered per square centimeter. These parameters should be assessed routinely as part of quality control programs used by radiation physicists. A code of practice was published in 1994 by the Radiation Therapy Committee of the American Association of Physicists in Medicine for the quality control of radiotherapy accelerators. The described practices are used routinely by radiation physicists. It would he prudent to ensure that an institution using a linear accelerator for blood irradiation follow these quality assurance guidelines and recommendations.

9. Confirming that irradiation occurred

It is important to have positive confirmation that the irradiation process has taken place. This is to identify whether an operator fails to initiate the electronically controlled irradiation process or when the irradiation process is not performed because of instrumentation malfunction. A radiation-sensitive indicator label has been developed specifically for this purpose by International Speciality Products,Wayne, NJ. The label containing a radiationsensitive film strip is placed on the external surface of the blood

component. Irradiation causes distinct visually observable changes: The appearance changes from clear red to opaque with obliteration of the word "NOT." When the label is placed on a blood component, there is a visual record that the irradiation process took place. The reliability of this type of indicator was documented recently in a multisite study.
Two versions of the indicator label have been manufactured. The difference is the range of radiation needed to cause a change in the radiationsensitive film. The ratings for these indicators are 1500 cGy or 2500 cGy. The ratings serve as an approximate guideline for the amount of absorbed radiation that will be needed to completely change the window from reddish to opaque with complete obliteration of the word "NOT." Because the indicator labels are designed for and are used to confirm that the irradiation process has occurred, we have concluded that the 1500 cGy label is the most appropriate tool to perform this quality control measure. This is based on the routinely observed pattern of dose distribution to a blood component in a canister of a free-standing irradiator. Despite a targeted central dose of 2500 cGy, there will be spots at which the dose will be less. If the theoretical dose map presented in Figure 2 is used as an example, there will be a spot that will receive only 1800 cGy. If the 2500 cGy-rated label were to be located on the external surface of a component, there may be minimal changes in the appearance of the radiation-sensitive film window.This would result in a judgment that the blood component was not irradiated, when in actuality it was treated satisfactorily.

Dose	
Linear accelerators	*Free standing irradiators*
2500 cGy to the center of an irradiation field with a minumum of 1500 cGy elsewhere.	2500 cGy to the central midplane of a canister with a minumum of 1500 cGy elsewhere.
Dose mapping	
Linear accelerators	*Free Standing irradiators*
Yearly dose mapping with an ionization chamber and a water phantom.More frequent evaluation of instrument conditions to ensure consistency of x-rays.	Routinely,once a year Cs-137 or twice a year Co-60 and after major repairs;the irradiation procedure should be tested using a fully filled canister with a dosimetry system to map the distrubition of the absorbed dose.
Correction for radioisotopic decay	
Cs-137; annually	Co-60; every 3 month
Turntable rotation(Free standing Cs-137 irradiators)	
Daily should be checked.	
Storage time (after irradiation)	
Red cells	*Platelets*
For up to 28 days;total storage time cannot exceed maximum stroga time for unirradiated red cells	No change due to the irradiation.

Table 4. Guidelines for irradiating blood components.

10. References

[1] Anderson KC, WeinsteinHJ: Transfusion-associated graftversus-host disease. New Eng J Med .1994;323:315-321

[2] Roberts GT, Luban NLC: Transfusion-associated graftversus-host disease, in Rossi EC, Simon TL, Moss GC, Goldis A(eds): Principles of Transfusion Medicine. Baltimore MD,Williams and Wilkins, 1996, pp 785-801

[3] Linden JV, Pisciotto PT: Transfusion-associated graftversus-host disease and blood irradiation. Transfus Med Rev .1992;6:116-123

[4] Anderson KC: Clinical indications for blood component irradiation, in Baldwin ML, Jefferies LC (eds): Irradiation of Blood Components, Bethesda, MD, American Association of Blood Banks, 1992, pp 31-49

[5] Brubaker DB: Transfusion-associated graft-versus-host disease, in Anderson KC, Ness PM (eds): Scientific Basis of Transfusion Medicine. Implications for Clinical Practice. Philadelphia, PA, W.B. Saunders Company, 1994, pp 544-573

[6] Williamson LM: UKguidelines for the irradiation of blood components. Transfus Sci .1995;16:135-137

[7] Davey RJ: Transfusion-associated graft-versus-host disease and the irradiation of blood components. Immunological Investigations .1995;24:431-434

[8] Kanter MH: Transfusion-associated graft-versus-host disease disease: Do transfusions from second-degree relatives pose a greater risk than those from first-degree relatives? Transfusion.1992;32:323-327

[9] McMilan KD, Johnson RL: HLA-homozygosity and the risk of related-donor transfusion-associated graft-versus-host disease. Transfus Med Rev.1993; 7:37-41

[10] Petz LD, Calhoun L, Yam P, et al: Transfusion-associated graft-versus-host disease in immunocompetent patients: Report of a fatal case associated with transfusion of blood from a second-degree relative, and a survey of predisposing factors. Transfusion .1993;33:742-750

[11] Williamson LM, Warwick RM: Transfusion-associated graft-versus-host disease and its prevention. Blood Reviews.1995; 9:251-261

[12] Ohto H, Anderson KC: Survey of transfusion-associated graft-versus-host disease in immunocompetent recipients. Transfus Med Rev .1996;10:31-43

[13] Davey RJ: The effect of irradiation on blood components, in Baldwin ML and Jefferies LC (eds): Irradiation of Blood Components, Bethesda, MD, American Association of Blood Banks, 1992, pp 51-62

[14] Fearon TC, Luban NLC: Practical dosimetric aspects of blood and blood product irradiation. Transfusion .1986.26:457-459

[15] Suda BA, Leitman SF, Davey RJ: Characteristics of red cells irradiated and subsequently frozen for long term storage. Transfusion.1993 33:389-392

[16] Miraglia CC, Anderson G, Mintz PD: Effect of freezing on the in vivo recovery of irradiated red cells. Transfusion .1994;34:775-778

[17] Crowley JR Skrabut EM, Valeri CR: Immunocompetent lymphocytes in previously frozen washed red cells. Vox Sang .1974;26:513-517

[18] Akahoshi M, Takanashi M, Masuda M, et al: A case of transfusion-associated graft-versus-host disease not prevented by white cell-reduction filters. Transfusion.1992;32:169-172

[19] Heim MU, Munker R, Saner H, et at: Graft-versus-host Kranldleit(GVH mit letalem ausgang nach der gabe von gefilterten erythrozytenkonzentraten(Ek). Infusionstherapie.1991; 18:8-9

[20] Hayashi H, Nishiuchi T, Tamura H, et al: Transfusion associated graft-versus-host disease caused by leukocyte fltered stored blood. Anesthesiology.1993;79:1419-1421
[21] Anderson KC: Leukodepleted cellular blood components for prevention of transfusion-associated graft-versus-host disease.Transfus Sci .1995;16:265-268
[22] Ramirez AM, Woodfield DG, Scott R, et at: High potassium levels in stored irradiated blood. Transfusion .1997;27:444-445
[23] Rivet C, Baxter A, Rock G: Potassium levels in irradiated blood. Transfusion.1989: 29:185
[24] Swann ID, Williamson LM: Potassium loss from leucodepleted red cells following "v-irradiation. Vox Sang .1996;70:117-118
[25] Strauss RG: Routine washing of irradiated red cells before transfusion seems unwarranted. Transfusion.1990; 30:675-677
[26] Luban NLC, Strauss RG, Hume HA: Commentary on the safety of red cells preserved in extended-storage media for neonatal transfusion. Transfusion.1991; 31:229-235
[27] Benson K, Marks AR, Marshall MJ, et al: Fatal graft versus-host disease associated with transfusions of HLAmatched,HLA-homozygous platelets from unrelated donors. Transfusion. 1994; 34:432-437
[28] Grishaber JE, Birney SM, Strauss RG: Potential for transfusion-associated graft-versus-host disease due to apheresis platelets matched for HLA class`, I antigens. Transfusion.1993; 33:910-914
[29] Wielding JU, Vehmeyer K, Dittman J, et at: Contamination of fresh-frozen plasma with viable white cells and proliferable stem cells. Transfusion .1994;34:185-186
[30] Bernvill SS, Abdulatiff M, Al-Sedairy S, et at: Fresh frozen plasma contains viable progenitor cells-should we irradiate.Vox Sang .1994;67:405
[31] Davey RJ, McCoy NC, Yu M, et al: The effect of pre-storage irradiation on post-transfusion red cell survival. Transfusion .1992;32:525-528
[32] Mintz PD, Anderson G: Effect of gamma irradiation on the in vivo recovery of stored red blood cells. Ann Clin Lab Sci .1993;23:216-220
[33] Moroff G, Holme S, Heaton A, et al: Effect of gamma irradiationon viability of AS-I red cells. Transfusion .1992;32(suppl):70S(abstr)
[34] Friedman KD, McDonough WC, Cimino DF: The effect of pre-storage gamma irradiation on post-transfusion red blood cell recovery. Transfusion.1991; 31:50S(abstr)
[35] Moroff G, Holme S, AuBuchon J, et al: Storage of red cells and platelets following gamma irradiation. Vox Sang.1994; 67:42, 1994 (Abstr, suppl 2)
[36] Moroff G, George VM, Siegl AM, et al: The influence of irradiation on stored platelets. Transfusion.1996;26:453-456
[37] Espersen GT, Ernst E, Christiansen OB, et at: Irradiated blood platelet concentrates stored for five days--evaluation by in vitro tests. Vox Sang .1988;55:218-221
[38] Duguid JKM, Cart R, Jenkins JA, et al: Clinical evaluation of the effects of storage time and irradiation on transfused platelets. Vox Sang.1991; 60:151 - 154
[39] Read EJ, Kodis C, Carter CS, et at: Viability of platelets following storage in the irradiated state. A paired-controlled study. Transfusion.1988;28:446-450
[40] Rock G, Adams GA, Labow RS: The effects of irradiation on platelet function. Transfusion .1988;28:451-455

[41] Sweeney JD, Holme S, Moroff G: Storage of apheresis platelets after gamma irradiation. Transfusion .1994;34:779-783
[42] Seghatchian MJ, Stivala JFA: Effect of 25 Gy gamma irradiation on storage stability of three types of platelet concentrates: a comparative analysis with paired controls and random preparation. Transfus Sci .1995;16:121-129
[43] Bessos H, Atkinson A, Murphy WG, et at: A comparison of in vitro storage markers between gamma-irradiated and non-irradiated apheresis platelet concentrates. Transfus Sci.1995; 16:131-134
[44] Anderson KC, Goodnough LT, Sayers M, et at: Variation in blood component irradiation practice: Implications for prevention of transfusion-associated graft-versus-host disease. Blood .1991;77:2096-2102
[45] Sprent J, Anderson RE, Miller JF: Radiosensitivity of T and B lymphocytes. II Effect of irradiation on response of T cells to alloantigens. Eur J Immuuol .1974;4:204-210,
[46] Valerius NH, Johansen KS, Nielsen OS, et al: Effect of invitro x-irradiation on lymphocyte and granulocyte function.Scand J Hematol .1981;27:9-18
[47] Pelszynski MM, Moroff G, Luban NLC, et at: Effect of irradiation of red blood cell units on T-cell inactivation as assessed by limiting dilution analysis: implications for preventingtransfusion-associated graft-versus-host disease. Blood.1994; 83:1683-1689
[48] Luban NLC, Drothler D, Moroff G, et at: The effect of irradiation on lymphocyte reactivity in platelctpheresis components assessed by limiting dilution analysis. Transfusion.1994; 34:66S(abstr)
[49] Rosen NR, Weidner JG, Bold HD, et al: Prevention of transfusion-associated graft-versus-host disease: selection of an adequate dose of gamma irradiation. Transfusion.1993; 33:125-127
[50] Center for Biologics Evaluation and Research, Food and Drug Administration: Recommendations regarding license amendments and procedures for gamma irradiation of bloodproducts. http://www.fda.gov/downloads/BiologicsBloodVaccines/GuidanceComplianceRegulatoryInformation/OtherRecommendationsforManufacturers/MemorandumtoBloodEstablishments/UCM062815.pdf
[51] Anderson G: Quality assurance of the irradiation process of blood components, in Baldwin JL, Jefferies LC (eds): Irradiation of Blood Components, Bethesda MD, American Association of Blood Banks, 1992, pp 63-75
[52] Masterson ME, Febo R: Pretransfusion blood irradiation Clinical rationale and dosimetric considerations. Med Phys.1992; 19:649-457
[53] Leitman SF: Dose, dosimetry and quality improvements of irradiated blood components. Transfusion.1993; 33:447-449
[54] Perkins JT, Papoulias SA: The effect of loading conditions on dose distribution within a blood irradiator. Transfusion .1994;34:75S(abstr)
[55] Moroff G, Luban NLC, Wolf L, et al: Dosimetry measurements after gamma irradiation with cesium-137 and linear acceleration sources. Transfusion.1993; 33:52S (abstr)
[56] Luban NLC, Fearon T, Leitman SF, et al: Absorption of gamma irradiation in simulated blood components using cesium irradiators. Transfusion.1995; 35:63S(abstr)

[57] Kutcher GJ, Coia L, Gillin M et al.: Comprehensive QA for radiation oncology: report of AAPM radiation therapy committee task group 40. Med Phys.1994; 21:581-618

[58] Nath R, Biggs PJ, Bova FJ, et al: AAPM code of practice for radiotherapy accelerators: report of AAPM radiation therapy task group no 45. Med Phys.1994; 21:1093-1121

[59] Leitman SF, Silberstein L, Fairman RM, et al: Use of a radiation-sensitive film label in the quality control of irradiated blood components. Transfusion.1992;32:4S (abstr)

Part 6

Examples for Different Quality Control Processes

19

R&D: Foundation Stone of Quality

Petr Košin, Jan Šavel and Adam Brož
Budweiser Budvar, N.C.
Czech Republic

1. Introduction

There are many definitions of quality. The oldest based on simple fulfilling of desired technical parameters developed into nowadays agreement, that the pivot of interest should be customer and his needs. Modern quality systems define quality as a degree of fulfilling of customers´ demands, or the degree of customers´ satisfaction by goods or services that he had paid for (Juran, 2000).

1.1 The role of quality in success on the market

Quality management systems have been developing since the beginning of industrial goods production. Producers of goods with higher quality had advantages for fight with competitors and high quality products were the main way how to satisfy customers. Although quality products are still necessary for success on the market, quality itself does not make nowadays business. Since the end of eighties new kind of companies appeared. These companies were orientated not only on quality goods production, but mainly on branding.

For these companies quality product was just one fourth of their marketing mix, whose management under certain brand was to satisfy customers. The resting parts of marketing mix are e.g. the price, place of purchase and accompanying services, or promotion quality. The original sense of brand was to label product with the place of origin and name of producer, which together had served as a guarantee of quality.

Due to sophisticated marketing methods can brand nowadays bring to customer whole battery of emotional values, which can sort customers into varying social groups. The satisfaction of customers is not brought only by the product quality, but also by emotions connected with social enlistment of brand (Klein, 2005; Olins 2009).

Brand management added to product new value, which allowed brand keepers to sell for much higher prices than would be the prices of their products without brand. Because not all customers could afford to pay extra price for emotional values and engaging to higher social classes, the byproduct of branding was the polarization of marked and origin of new distribution channels selling products without strong brands. These products are sold for much lower prices and usually have good enough quality to satisfy demands of its users. These so called cheap brands e.g. made up 40 % of German beer marked in 2005 (Verstl, 2005).

Brand management is also the reason why contact of quality department with customers is nowadays mediated by marketing department and quality improvement is often managed by marketing manager.

1.2 Quality improvement

Quality manager can never be satisfied with quality. It is because quality is not static; it is developing together with the development of customers´ needs. The process of developing quality is called quality improvement and it is integral part of modern quality management. As any other management systems, the management of quality improvement is based on the flow of information. There are several model systems for information flow in quality improvement management; of the most famous is Deming´s PDCA cycle (fig. 1) or Juran's quality spiral (fig. 2).

These model systems have in common four basic steps, which also represent four levels where quality is managed in practice. The first level of quality management called "plan" or "product development" step is usually secured by the R&D department or by external consulting expert. This step includes designing of the product with all technical parameters and proposing of production processes with all operation steps and control points.

The second step ("Do" or "Production & process control), usually secured by production department in close cooperation with quality control department, includes production of products by production processes, which are carefully operated by feedback regulation in originally proposed control steps.

The third step where quality is managed is the "Check" or "Final inspection". This step is usually secured by the Quality control department and sampling or evaluation of the results are usually planed and processed with the help of statistical tools, like control charts or histograms.

The last step of information flow at the quality improvement management ("Act" or "Market research") is secured by quality assurance or marketing department. At this step customers´ satisfaction with product is measured. Method can be common market research, like statistic study with questionnaires, or with direct interviews.

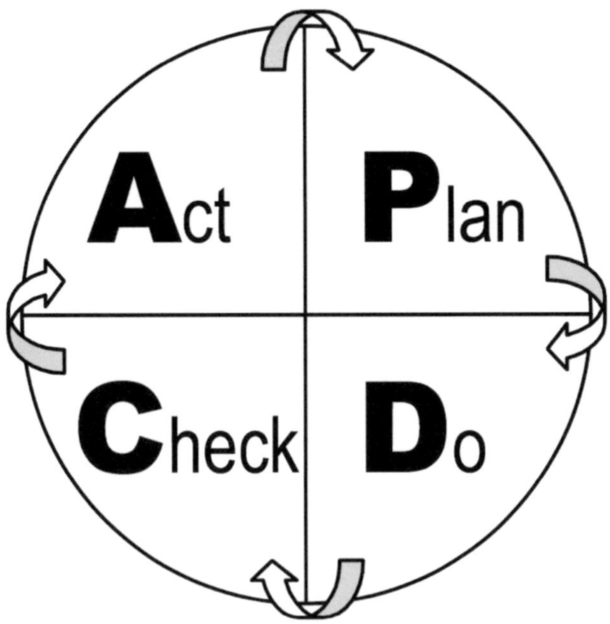

Fig. 1. PDCA cycle.

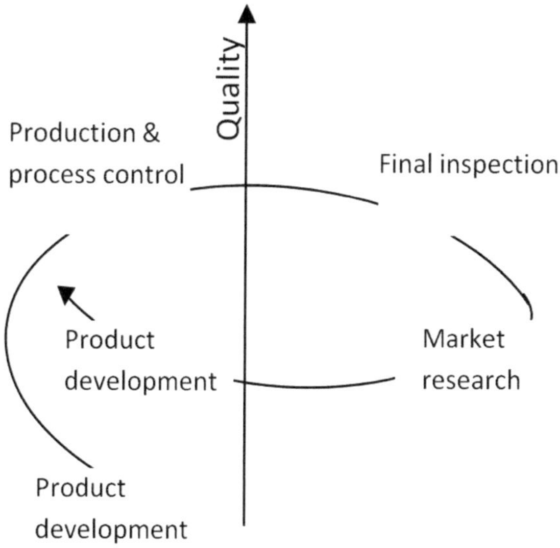

Fig. 2. Juran´s quality spiral.

1.3 Quality characteristics

Quality characteristics can be defined technically as inherent property of product that serves for identification, description and differentiation of product form other products and has a quantity and unit. Customer orientated definition of quality characteristic would be a property of product, that satisfy customer.

Quality characteristics can be divided into two groups: real characteristics and measurable attributes. Real characteristics directly correspond to the customer orientated definition of quality and are the reason why customers buy selected product. These characteristics should be evaluated in the fourth step of PDCA or Juran´s quality spiral. Their disadvantage is problematic measurement and evaluation and that is why real characteristics are usually translated into measurable attributes in the first step of quality improvement process.

Measurable attribute suit more the technical definition of quality and although not corresponding directly to customers' satisfaction they can be quite easily measured and evaluated during production and by feedback effect serve in the second and third step of quality management.

2. Case study: beer foam stability

The case study of this chapter for illustration of quality improvement in practice will be beer foam stability. It is a measurable attribute, which closely describes real quality characteristic called foam appearance. Foam appearance is one of the most important quality parameters of beer, because it is a visual parameter and visual parameters are ease to evaluate by almost all customers, who are much surer by what they see than what they taste.

Foam stability is not the only attribute describing foam appearance; the others are e.g. foam density, creaminess, color, or ability to cling on beer glass. Although these technical parameters have their meaning for foam appearance, stability is the most important parameter because when foam is not stable, it disappears and there is nothing to judge.

Assignment for this study is was to improve beer foam stability without changing any other beer quality and production parameters, which most often include raw materials. Complexity of this quality parameter much differs from common example situations like screw production.

2.1 Procedure of foam quality improvement

The general procedure of quality improvement has several steps. Whole process starts with bibliographic research, because many of the basic questions have been solved before by someone else.

The second step should be the verification of bibliographic result under the specific conditions of the company. With good luck this can be the end of quality improvement.

If bibliographic results do not help, the third step should be the start of own primary research. There cannot be given general instruction on this step, but there is one way that occasionally helps and was useful also in the case study of this chapter. It is to develop your own analytical method, of which results closely corresponds to customers sensation of the quality parameter and simultaneously can be used all over whole production line to evaluate how the quality attribute develops during production.

The fourth step can then be to use the new analytical method to find weak points of the production line and find a way how to control these processes to improve the quality problem. Results of the method can then be used for feed-back regulation of selected processes parameters in second step of the quality improvement information flow described by PDCA or Juran´s quality spiral.

The last step of the research would be identical with the third or fourth step of quality management. With the help of sophisticated statistical tools should be precisely evaluated the extent of quality improvement and the economical balance of quality profits and costs.

2.2 Bibliographic search

The research usually starts with bibliographic search. In many cases the same problem has already been discussed either in academic or applied research.

Academic sources

Academic research offers several solutions for foam quality improvement, mostly based on reductionist analytical approach. The idea is that foam stability can be increased by addition of foam stabilizing material to beer. There have been described several foam stabilizing substances, but less methods how to increase the content of these and not change beer taste or composition. Of the most discussed are bitter acids and proteins (Evans, 2002).

All of hop bitter acids can increase foam stability, but the most effective are chemically reduced derivates. Their production stars with extraction of α-bitter acids from hop by organic solvent or supercritical CO_2. The second step is isomerization of α-bitter acids in alkali and high temperature conditions and the third step chemical reduction of iso-α-bitter acids into di-, tetra- or hexa-hydro-iso-α-bitter acids. Most often discussed are tetrahydro-iso-α-bitter acids, produced under the brand Tetrahop.

Tetrahop is used in downstream processes as additive in milligrams per liters of beer. There are several problem of this way of improving foam quality. Major is that although foam stability is increased, foam structure at the end of foam collapse has unnatural appearance resembling polystyrene foam. Next problem is harsh character of Tetrahop's taste, which is far away from fine taste of natural hops. Probably the least important problem is that the chemical preparation of Tetrahop collides with Reinheitsgebot, German beer purity law

saying that beer can be only made from water, barley malt and hops. Improving foam quality by simple increase of natural hop components would have negative effect in change of bitterness intensity, one of the most sensed sensory attribute of beer.
The role of proteins in foam stability has been the most studied part of foam quality in academic research. There have been described several proteins that influence foam quality, mainly hydrophobic proteins like protein Z, or lipid transfer proteins (LTP). Protein Z represents proteins with high molecular weight (relative molecular weight 35 000 - 50 000) and LTP have relative molecular weight 5 000 - 15 000. Proteins, which together with bitter acids and ions build up the framework of foam bubble walls, come to beer from malt.
A lot of studies on which malt contains more of these foam promoting proteins were driven by the idea, that change of malt specifications could be a way for a brewer how to fix problems with foam. The problem of this approach is that changing malt specifications can substantially change some of the other important parameters of beer, e.g. color, fermentability and final degree of attenuation, or the essential character of beer taste, which is hidden in the unfermented remainder of malt in the beer body.
Although foam stability has been in focus of academic research for quite a long time, there have not been found a practical recipe how to improve foam quality and not change any of the other beer parameters, including beer raw materials and composition.

Applied sources

There are far less papers written from applied research compared to academic research. The reason is not only the evaluation of academic research quality by the quantity of published papers, but also historic transfer of applied research from goods producers to service and suppliers companies, who more carefully guard their knowhow and do not publish much of technical papers.
Although it is quite hard to come across this kind of publication, they are of great use because they usually look for practical solutions. Contrary to academic research, which usually looks for answers on questions "how does it work", applied research usually solves questions concerning what can one do to economically solve a problem.
For our case study of foam stability can be found sporadic publications recommending some practical solutions like optimization of the malt grinding, correct choice of lauthering tun, sufficient separation of sediment after wort boiling, or consistent rinsing of bottles at the end of washing (Haukeli, 1993).

3. Improvement of foam stability

The assignment of the research was to improve foam without changing any other quality characteristics, especially beer appearance and taste, which is secured by constant specifications of raw materials. That is why trials with alternative malt specification, hop dosage or use of any additive to beer as discussed in the academic research was excluded from the design of this study.

3.1 Foam stability measurement

Foam collapse can be divided into three stages from the macroscopic point of view. The first is the drainage of beer out of wet foam, where significant upward movement of beer-foam interface can be observed. The second stage is the collapse of dry foam and is accompanied by significant decrease of foam surface. The third stage is the break-up of the last foam layer, which results in the appearance of a "bald patch" on the surface of the beer.

This division corresponds to measurement strategy focused on the second stage of foam decay. The collapse of the foam accordance to first order kinetic equation is usually expressed as the time dependency of beer volume remaining in dry foam after initial beer drainage.
Kinetic equation can also connect the first two phases of foam collapse as expressed in formula (1),

$$c = c_{\infty} - a_0 \cdot \frac{k_2}{k_2 - k_1} \cdot e^{-k_1 \tau} + (a_0 \cdot \frac{k_1}{k_2 - k_1} - b_0) \cdot e^{-k_2 \tau} \tag{1}$$

where c is the beer volume or its height under the foam, c_{∞} is the total volume of beer after complete foam decay, a is volume of beer bound in dry foam, b is beer freely present in the foam and τ is time. The constants k_1, k_2 describe the foam collapse in the first and second stage of decay, index 0 indicates the beginning of foam degradation (Savel, 1986).
How customers evaluate foam quality was uncovered by qualitative research at which 30 random customers were asked about their satisfaction with beer foam on the beer they were drinking in pubs or bars in the Czech Republic. In contrast to similar investigations, interviewees were not asked a long series of questions about foam quality or served any adjusted beer samples. The intention was to discretely interview the drinkers in their normal pub or bar drinking situation and gauge their opinion of the foam on the beer they were being served. The only question asked by the interviewers during drinking was the unforced question "is everything OK or not with the foam?" At this point, according to the interviewee's opinion, if there was something wrong with the foam, we visually evaluated the stage of foam collapse, in particular noting the presence of a "bald patch" in the foam, on the surface of the beer.
According to this qualitative assessment of customer perception of foam quality, customers in did not pay much attention to the foam until a problem with the foam is perceived. The beer was seen as problem free so long as there was a sufficient amount of foam to cover the beer surface in the glass. Customers start to be concerned about the foam quality in their glass only once they perceive that there was something wrong with the foam in their glass. This was at the end of foam collapse, when bald patches start to appear on the beer surface. At this point, approximately a quarter of the customers started to pay attention to the quality of beer foam in their glasses. The other three quarters of customers did not have any problems with the foam quality, even at this point. As the break-up of the last foam layer proceeded to produce a substantial bald patch on the beer surface, more customers started to be concerned with foam quality. Once the beer surface was almost completely bald, almost all customers commented that the foam quality was not satisfactory. Thus it can be concluded that for beer drinkers, the early appearance of this bald patch indicates a poor quality beer.
Close to this sensation of foam quality is a method for foam stability measurement called pouring test, which measures foam stability through whole collapse curve and includes the last collapse stages where bald patch appears. It is based on pouring atemperated (8 °C) beer from the bottle to a standard tasting glass and time from pouring to the first bald patch larger than 5 mm appearance is recorded. Although this test is principally very close to the customer sensation of foam quality, it has a disadvantage of low reproducibility.
One of the most spread methods among brewing laboratories is a method called NIBEM. This method is based on recording of the speed of downward movement of foam surface in

the second stage of foam collapse. This method has much higher reproducibility than the pouring test, but is quite far away from the customer perceived foam stability, as can be seen from the low correlation with pouring test (fig. 3).
Disadvantage of both of these methods, NIBEM and pouring test, is that it cannot be used to measure foam stability of samples that do not contain sufficient amount of CO_2 to create foam. NIBEM is slightly less sensitive to CO_2 content of sample than pouring test, because foam is created by flushing of the sample through the jet.

3.2 Matrix foaming potential

The new method for foam stability measurement, which was optimized and tested for foam stability improvement, is called the matrix foaming potential (MFP) and is measured by Foam stability tester type FA by 1-CUBE, Havlickuv Brod, Czech Republic (fig. 4). Foam is created by introducing of a gas into liquid sample and mixing with stirrer. By the combination of gas type, gas flow rate and revolution speed of the mixer there can be prepared foam of various structures, eg. by introducing 0,25 mL/min of air and mixing at 1200 RPM creates very fine foam resembling foam created on draught beer, or introducing 0,5 ml/min of air and mixing at 900 RPM creates medium coarse foam resembling foam on beer poured from the bottle (fig. 5).
Foam stability is evaluated as a time from the end foam generation to the decrease of foam surface over a set distance, which is a distance of electrodes that are in the place of measurement. The MFP value expressed in seconds covers the height of created foam under standard conditions, which corresponds to foaming ability of the sample, time to foam drainage in the first stage of foam collapse and the whole second stage of foam collapse.
As can be seen from the measurement principal, MFP can be used all over the whole beer production line, because even samples without CO_2 can be evaluated. Samples that contain CO_2 have to be degassed prior to the measurement. The MFP measurement has lower reproducibility due to various reasons, e.g. the temperature sensitivity (fig. 6). Regardless the reproducibility this method is much close to real foam quality as sensed by consumers, as can be seen from satisfactory correlation with pouring test (fig. 7).

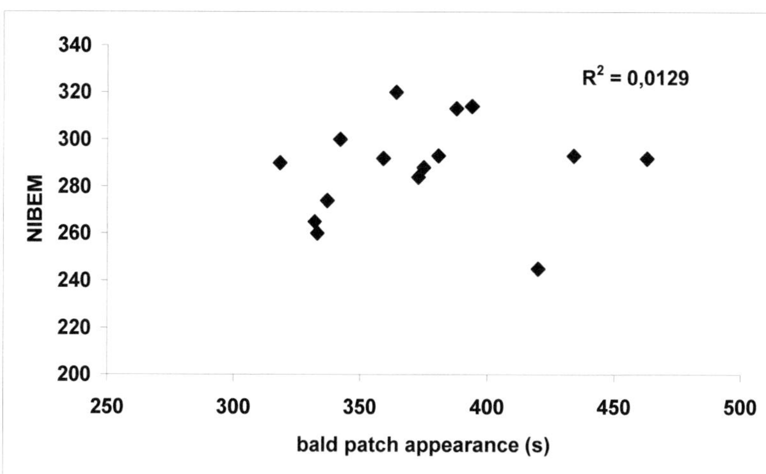

Fig. 3. Scatter plot and regression analysis of NIBEM with customer perceived stability measured by the pouring test.

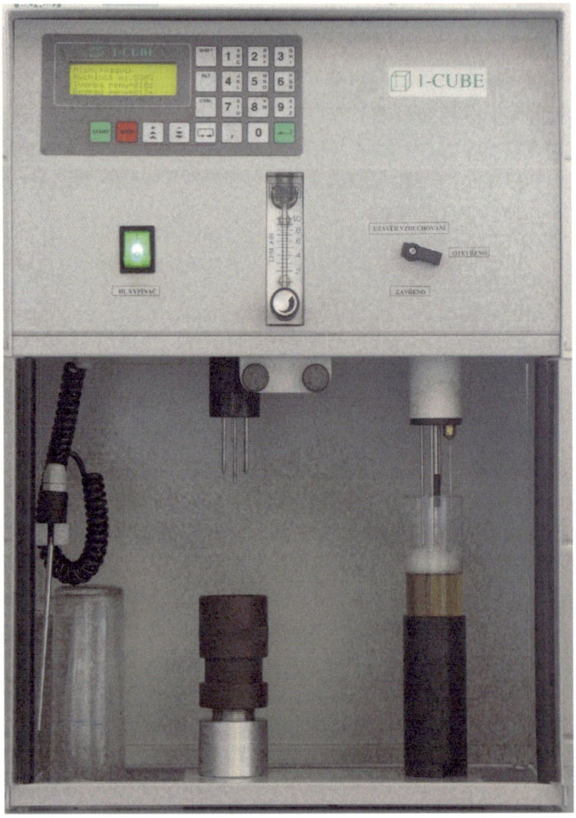

Fig. 4. Foam stability tester type FA by 1-CUBE.

Fig. 5. Fine (right) and medium coarse foam.

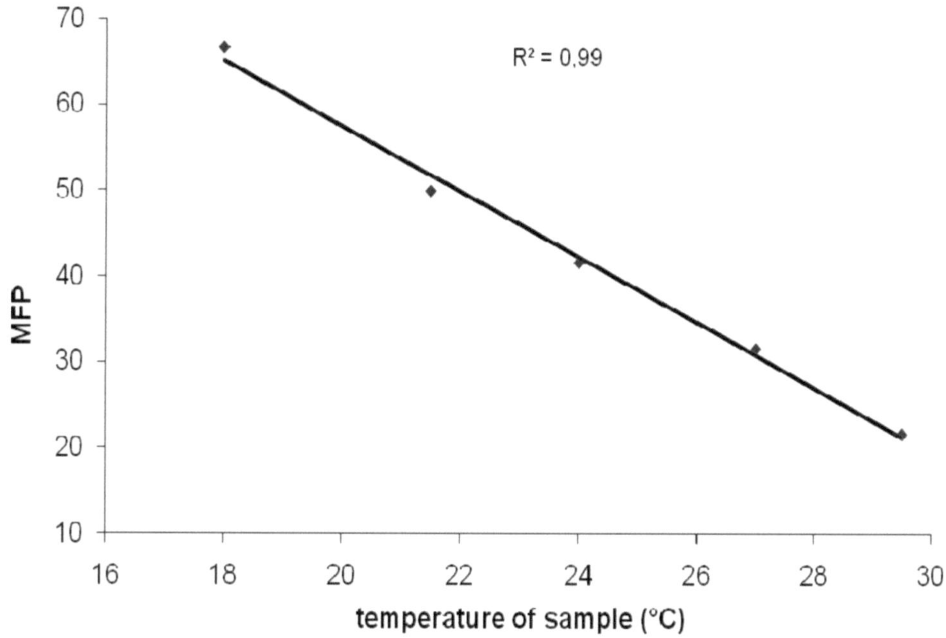

Fig. 6. Temperature dependance of Matrix foaming potential.

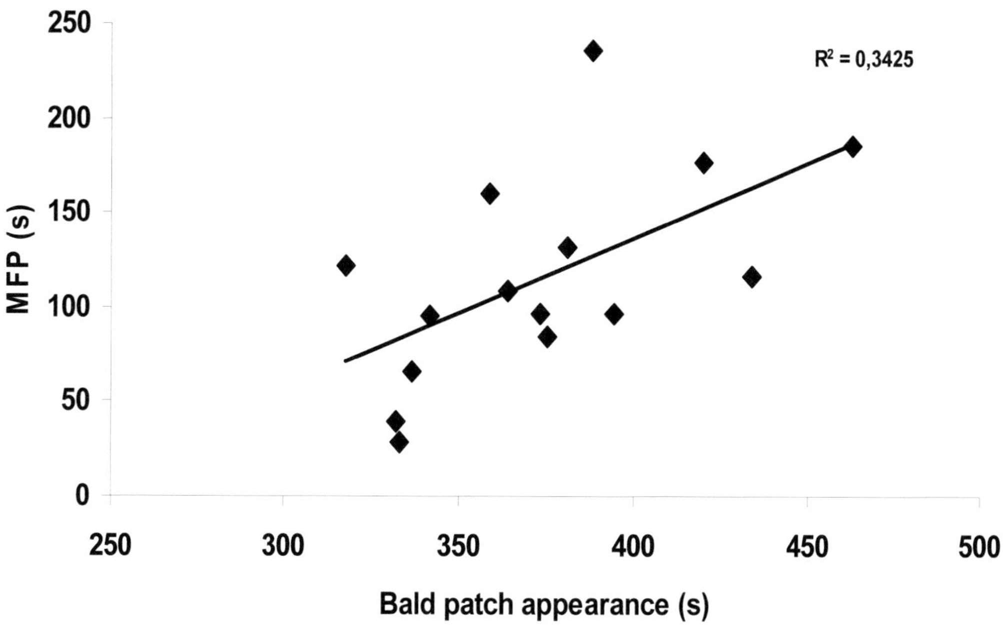

Fig. 7. Correlation of pouring test with Matrix Foaming Potential (MFP).

3.3 Foam positive and negative substances

As discussed above, brewers or researchers looking to improve foam quality typically take a reductionist analytical approach. Accordingly, the quality of beer foam generated is tried to be evaluated by measuring the content of foam positive components in beer, and then attempting to modify the brewing process to increase the content of these foam positive compounds to improve foam quality. Most often targeted with such an approach are foam positive components including protein Z, LTP1 and other proteins, and iso-α-acids or their reduced forms. Much more infrequently, the role of foam negative components such as lipids is considered in the technical literature.

It was observed that beer, even with the lowest content of proteins, could be foamed to 100 % of volume of relatively stable foam by simple foaming technique (Fig 8). A simple approach was to correlate the content of foam positive proteins assessed by the Bradford Coommassie blue binding assay (CBB) with foam stability measured by both NIBEM value and by pouring the beer to a glass from the bottle and measuring the time to bald patch appearance (Fig 9). This experiment was conducted with 15 brands of commercial lagers and showed no association between the level of foam positive proteins in beer or its foam stability with NIBEM (Fig 9A), although there was some association with bald patch formation (Fig 9B) although the slope was relatively low. On the basis of these results it was questioned whether the content of foam positive compounds was as important for the beer foam quality of beer as found in previous studies.

This suggests that the content of foam positive proteins/components may not be limiting with respect to foam quality, a similar conclusion that can perhaps be drawn from the beer dilution experiments of Roberts over 30 years ago (Roberts, 1978).

Fig. 8. whole volume of low protein beer converted into foam.

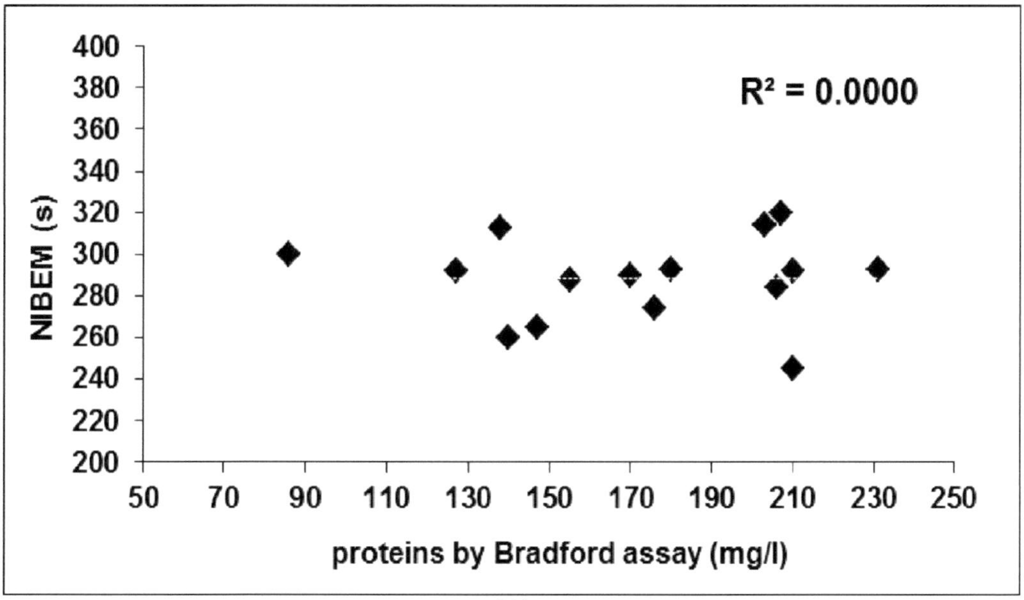

(A)

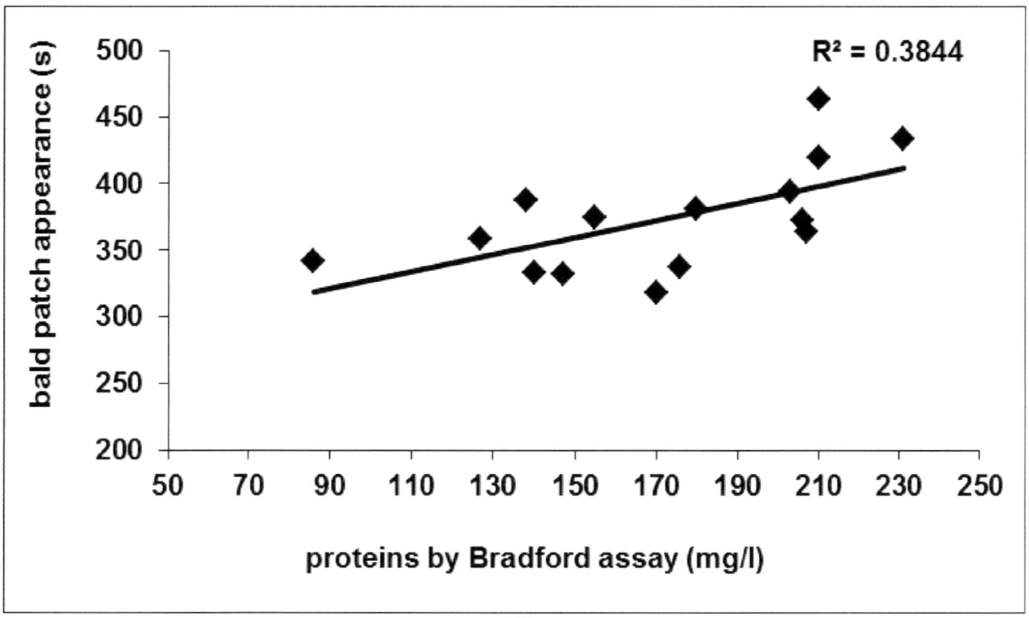

(B)

Fig. 9. Scatter plot and regression analysis of protein content of beer with NIBEM value (A), and time to bald patch appearance (B).

Previous investigations using a "foam tower" have shown that hydrophobic and foam positive components such as LTP1 and iso-α-acids are preferentially concentrated in the foam. To study if the content of beer foam positive proteins/components were limiting, a serial re-foaming experiment as depicted in figure 10 was designed. Degassed beer, created foam by stirrer as in MFP measurement with foaming time kept constant so as to measure the quantity of foam produced. The foam and beer phases were separated by pouring beer from under the foam, refilled the beer phase to original volume by "fresh" degassed beer to keep standard foaming conditions and again created foam with the mixer. The refilled amount was less than 10 % of the total volume. This cycle of foaming and separation was repeated 15 times.

The basic premise of the experiment was that if the content of foam positive proteins/components content was limiting in beer foam, foam capacity (amount of generated foam) and foam stability would decrease with sample order number in the experiment as these foam active components were concentrated in the foam and depleted from the beer. Thus foam positive proteins/components would migrate and concentrate in the foam phase in the earlier foaming and separation steps and so that there would not be a sufficient amount in beer phase in the later steps to generate sufficient amounts of stable foam.

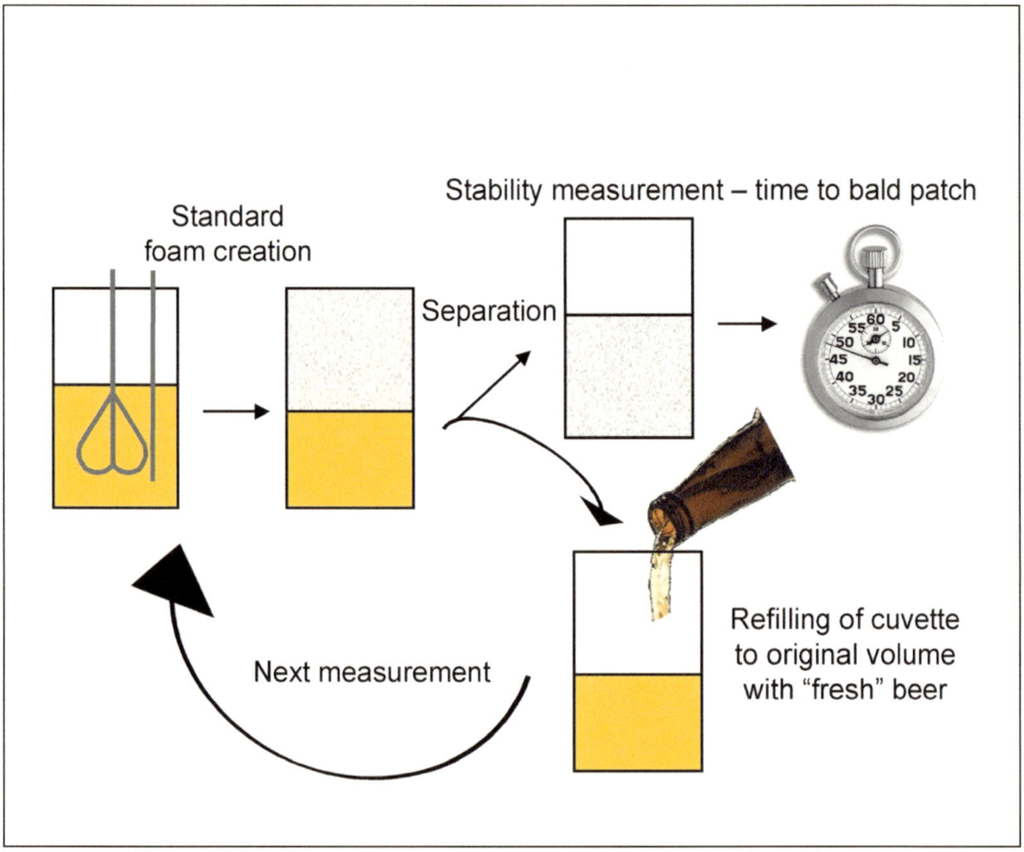

Fig. 10. Schematic of the design of experiment for the serial re-foaming of beer. Fifteen iterations of re-foaming of the beer were undertaken.

Figure 11 clearly shows that with serial re-foaming, foam capacity throughout the experiment was unchanged and foam stability, in terms of time till bald patch formation, was substantially increased, being five times higher at the end of experiment than at the beginning. It follows that foam stability was not just determined by the level of foam positive compounds, but it was more the result of compromise or balance between foam positive and negative components. Moreover, as the foam stability increased with re-foaming, it was apparent that both negative and positive foam components were presumably concentrated in the foam, thus separated from the beer to be re-foamed in the next cycle. This unexpected and contrary result could be explained by the following hypotheses. Firstly, Bamforth proposed that "hydrolyzed hordein appears to selectively enter beer foams at the expense of the more foam-stabilising albuminous polypeptides" such as protein Z (Bamforth, 2004). As such, as the level of hydrolyzed hordein is depleted relative to the albuminous polypeptides, foam stability would be seen to improve. However, Lusk et al. found in their foam tower experiments, as the content of LTP1 was depleted, the foam became less "creamy" and contained coarse bubbles, that were not observed in this experiment (Lusk, 1999). Secondly, as LTP1 is concentrated in beer foam and is thought to play a lipid-binding role in beer, both the LTP1 and the foam stabilizing lipids would be removed with the separated foam. An improvement in foam stability would occur if the level of lipids were limiting in the beer relative to LTP1, other lipid binding components and foam positive proteins/components.

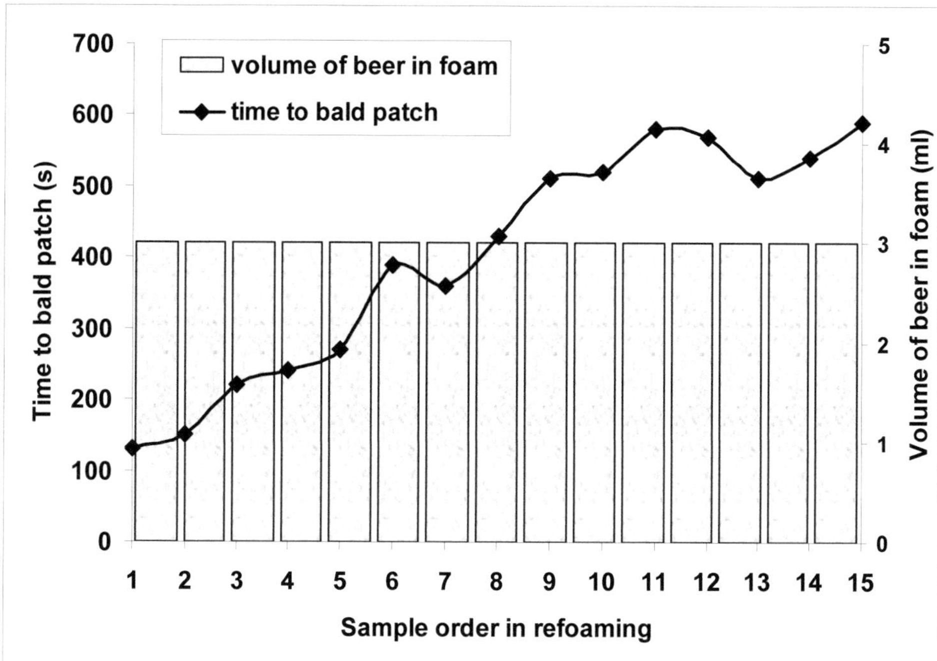

Fig. 11. Results of foaming capacity and foam stability during a serial refoaming experiment with 15 iterations. The volume of beer foam formed was exactly the same for each refoaming iteration, when using a constant time of foaming.

3.4 Practical approach to foam stability

The insights gained from these experiments recommend several practical approaches to foam quality improvement in commercial production. These are based on the premise that by measuring the MFP during and within each production stage, critical points can be identified in the process that reduce foam stability and indicates process parameters that can be modified to improve foam stability, particularly the limiting of the inclusion of foam negative components. One example was to apply MFP measurement during the course of lautering and sparging process. Figure 12 shows, that extended sparging was one of the steps that reduces foam stability. It has long been known that although extended sparging recovers more extract, it also results in the extraction of increasing amounts of undesirable substances such as polyphenols, husk bitter substances, etc , and foam negative materials. Similarly, the MFP analysis was applied during the course of main fermentation (Fig 13). During fermentation, foam stability was decreased to almost a third.

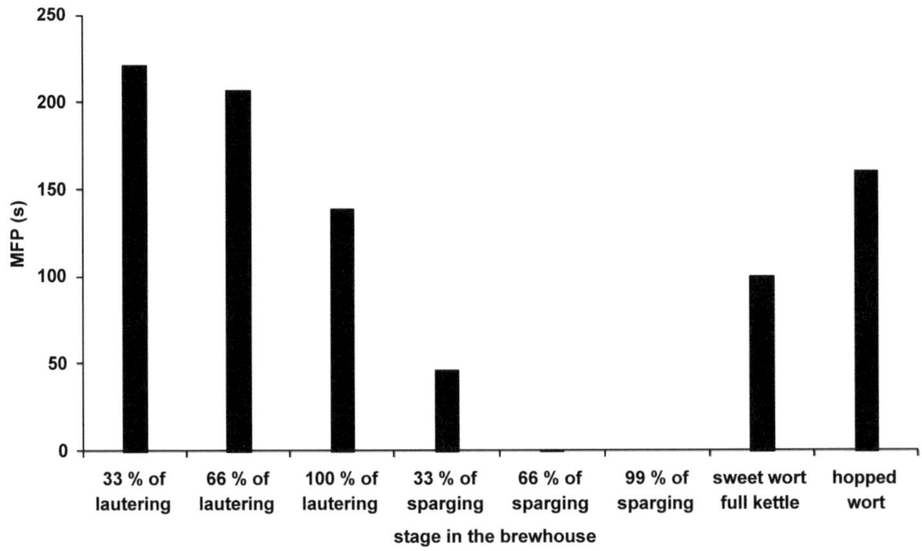

Fig. 12. Matrix foaming potential of wort samples taken during the course of lautering and sparging.

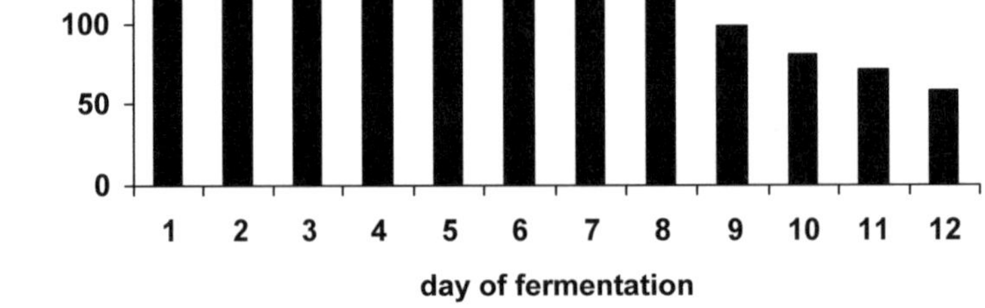

Fig. 13. Matrix foaming potential of "beer" samples during the course of main fermentation.

By this approach there were found several control points for foam stability, of which some fulfilled the demand for not changing of any other beer quality parameter, including raw materials. The effectiveness of these new control points has to be validated by statistical methods, which also serve for control of constancy of other quality parameters after setting new control points.
The most suitable statistic tool is regulation chart, which illustrates the change of foam stability in figure 14 by NIBEM value. Foam stability increased from values around lower specification limit into optimal central zone for NIBEM, MFP increased approximately three times.

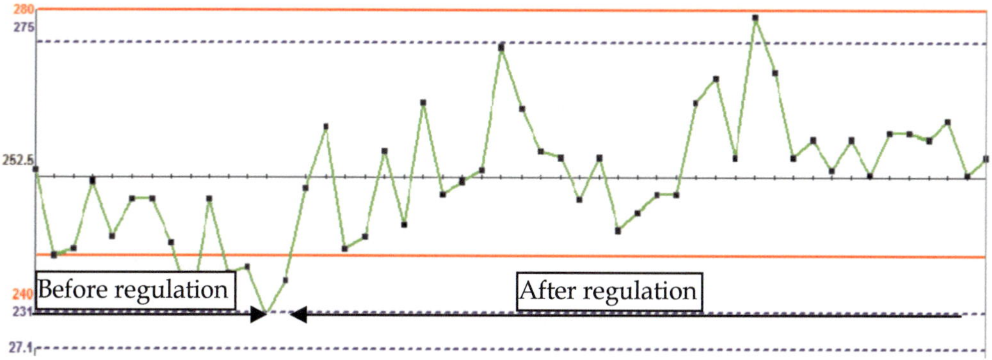

Fig. 14. Regulation chart of foam stability measured by NIBEM during process optimization.

4. Conclusion

Integral part of modern quality management is improving quality. For the task of improving beer foam quality, more successful strategy appeared to be employing own R&D, compared to external consulting expert, even with deep knowledge of bibliographic results on given topic.
The quality improvement procedure included optimization of new method with results close to customer sensation of quality, which could be used even for evaluation of intermediate product all over the production line. By this method weak points were discovered and the success of new regulation was evaluated by control charts.

5. References

Bamforth, C.W. (2004). A critical control point analysis for flavour stability of beer. *Master Brewers Association of the American Technical Quarterly*, Vol.41, (2004), pp. 97-103

Evans, E. & Sheehan, M. C. (2002). Don´t be fobbed off: The substance of beer foam - a review. *Journal of the American Society of Brewing Chemists*, Vol.60, No.2, (2002), pp. 47-57

Haukeli, A. D.; Wulff, T. O. & Lie, S. (1993). Practical experiments to improve foam stability, *Proceedings of 1993 24th European Brewery Convention Congress*, p. 40, Oslo, Norway, 1993

Juran, J.M. & Godfrey, A.B. (2000). *Juran´s quality handbook, fifth edition*, McGraw Hill, ISBN 0-07-116539-8, Singapore, Singapore

Klein, N. (2005). *Bez loga*, Dokořán, ISBN 80-7363-010-9, Prague, Czech Republic

Lusk, L.T.; Ting, P.; Goldstein, H.; Ryder, D. & Navarro, A. (1999). Foam tower fractionation of beer proteins and bittering acids, In: *European Brewery Convention Monograph*, 166 - 187, ISBN 3-418-00774-0, Amsterdam, Netherlands

Olins, W. (2009) *O Značkách*, Argo, ISBN 978-80-257-0158-4, Prague, Czech Republic

Roberts, R.T.; Keeney, P.J. & Wainwright, T. (1978). The effects of lipids and related materials on beer foam. *Journal of the Institute of Brewing*, Vol.84, (1978), pp. 9-12

Verstl, I. (2005). The servant of two masters or how to keep your customers satisfied : the European brewing industry, the Zeitgeist and Tradition, *Proceedings of 2005 30th European Brewery Convention Congress*, p. 171, Prague, Czech Republic, May 14-19, 2005

Šavel, J. (1986). Dva modely rozpadu pivní pěny, *Kvasný Průmysl*, Vol.32, No.4, (1986), pp. 76-78

20

Herbal Drug Regulation Illustrated with Niprifan® Antifungal Phytomedicine

Sunday J. Ameh[1], Obiageri O. Obodozie[1], Mujitaba S. Abubakar[2], Magaji Garba[3] and Karnius S. Gamaniel[1]

[1]Department of Medicinal Chemistry and Quality Control, National Institute for Pharmaceutical Research and Development (NIPRD), Garki, Abuja,
[2]Department of Pharmacognosy and Drug Development, Ahmadu Bello University, Zaria,
[3]Department of Pharmaceutical and Medicinal Chemistry, Ahmadu Bello University, Zaria,
Nigeria

1. Introduction

Quality system is defined as arrangements, procedures, processes and resources; and the systematic actions necessary to ensure that a manufactured product will meet given specifications. On the other hand, quality control is defined as measures taken, including sampling and testing, to ensure that raw materials, intermediates, packaging materials and finished goods conform to given specifications. Quality specification refers to a written procedure and requirements that a raw material, intermediate or finished good must meet for approval. On the other hand, standard operating procedure (SOP) refers a written procedure, giving step-by-step directions on how a particular operation is to be carried out. Quality manual means, a document that describes the various elements of the system used in assuring the quality of results or products generated by a laboratory or factory. The term quality assurance refers to the totality of all the arrangements made with the objective of ensuring that products are of the quality required for their intended use. Good manufacturing practice (GMP), on the other hand, is that aspect of quality control that deals directly with manufacturing and testing of raw materials, intermediates and finished goods to ensure a product of consistent quality. Essentially, GMP involves two types of control - analytical and inspection, and both require: i) clear instructions for every manufacturing process; ii) a means of controlling and recording every manufacturing process; iii) a means of ensuring that the complete history of a batch can be traced; iv) a mechanism for recalling any batch of product from circulation; v) a system for attending to complaints on quality of product or service; and vii) a programme for training operators to carry out and to document procedures. The foregoing definitions and description of GMP conform to those of WHO (2000). It is also clear from the foregoing that GMPs are not prescriptive instructions on how a manufacturer can produce, but are rather a series of principles that must be observed for quality products, services or results to emerge. Invariably, GMPs are approved and enforced by an appropriate National Agency, but the onus of preparing and

executing GMPs rests with the manufacturer. In Nigeria (population ~ 150 million), GMPs are enforced by NAFDAC - established by decree in 1992/93.
The tests carried out for this study were according to official procedures - mostly BP (2004) and WHO (1998). The results are discussed within the context of requirements for herbal drug regulation as per WHO, EMEA and NAFDAC. It is noted that herbal drug regulation in Nigeria (Table 1) as compared that in Europe (Table 2) is paradoxically hampered not by the rigor and "stringency" of rules, but by the fact that the rules are only merely cumbersome, being neither adequate nor enforceable (Table 3), unlike those of EMEA.

S/No.	**Regulatory aspect**	**Requirement**
1	Legal status of applicant, who may be: Manufacturer. Marketer. Distributor.	The applicant must be certified by the Corporate Affairs Commission as a registered business in Nigeria. A marketer or distributor must show evidence of Power of Attorney issued by the manufacturer.
2	Analytical status of the product for registration.	The product must : Have a certificate of analysis. Be accompanied by a dossier with information on: Ingredients. Method of analysis. Stability. Dosage. Safety, among others.
3	Pre-registration inspection of premises.	Manufacturing, storage and distribution premises must be GXP compliant. Marketers must provide convincing evidence of GDP and GSP.
4	Post marketing surveillance plan/ report	Applicant may be required to provide a plan for reporting on: The use of the product. Any adverse reactions.

The above information were drawn from NAFDAC's leaflets and website: www.nafdacnigeria.org

Table 1. NAFDAC requirements for registering herbal medicines.

Essentially, the NAFDAC rules seem perhaps affected, or rather made cumbersome without being truly rigorous or "stringent" as actually claimed in the Agency's website. It is further noted that, like in China, where the head of the drug regulatory agency was sentenced to death for corruption (Gross and Minot, 2007), drastic actions, including the wholesale reorganization of NAFDAC management, had to take place in 2000 to straighten things out. The high frequency of confiscation and public destruction of counterfeit products by NAFDAC strongly testifies to the inadequacy of the rules and policies guiding drug regulation in Nigeria. Unfortunately, this worrisome state of affairs is equally true of many countries, as stated in the case of china. There is thus, the need for drug regulatory agencies in these countries to brace up. The aim of this article therefore, is to further an earlier advocacy (Ameh et al., 2010a) that includes alerting and encouraging Drug Regulatory Agencies, Health Ministries, and Parliamentary Health Committees, especially those in developing countries, to enact laws and evolve policies that will better regulate the

S/No.	Regulatory aspect	Requirement
1	**Product information**: Summary of product characteristics	Name of the product. Strength. Dosage form. Quantity of active ingredient (Example: 25 mg *P. guineense*). List of excipients (Example: *P. guineense, E. caryophyllata* etc., etc). Shelf life. Posology and method of administration. Indications. Contraindications/ special warnings. Precautions for use. These data are used as the basis for inserts, packaging, or advertisement. Inserts must pass "readability testing."
2	**Quality control data**: Refer to GMP requirements for production.	Production must be in a GMP compliant facility Drug must be produced with validated/ reproducible formula method. There must be a finished product specification. The product must be manufactured at least on pilot scale and three batches used for stability studies. Stability studies should be carried out on the product packaged in the container proposed for marketing. A summary of the stability studies undertaken must be provided. From stability data shelf life/ storage precautions should be proposed. A quality dossier must be provided for raw and finished materials The product must be produced from herbs that have been cultivated and harvested in accordance with GACP. 10. The starting material must be evaluated for risk of any environmental contamination.
3	**Safety data requirements**: Refers to safety pharmacology, including animal and human studies	Published animal or human studies. Review of any potential interactions with other drugs, side effects, and any proposed contraindications/ precautions in the product information. Recognized monographs on the material or product with information on safety. Any information concerning special groups such as children, the elderly or pregnant women. It is interesting to note that in the US, where herbal medicines are regulated as dietary supplements, manufacturers are not required to prove safety or efficacy, but the FDA can withdraw a product from sale if it proves harmful.
4	**Traditional use evidence**: Refers to history and prevalence.	There is no requirement to prove efficacy. Instead data must provide reference that the product has been in use as medicine for 30 years or more. Of 30 or more years, the last 15 must be in Europe. The data must be presented in a special format, called: *Common Technical Document Format*.

The Table was drawn based on data gathered from references including (DSHEA 1994; Goldman, 2001; De Smet, 2005; Ann Godsell Regulatory, 2008).

Table 2. EMEA requirements for registering herbal medicines.

S/No.	**Extra requirement**	**Remark**
1	Five (5) copies of the product dossier.	Probably unreasonable
2	Three (3) packs of the products samples.	Probably reasonable
3	Notarized original copy of the duly executed Power of Attorney from the product manufacturer.	Clearly unreasonable for all categories of applicants
4	Certificate of Manufacture issued by the competent health or regulatory authority in country of origin and authenticated by the Nigerian Mission in that country. Where there is no Nigerian mission, The British High Commission or an ECOWAS country Mission will authenticate.	Probably unreasonable for all categories of applicants
5	If contract-manufactured, Contract Manufacturing Agreement, properly executed and notarized by a Notary Public in the country of manufacture.	Clearly unreasonable for all categories of applicants
6	Current World Health Organization Good Manufacturing Practice Certificate for the manufacturer, authenticated by the Nigerian Mission.	Clearly unreasonable for all categories of applicants
7	Certificate of Pharmaceutical Products (COOP) duly issued and authenticated.	Clearly unreasonable for all categories of applicants
8	Current Superintendent Pharmacists license to practice issued by the Pharmacists Council of Nigeria (PCN).	Only probably reasonable
9	Premises Registration License from PCN	Only probably reasonable
10	Certificate of Registration of brand name with trademark registry in the Ministry of Commerce here in Nigeria; Letter of invitation from manufacturer to inspect factory abroad, stating full name and location of plant.	Probably unreasonable for all categories of applicants
11	The applicable fee payable only if documents are confirmed to be satisfactory.	Likely to be abused if the amount is high. The fee should be a token amount paid by all applicants
12	Nutraceuticals, medical devices and other regulated drug products have similar requirements, with minor variations. Specific details can be obtained from NAFDAC.	A sketch of the minor variations should be provided in print no matter how brief. Any information provided by NAFDAC should be printable for sake of transparency.

The information on NAFDAC were drawn from leaflets and NAFDAC's website (2010): www.nafdacnigeria.org/ The remarks are informed by current affairs and public perception of NAFDAC's role and activities including the wholesale reorganization of its Management in 2000.

Table 3. NAFDAC's extra requirements for registering herbal medicines.

production, distribution and use of herbal drugs. This is in view of the ever increasing use of herbs notably after the Alma-Ata Declaration (Ameh et al, 2010b) which paved the way for the stupendous growth of herbal drug use worldwide, particularly in North America where that growth had been stymied by the Flexner Report of 1910. That Report, which coincided with Paul Ehrlich's introduction of Salvarsan and the term "chemotherapy" in 1909, had favoured chemical medicine over herbal (Pelletier, 2006). Furthermore, apart from the said growing use in the West, it is held that some 80 % of the populace in many developing countries still relies predominantly on herbs and other alternative remedies (WHO, 2008). Indeed, in some parts of Africa, for example, Ethiopia, a dependence of up to 90% has been claimed (BBC, 2006).

2. Experimental

The study applied official procedures - mainly WHO (1998) and BP (2004) to: evaluate the quality parameters of the raw materials and their extracts; and the changes in these parameters during dark, dry storage in capped glass bottles under tropical room temperature and humidity (RTH) as obtain in a typical Nigerian Traditional Apothecary (NTA). The parameters evaluated were: appearance, loss on drying, ash values, extractability, solubility, pH, TLC features, light absorption and foaming index. Basic morphological studies were carried out as per WHO (1998). Appropriate phytochemical tests were also conducted by official methods as described elsewhere (Ameh et al., 2010c; 2010d).

2.1 Treatment and sampling of material

The aerial parts of *Mitracarpus scaber* obtained during the months of October and November from the botanical garden of the National Institute for Pharmaceutical Research and Development (NIPRD) were air-dried in a well-ventilated shade, designed for drying medicinal plant materials. The materials were subsequently comminuted to coarse powder with a grinding machine. The procedure for sampling was as per WHO (1998) as had been described in detail earlier (Ameh et al., 2010c). Three (3) original samples from each batch or container were combined into a pooled sample and subsequently used to prepare the average sample. The average sample was prepared by "quartering" the pooled sample as follows: each pooled sample was mixed thoroughly, and constituted into a square-shaped heap. The heap was then divided diagonally into 4 equal parts. Any 2 diagonally opposite parts were taken and mixed carefully. This step was repeated 2 to 4 times to obtain the required quantity of sample. Any material remaining was returned to the batch. The final samples were obtained from an average sample by quartering, as described above. This means that an average sample gave rise to 4 final samples. Each final sample was divided into 2 portions. One portion was retained as reference material, while the other was tested in duplicate or triplicate. The samples for stability study were stored at room temperature and humidity (RTH) in capped glass bottles and placed in a shelf protected from light.

2.2 Macroscopic examination and phytochemical tests on the fresh and air-dried materials

The procedures adopted were as per WHO (1998). Shape and size were determined with the aid of a ruler and a pair of calipers. Diffuse day light was used on the untreated sample to determine its colour. The texture and surface/ fracture characteristics of the untreated sample were examined, where necessary, with x10 magnification hand lens to reveal the characteristics of cut surfaces. The material was felt by touch, to determine if it was soft or hard. Or was bent and ruptured, to obtain information on brittleness and appearance of fracture planes - whether it was fibrous, smooth, rough or granular. Odour was determined by placing a pinch in a 50-ml beaker, and then slowly and repeatedly the air above the material was inhaled. If no distinct odour was perceived, the material was crushed between the thumb and index finger, and inhaled as above. The strength of the odour was determined as: odourless, weak, distinct, or strong. The sensation of the odour was determined as: aromatic, fruity, rancid, etc. etc. When possible, the odour was compared with that of a defined substance, such as menthol, sulphur dioxide, eugenol, etc. etc. Taste: In tasting the material, as recommended by our experience with the material, the following procedure was applied: a pinch of the material was mixed with water and savored, or chewed without swallowing, to determine the strength and the sensation of the taste. The strength is recorded as: tasteless, weak, distinct, or strong; and the sensation, as: sweet, sour, saline, or bitter. Phytochemical tests for tannins, saponins, terpenoids, anthraquinones and alkaloids were carried out on samples by procedures as described in detail elsewhere (Ameh et al., 2010c, d).

2.3 Loss on drying

This was carried out using a minimum of 0.5 - 1.0 g of material. Drying was effected in a Lindberg/Blue M gravity-convention oven maintained at 105-110 ^{0}C, for 3 h, after which the sample was allowed to cool to room temperature in a desiccator, and subsequently weighed. The time interval from the oven to point of weighing was usually about 30 minutes. The results are expressed as a range or as mean ± standard deviation.

2.4 Evaluation of extractive matter

About 4 g of accurately weighed coarsely powdered, air-dried sample was transferred into a glass-stoppered, 250-ml reflux conical flask, followed by the addition of 100 ml of solvent. The flask was weighed along with its contents, and recorded as W1. The flask was well shaken, and allowed to stand for 1 h. Subsequently a reflux condenser was attached to the flask, and gently brought to boiling and maintained thereat boiled 20 - 60 minutes depending upon the solvent. The mixture was subsequently cooled and weighed again. The weight was recorded as W2, and then readjusted to W1 with the solvent. The flask was shaken well once again and its contents rapidly filtered through a dry filter paper. By means a pipette, 25 ml of the filtrate was transferred to a previously dried and tarred glass dish and then gently evaporated to dryness on a hot plate. Subsequently, the dish was dried at 105 °C for 1-6 hours, cooled in a desiccator for 30 min, and weighed. The extractable matter was calculated as %w/w of the air-dried sample.

2.5 Determination of solubility of material in a given solvent – Methods I and II

The solubility of a material was determined at room temperature ~ 25°C and expressed in terms of "parts', representing the number of milliliter of solvent, in which 1 g of the material is soluble. Vials of appropriate sizes: ~4-ml, ~ 12-ml and ~20-ml capacities were used. The mixtures were thoroughly shaken for at least 30 min before inspection for un-dissolved solute. In methods I, each vial received 100 mg of sample and the volume of solvent indicated. In method II, a vial received 100 mg and increasing volumes of solvent. The methods give the same results.

2.6 Light absorption and thin layer chromatography (TLC)

UV-VIS Spectrophotometer (Jenway or Shimadzu) and quartz 1-cm cells were used for the study. Solutions of herb and extract were made by thoroughly mixing 1 part of the solute and with 100 parts of solvent methanol: water 80:20, v/v filtering, and diluting the filtrates by 150x with the same solvent. Absorbencies were measured at λ227 nm, using the solvent as the blank. Florescent, precoated plates were used for normal phase TLC, utilizing silica K6, and hexane: ethylacetate as mobile phase. Solutions of analytes were prepared and applied as follows: To 1 mg of the analyte, 2 drops of ethanol were added and mixed well (~1 %w/v solution). The plates used were 5 cm wide x 20 cm long. With a ruler and a pencil, a distance of 5 mm was measured from the bottom of the plate, and a line of origin was lightly drawn across the plate, without disturbing the adsorbent. The analyte was applied to the origin as a 1 µl droplet. The spot was allowed to dry. Subsequently, the plate was developed in a developing tank saturated with the vapour of the solvent system to be used as mobile phase. The level of the solvent in the tank was adjusted to a level 2 to 3 mm below the line of origin on the plate. The plate was considered developed when the solvent front reached a predetermined line, not less than 5 mm below the top of the plate. The air-dried plate was visualized using a viewing cabinet (Cammag) and a UV-lamp (Cammag - equipped to emit light at 254 or 366 nm). The resulting chromatogram was photographed and subsequently drawn to scale.

2.7 Determinations of pH of preparations – herb and the dry extract

Determination of pH values was with a Jenway pH Meter. Standard pH solutions: 4, 7 and 10; and freshly distilled water were used for the study.

2.8 Determination of foaming indices of preparations – herb and the dry extract

Decoctions of plant materials foam due to the presence of saponins. This ability is measured as foaming index, and is an important quality control parameter. The requirements for the test include: conical flasks (500-ml); volumetric flasks (100-ml); test tubes (16cm x 16mm); ruler; and stop-clock. The procedure was as follows: Exactly 1.0 g of powdered material was accurately transferred into a 500-ml conical flask containing 100 ml of boiling water, and maintained at moderate boiling for 30 minutes. The mixture was then cooled and filtered into a 100-ml volumetric flask. The volume was made up to 100 ml with water. Successive portions of 1 ml, 2 ml, 3 ml etc up to 10 ml of the filtrate was poured into ten stoppered tubes having the following dimensions: height, 16 cm; diameter, 16 mm. Subsequently, the volume of each tube was adjusted to 10 ml with water, stoppered and shaken in lengthwise motion for 15 seconds, at 2 shakes per second. The tubes were allowed to stand for 15 minutes, and the height of the foam in each tube was measured. The results were assessed as follows:

Foaming index is ≤ 100, if the height of foam in all the tubes is less than 1 cm.

If a height of 1 cm is obtained in any tube, the volume [V] of the decoction in that tube, is used to determine the foaming index, as = 1000/V.

But if the tube above is the first or the second in the series, prepare an intermediate dilution to obtain a more precise result.

If the height of the foam is > 1 cm in every tube, the foaming index is over 1000.

To obtain a more precise result, repeat the determination using a new series of dilutions of the decoction. Note the tube in which the height of foam is 1 cm, and the volume [V] of the decoction therein, and calculate the foaming index, as = 1000/V. Results are expressed as expressed as a quantity [Q] per ml or as [Q]ml^{-1}.

3. Results

3.1 Results of botanical examination / phytochemical tests on the herb and extract

The key botanical and phytochemical characteristics of *Mitracarpus scaber* Zucc (Family: Rubiaceae) with Voucher specimen number: NIPRD/H/4208, preserved in the Institute's Herbarium are indicated in Table 4. The plant grows erect, up to 55 cm high, usually branched; the leaves are lanceolate, 2-4 cm long, with the upper surface bearing minute hairs. The plant manifests dense clusters of inflorescence, 6-14 mm across, with minute white flowers. The fruits are dehiscent capsules, about 0.5-1 mm long. Both the fresh plant and air-dried weed are practically odourless but possess a slight warm taste. Tannins, saponins and anthraquinones were detected in the weed. The extract also contained tannins and anthraquinones but not saponins. Tests for alkaloids were negative for the weed and extract.

Characteristic	Live Sample	Air-dried Sample
General appearance	 M. scaber is an annual, with erect stems, up to 55 cm high, and often branched. The leaves are lanceolate, 3-5 cm long, with the upper surface scabrous. The inflorescence consists of clusters of small white flowers. The fruits are dehiscent capsules, up to 1 mm long. The plant is of the Family, Rubiaceae, reproduces by seeds, and is found in the tropics.	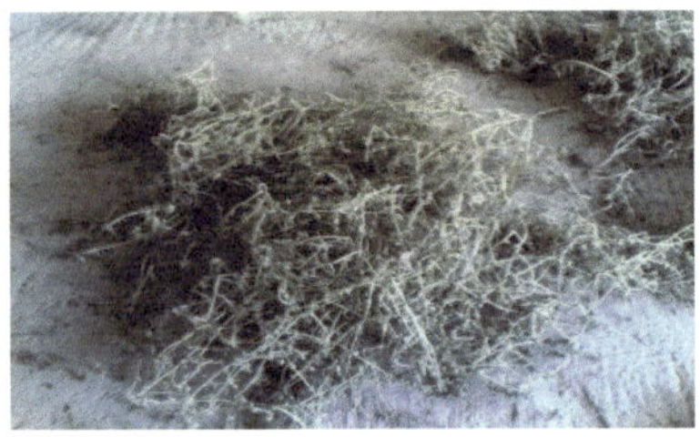 The air-dried sample consists of brownish green twigs and other parts that can readily be ground in a mortar or comminuting machine. The air-drying process takes about a week during the months of October to December, at NIPRD Herbarium, Abuja. The extracts obtained with various solvents yield a black, odorless and sticky mass.
Odor	Odourless	Odourless
Taste	Very slightly warm	Slightly warm
Phytochemicals	Tannins, saponins and anthraquinones were detected. The tests for alkaloids were negative.	Tannins and anthraquinones were detected. The tests for alkaloids and saponins were negative.

The samples described above were obtained from the NIPRD Botanical Gardens at Idu Industrial Area, Idu, Abuja, Federal Capital Territory, Nigeria.

Table 4. Some key characteristics of *Mitracarpus scaber* and its aerial parts.

3.2 Results of physicochemical tests on the herb and extract of *Mitracarpus scaber*

Typical results of loss on drying (LOD as %w/w was 10.29 ± 1.81 for the herb; and 15.86 ± 0.72 for the extract) and total ash (TA as %w/w was 12.44 ± 2.95 for the herb; and 0.40 ± 0.09 for the extract) are shown in Table 5. The Table also shows that a 5%w/v mixture of the herb had pH of 5.7 ± 0.3, while that of the extract was 6.9 ± 0.3. The herb in water foamed slightly (that is: Foaming Index [FI] ≤ 100), but the extract did not foam at all (that is: FI = 0). Table 5 further shows that the herb in methanol/water (80/20: v/v) had an A1%1cm of 325.8 ± 15.6, while that of the extract was 349.5 ± 14.1. The Table also shows the extractabilities of the herb in various solvents. The extractability results expressed as (%w/w) were as follows: acetone, 6.89 ± 0.89; water, 28.37 ± 1.77; ethanol, 11.72 ± 0.81; ethylacetate, 14.02 ± 1.89; hexane, 4.11 ± 0.47; and methanol, 15.11 ± 1.07. The extractability of the air-dried weed was highest in water and least in hexane. Among the organic solvents, extractability was lowest in hexane, and highest in methanol, followed by ethylacetate, ethanol and acetone.

Parameter (Mean ± SD)	Air-dried Herb	Ethylacetate Extract
Loss on drying (LOD: % w/w)	10.29 ± 1.81 (n=12)	15.86 ± 0.72 (n=12)
Total ash (TA: % w/w)	12.44 ± 2.95 (n=12)	0.40 ± 0.09 (n=11)
pH of 5 % w/v in water	5.7 ± 0.3 (n=5)	6.9 ± 0.3 (n=5)
Foaming Index (FI: as ml^{-1})	Slight foam. FI ≤ 100 (n=5)	No foam. FI = 0 (n=5)
A 1%1cm at λ227 nm (MeOH/ H_2O: 80/20 v/v)	325.8 ± 15.6 (n=5)	349.5 ± 14.1 (n=5)
Extractability (% w/w) in: **Solubility** (ml/g) in:	Extractive value (n=8-12) -	- Solubility (n=2-6)
Water	28.37 ± 1.77	>10^3
Methanol	15.11 ± 1.07	25.0 ± 0.0
Ethylacetate	14.02 ± 1.89	25.0 ± 5.0
Ethanol	11.72 ± 0.81	25.0 ± 0.0
Acetone	6.89 ± 0.89	15.0 ± 0.0
Hexane	4.11 ± 0.47	55.0 ± 5.0

The LOD results were validated by concurrent determination of the LOD of copper sulphate, which result (mean ± SD) was 36.12 ± 0.19 %. The results prove that the extract was quite hygroscopic. The low TA results for the extract but not the herb probably suggests a high presence of high bio-minerals. The high water extractability result agrees with the high TA of the herb, and the fact that hexane, the least polar solvent, produced the lowest extractability result. Among the organic solvents the solubility of the extract was least in hexane (55 ml/g), but higher in ethanol, ethylacetate and methanol (15-25 ml/g). The extract was practically insoluble in water (>10^3ml/g). The colour of the solution obtained from the herb using different organic solvents was clear and greenish-brown in each case, but that obtained with water was yellowish brown, and slightly cloudy, with no tinge of green.

Table 5. Various physicochemical parameters of herb and extract of *Mitracarpus scaber.*

3.3 Results of thin layer chromatographic (TLC) studies on the herb and extracts of *M. scaber*

Figure 1 is a normal phase TLC of the herb and extract developed with hexane-ethylacetate. The Figure indicates the following: The herb in acetone (A) or ethanol (C) yielded 5 identical principal spots, while the herb in water (B) yielded only 2 principal spots – Rf1 and Rf4. The herb in ethylacetate (D) or hexane (E) yielded 3 spots, while the herb (F) in methanol yielded 4. The dry hexane extract (G) re-dissolved in hexane yielded 7 spots, while the ethylacetate extract (H) re-dissolved in ethylacetate yielded 10. Notably, Rf4 was present in all the chromatograms, while Rf2 and Rf3 were present only in the H chromatogram. On the other hand Rf1 was present only in the B, G and H chromatograms.

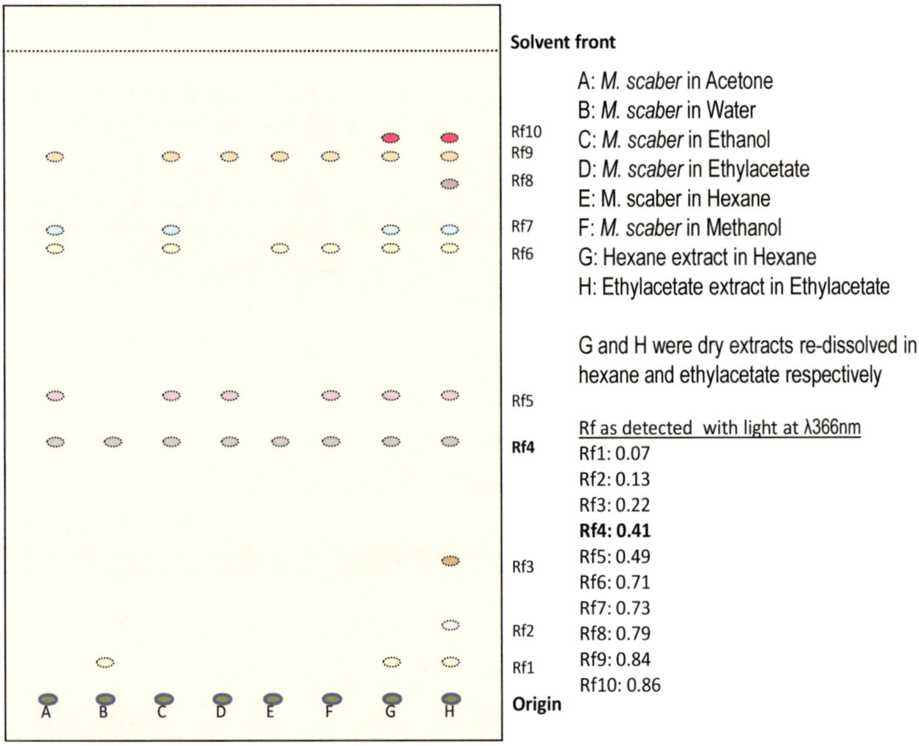

The above diagram is of a normal phase TLC (K5 Silica, using hexane: ethylacetate at 60:40 v/v as mobile phase) of samples of samples of herb and extracts in various solvents as follows. Types of samples/ solvents: A: Herb in Acetone, B: Herb in Water, C: Herb in Ethanol, D: Herb in Ethylacetate, E: Herb in Hexane, F: Herb in Methanol, G: Hexane extract in Hexane, H: Ethylacetate extract in Ethylacetate. The samples in G and H were dry extracts re-dissolved in hexane and ethylacetate respectively. Rf as detected at λ366nm: Rf1: 0.07, Rf2: 0.13, Rf3: 0.22, Rf4: 0.41, Rf5: 0.49, Rf6: 0.71, Rf7: 0.73, Rf8: 0.79, Rf9: 0.84, Rf10: 0.86. Descriptions/interpretations: The herb in acetone (A) or ethanol (C) yielded 5 identical principal spots, while the herb in water (B) yielded only 2 principal spots – Rf1 and Rf4. The herb in ethylacetate (D) or hexane (E) yielded 3 spots, while the herb (F) in methanol yielded 4. The dry hexane extract (G) re-dissolved in hexane yielded 7 spots, while the ethylacetate extract (H) re-dissolved in ethylacetate yielded 10. Notably, Rf4 was present in all the chromatograms, while Rf2 and Rf3 were present only in the H chromatogram. On the other hand Rf1 was present only in the B, G and H chromatograms.

Fig. 1. Diagrammatized normal phase TLC of *M. scaber* extracts showing up to ten principal spots.

Months of storage in capped glass bottles at RTH	Appearance		Extractability of herb in different solvents			Solubility of extract in different solvents			Loss on drying (mean ± SD) %w/w	
	Herb	Extract	Ethyl-acetate	n-C_6H_{14}	H_2O	Ethyl-acetate	n-C_6H_{14}	H_2O	Herb	Extract
0	Wrinkled, brownish green/ grey leaves and twigs	Sticky mass becoming slightly gritty upon exposure	14.0 ± 1.8	4.2 ± 0.5	36.4 ±2.2	25.0 ± 5.0	55.0 ± 5.0	>10^3	10.51 ± 0.74[a] (7)	14.07 ± 1.73[b] (7)
3	Practically unchanged from above	Practically unchanged from above	15.3± 2.2	4.3 ± 0.5	33.9 ± 2.4	25.0 ± 5.0	45.0 ± 5.0	>10^3	9.11 ± 0.67[a] (6)	12.84 ± 1.18[b] (6)
9	Practically unchanged from above	Practically unchanged from above	14.4 ± 1.9	4.1 ± 0.5	30.2 ± 2.4	25.0 ± 0.0	55.0 ± 0.0	>10^3	10.44 ± 0.51[a] (6)	13.14 ± 1.11[b] (7)
21	Practically unchanged from above	Practically unchanged from above	14.7 ± 1.9	4.1 ± 0.5	30.7± 2.2[a]	25.0 ± 5.0	50.0 ± 5.0	>10^3	10.87 ± 0.73[a] (7)	14.84 ± 0.99[b] (5)
39	Practically unchanged from above	Practically unchanged from above	15.0 ± 2.0	4.3 ± 0.5	27.8 ±2.0[a]	25.0 ± 0.0	55.0 ± 0.0	>10^3	8.96 ± 0.59[a] (5)	11.44 ± 0.86[b] (6)

The results show that storage of both the herb and the extract in capped glass bottles at room temperature and humidity for up to 39 months, produced no consistent or statistically significant change in moisture content. Both ([a]) and ([b]) indicate that the figures bearing them do not differ statistically from one another ($P > 0.05$).

Table 6. Effect of storage on herb and ethylacetate extract of *M. scaber* as evaluated by appearance, extractability, solubility and loss on drying.

3.4 Effect of storage on herb and ethylacetate extract of *M. scaber* as evaluated by appearance, extractability, solubility and loss on drying

Table 6 shows that the general appearance of the herb as wrinkled, brownish green/ grey leaves and twigs remained essentially unchanged up to the 39th month of storage. However, the extractability of the herb in water fell slightly but significantly as from after the 3rd month of storage. Table 6 also shows that neither the solubility profile nor the appearance of the extract and the solutions made from them in different solvents changed with storage. The same Table 6 shows that storage of the herb and the extract in capped glass bottles at room temperature and humidity (RTH) for up to 39 months produced no consistent or statistically significant changes in moisture content.

3.5 Effect of storage on light absorption and TLC features of the herb and extract

Table 7 presents the effect of storage on light absorption and TLC features of the herb and extract. It shows the following: that the difference in absorbance between 0th month and the 21st/ 39th months was insignificant for the herb ($P > 0.05$). By contrast, the corresponding difference for the extract was significant ($P < 0.05$). Table 7 also shows that the number of TLC spots observed for the herb and extract at every stage of storage was unchanged up to the 39th month.

Months of storage in capped glass bottles at RTH	Herb							Extract		
	Abs. at λ227 nm	Types of solvent/ number of TLC spots						Abs. at λ227 nm	TLC spots	
		A	B	C	D	E	F		G	H
0	2.172 ± 0.104 (5)	5	2	5	3	3	4	2.330 ± 0.094 (5)	7	10
3	2.221 ± 0.114 (6)	5	2	5	3	3	4	2.174 ± 0.107 (5)	7	10
9	2.144 ± 0.098 (5)	5	2	5	3	3	4	2.104 ± 0.070 (5)	7	10
21	2.322 ± 0.117[a] (5)	5	2	5	3	3	4	2.039 ± 0.104[b] (5)	7	10
39	2.233 ± 0.114[a] (5)	5	2	5	3	3	4	2.084± 0.111[b] (6)	7	10

For the herb, the difference in absorbance between 0th month and the 21st/ 39th months, denoted by (a), was insignificant ($P > 0.05$). By contrast, for the extract, the difference in absorbance between the 0th or 3rd month and the 21st or 39th month, denoted by (b), was significant ($P < 0.05$). Notably, the number of TLC spots observed for both the herb and extract remained unchanged up to the 39th month.

Table 7. Effect of storage on light absorption and TLC characteristics of herb and extract of *M. scaber*.

3.6 Effect of storage on pH and foaming indices of herb and ethylacetate extract of *M. scaber*

Table 8 shows that the pH of a 5 %w/v mixture of the herb or extract in water did not change significantly with storage for up to 39 months. However, although the 5 %w/v mixture of the fresh plant material in water did foam slightly; this property diminished rapidly, and was totally lost after the 3rd month of storage. By contrast, the ethylacetate extract never foamed at any stage of storage.

Months of storage in capped glass bottles at RTH	pH of mixture (5% w/v)		Foaming index	
	Herb in water	Ethyl acetate extract in water	Herb in water	Ethylacetate extract in water
Within 1st day of harvest or preparation	5.7 ± 0.3 (5)	6.9 ± 0.3 (5)	Foam: Slight ≤ 100 (5)	Foam: Nil (5)
0	5.9 ± 0.2 (6)	6.1 ± 0.3 (5)	Foam: Slight < 100 (5)	Foam: Nil (6)
3	5.6 ± 0.2 (5)	6.2 ± 0.4 (6)	Foam: Slight < 100 (5)	Foam: Nil (5)
9	5.4 ± 0.2 (5)	6.1 ± 0.3 (5)	Foam: Nil (6)	Foam: Nil (5)
21	5.9 ± 0.3 (7)	6.7 ± 0.3 (5)	Foam: Nil (5)	Foam: Nil (5)
39	5.9 ± 0.3 (5)	6.5 ± 0.3 (5)	Foam: Nil (5)	Foam: Nil (5)

Both (a) and (b) indicate that the pH of the 5 %w/v aqueous mixtures at every stage fell within the mean values of 5.8 ± 0.2a and 6.4 ± 0.4b - they indicate that any deviations from these mean values were insignificant ($P > 0.05$). The freshly harvested samples foamed measurably, but the ability was totally lost after the 3rd month of dry storage. In all cases however, the dry ethylacetate extract was virtually insoluble in water, and did not foam at all.

Table 8. Effect of storage on pH and foaming indices of the fresh plant material, the dry herb and the ethylacetate extract of *M. scaber*.

4. Discussions

The aim of this study was to apply official methods of WHO (1998) and BP (2004) to study the key quality attributes of the air-dried weed and the ethylacetate extract of *Mitracarpus scaber*, for the purpose of quality control, GMP production and registration of Niprifan by NAFDAC. WHO had defined "Herbal Substance" as "Material derived from the plant(s) by extraction, mechanical manipulation, or some other process" (WHO, 2005). Thus, either the ethylacetate extract, or even the comminuted, air-dried vegetable matter, may rightly be termed the "Herbal Substance" of Niprifan. Since the advent of the Alma-ata Declaration in 1978, many developing countries opted to adopt the WHO model in developing their National Traditional Medicine, especially phytotherapy (Ameh et al., 2010b). NIPRD's adherence to the WHO model had resulted in the sickle cell drug - Niprisan, developed from Yoruba Traditional Medicine (Wambebe et al., 2001). It is generally held that in most countries, especially in Africa, the populations depend greatly on herbal remedies, up to 90 % in some instances like Ethiopia (BBC, 2006). Such high dependence calls for a system or mechanism for harnessing and optimizing all or most of such plant resources. That means that every effort must be made to obtain maximum benefits from them. One way to do this is to standardize the raw materials used in producing the remedies, by studies such as this one. Such studies will at least help to minimize waste, and even lead ultimately to conservation of endangered plants. Indeed, efforts at conservation are more likely to succeed when the value of what is to be conserved is proven.

Our immediate interest however, is in the need to entrench the use of these resources by taking appropriate actions, which, in this case is - an application to NAFDAC to consider the registration of Niprifan, based on folkloric use evidence, pertinent literature, and the experimental data provided in this study. These three lines of evidence can be summarized as follows. At the peak of British colonialism in Africa considerable effort was made to harness the continent's wealth in herbal traditions. Thus at as far back as the 1930s a team of British scientists had combed the entire West Africa to research traditional herbal remedies. Thus, for *Mitracarpus scaber,* Hutchinson and Dalziel (1948) reported a number of findings that have subsequently been confirmed by work in NIPRD and elsewhere (Benjamin et al., 1986; Irobi and Daramola, 1994; Cimanga et al., 2004; Abere et al., 2007a, 2007b). These include the following: that *M. scaber* was widely distributed and used topically in all of West Africa for various skin infections; and orally for various internal conditions. Among the traditional indications mentioned, and which have since been confirmed by NIPRD's Ibrahim Muaazzam (ethnobotanist and consultant on TM) are: leprosy, lice, ringworm, eczema and craw-craw. Currently, the plant is used orally for sore throat, for which purpose it is wholly macerated in water.

Among the vernacular names of *M. scaber* are: Hausa (*goga masu*); Fulani (*gadudal*); Yoruba (*irawo-ile*); and Ibo (*obu obwa*). Professor Ogundiani (2005) in his inaugural address at the University in Ile-Ife commented on Niprifan,

stressing the antimicrobial potency of *M. scaber*. Ogundiani, as stated in the lecture, had been unaware of the NIPRD's work on Niprifan, until shortly before the inaugural, since that work, led by Professors Wambebe, Okogun and Nasipuri, had been unpublished. Therefore, in this paper we elect to present not only these historical antecedents, but also to furnish the results of our evaluation of the key quality variables of the herb and extract of *M. scaber,* with a view to advancing the registration of Niprifan (for skin infections) by NAFDAC. The results here presented probably suffice for quality control and GMP production, particularly if more emphasis is placed on technical requirements than on bureaucracy. It must be remarked at this juncture that NAFDAC only belatedly recognized the sickle cell drug, Niprisan, after the US-FDA and EMEA had granted it orphan status (Pandey, 2003). One may wonder - What a paradox! Why should the US and Europe that need herbal drugs far less than Nigeria be keener in their regulation? Therefore, from the foregoing, it seems that the key to this Nigerian enigma lies not in the technical but in the non-technical differences between NAFDAC and EMEA as depicted in Tables 1-3. The said differences which hinge on NAFDAC's extra requirements (Table 3) suggest that NAFDAC needs to re-strategize for efficient discharge of its Mandate. For example, despite the widespread use of herbal medicines in Nigeria and the Federal Policy on TM (2007), NAFDAC is not known to have "fully registered" a single herbal medicine since its creation in 1992/3, whereas it should. This is the puzzle this article had hoped to address.

5. References

Abere , T. A.; Onyekweli, A. O. & Ukoh, G. C. (2007a). *In vitro* Antimicrobial Activity of the Extract of Mitracarpus scaber Leaves Formulated as Syrup. *Tropical Journal of Pharmaceutical Research* Vol. 6, No. 1, pp. 679-682.

Abere, T. A.; Onwukaeme, D. N. & Eboka, C. J. (2007b). Pharmacognostic evaluation of the leaves of *Mitracarpus scaber* Zucc (Rubiaceae). *Tropical Journal of Pharmaceutical Research* Vol. 6, No. 4, pp. 849-853.

Ameh, S. J.; Obodozie, O. O.; Inyang, U. S.; Abubakar, M. S. & Garba, M. (2010a). Current phytotherapy - an inter-regional perspective on policy, research and development of herbal medicine. *Journal of Medicinal Plants Research* Vol. 4(15), pp 1508-1516, 4 August, 2010.

Ameh, S. J.; Obodozie, O. O.; Inyang, U. S.; Abubakar, M. S. & Garba, M. (2010b). Current phytotherapy - A perspective on the science and regulation of herbal medicine. *Journal of Medicinal Plants Research*; Vol. 4(2): 072-081.

Ameh, S. J.; Tarfa, F. D.; Abdulkareem, T. M.; Ibe, M. C.; Onanuga, C. & Obodozie, O. O. (2010c). Physicochemical Analysis of the Aqueous Extracts of Six Nigerian Medicinal Plants. *Tropical Journal of Pharmaceutical Research,* 9 (2): 119-125.

Ameh, S. J.; Obodozie, O. O.; Inyang, U. S.; Abubakar, M. S. & Garba, M. (2010d). Quality Control Tests on Andrographis paniculata Nees (Family: Acanthaceae) - an Indian 'Wonder' Plant Grown in Nigeria. *Tropical Journal of Pharmaceutical Research,* Vol. 9, No 4, pp. 387-394

Ann Godsell Regulatory (2008). Pharmaceutical Good Manufacturing Practice for Herbal Drug Substances 2008 [cited 2010 April 8]. Available online at: http://www.pharmaceutical-int.com/article/category/treatment-herbal-medicines

BBC News (2006). Can herbal medicine combat Aids? Wednesday, 15 March, 13:10 GMT. http://newsvote.bbc.co.uk/mpapps/pagetools/print/news.bbc.co.uk/2/hi/Africa/4793106.stm .

Benjamin, T. V.; Anucha, T. C. & Hugbo, P. G. (1986). An approach to the study of medicinal plants with antimicrobial activity with reference to *Mitracarpus scaber*. *In*: Sofowora, A. (Ed.) *The State of medicinal Plants Research in Nigeria,* pp. 243-245, Ibadan University Press, ISBN 978-30285-0-2, Ibadan, Nigeria.

Bisignano, G.; Sanogo, R.; Marino, A.; Angelo, V.; Germano, M.; De Pasquale, R. & Pizza, C. (2000). Antimicrobial activity of Mitracarpus scaber extract and isolated constituents. *Letters in Applied Microbiology,* Vol. 30, pp. 105-108. doi:10.1046/j.1472-765x.2000.00692.x.

Cimanga, R. K.; Kambu, K.; Tona, L.; Bruyne, T.; Sandra, A.; Totte, J.; Pieters, L. & Vlietinck, A. J. (2004). Antibacterial and antifungal activities of some extracts and fractions of Mitracarpus scaber Zucc. (Rubiaceae). *Journal of Natural Remedies,* Vol. 4, No. 1, pp. 17-25

De Smet, P. N. (2005). Herbal medicine in Europe - relaxing regulatory standards. *New England Journal of Medicine,* Vol. 352, No. 12, pp. 1176-78.

DSHEA (1994). Dietary Supplements Health Education Act of 1994 [cited 2010 April 8]. Available at: http://fda/Food/DietarySupplements/ucm109764.htm

Gbaguidi, F.; Accrombessi, G.; Moudachirou, M. & Quetin-Leclercq, J. (2005). HPLC quantification of two isomeric triterpenic acids isolated from Mitracarpus scaber and antimicrobial activity on Dermatophilus congolensis. *Journal of Pharmaceutical & Biomedical Analysis,* Vol. 39, No. 5, pp. 990-995.

Germanò, M. P.; Sanogo, R; Costa, C; Fulco, R; D'Angelo, V.; Viscomi, E. G. & de Pasquale, R. (1999). Hepatoprotective Properties in the Rat of *Mitracarpus scaber* (Rubiaceae). *Journal of Pharmacy & Pharmacology,* June 1999. Vol. 51, No. 6, pp. 729-734.

Goldman, P. (2001). Herbal medicines today and the roots of modern pharmacology. *Annual Internal Medicine,* Vol. 135, No. 8, Pt 1, pp. 594–600.

Gross, A. & Minot, J. (2007). Chinese Manufacturing: Scandals and Opportunities. Published in MX, November/ December, Pacific Bridge Medicals. http://www.pacificbridgemedicals.com/

Houghton, P. J.; Ibewuike, J. C.; Mortimer, F.; Okeke, I. N. & Ogundaini, A. O. (2002). Antibacterial quinones from Mitracarpus scaber. *Journal of Pharmacy & Pharmacology,* 52 (Suppl.) 261. www.kcl.ac.uk/content/1/c6/01/66/69/PUB82008.pdf

Hutchinson, J. & Dalziel, J. M. (1948). *The flora of West Tropical Africa.* Crown Agents for the Colonies. 4 Millbank, Westminster, London, SW1., London. Vol 11. 1948. p. 222.

Irobi, O. N. & Daramola, S. O. (1994). Bactericidal properties of crude extracts of Mitracarpus villosus. *Journal of Ethnopharmacology,* Vol. 42, No. 1, pp. 39-43.

NAFDAC's website (2010). Website of the National Agency for Food and Drug Administration and Control. *www.nafdacnigeria.org/*

Ogundaini, A. O. (2005). From Greens into Medicine: Taking a Lead From Nature. Inaugural Lecture Series 176. Ile-Ife. Tuesday 26, April. Obafemi Awolowo University Press Limited

Okunade, A. L.; Clark, A. M.; Hufford, C. D. & Oguntimein, B. O. (1999). Azaanthraquinone: an antimicrobial alkaloid from Mitracarpus scaber [letter] *Planta Medica,*Vol. 65, No. 5, pp. 447-8.

Pandey, R. C. (2003). Xechem's Sickle Cell Drug, NIPRISAN – HEMOXIN Granted Orphan Drug Status by the FDA NEW BRUNSWICK, N.J.-- (BUSINESS WIRE)--Sept. 2, 2003--Xechem International, Inc. (OTC BB: XKEM).

Pelletier, K. R. (2009). Guide to Herbal Medicine (cited 2009 June 2). Available from: http://www.ca.encarta.msn.com/sidebar_701509401/Guide_to_Herbal_Medicine.html

Wambebe, C; Khamofu, H.; Momoh, J.; Ekpeyong, M.; Audu, B.; Njoku, B. S.; Nasipuri, N. R.; Kunle, O. O. et al. (2001). Double-blind, placebo-controlled, randomized cross-over clinical trial of NIPRISAN in patients with sickle cell disorder. Phytomedicine, 8(4):252-61.

WHO (1998). Quality control methods for medicinal plant materials. World Health Organization, WHO, Geneva, pp. 1-115.

WHO (2000). General guidelines for methodologies on research and evaluation of traditional medicine (Document WHO/EDM/ TRM/2000.1). World Health Organization, WHO, Geneva. pp. 1-184.

WHO (2005) APPENDIX 2 Glossary of key terms. In: Information needed to support Clinical Trials of herbal products. TDR/GEN/Guidance/05.1Operational Guidance: Special Programme for Research and Training in Tropical Diseases, WHO, Geneva, pp. 1-16.

21

Procedures for Evaluation of Slice Thickness in Medical Imaging Systems

Giuseppe Vermiglio, Giuseppe Acri, Barbara Testagrossa,
Federica Causa and Maria Giulia Tripepi
Environmental, Health, Social and Industrial Department – University of Messina Italy

1. Introduction

The main goal of a medical imaging system is to produce images to provide more accurate and timely diagnoses (Torfeh et al., 2007). In particular, Computed Tomography (CT), Magnetic Resonance Imaging (MRI), and Ultrasound (US) are resourceful tools in medical practice, and in many cases a life saving resource when rapid decisions are needed in the emergency room (Rehani et al., 2000). The above diagnostic techniques are based on the evaluation of high resolution images from technologically sophisticated equipment. Individual images are obtained by using several different electronic components and considerable amounts of data processing, which affect the quality of images produced and, consequently, make the diagnostic process more complicated (Torfeh et al., 2007).

In this context, to guarantee a consistent image quality over the lifetime of the diagnostic radiology equipment and to ensure safe and accurate operation of the process as a whole, it is necessary to establish and actively maintain regular and adequate Quality Assurance (QA) procedures. This is significant for computer-aided imaging systems, such as CT, MRI and US. The QA procedure should include periodic tests to ensure accurate target and critical structure localization (Mutic et al., 2003). Such tests are referred to as Quality Controls (QCs). They hold a key role within the QA procedure because they enable complete evaluation of system status and image quality (Chen et al., 2004; Vermiglio et al., 2006).

Importantly, QCs permit the identification of image quality degradation before it affects patient scans, and of the source of possible equipment malfunction, pointing to preventive or immediate maintenance requirements. Thus, image QC is crucial to ensure a safe and efficient diagnosis and treatment of diseases (Rampado et al., 2006; Torfeh et al., 2007). For this reason, periodic QCs have been recommended by manufacturers and medical physicists' organizations to test the performance of medical imaging systems. Protocols for QCs and QA in medical imaging systems have been produced by several professional groups (AAPM - NEMA) (Goodsitt et al., 1998). This highlights the extensive role of QA programs, including QC testing, preventive maintenance, etc. (Rampado et al., 2006).

In any clinical imaging study, it is important to have accurate confirmation of several physical characteristics of the medical imaging device. In particular, the slice thickness accuracy represents an important parameter that should be estimated during QC procedures, not only because the signal to noise ratio varies linearly with the slice thickness, but also because clinical image resolution is strongly affected by partial volume effects, thus reducing clinical image quality with increasing slice thickness (Narayan et al., 2005). In addition, to determine the FWHM the AAPM procedure involves the evaluation of a line profile of the slice. So, during QC procedures many and different available test objects are used to assess different physical characteristics of the medical imaging device, including slice thickness, spatial resolution, dark noise, uniformity, etc. AAPM Reports No 1 and No 28 state that the slice thickness can be evaluated from the measure of the full width at half maximum (FWHM) of the response across the slice (Judy et al., 1977; Price et al., 1990). In particular, for a high-accuracy measurement of slice thickness, several test objects inserted in multipurpose phantoms can be used, most of which utilize inclined surfaces (plane, cone or spiral). A typical test object for the slice thickness evaluation is the crossed high signal ramps oriented at a fixed angle (Price et al., 1990).

Whereas in US equipment the slice thickness is typically not measured (Skolnick, 1991), most CT and MRI scanners adopt specific and automated procedures that require the use of dedicated phantoms, coupled with a dedicated imaging software that, however, does not always include line profile tools.

Standard slice thickness accuracy evaluation methods consist of scan explorations of phantoms that contain different specific patterns. These methods are based on manual scans with graphics tools or, alternatively, on automatic scans utilizing encoded masks to determine the Region Of Interest (ROI) for quantization (Torfeh et al., 2007). Therefore, a variety of different phantoms presently exists, but each requiring a specific QC protocol.

Further, even the newest medical imaging software do not allow a direct measurement of the slice thickness accuracy with CT and MRI scanners, but require a complicated procedure to be performed by specialized technicians authorized to enter in the SERVICE menu of medical devices.

To reduce complications and provide a versatile and unique QC procedure to estimate slice thickness accuracy, a novel dedicated phantom and associated procedure is proposed here that is easy to implement and that can be used on both CT and MRI scanners.

Such phantom can be used either with already existing dedicated software or with to this aim dedicated LabView-based tools, to readily measure the slice thickness in real time and/or post-processing operation. The reliability of the innovative technique proposed here has been evaluated with respect to previously validated procedures by conducting statistical analysis, as discussed in detail in the following sections.

Further, this novel technique is suitable also for the evaluation of the elevation resolution in US scanners, and easier to perform than standard techniques. This chapter is structured as follows: a review of the materials and methods commonly used in CT, MRI and US imaging systems, and the novel and versatile methodology proposed here for slice thickness measurements are presented in Section 2. The results obtained from the application of the proposed novel methodology to the three different imaging techniques are discussed in Section 3. The conclusions are drawn in Section 4.

2. Materials and methods

In this section the commonly used methods for determining the slice thickness accuracy in CT, MRI and US scanners are presented. In addition, a novel procedure that uses a dedicated phantom and the following image elaboration by employing LabView based software is also proposed.

2.1 CT scanners

The slice thickness is evaluated by measuring the width (FWHM) of the image of one or more Aluminium ramps at the intersection of the ramp(s) with the scan plane (CEI, 1998), as illustrated schematically in Fig. 1. The sensitivity profile measures the system response to an attenuating impulse as a function of the z-axis position, through the slice plane. The sensitivity profile is a function of pre- and post-patient collimation, and appears as a blurred square wave. The FWHM of this blurred square is defined as the nominal slice width. The slice thickness of a CT scanner is determined by focal spot geometry as well as pre-patient and detector collimation and alignment.

An customary method of monitoring equipment performance is to measure the parameters of interest using test objects. For example, Goodenough et al. proposed an approximate measure of beam width by using a series of small beads positioned across the beam width. However, this phantom is difficult to use for a precise quantitative measure because of the uncertainty on the alignment of the beads with respect to the beam. Indeed, the beam width can be measured directly from a beam profile plot only if care is taken to ensure that the Aluminium piece is oriented at 45 degrees across the width of beam (Judy et al., 1977, as cited in Goodenough et al., 1977).

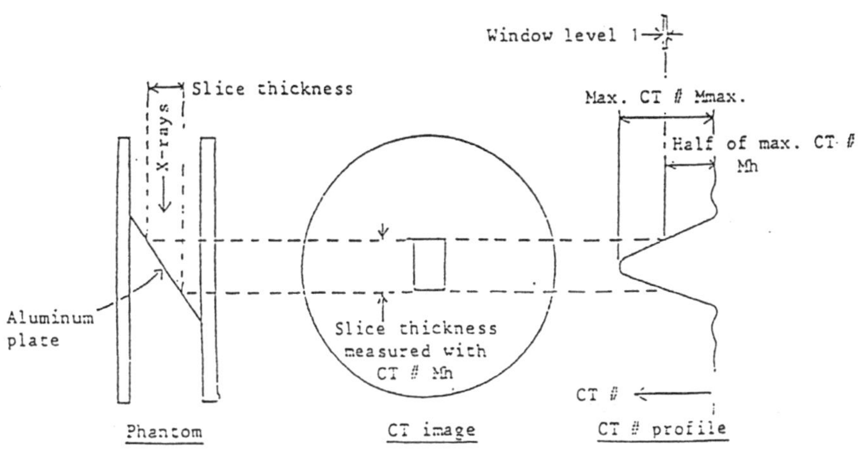

Fig. 1. Principle of slice thickness measurement (Philips, 1997).

A typical performance phantom (Philips, 1997) uses Aluminium plates slanted 26.565 degrees and which are across each other, as shown in Fig. 2. With X-rays irradiated to the Aluminium plates, the axial length of each Aluminium can be measured in CT image. With this method, it is possible to obtain an accurate measurement even if the intersection of the Aluminium plates is not aligned with the X-ray beam, by averaging the two measurements L_a and L_b. The slice thickness L is calculated as follows:

$$L = \left(\frac{L_a + L_b}{2}\right) \cdot \tan 26.565° = \frac{L_a + L_b}{4} \tag{1}$$

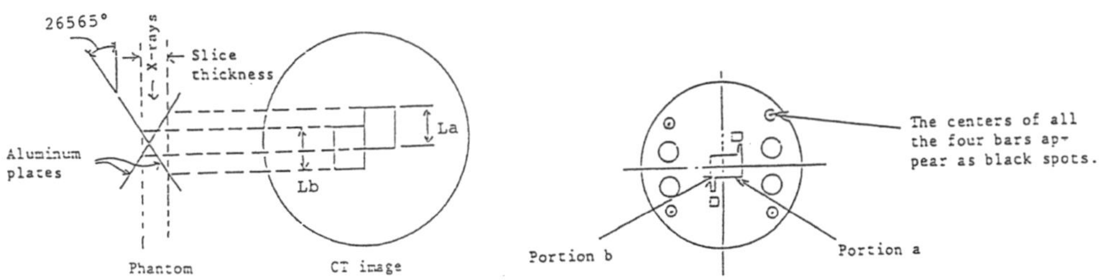

(a) (b)

Fig. 2. CT image of the section dedicated to the measurements of slice thickness of a typical performance phantom: (a) layout of the measurement methodology; (b) schematic of resulting image with the centres of the four bars appearing as black spots (Philips, 1997).

Another performance phantom used for CT scanners (Fig. 3) is a poli methyl methacrilate (PMMA) box presenting a pattern of air filled holes drilled 1 mm apart and aligned in the direction of the slice thickness (perpendicular to the scan plane). Each visible hole in the image represents 1 mm of beam thickness (General Electric [GE], 2000).

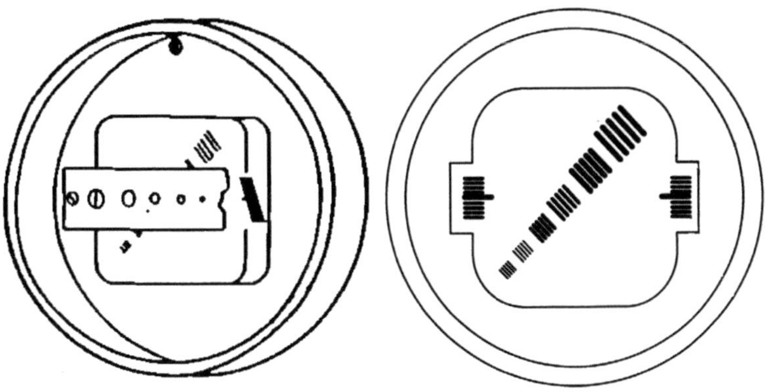

(a) (b)

Fig. 3. Phantom with air-filled holes for slice thickness measurement in CT scanners: (a) schematic of the phantom, (b) schematic of the calibration image (General Electric [GE], 2000).

To determine the slice thickness, the image is displayed at the recommended window level and width. The number of visible holes (representing air-filled holes) is counted. Holes that appear black in the image represent a full millimetre slice thickness. Holes that appear grey count as fractions of a millimetre; two equally grey holes count as a single 1 mm slice thickness.

2.2 MRI scanners

The technique of MRI differs from X-ray CT in many ways, but one of the most interesting is perhaps that the slice is not determined primarily by the geometry of the scanning apparatus but rather by electronic factors, namely the spectrum of radio frequency pulse and the nature of the slice selection gradient (Mc Robbie et al., 1986). The slice profile and width of a 2D imaging technique such as MRI is a very important feature of its performance.

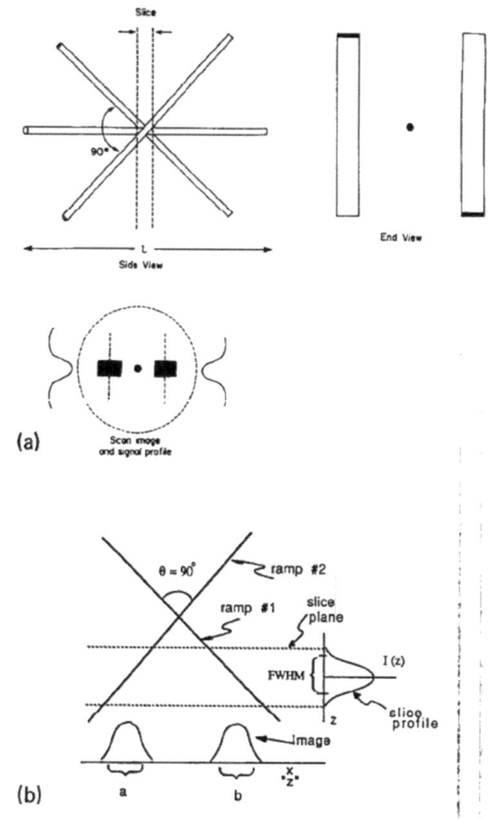

Fig. 4. High signal ramp phantoms: (a) A typical slice-thickness phantom consisting of a set of crossed thin ramps. A ramp crossing angle of 90° yields an angle of 45° between the ramp and the image plane. The phantom length (L) should be greater than twice the maximum slice thickness. An alignment rod placed between the two ramps defines the point where the two ramps cross. When the slice is properly aligned through the intersection of the ramps the images of the ramps and of the rod will be aligned. (b) The slice sensitivity profile is directly proportional to the image intensity profiles if the image plane is perpendicular to the alignment rod (Price et al., 1990).

Since the signal obtained is directly proportional to the thickness of the slice, an inaccurate slice width can lead to a reduced Signal-to-Noise Ratio (SNR) (Lerski, 1992). Partial volume effects can significantly alter sensitivity and specificity. Quantitative measurements such as relaxation time T_1 and T_2 values, are also greatly influenced by slice thickness. Inaccuracies in slice thickness may result in inter-slice interference during multi-slice acquisitions, and invalid SNR measurements (Och et al., 1992). The slice profile, ideally rectangular, may contain side lobes which can produce very confusing effects (Lerski, 1992). In addition, gradient field nonuniformity, radio frequency field nonuniformity, nonuniform static magnetic field, noncoplanar slice selection pulses between excitation and readout, T_R/T_1 ratio (where T_R represents the repetition time), and radio frequency pulse shape and stimulated echoes can also affect the slice thickness accuracy (Price et al., 1990).

A variety of phantoms have been designed to evaluate slice thickness. All are some variation of an inclined surface. These may include wedges, ramps, spirals, or steps. A typically used phantom is the crossed high signal ramps.

High signal ramp (HSR) phantoms generally consist of opposing ramp pairs oriented at a fixed angle θ (Fig. 4). The HSR's should be thin (ideally infinitesimally thin) to quantify the slice profile accurately. In general, the thickness of a (90°) HSR oriented at 45° respect to the image plane should be < 20% of the slice profile FWHM (i.e., for 5-mm slice it is necessary to use a l-mm ramp) to obtain a measurement with < 20% error.

The FWHM is the width of the slice profile (SP) at one-half of the maximum value. In this case, the SP should be obtained for each ramp. The FWHM then becomes

$$L = FWHM = \frac{(a+b)\cos\theta + \sqrt{(a+b)^2 \cos^2\theta + 4ab\sin^2\theta}}{2\sin\theta} \tag{2}$$

where a and b refer to the FWHM of the intensity profiles measured for ramp 1 and ramp 2, respectively. Note that for θ=90° then Eq. 2 is simplified to:

$$L = FWHM = \sqrt{ab} \tag{3}$$

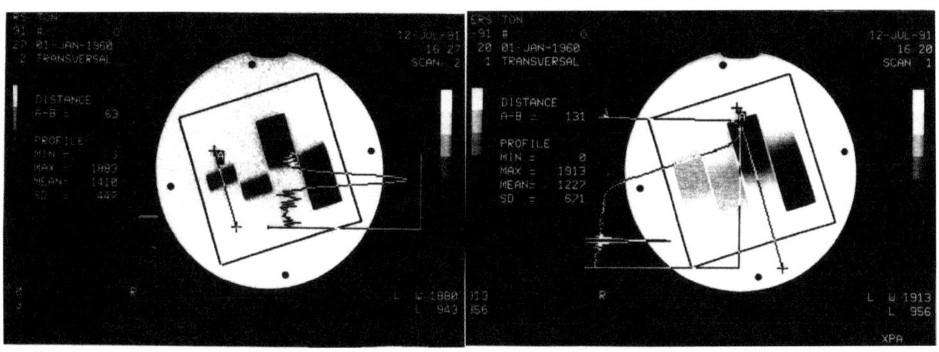

(a) (b)

Fig. 5. Example of CT scanner image obtained from the assessment of the EUROSPIN phantom (Lerski, 1992).

The EUROSPIN test phantom contains two sets of structures that may be used for slice profile and width measurement. Pairs of angled plates are used to obtain a direct measurement. The additional pairs of wedges are used to calibrate especially thin slices. Typical examples of images obtained for slice width are presented in Fig. 5(a and b). The dark bands on the left hand side of Fig. 5a represent a projection of the slice profile from the angled plates; the shaded region on the right hand side of the Fig. 5b represents the projection of the profile at the wedge.

2.3 US scanners

Ultrasound image resolution depends on beam width in the scan and elevation (section thickness) planes (Skolnick, 1991). On US scanners slice thickness evaluation or elevational resolution is useful to understand some of the problems due to partial volume effect. Section thickness is significantly more complicated to check. This characteristic of the ultrasound

beam depends on the focusing effect in the elevation direction, which is perpendicular to the scanning plane (Richard, 1999). In linear, curved linear and phased array sector probes, focus is controlled electronically, but in the elevation plane it is determined mechanically by the curvature of the crystals. The beam in the scan plane can be sharply focused only in a narrow focal range. Thus, beam profiles in the scan plane are not indicative of beam profiles in the elevation plane. As with lateral and axial resolution, elevational resolution can be measured indirectly with anechoic spherical objects or cylindrical plug phantoms. Slice thickness focusing can also be evaluated qualitatively by scanning the anechoic cylindrical objects in an ultrasound QC test phantom with the scan plane along the lengths of the cylinders (e.g., perpendicular to the usual scan direction). Quantitative assessment can be achieved by using an "inclined plane" phantom (Fig. 6) (Goodsitt et al., 1998).
The methodology used with the inclined plane phantom consists in obtaining the elevation beam profile, finding the depth where the image is narrowest. By focusing on that plane, the thickness of the image is measured at the focal plane.

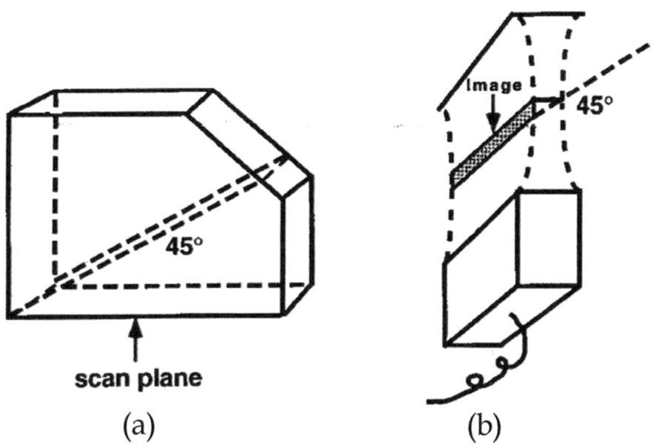

Fig. 6. Phantom used for beam width measurement in the elevation plane: (a) schematic of the phantom; (b) schematic of the procedure to obtain the image at the beam waist (Goodsitt et al., 1998).

This technique enables measurement of the elevation dimension of the beam only at a single depth. To determine the entire profile in the elevation plane, the probe must be moved horizontally along the surface of the phantom to make a series of measurements, with the beam intersecting the inclined plane at different depths (Skolnick, 1991).
Slice thickness focal range, thickness and depth should be recorded on the US unit for each commonly used transducer. Any significant variation from those reference values may indicate a detachment of the focusing lens.

2.4 The novel procedure

A novel and versatile methodology is proposed here to determine the slice thickness accuracy using a novel phantom. The methodology can be applied to any image scanning technique, including CT, MRI and US scanning. The methodology consists of two steps: 1) acquisition of images of the phantom; 2) image elaboration by using the dedicated LabView-based software. To test the proposed procedure and to obtain detailed information about the

quality of the obtained results, the acquired and processed images were compared with those obtained by elaborating the same phantom images using commercial software following already validated procedures (Testagrossa et al., 2006).
The novel proposed dedicated phantom consists of a poli-methyl-methacrilate (PMMA) empty box (14.0 cm x 7.5 cm x 7.0 cm) diagonally cut by a septum at 26 degrees (Fig. 7). The PMMA septum is 2.0 mm thick and it divides the box into two sections, thus reproducing both single and double wedges. The two sections can be filled with the same fluid or with fluids of different densities. In particular, to determine the slice thickness accuracy the PMMA box was filled with two different fluids (water and air) for assessment in a CT scanner. To perform the same assessment with an MRI scanner, water was replaced with a $CuSO_4\ 5H_2O+H_2SO_4$+1ml/l antialga (ARQUAD) liquid solution (T_1=300 ms, T_2=280ms). For US systems the upper wedge was filled with ultrasound gel, as conductive medium.
In addition, a spirit level is used to verify the planarity of the phantom with respect to the beam and the patient couch.

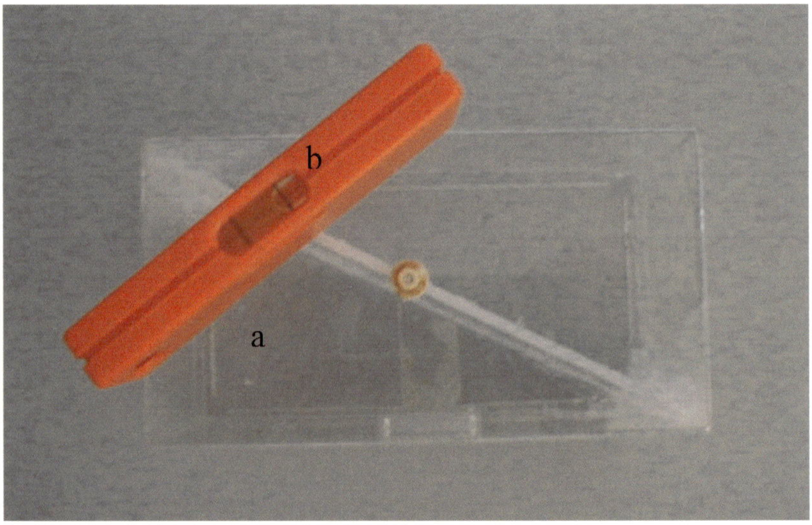

Fig. 7. The novel proposed PMMA phantom for the evaluation of the slice thickness accuracy: (a) proposed for the evaluation of the slice thickness accuracy. (b) the spirit level used to verify the planarity of the phantom with respect to the beam and the patient couch is also shown for completeness.

The test procedure followed for both CT and MRI devices consists of four steps:

1. Placing the slice thickness accuracy phantom in the scanner head holder.
2. Adjusting level position of the phantom if necessary.
3. Moving/positioning the phantom in the gantry aperture.
4. Scanning the phantom with a single slice using the desired slice width available

The phantom images were acquired using standard Head and Body protocols, shown in Table 1 and Table 2 for CT and MRI medical devices, respectively. After, the phantom images were acquired, elaborated and analyzed, they were stored and/or transmitted to a printer. The stored ones were further transferred to a dedicated workstation, whereas the printed ones were acquired by the same workstation using a VIDAR Scanner, for the next elaboration.

Scan Parameters	Head Protocol	Body Protocol
kV	120	120
mA	100	45
Scan Time (s)	3	1
Field of View (mm)	230	360
Reconstruction Matrix	512	512
Filter	None	None

Table 1. Standard Protocol for testing CT medical devices.

Scan Parameters	Head Protocol	Body Protocol
Coil type	Head	Body
Scan mode	SE	SE
Scan technique	MS	MS
Slice orientation	Transversal	Transversal
Numbers of echoes	2	3
Field of View (mm)	250	250
Repetition Time (ms)	1000	1000
Scan matrix	256	256
Water fat shift	1.3	Maximum

Table 2. Standard Protocol for testing MRI medical devices.

The methodology developed here for data elaboration, utilizes a purposely developed LabView-based slice thickness measurement software. LabView is a graphical programming language that uses icons instead of lines of text to create applications. In contrast to text-based programming languages, where instructions determine program execution, LabView uses data flow programming, where the flow of data determines execution (National Instruments [NI], 2003). The software is compatible with both non-standard and standard image formats (BMP, TIFF, JPEG, JPEG2000, PNG, and AIPD) (Vermiglio et al., 2008). To evaluate the slice width the FWHM of the wedge, expressed in pixels, is measured and calibrated with respect to the effective length of the PMMA box, expressed in mm. The result is displayed in real-time at the user interface, known as the software Front Panel (Fig. 8), with the advantage of providing a complete set of data with a user-friendly interface.
By plotting the radiation profile obtained as a system response to an attenuating impulse, as a function of position (z axis), that is through the slice plane, it is possible to estimate the slice thickness accuracy of the acquired image utilising the developed software. This is referred to as the sensitivity profile. Gaussian smoothing is applied to smooth out the sensitivity profile and permit a clearer estimate of the desired FWHM. To evaluate the slice thickness , in real time, the software utilises the following equation:

$$ST = FWHM \cdot \tan(26°) \tag{4}$$

To test the proposed procedure, results are compared with those obtained by elaborating the same phantom images using commercial software, in particular Image-Pro Plus software from Media Cybernetics (Sansotta et al., 2002; Testagrossa et al., 2006).

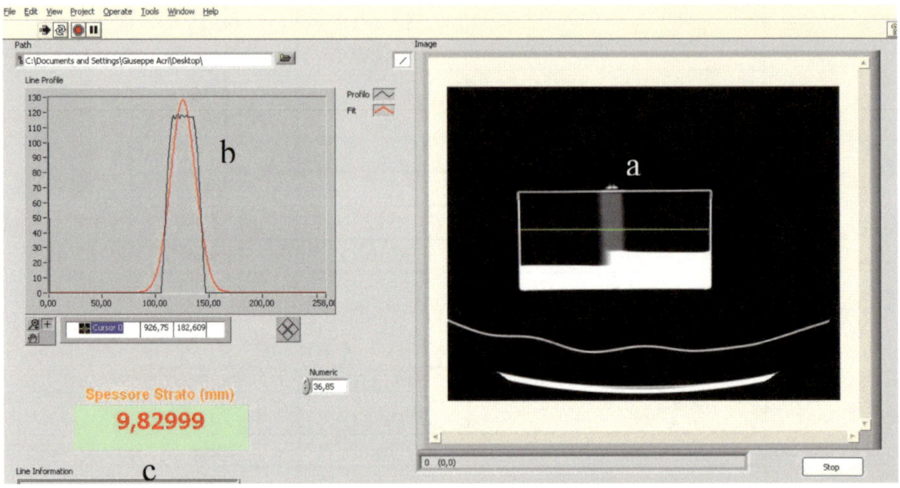

Fig. 8. Front Panel of the dedicated slice thickness LabView software showing (a) X-ray phantom section, (b) the detected line profile and corresponding Gaussian fit, and (c) the resulting slice thickness value. All steps are performed in real time.

The measurements presented here have been conducted on several CT and MRI devices in an extended study from 2006 to 2010. In addition, a statistical analysis was conducted on the resulting datasets to further validate the proposed methodology. The chosen statistical method is the variance analysis, through Fisher's exact test (F-test), to assess if a significant difference exists between datasets obtained following different procedures (C.A. Markowski & E.P. Markowski, 1990). The F-test is useful when the aim of the study is the evaluation of the precision of a measurements technique. In fact, the variance analysis consists in the factorisation of the total variance into a set of partial variances corresponding to different and estimated variations. The statistical analysis was conducted both on CT and MRI datasets.

For CT scanners three different datasets of slice thickness measurements were considered. The first dataset of 16 measurements was done on a reference (RF) value of 10 mm The second dataset of 14 measurements, on a RF value of 5 mm. The third dataset of 10 measurements, on a RF value of 2 mm. The data was obtained using two different procedures.

For MRI scanners slice thickness measurements were done on a RF value of 10 mm. In this case three different procedures were compared and 24 measurements in total were obtained on different MRI systems.

3. Results

The slice thickness measurements results using the novel proposed methodology and the statistical data analysis are presented in this section for CT, MRI and US systems.

3.1 CT medical devices

A two-dimensional image of the wedge of the dedicated phantom of Fig. 7 was acquired using the Standard Head Protocol described in Table 1. The X-Ray image of the phantom is presented in Fig. 9. In this case the line profile tool was available and the related trend was shown in the same figure.

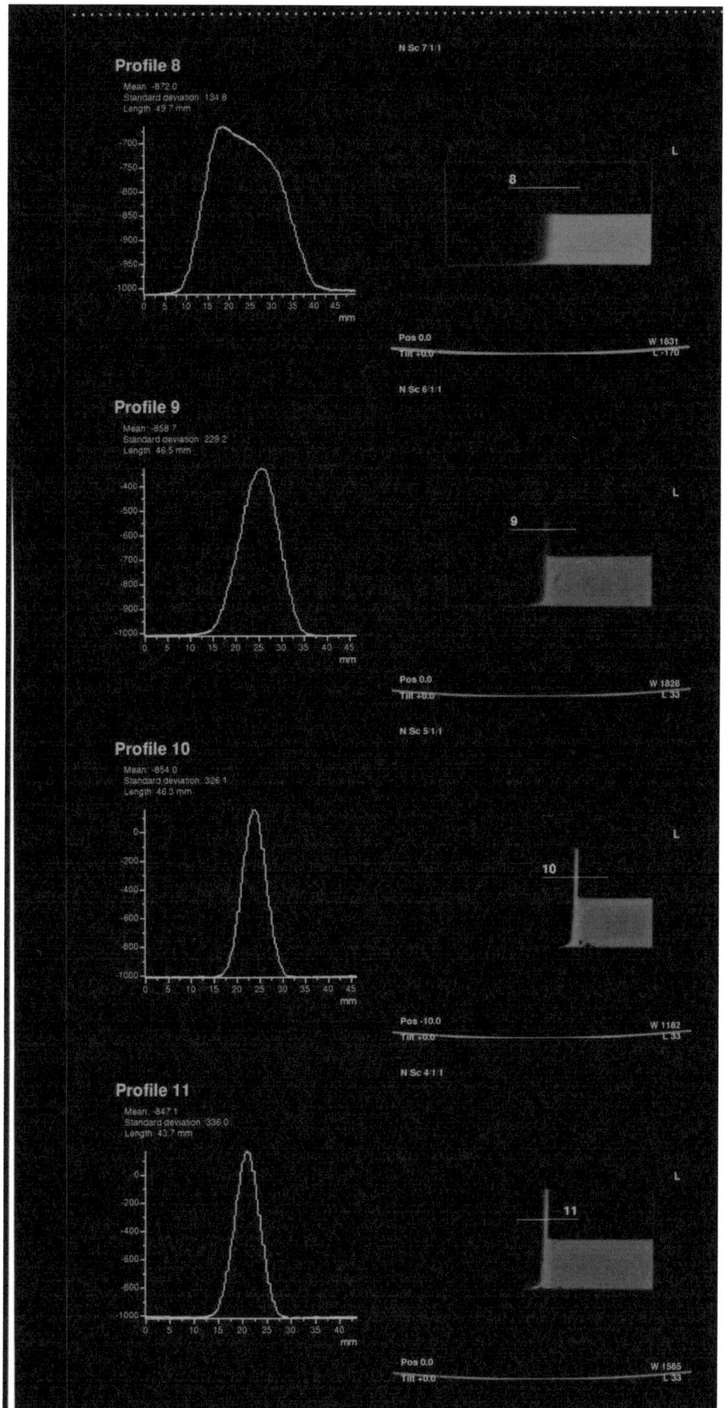

Fig. 9. CT scanner in-built software results displaying the line profiles obtained for different readings of the X-ray image of the dedicated phantom.

The results obtained by measuring the slice thickness accuracy with different CT scanners and by employing the in-house developed LabView (LV) program and the commercial Image Pro Plus (IPP) software are compared with the corresponding 10 mm RF value in Table 3. In the same table mean values and standard deviations are also reported for both procedures.

RF Value (mm)	LV (mm)	LV mean value and standard deviation (mm)	IPP (mm)	IPP mean value and standard deviation (mm)
10	8.61	9.69±0.68	8.07	9.64±1.26
	9.30		10.33	
	9.44		9.44	
	10.80		10.04	
	9.50		8.62	
	10.18		12.00	
	10.26		8.54	
	9.45		10.10	

Table 3. Slice thickness accuracy results obtained with LV and IPP for a 10 mm RF value, using different types of CT scanners.

The data of Table 3 are presented in graphical form in Fig. 10. The slice thickness accuracy results (blue: IPP; red: LV) and deviation from the RF value (blue: IPP; red: LV) obtained using the IPP and LV procedures are presented for the 10 mm RF value.

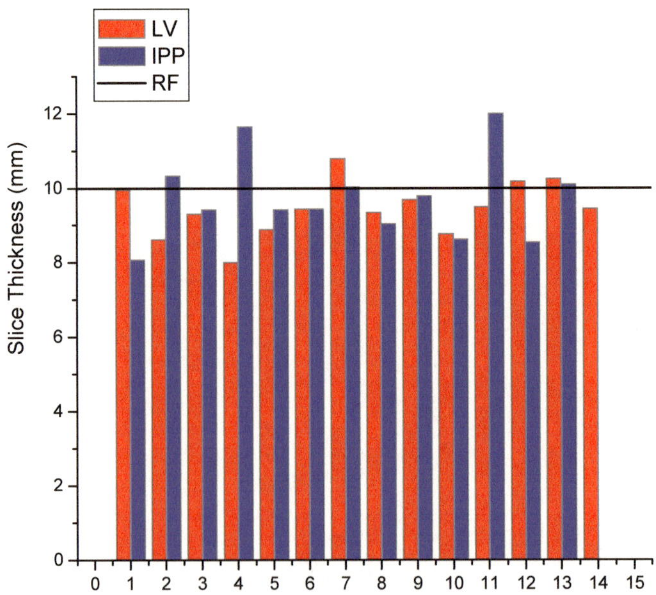

Fig. 10. (a)

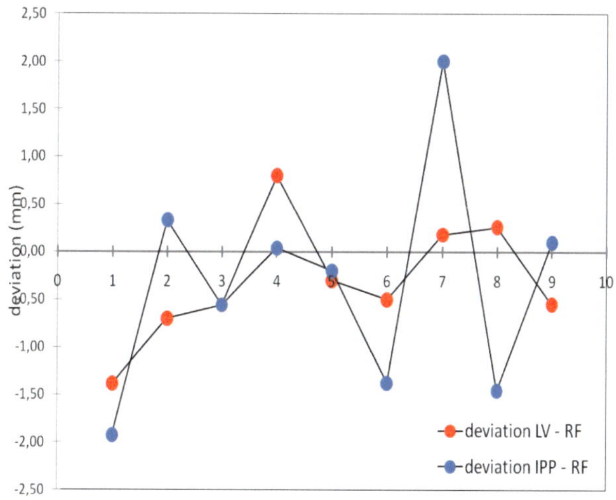

(b)

Fig. 10. CT slice thickness accuracy obtained for a 10mm RF value: (a) data set (blue: IPP; red: LV); (b) deviation from the RF value (blue: IPP; red: LV).

From the results presented in Table 3, it is observed that the mean values calculated for both the IPP and LV procedures, are comparable. However, the standard deviation obtained from the two different procedures is considerably different, with the LV procedure providing a narrower deviation, and hence more accurate results of the performed measurements.

In Table 4, the slice thickness accuracy results obtained with IPP and LV are compared with the corresponding 5 mm RF value. Also in this case, for the sake of completeness, the respective mean values and standard deviations obtained from the IPP and LV datasets are reported.

RF Values (mm)	LV (mm)	LV mean value and standard deviation (mm)	IPP (mm)	IPP mean value and standard deviation (mm)
5	4.50	4.73±0.25	4.03	4.83±0.41
	4.75		4.82	
	4.80		4.83	
	4.36		5.36	
	4.89		4.85	
	4.71		4.80	
	5.13		5.13	

Table 4. Slice thickness accuracy results obtained with LV and IPP for a 5mm RF value, using different types of CT scanners.

Also for the 5 mm RF value case, the mean values obtained from LV and IPP datasets are comparable, with a slightly better estimate for the IPP procedure. However, the standard deviations obtained from the two procedures are significantly different. As in the previous case, the LV procedure provides a narrower deviation, thus enabling a more accurate measurement.

The data of Table 4 are reported in graphical form in Fig 11, where the slice thickness accuracy results obtained with the two procedures (blue: IPP; red: LV) and their deviation from the 5 mm RF value are presented.

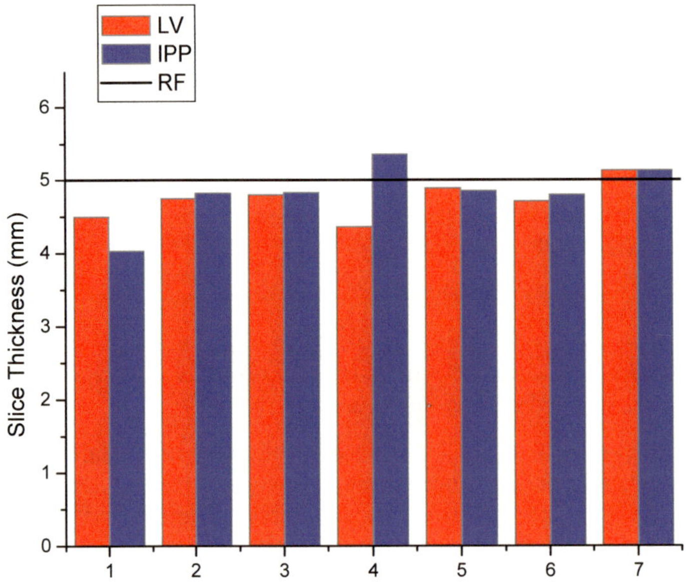

(a)

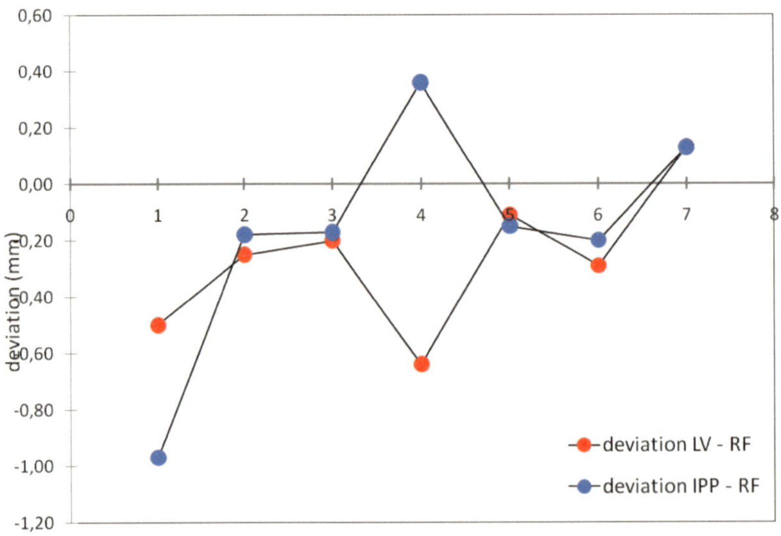

(b)

Fig. 11. CT slice thickness accuracy obtained for a 5 mm RF value: (a) sample set (blue: IPP; red: LV); (b) deviation from the RF value (blue: IPP; red: LV).

Finally, in Table 5, the results obtained by measuring the slice thickness accuracy by employing the IPP and the LV procedures are compared with the corresponding 2 mm RF value and mean values and standard deviation are also indicated. From the analysis of the data of Table 5, it can be observed that the mean value calculated by the LV dataset is significantly closer to the RF value than that calculated from the IPP dataset. This further supports the validity of the proposed technique.

RF Values (mm)	LV (mm)	LV mean value and standard deviation (mm)	IPP (mm)	IPP mean value and standard deviation (mm)
	2.66		2.79	
	3.00		2.75	
2	2.06	2.32±0.48	2.83	2.79±0.16
	1.96		2.56	
	1.95		3.00	

Table 5. Slice thickness accuracy results obtained with LV and IPP for a 2 mm RF value, using different types of CT scanners.

The data of Table 5 are represented in graphical form in Fig 12, where the slice thickness accuracy results obtained using IPP and LV (blue: IPP; red: LV) and their deviation from the 2 mm RF value (blue: IPP; red: LV) are presented.

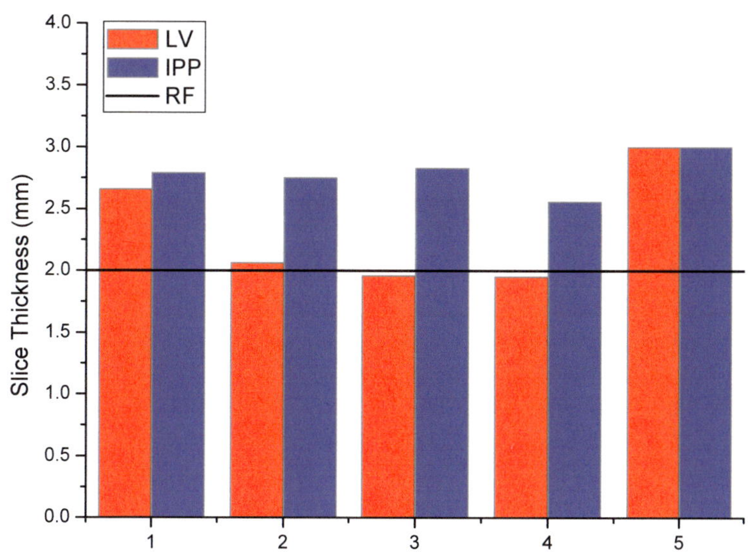

Fig. 12. (a)

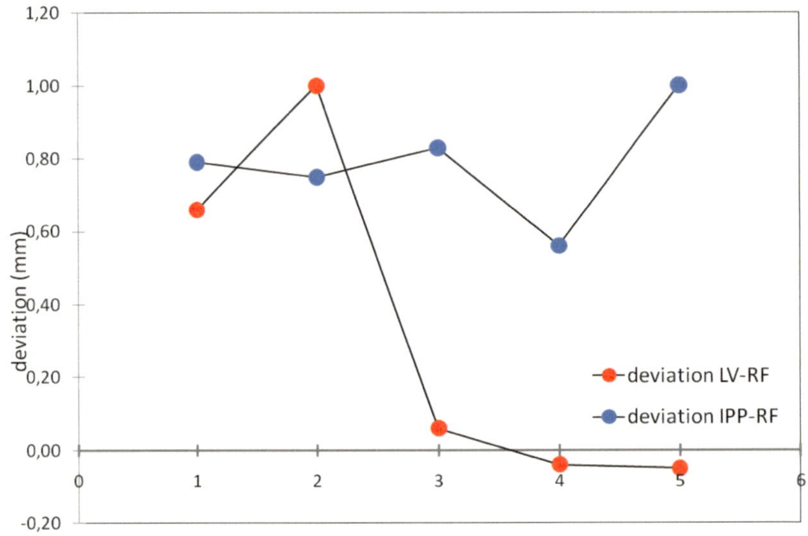

(b)

Fig. 12. CT slice thickness accuracy obtained for a 2 mm RF value: (a) sample set (blue: IPP; red: LV); (b) deviation from the RF value (blue: IPP; red: LV).

Statistical analysis conducted on the datasets shown in Tables 3-5 yielded the F-values reported in Table 6. Such F-values indicate that there is no significant statistical variation between the two different procedures, thus validating the methodology.

RF Values (mm)	F_C	F_T
10	0.011	4.54
5	0.332	4.67
2	4.18	5.12

Table 6. F values calculated for different RF values (F_C). These F-values were compared to the tabulated ones (F_T) for the P=0.05 confidence level.

3.2 MRI medical devices

Two-dimensional images of the wedge of the dedicated phantom were acquired using the Scan Head Protocol reported in Table 2. A typical MRI image of the PMMA box and the corresponding line profile calculated with the in-built MRI software are shown in Fig. 13. The preliminary results obtained using the IPP and LV procedures discussed above applied to MRI are reported in Table 7 for a RF value of 10 mm. In this case the slice thickness accuracy measured using an in-built MRI software (CS) is also included. The corresponding mean value and standard deviation of the three datasets are also reported.
Also in this case, the mean values calculated from the three different datasets of measurements are comparable between them. However, whereas the standard deviations obtained from CS and LV procedures are comparable, those obtained from IPP and LV are significantly different. The LV procedure provides a narrower deviation with respect to that obtained with IPP, which gives evidence for a more accurate measurement.

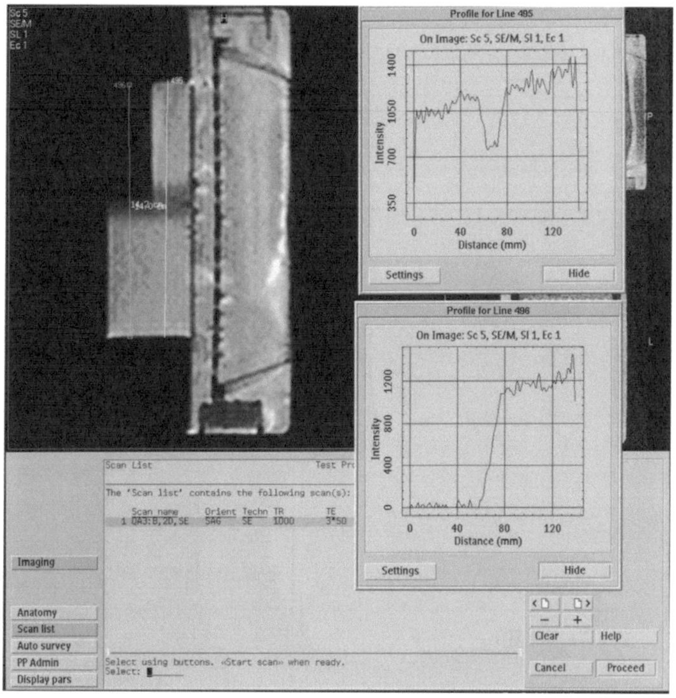

Fig. 13. MRI PMMA box image and corresponding line profile appearing on the MRI display. It is possible to notice the related line profile as obtained directly at the equipment console.

The slice thickness accuracy data of Table 7 are presented in graphical form in Fig. 14(a), where the slice thickness accuracy determined directly at the equipment console and that measured using the IPP and the LV procedures are compared. In Fig. 14(b) deviations from the RF value are presented for the three sets of data.

RF value (mm)	CS (mm)	CS mean value and standard deviation (mm)	LV (mm)	LV mean value and standard deviation (mm)	IPP (mm)	IPP mean value and standard deviation (mm)
10	10.1	10.05±0.17	10.3	10.05±0.15	9.80	9.96±0.31
	9.90		9.92		9.85	
	10.3		10.2		9.80	
	9.80		10.0		10.5	
	10.1		9.94		10.1	
	10.2		10.17		10.3	
	9.90		9.95		9.60	
	10.1		9.90		9.90	

Table 7. Comparison between the slice thickness accuracy obtained from the MRI in-built software (CS), the Image Pro Plus (IPP) procedure and the dedicated LabView (LV) software for a 10.0 mm Reference (RF) value. In the same table mean values and corresponding standard deviations for the three different procedures are also reported.

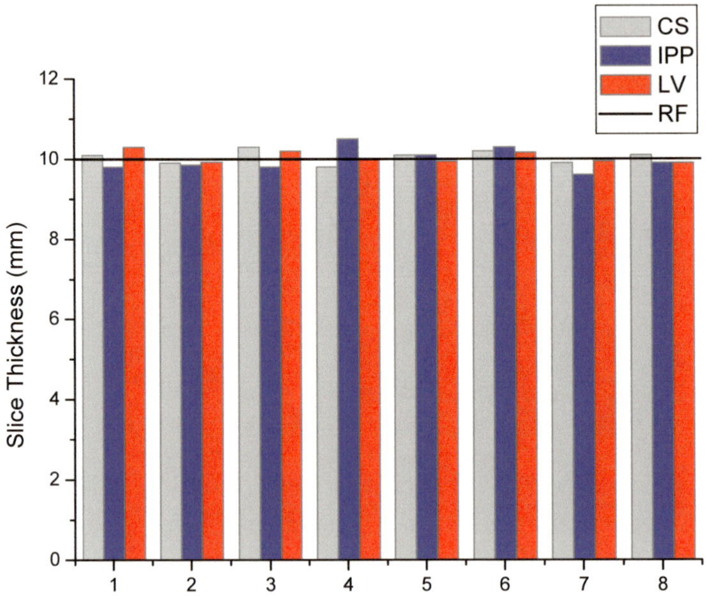

(a)

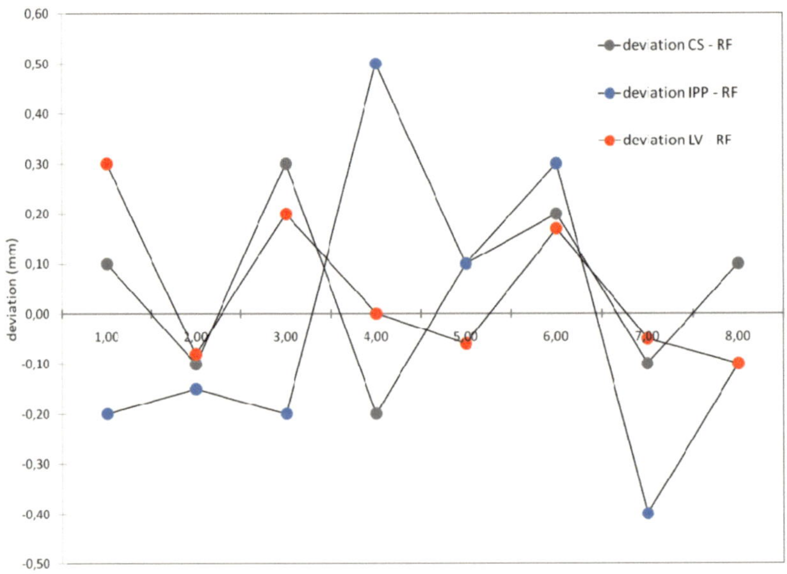

(b)

Fig. 14. MRI slice thickness accuracy obtained for a 10 mm RF value: (a) data obtained directly from the MRI scanner (CS) and measured using the IPP and the LV procedures (grey: CS; blue: IPP; red: LV); (b) deviation from the RF value (grey dots: CS; blue dots: IPP; red dots: LV).

The F-test is then used to verify that the three datasets are comparable (Table 8). As evident from the data of Table 8, the F values calculated for the three sets of data are significantly smaller than that tabulated (F_T) for a P=0.05 confidence level in all cases. Therefore, there is no significant statistical difference between the three different procedures, thus validating the novel LV procedure proposed here also for MRI imaging.

Datasets	F_C	F_T
CS-IPP	0.836	
CS-LV	0.00165	4.54
IPP-LV	0.513	

Table 8. Results of the F-test calculated comparing the CS, IPP and LV datasets (F_C). These F-values were compared to the tabulated ones (F_T) for the P=0.05 confidence level.

3.3 US scanners

The novel PMMA phantom proposed here (Fig. 7) was utilised also to test the elevation resolution in ultrasound systems. Preliminary measurements have been conducted on US probes to evaluate if it is possible to measure the beam width in the elevation plane. To correctly determine the elevation beam profile, it was necessary to slightly modify the phantom. In particular, the septum position was modified and, for the beam width evaluation, the inclined plane was oriented 28 degrees to the top and bottom surface, so the probe intersects the inclined plane at 28 degrees. The upper side of the inclined surface was filled with ultrasound gel, as conductive medium. The echoes reflected from the inclined plane are displayed as a horizontal band. Because the beam intersects the plane at 28 degrees, the beam width in the elevation plane can be calculated as follows:

$$ST = FWHM \cdot \tan(28°) \tag{5}$$

The above technique enables the measurement of the elevation dimension of the beam at a single depth only. To determine the entire profile in the elevation plane, the probe must be moved horizontally along the surface of the phantom to make a series of measurements, with the beam intersecting the inclined plane at different depths. With the use of the novel phantom described here, the resolution in the elevation plane is completely independent from the lateral resolution in the scanning plane. The new phantom proposed here is easy to use and does not require any additional equipment and the results are immediately displayed on the screen of the dedicated PC.

4. Conclusion

Slice thickness represents an important parameter to be monitored in CT, MRI an US. Partial volume effects can significantly alter sensitivity and specificity. Especially for MRI, quantitative measurements such as T_1 and T_2, are also greatly influenced by the accuracy of the slice thickness. Inaccuracies in the measurement of this parameter may result in inter-slice interference in multi-slice acquisitions, leading to invalid SNR measurements. In addition, for the US scanners, significant differences in image resolution can occur because of variations in the beam width of the elevation plane. So, it is important to know the profiles of the elevation planes of various probes to choose the probe that offers resolution

in the elevation plane that is most appropriate for the particular clinical application. Moreover, slice thickness accuracy is an important element also in the QC program for CT scanner, because quantitative CT analysis is heavily dependent on accuracy of slice thickness. Therefore, to provide an adaptable, reliable and robust procedure to evaluate the slice thickness accuracy a novel, dedicated phantom and corresponding LabView-based procedure have been proposed here, up to date applied to both CT and MRI devices. The new PMMA box proposed here to be associated with the dedicated software enables an innovative, accurate, easily applicable and automated determination of this parameter.
The carried on studies have utilised the 2 mm septum because the z axis dimensions for rotating anode x-ray tube foci are typically less than this value, but is our intention reduce the septum width, in order to evaluate the slice thickness also for the helicoidal CT scanners, that can reach slice widths smaller than 2 mm.
The accuracy (standard deviation) obtained with the novel procedure proposed here is significantly higher than that obtained with other procedures (e.g., in-built software, IPP).
This new slice thickness accuracy procedure is proposed as an alternative to the commonly adopted ones, which are typically complicated by the use of ad-hoc software and phantoms distributed by manufacturers and specific to the medical equipment. The proposed method employs a novel universal phantom, coupled with a dedicated LabView-based software that can be used on any CT and MRI scanner in a quick, simple and reproducible manner.
The readiness and applicability of the proposed procedure has been validated by quantitative tests using several different medical devices and procedures. In all cases the results obtained using the novel proposed procedure were statistically compatible with other commonly used procedures, but provided a very immediate determination of the slice thickness for both CT and MRI equipments, thus confirming the flexibility of the described method, its simplicity and reproducibility as an efficient tool for quick inspections.
The same procedure should be suitable also for determining elevation accuracy on US scanner: in fact, preliminary results confirmed that the novel phantom, opportunely modified, when coupled with the LabView dedicated software, allowed measurements of the section thickness

5. References

CEI EN 61223-2-6. (1997). Evaluation and routine testing in medical imaging departments. Part 2-6: Constancy tests – X-ray equipments for computed tomography. CEI, (Ed.), pp. 1-26

Chen, C.C.; Wan, Y.L.; Wai, Y.Y. & Liu, H.L. (2004). Quality assurance of clinical MRI scanners using ACR MRI phantom: preliminary results. *Journal of Digital Imaging,* Vol. 17, No. 4 (December 2004), pp. 279-284, ISSN 0897-1889

General Electric Medical System. (2000). Quality Assurance. In: *CT HiSpeed DX/i Operator manual Rev. 0.* General Electric Company (Ed.). Chapter 6, pp. 1-28

Goodsitt, M.M.; Carson, P.L.; Witt, S.; Hykes, D.L. & Kofler, J.M. (1998). Real-time B-mode ultrasound quality control test procedures. Report of the AAPM ultrasound task group No. 1. *Medical Physics,* Vol. 25, No. 8 (August 1998), pp. 1385-1406, ISSN 0094-2405

Judy, P.F.; Balter, S.; Bassano, D.; McCollough, E.C.; Payne, J.T. & Rothenberg, L. (1977). Phantoms for performance evaluation and quality assurance of CT scanners.

AAPM Report No. 1 (American Association of Physicist in Medicine, Chicago, Illinois, 1977)

Lerski R.A. & Mc Robbie D.W. (1992). EUROSPIN II. Magnetic resonance quality assessment test objects. Instruction for use. Diagnostic Sonar LTD (January 1992), pp. 1-79

Markowski, C.A. & Markowski, E.P. (1990). Conditions for the effectiveness of a preliminary test of variance. *The American Statistician,* Vol. 44, No. 4 (November 1990), pp. 322-326, ISSN 0003-1305

Mc Robbie, D.W.; Lerski, R.A.; Straughan, K.; Quilter, P. & Orr, J.S. (1986). Investigation of slice characteristics in nuclear magnetic resonance imaging. *Physics in Medicine and Biology,* Vol. 31, No. 6 (June 1986), pp. 613-626, ISSN 0031-9155

Mutic, S.; Palta, J.R.; Butker, E.K.; Das, I.J.; Huq, M.S.; Loo, L.N.D.; Salter, B.J.; McCollough, C.H. & Van Dyk, J. (2003). Quality assurance for computed-tomography simulators and the computed tomography simulation process. Report of the AAPM radiation therapy committee task group No. 66. *Medical Physics,* Vol. 30, No. 10 (October 2003), pp. 2762-2792, ISSN 0094-2405

Narayan, P.; Suri, S.; Choudhary, S.R. & Kalra, N. (2005). Evaluation of Slice Thickness and Inter Slice Distance in MR scanning using designed test tool. *Indian Journal of Radiology & Imaging,* Vol. 15, No. 1 (February 2005), pp.103-106, ISSN 0971-3026

National Instruments. (2001). LabView User Manual. National Instruments Corporation (Ed.). Available from
http://www.ni.com/pdf/manuals/320999d.pdf

Och, J.G.; Clarke, G.D.; Sobol, W.T.; Rosen, C.W. & Ki Mun, S. (1992). Acceptance testing of magnetic resonance imaging systems: Report of AAPM Nuclear Magnetic Resonance Task Group No.6. *Medical Physics,* Vol. 19, No. 1 (January-February 1992), pp. 217-229, ISSN 0094-2405

Philips. (1997). Performance Phantom C instruction manual. In: Tomoscan CX/Q technical documents. pp. 1-18

Price, R.R.; Axel, L.; Morgan, T.; Newman, R.; Perman, W.; Schneiders, N.; Selikson, M.; Wood, M. & Thomas S.R. (1990). Quality assurance methods and phantoms for Magnetic Resonance Imaging. Report of the AAPM Nuclear Magnetic Resonance Task Group No. 28. *Medical Physics,* Vol. 17, No. 2 (March-April 1990), pp. 287-295, ISSN 0094-2405

Rampado, O.; Isoardi, P. & Ropolo, R. (2006). Quantitative assessment of computed radiography quality control parameters. *Physics in Medicine and Biology,* Vol. 51, No. 6 (March 2006), pp. 1577-1593, ISSN 0031-9155

Rehani, M.M.; Bongartz, G.; Golding, S.J.; Gordon, L.; Kalender, W.; Albrecht, R.; Wei, K.; Murakami, T. & Shrimpton, P. (2000). Managing patient dose in computed tomography. *Annals of ICRP,* Vol. 30, No. 4 (December 2000), pp. 7-45, ISSN 0146-6453

Richard, B. (1999). Test object for measurement of section thickness at US. *Radiology,* Vol. 211, No. 1 (April 1999), pp. 279-282, ISSN 0033-8419

Sansotta, C.; Testagrossa, B.; de Leonardis, R.; Tripepi, M.G. & Vermiglio, G. (2002). Remote image quality and validation on radiographic films. *Proceedings of the 7th Internet World Congress for Biomedical Sciences, INABIS 2002,* Available from
http://www.informedicajournal.org/a1n2/files/papers_inabis/sansotta1.pdf

Skolnick, M.L. (1991). Estimation of Ultrasound beam width in the elevation (section thickness) plane. *Radiology*, Vol. 180, No. 1 (July 1991), pp. 286-288, ISSN 0033-8419

Torfeh, T.; Beaumont, S.; Guédon, J.P.; Normand, N. & Denis, E. (2007). Software tools dedicated for an automatic analysis of the CT scanner Quality Control's Images, In: *Medical Imaging 2007: Physics of Medical Imaging. Proceedings of SPIE*, Vol. 6510, J. Hsien & M.J. Flynn (Eds.), 65104G, ISBN 978-081-9466-28-0, San Diego, California, USA, March 6, 2007

Testagrossa, B.; Novario, R.; Sansotta, C.; Tripepi, M.G.; Acri, G. & Vermiglio, G. (2006). Fantocci multiuso per i controlli di qualità in diagnostica per immagini. *Proceedings of the XXXIIIth International Radio Protection Association (IRPA) Conference*, IRPA, ISBN 88-88648-05-4, Turin, Italy, september 20-23, 2006

Vermiglio, G.; Testagrossa, B.; Sansotta, C. & Tripepi, M.G. (2006). Radiation protection of patients and quality controls in teleradiology. *Proceedings of the 2nd European Congress on Radiation Protection «Radiation protection: from knowledge to action»*. Paris, France, May 15-19, 2006, Available from http://www.colloquium.fr/06IRPA/CDROM/docs/P-121.pdf

Vermiglio, G.; Tripepi, M.G.; Testagrossa, B.; Acri, G.; Campanella, F. & Bramanti, P. (2008). LabView employment to determine dB/dt in Magnetic Resonance quality controls. *Proceedings of the 2nd NIDays*, pp. 223-224. Rome, Italy, February 27, 2008

22

Nursing Business Modeling with UML: From Time and Motion Study to Business Modeling

Sachiko Shimizu et al.[1]
Osaka University
Japan

1. Introduction

A nurse is an autonomous, decentralized worker who recognizes goals, his or her environment, the conditions and actions of patients and other staff members, and determines his or her own actions. Put another way, the nurse makes decisions flexibly in the midst of uncertainty. Because of this, nursing work differs from individual nurse to nurse, and understanding this process theoretically is considered to be difficult.

Concerning nursing work analysis, research has been done on task load (time required for tasks). However, there has been scant academic research on work processes in nursing compared with research that has accumulated in other industrial fields, including research on structuralizing work, i.e., defining and visualizing work processes. To improve work processes, it is necessary to understand and clarify work as a chain of theoretically related activities.

Thus in this study, using time and motion study techniques, a method used to measure jobs, we clarify the structure of the work of transporting patients by nurses. We also attempt to visualize it. We use objected-oriented modeling to express the operation visually.

2. From time and motion study to business modeling

Time and motion study is a method that actually measures the movements of a particular person. Its results can be applied not only to measuring the work load of nurses (Van de Werf et al., 2009; Were et al., 2008;Hendrich et al.,2008) and analyzing the workflow(Tang et al., 2007), they can also be used as basic data for task scheduling(Yokouchi et al., 2005) and efficient arrangement of personnel. In addition, the results are being used as indicators to evaluate changes in a hospital brought about by systems deployed (Yen et al., 2009), such as an electronic medical record (EMR) system. Thus many time and motion studies of hospitals have been conducted both within Japan and without.

Specifically, a time and motion study is defined as a study that records the time of occurrences of tasks through continuous observation. A type of measuring technique similar

---

[1]Rie Tomizawa, Maya Iwasa, Satoko Kasahara, Tamami Suzuki, Fumiko Wako, Ichiroh Kanaya, Kazuo Kawasaki, Atsue Ishii, Kenji Yamada and Yuko Ohno
Osaka University, Japan

to the time study is work sampling, which seeks to comprehend a job by sampling its conditions at predetermined time intervals. Work sampling cannot comprehend a job in its entirety, but it lessens the burden on the measurer. It also makes it possible for the worker himself or herself to record time. In contrast, a time and motion study comprehends the job in its entirety, but the burden on the measurer is great. The differences in results between the two methods have been observed to be large for jobs in which there were few events(Finkler et al., 1993). Currently, the results that come from measuring a job through continuous time and motion observation are said to be the gold standard.

While the breadth of research that utilize measurement results from time and motion studies encompasses all nursing work, individual studies have been limited to examining the amount of work for individual caring assignments, such as cleaning a patient, feeding a patient, and taking care of a patient's toilet needs. There have been especially few studies that evaluate the work amount of a job by focusing on the job and clarifying its work process. While not on concerned with nursing work, the only such study conducted so far in the medical field was visualizing and understanding the amount of work involved in the process of registering cancer patients by Shiki et al. (Shiki et al., 2009). They proposed the method of "time-process study," a method to visualize tasks by adding time information to the process. However, because both the process and amount of work were estimated through interviews, the results can be said to be lacking in objectivity. Thus our study uses the time and motion study method, which actually measures a task. We focus on the job of transporting patients and clarifying its process. We also study the possibility of a method to visualize the work process using the clarified process and time information.

Transporting patients is an operation that is often performed outside hospital wards. It is both physically and mentally demanding of nurses. This job should also be scrutinized because it reduces the number of nursing staff inside the wards, as nurses go outside the wards in order to safely transport patients.

3. Methods

3.1 Study setting

We carried out a time and motion study of nursing work related to transporting patients in four hospital wards of a cardiovascular treatment facility. We tracked our subjects, who were nurses, nursing assistants, and medical clerks, from the time of the start of a task until its end, and recorded the task actions. The record of a task action included the content of the action, the time of its start and end, the person who was the target of the action, and the location of the action. The four wards of the treatment facility consisted of the cardiac failure ward, arrhythmia ward, cardiomyopathy/pulmonary hypertension ward, and cerebral vascular and metabolism ward. The destinations of patient transport included exam rooms for CT, X-ray, MRI, echocardiography, respiratory function testing, cardiac rehabilitation, neurological rehabilitation, cardiac catheterization investigation, and dialysis.

3.2 Business modeling with UML

From the time and motion study records we obtained, we created a use case diagram and activity diagram. Use case diagrams and activity diagrams are types of diagrams created using Unified Modeling Language (UML). UML is the de facto standard objected-oriented modeling language, and was developed for software development. In recent years,

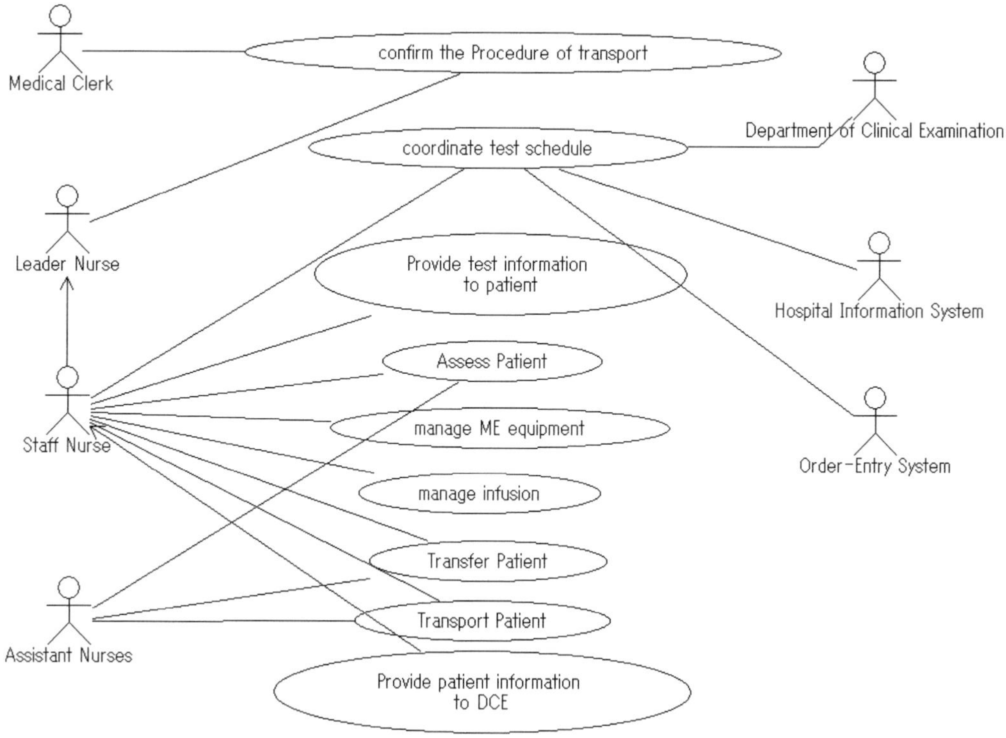

Fig. 1. Functional diagram-patient transports system.

however, its use for business modeling has been suggested (Eriksson and Penker, 2000). The reason is that the structure of a job can considered oriented-oriented in nature. The content of a job can be treated as exchanges of messages between objects, such as materials and users. Thus UML as a descriptive method can allow one to intuitively understand the job. In this study, we elucidated the functional aspect of the operation of transporting patients. We also used an activity diagram to visualize the work process of transporting patients. Finally, we discussed the work load and its time efficiency by adding time information to the activity diagram. This study was approved by the ethics committee of the hospital we studied.

4. Results

From the time and motion study, we observed and recorded 213 jobs of transferring patients. Overall, the number patient transfer assignments recorded was 3,775. Of these records, 387 records were not jobs related to transporting patients, so they were removed from our analysis.

A use case diagram extracted from the results of the time and motion study is shown in Figure 1. There were seven types of actors involving in transporting patients: nurses, head nurses, medical clerks, nursing assistants, the central medical examination department, the order entry system, and the hospital information system. The nurses were divided into two groups: head nurses, who had the responsibility of being in charge of nursing duties, and staff nurses, who received patients and provided care for them. The head nurse and the medical clerk received communication about the transport of a patient, and confirmed the predetermined method of transport care. In addition, the head nurse made adjustments such as changing the transport personnel and finding appropriate personnel. Of the tasks related to transport care, the nurse and nursing assistant handled tasks that had direct bearing on the patient. In the hospital of this study, patients undergoing oxygen therapy, patients being monitored by EKG, and patients undergoing transfusion were the responsibility of nurses, not nursing assistants.

Task	TOT	Frequency	Median	Range
T01 Coordinate time for examination	0:33:27	28	58	(5-273)
T02 Confirm schedule of examination	0:05:24	10	29	(4-100)
T03 Accept call for examination	0:31:30	45	34	(1-324)
T04 Look for patient record	0:04:32	11	18	(2-64)
T05 Check bed rest level	0:09:11	10	36	(6-186)
T06 Identify care-giver	0:00:58	3	21	(4-32)
T07 Prepare map	0:08:27	20	23	(3-70)
T08 Prepare patient consultation card	0:14:37	31	18	(1-108)
T09 Prepare patient record	0:28:41	42	31	(5-187)
T10 Find care-giver	0:01:59	3	42	(16-60)
T11 Find patient	0:07:33	11	17	(4-116)
T12 Wait for care-giver	0:00:21	1	21	(21-21)
T13 Relay examination information to patient	0:29:55	43	34	(1-144)
T14 Hand necessary materials to patient	0:00:21	3	6	(2-13)
T15 Change care-giver assignment	0:00:37	1	37	(36-36)
T16 Relay exam information to nurse	0:26:48	38	21	(1-384)
T17 Prepare film	0:00:44	2	22	(15-29)
T18 Prepare materials to be brought	0:04:02	3	38	(6-198)
T19 Prepare transport care equipment	0:22:38	46	20	(1-139)
T20 Carry transport care equipment	0:21:27	40	26	(1-88)
T21 Assess situation	0:24:48	17	26	(2-382)
T22 Confirm patient name	0:02:45	10	16	(6-30)
T23 Prepare to move ME devices	0:13:50	19	31	(7-237)
T24 Prepare to move medical supplies	0:16:43	23	42	(2-117)
T25 Assist in excretion	0:05:16	5	52	(10-152)
T26 Assist in changing of clothes	0:12:35	19	25	(10-127)
T27 Prepare for transfer	0:10:22	13	29	(5-199)
T28 Carry patient	1:46:59	83	43	(3-707)
T29 Transport patient	9:15:49	109	292	(1-866)
T30 Go through reception procedures	0:08:56	34	9	(1-90)
T31 Hand-over patient	0:01:55	8	13	(2-34)
T32 Hand-over necessary supplies	0:10:31	30	15	(1-89)
T33 Relay information	0:33:09	31	63	(3-156)
T34 Prepare for examination	0:27:16	26	32	(1-370)
T35 Assist in examination	0:42:01	41	28	(6-255)
T36 Standby at destination	1:57:19	35	92	(1-1612)
T37 Receive patient	0:06:37	7	20	(6-208)

Task	TOT	Frequency	Median	Range
T38 Reattach ME devices	0:41:25	18	82	(6-766)
T39 Reattach medical supplies	0:21:23	14	69	(2-396)
T40 Secure consultation card	0:04:35	23	9	(1-44)
T41 Secure patient record	0:23:02	30	19	(1-560)
T42 Clear away film	0:00:28	4	5	(3-16)
T43 Clear away transport care equipment	0:25:52	40	34	(2-115)
T44 Clear away map	0:01:54	5	11	(1-78)
T45 Finish clean up	0:13:24	15	33	(1-159)
T46 Record the transfer	0:11:10	11	32	(3-247)
M Move	4:36:03	119	95	(2-1068)

TOT: time on task.

Table 1. Identified tasks and their descriptive statistics.

The dynamic aspect of transporting patients is shown as an activity diagram (see Figure 2). The head nurse, who is in charge of communication in the hospital ward, and the medical clerk receive a call for a patient from the central medical examination department. They confirm the bed rest level of the patient from his or her chart. If the patient can walk outside the ward by himself or herself (self-reliant), the person in charge of communication prepares the chart, the patient's exam ticket, and the map to the exam room. He or she searches for the patient, relays the call for examination to the patient, and hands over necessary items. If the bed rest level is escort (transport in a wheelchair) or litter care (transport on a stretcher), the person in charge of communication searches for the transport personnel and hands over the exam call. The transport personnel prepare the patient's chart, the exam ticket, and the instrument for transport care such as a wheelchair or stretcher, and move to the patient's location. They relay the exam call to the patient, and assess the patient's conditions to determine if transport is possible. If the transport personnel determine that the patient can be transported, he/she/they prepare oxygen or transfusion devices for transport, and perform excrement care and assist the patient in changing clothes. Next, the transport personnel move the patient from the bed to the transport instrument, and transport the patient to the exam room. After the patient arrives in the examination room, the transport personnel notify the exam receptionist of the patient's arrival, hand over the patient, and hand over items brought along, such as the patient chart and the exam ticket. If the exam takes only a short time, e.g. in the case of an x-ray exam, the transport personnel wait in the exam room, assist with preparing the patient for examination, and assists in the examination. If the exam takes a longer period of time, the transport personnel return to the hospital ward and perform other tasks. When communication comes from the examination room, the transport personnel receive the message and move to the exam room. After the exam has completed, the transport personnel receive the call from the patient, transfer the patient to the transport instrument, transport him or her back to the ward, and again move him or her to the hospital bed. The transport personnel prepare medical electronic equipment and medical devices attached to the patient so that subsistence in bed is possible. After assessing the patient's conditions, the transport personnel puts away the items brought along, such as the exam ticket and the patient chart, and record the transport. As shown in the activity diagram, we clarified that the process of transporting a patient was composed of 47 tasks.

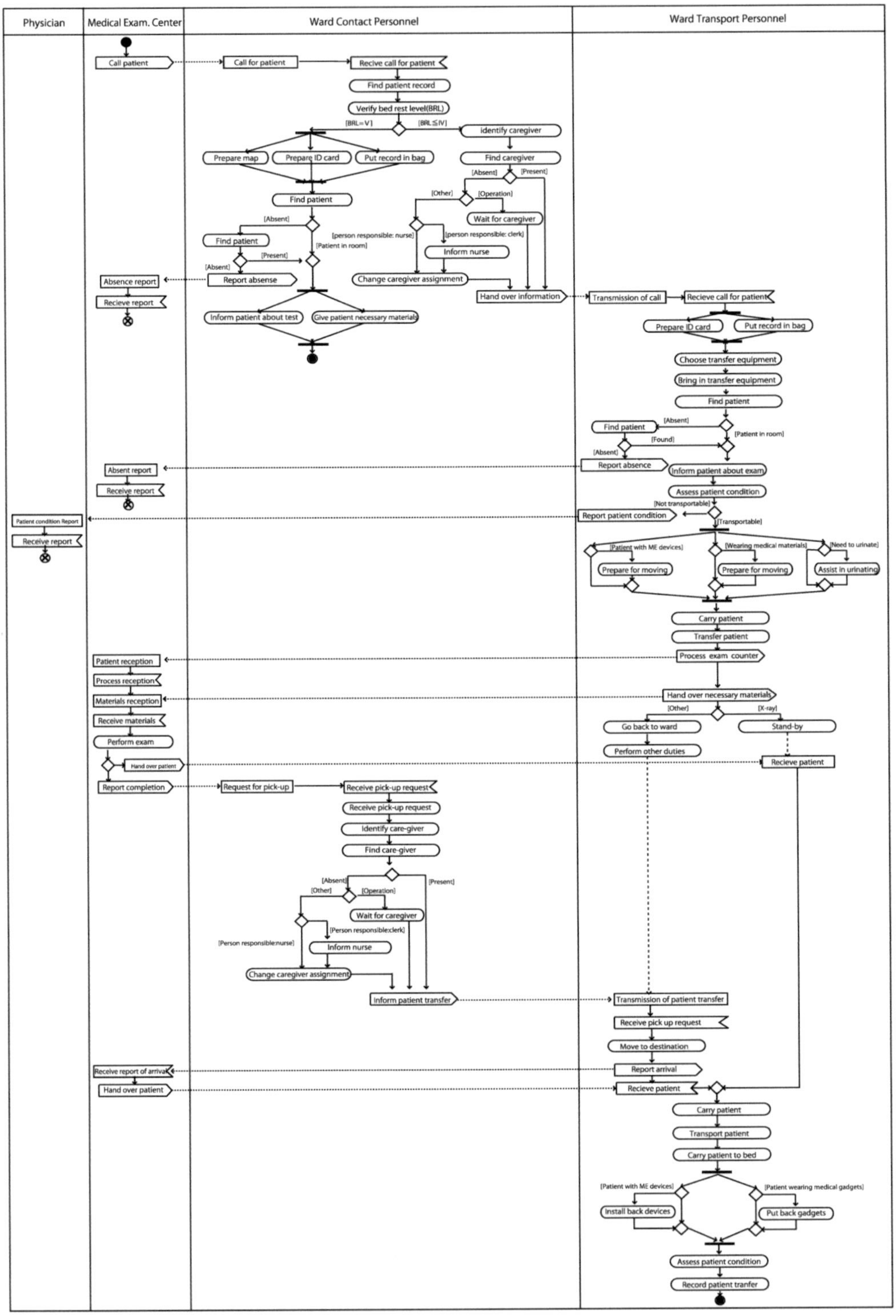

Fig. 2. Dynamic diagram-patient transports.

Table 1 shows the total time on task during a day in the four wards for each of the 47 tasks shown in the activity diagram. Also shown are the number of occurrences of each task, the median value, and the range. The task that took up the most total time was "T29 Transporting patient" (9:15:49). It took about 5 minutes on average for the nurse(s) to transport a patient. Of the 213 patient transport jobs observed, 109 actually involved transporting the patient. Patient transport jobs that did not involve transport were only those to support self-reliant patients and to adjust the scheduled time of exams. After T29, the task that took the most time was "T36 Standing by at the destination" (1:57:19), followed by "T28 Transferring the patient" (1:46:59). On the other hand, there were few occurrences of tasks related to searching for or changing transport personnel, such as "T06 Identifying care provider," "T12 Waiting for care provider," and "T15 Changing care provider." Comparing the coefficient of variance, we found that the coefficient of variance for "T41 Putting patient chart away," "T16 Conveying exam information to nurse," "T36 Standing by at destination," and "T21 Assessing conditions" was high. On the other hand, the coefficient of variance of "T29 Transporting patient" and "T43 Putting instruments for transport care" was relatively low.
The time on task for each type of task is shown in Table 2. Direct tasks are those that deal directly with the patient. Indirect tasks are tasks carried out without direct contact with the patient, including preparatory tasks for direct tasks and cleaning tasks. Direct tasks, which involve transporting the patient, made up about 60 percent of all tasks, and indirect tasks made up 14 percent of all tasks.

Task category	No. of task	Time on Task	(%)
Indirect care	21	3:56:23	(14.1)
Direct care	21	16:08:27	(58.0)
Communication	2	0:59:57	(3.5)
Waiting	1	1:57:19	(7.0)
Record	1	0:11:10	(0.6)
Move	1	4:36:03	(16.5)
Total	47	27:49:19	(100.0)

Table 2. Time on task by each task category.

5. Discussion

First, we clarified the location and roles of persons in charge of tasks by making use of time and motion study data to visualize the object-oriented work process. From a functional point of view, the main persons in charge of the job of transporting patients were nurses. However, we understood that medical clerks participated in coordinating communication and that nursing assistants participated in transporting patients who did not need custody or attachment of medical electronic or transfusion devices. We understood that while medical clerks received communication about exams and confirmed the method of transport care on the patient chart, they did not have privilege to change the transport personnel or delegate the task, so they turned the task over to lead nurses. Furthermore, in the case of self-reliant patients, the person in charge of communication in a ward had the responsibility of transmitting the exam information to the patient regardless of whether he or she was a medical clerk or nurse. Furthermore, in the case of patients who needed wheelchair or stretcher transport, the person in charge of communication had the responsibility of sending information about the exam call to the transport personnel after receiving the communication about the exam. Our study showed that if the person in charge of communication was a medical clerk, he or she turned the task over the head nurse, because he or she did not have the privilege to change the care provider. The task that took the most time in this process was "Conveying exam information to the patient," followed by "Preparing patient chart" and "Preparing exam ticket." Use of the exam ticket was limited to outpatient exams of hospitalized patients and during the medical exam, so the repositories of the tickets were fixed. In contrast, because patient charts were used for a variety of purposes by physicians, nurses, medical clerks, and many other hospital employees, search for the charts took place, and the time required to prepare the charts grew longer. After information was conveyed to the transport personnel by the person in charge of communication, the transport personnel handled all responsibilities, including the final task of recording the transport.

Second, we understood the divergence between the work process specified in the hospital procedures manual and the actual work process. The manual used in the hospital of our study did not specify tasks such as "Searching for the patient," "Searching for the transport personnel," "Changing the transport personnel," "Preparing the exam (in the exam room)," and "Assisting in the exam." This reason is that the work procedures manual contains standard procedures. Irregular events and redundant tasks that should be kept in mind were not included. Also, the procedures manual was written to describe work procedures for individual nurses, so the location and role of workers described above were not clarified.

Third, from the work process diagram based on actual work records collected by this study and by adding time information to the process, we understood the efficiency with which tasks were carried out. By understanding the time used for each task and the variability of time, we clarified the time element that makes up the care of transporting patients. In the future, we seek to understand in detail how time on task changes depending on constraints.

Fourth, our study suggests that the data can be used for risk analysis. Our study extracted 47 tasks that made up the transport of patients, and listed their sequential order from time study records. Through our study, we clarified the input and output of each task, as well as the frequency of irregular events. Irregular events such as "Searching for the patient" and "Searching for the nurse" can be considered risks recorded by this study that prevent the work goal from being achieved. Although not carried out in this study, each task can be scrutinized to clarify factors that hinder each of their output. Doing this can draw out the risks associated with the work of transporting the patient, and produce discussions about concentrating risks and avoiding risks.

6. Future outlook

In this study, the structure of the work of transporting patients was visualized. The study suggests that the work of transporting patients has great differences in the objects, the process, and time efficiency depending on the conditions of the patients, type of exam, and occurrence of the work. Also, because many work occurrences were irregular and required quick responses, we learned that nurses must make adjustments with other tasks while at the same time accomplishing the task of transporting patients.
This study showed the usefulness of time and motion study for clarifying not only work load but also work structure and work processes. In the future, we seek to confirm the applicability of this study by conducting similar studies based on other jobs and records of jobs in several other facilities.

7. Acknowledgment

We wish to express our sincere gratitude to all the nurses, nursing assistants, and medical clerks who cooperated with us in this study. We also received assistance from research seminar students of the Osaka University Graduate School of Medicine, Division of Health Sciences. They conducted the time and motion study measurements. Once again, our deep gratitude to all involved.

8. References

Eriksson , H.E., Penker, M. (2000). *Business modeling with UML.* Wiley, New Jersey, 9780471295518.

Finkler, S.A., Knickman, J.R., Hendrickson, G. et al.(1993). A Comparison of work-sampling and time and motion techniques for studies in health services research. *Health Service Research,* Vol.28, No.5, pp.577-597, 0017-9124.

Hendrich, A., Chow, M.P., Skierczynski, B. A. et al.(2008). A 36-Hospital Time and Motion Study: How Do Medical-Surgical Nurses Spend Their Time? *The Permanente Journal,* Vol.12, No.3, pp.25-34, 1552-5767.

Shiki, N., Ohno, Y., Fujii, A. et al.(2009).Time process study with UML a new method for process analysis. *Methods of informatics in Medicine,* Vol.48, No.6, pp.582-588, 0026-1270.

Tang, Z., Weavind, L., Mazabob, J. et al.(2007).Workflow in intensive care unit remote monitoring: a time and motion study. *Critical Care Medicine,* Vol.35, No.9, pp.2057-2063, 0090-3493.

Van de Werf, E., Lievens, Y., Verstraete, J. et al.(2009).Time and motion study of radiotherapy deliverly: economic burden of increased quality assurance and IMRT. *Radiotherapy and Oncology* ,Vol.93, pp.137-140, 0167-8140 .

Were, M.C., Sutherland, J.M., Bwana, M. et al.(2008).Patterns of care in two HIV continuity Clinics in Uganda, Africa:a time motion study. *AIDS care,* Vol.20, No.6, pp.677-682, 0954-0121.

Yen, K., Shane, E.L., Pawar, S.S. et al.(2009).Time motion study in a pediatric emergency Department before and after computer physician order entry. *Annals of emergency medicine,* Vol.53, No.4, pp.462-468, 0196-0644.

Yokouchi, M., Ohno, Y., Kasahara, S. et al.(2005). Development of Medical Task Classification for Job Scheduling. *Medical and Biological engineering,* Vol.43, No.4, pp.762-768, 1347-443X.

23

Practical Quality Control: the Experiences of a Public Health Laboratory

Francesc Centrich[1], Teresa Subirana[1],
Mercè Granados[2] and Ramon Companyó[2]
[1]*Laboratori de l'Agència de Salut Pública de Barcelona;*
[2]*Universitat de Barcelona*
Spain

1. Introduction

In the 1930's W.A. Shewhart pioneered the application of statistical principles to the quality control (QC) of production processes, eventually publishing the landmark book "Economic Control of Quality of Manufactured Products" (Shewhart, 1931). In this book, he states that a phenomenon is *under control* if its future variation can be predicted (within limits) based on previous experience. This is precisely the idea behind the control charts used in measurement processes—specifically, for chemical analysis. The International Organization for Standardization (ISO), in its standard ISO 9000 (ISO, 2005a), defines *quality control* as "the part of quality management focused on fulfilling quality requirements". According to the standard, quality management also includes quality planning, quality assurance and quality improvement. The above definition is rather vague, because quality management systems based on the ISO 9000 family of standards can be applied to any kind of organization regardless of its field of activity, its size or whether it is from the public or private sectors. Testing laboratories typically distinguish between *internal* and *external* QC. In this context, the International Union of Pure and Applied Chemistry (IUPAC, 1998) gives a definition of internal QC that is well-suited to an analytical laboratory: "the set of procedures undertaken by laboratory staff for the continuous monitoring of operation and the results of measurements in order to decide whether results are reliable enough to be released". Although the aforementioned document does not formally define *external QC,* it does mention that *external control* may be done by submitting blind samples to the measuring laboratory. This activity can be organized in the form of a collaborative test.
The aim of these QC activities is to verify that the quality parameters of an analytical method ascertained in the method validation are maintained during its operational lifetime. Thus, method validation or revalidation tasks are periodic activities that end with a validation report, whereas QC activities are recurrent activities implemented in routine work. Apart from the use of fully validated methods, QC assumes the use of properly maintained, verified and calibrated equipment, reagents and consumables with the proper specifications; standards with well-established traceability; and qualified technicians working in suitable environmental conditions. However, fulfilling all these requirements is not enough to ensure the delivery of appropriate quality results over time: a laboratory's capacity to produce technically correct results must be continuously monitored. Indeed, according to Thompson *et al.* (Thompson & Lowthian, 1993), QC is the only quality

management measure that provides a high level of protection against the release of inaccurate data. The authors demonstrate a significant relationship between the efficacy of a laboratory's QC and its subsequent performance in proficiency tests. They also consider that the implementation of QC activities and the participation in proficiency tests are two sides of the same coin: a laboratory's commitment to quality.

Once a laboratory has implemented a method in its routine work, is performing adequate QC, has taken any appropriate corrective and/or preventive actions, and its staff has acquired sufficient expertise, it may consider including this method in its scope of accreditation. Figure 1 shows these activities in the context of the operational lifetime of an analytical method.

This chapter was written to explain, in practical terms, the QC activities and management at an analytical laboratory—namely, the Chemical Analysis Service at the Laboratory of the Public Health Agency of Barcelona (*Laboratori de l'Agència de Salut Pública de Barcelona*; hereafter, *LPHAB*).

2. History and present context of the LPHAB

The LPHAB has its origin in the Municipal Laboratory of Barcelona, a microbiology laboratory created in 1886 to provide support to the sanitary authorities in their efforts to prevent rabies; since its inception, the Municipal Laboratory of Barcelona was a reference laboratory in Spain. In 1907, owing to its ever-increasing activities, it was given a new structure that led to creation of a section dedicated to chemical analysis of foods, with the then innovative objective of studying health problems attributable to the presence of hazardous chemicals in foods.

From the 1950's onwards, the section on chemical analysis of foods underwent major development. This stemmed from advances in knowledge on food chemistry and was catalyzed by various international food crises caused by chemical pollutants such as mercury and methanol. A case of widespread food poisoning in Spain in 1981, traced to denatured rapeseed oil, triggered the modernization of many Spanish public health laboratories, including the Municipal Laboratory of Barcelona. The Laboratory's equipment was soon updated, and its organization and management were overhauled. These changes enabled the Municipal Laboratory of Barcelona to face new analytical challenges. In addition to assessing the nutritional properties of food, it also focused on detection and determination of additives, residues and contaminants in food. The Municipal Laboratory of Barcelona began serving customers outside of the municipal administration; the challenge of providing these customers with the data they sought at specific analysis costs and response times proved highly stimulating. By the year 2000, it had analyzed 20,000 samples. In 2003 the Municipal Laboratory of Barcelona merged with the Public Health Laboratory of the Autonomous Government of Catalonia (*Generalitat de Catalunya)* in Barcelona to form the LPHAB. This union led to significant investments in instrumentation and to the recruitment of new staff; consequently, the newly formed LPHAB became one of the strongest laboratories in Spain for food analysis.

The LPHAB currently comprises four departments: two technical departments (the Chemical Analysis Service [CAS] and the Microbiological Analysis Service) and two management & support departments (the Quality Assurance Unit [QAU] and the Logistics & Services Unit). It presently employs 65 people, 31 of which work in the CAS (11 senior technicians and 20 mid-level technicians and support staff). The CAS encompasses four areas: two dealing with applications (food analysis and environmental analysis) and two dealing with analytical techniques (spectroscopic analysis and chromatographic analysis).

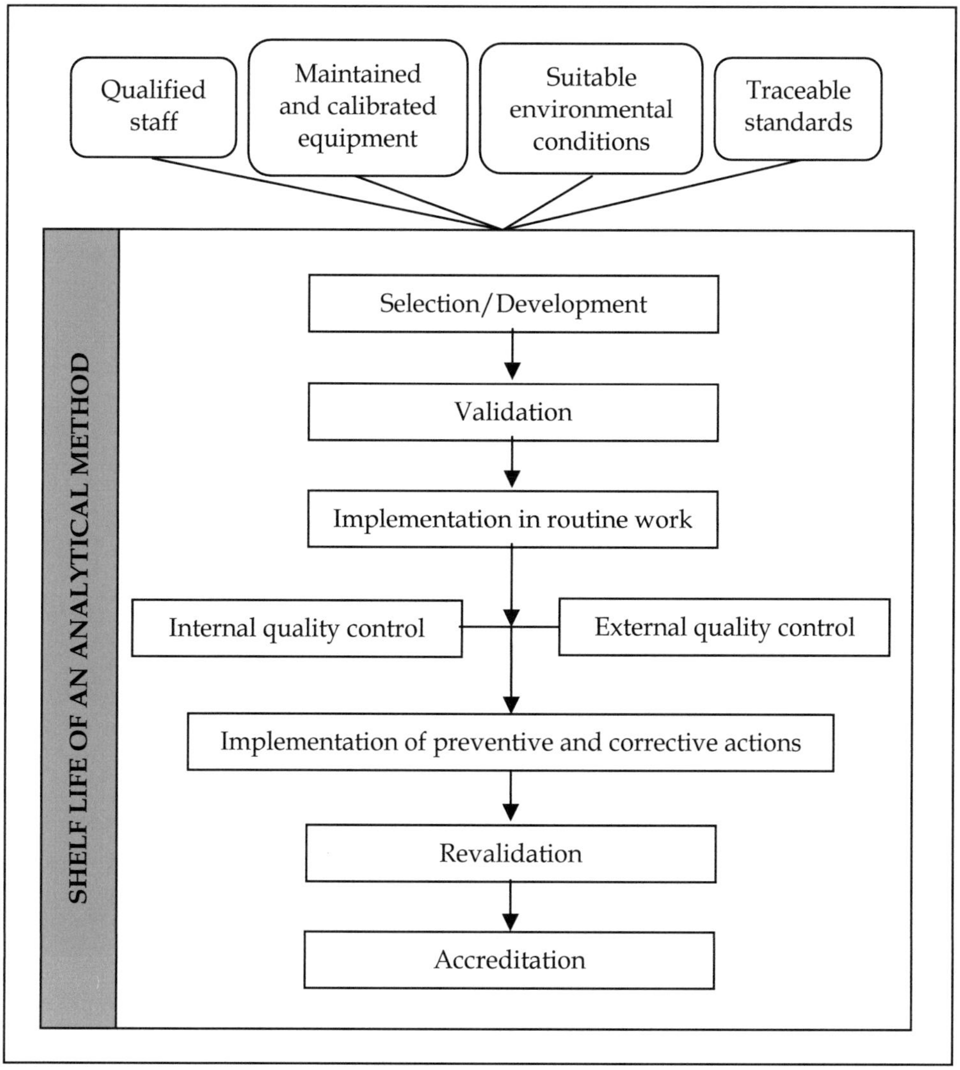

Fig. 1. Activities that determine the reliability of test results.

The LPHAB features a broad array of state-of-the-art equipment: roughly 500 instruments, including those for sample treatment, chromatography and spectroscopy. These include various gas and liquid chromatographs coupled to tandem mass spectrometry plus two inductively coupled plasma spectrometers, one equipped with photometric detection, and the other, with mass spectrometry detection. The LPHAB also uses a laboratory information management system (LIMS).

To date, the CAS has implemented about 110 analytical methodologies included in the scope of accreditation according to the requirements of the ISO 17025 standard (ISO, 2005b). In 2010, the CAS portfolio included approximately 1,800 different determinations, 1,400 of which correspond to its scope of accreditation. Moreover, the flexible scope includes some 55 analytical methods, grouped according to instrumental techniques, and numerous analytes.

In 2010 the LPHAB tested 32,225 samples, for which it performed some 550,000 determinations. Roughly half of these samples were food samples, and the other half, environmental samples (chiefly, potable water and filters for atmospheric control). The LPHAB's main customers are the Public Health Agency of Barcelona (which owns it), and the inspection bodies of the Catalonian and Spanish governments.

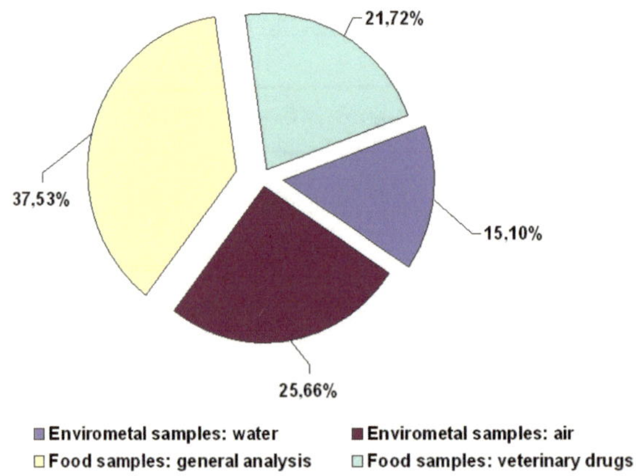

Fig. 2. Breakdown of analyses performed at the LPHAB in 2010, by sample type.

In 2010, the LPHAB's budget, excluding staff costs, was €1.2 million. This includes consumables, gases, reagents, culture media, equipment maintenance, participation in proficiency testing, and small investments. Its revenue contracts and invoices totaled €7 million.

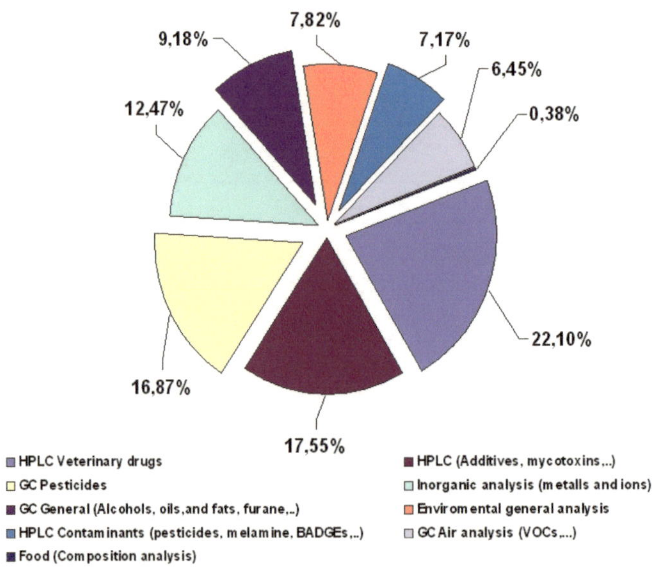

Fig. 3. Breakdown of analyses performed at the LPHAB in 2010, by analytical technique

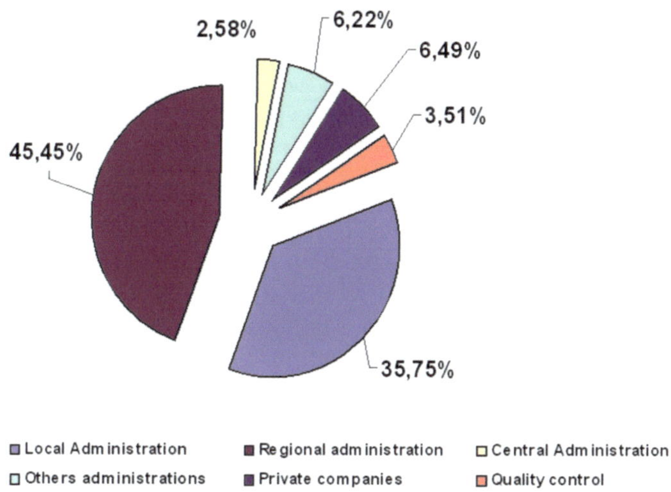

Fig. 4. Breakdown of the LPHAB's customers

The LPHAB performs research on developing and improving analytical methodology, both on its own and in collaboration with various universities. Its staff members often participate as experts in training courses organized by universities or government bodies, and some of its senior technicians are regularly asked by the Spanish Accreditation Body to participate as technical experts in laboratory accreditation audits for the food sector. Lastly, the LPHAB regularly hosts university or vocational students for training stays and internships.

3. QC within the framework of the ISO/IEC 17025 standard

Since it was issued in 2005, the ISO/IEC 17025 standard (ISO, 2005b) has been the international reference for accreditation of the technical competence of testing and calibration laboratories. The requirements of ISO/IEC 17025 (ISO, 2005b) concerning QC are concisely set out in Section 5.9 of the Standard, entitled "Assuring the Quality of Test and Calibration Results". Briefly, the Standard states that QC activities are mandatory and dictates that their results must be recorded. It also mentions the most frequent internal QC and external QC activities, without excluding other possible activities:

"The laboratory shall have QC procedures for monitoring the validity of tests and calibrations undertaken. The resulting data shall be recorded in such a way that trends are detectable and, where practicable, statistical techniques shall be applied to the reviewing of the results. This monitoring shall be planned and reviewed and may include, but not be limited to, the following:

a) Regular use of certified reference materials and/or internal QC using secondary reference materials;

b) Participation in proficiency test or proficiency-testing programs;

c) Replicate tests or calibrations using the same or different methods;

d) Retesting or recalibration of retained items;

e) Correlation of results for different characteristics of an item."

The standard goes on to state that the results of the monitoring activities performed must be analyzed and that appropriate measures should be taken:

"QC data shall be analyzed and, where they are found to be outside pre-defined criteria, planned action shall be taken to correct the problem and to prevent incorrect results from being reported."

4. Legislative requirements

Food safety and environmental protection are top priorities in the EU, which has implemented widespread legislation to support its policies in these fields. Noteworthy examples include Regulation (EC) No 178/2002, which establishes a legal framework for food; Directive 2000/60/EC, which establishes a framework for actions in the EU's water policy; and Directive 2008/50/EC, which outlines measures on ambient air quality. Currently, there is also a proposal for a framework Directive to create common principles for soil protection across the EU.

The EU has high standards for food safety and environmental protection. For instance, Regulation (EC) No 1881/2006 defines maximum levels for certain contaminants (*e.g.* mycotoxins, dioxins, heavy metals, and nitrate) in foodstuffs; Regulations (EU) No 37/2010 and (EC) No 830/2008 stipulate maximum residue levels of pharmacologically active substances or pesticides, respectively, in foodstuffs; and Directive 98/83/EC defines values for several microbiological and chemical parameters for water intended for human consumption. Regarding the environment, Directive 2008/50/EC defines objectives for ambient air quality and establishes limits on the concentration levels of air pollutants; Water Framework Directive 2000/60/EC presents a list of 33 priority pollutants based on their substantial risk; and Directive 2008/105/EC establishes environmental quality standards for these 33 pollutants.

Laboratories in charge of official controls provide essential support for these policies, by proficiently monitoring environmental and food samples. These laboratories should be equipped with instrumentation that enables correct determination of maximum levels as stipulated by EU law. According to Regulation (EC) No 882/2004, the laboratories designated for official controls in feed and food samples must operate and be assessed and accredited in accordance with ISO/IEC 17025 (ISO, 2005b). Likewise, Directive 2009/90/EC establishes that laboratories that perform chemical monitoring under Water Framework Directive 2000/60/EC must apply quality management system practices in accordance with the ISO/IEC 17025 standard or an equivalent standard accepted at the international level. Moreover, the laboratories must demonstrate their competence in analyzing relevant physicochemical parameters or compounds by participating in proficiency testing programs and by analysis of available reference materials representatives of the monitored samples. In Spain, Royal Decree 140/2003 stipulates that laboratories designated for official controls of water intended for human consumption that analyze more than 5,000 samples per year must be accredited in accordance with ISO/IEC 17025 (ISO, 2005b), and that other laboratories, if they are not accredited as such, must be at least certified according to ISO 9001 (ISO, 2005a).

There has been a shift from using official analytical methods to a more open approach that allows the laboratories involved in official controls to use validated analytical methods that have been proven to meet established performance criteria. Thus, different scenarios are presently possible: in very few cases, such as Commission Regulation (EEC) 2676/90, on the analysis of lead in wine, the method is defined; more frequently, as in Directive 2008/50/EC on air quality or in Directive 98/83/EC on water intended for human consumption, although methods are explicitly specified, laboratories are allowed to use alternative methods, providing they can demonstrate that the results are at least as reliable as those produced by the specified methods. Another approach is that of Decision 2002/657/EC, concerning analytical methods for the analysis of residues and contaminants in food products, which establishes the performance criteria for methods. Directives 2009/90/EC and 98/83/EC establish analogous analytical method criteria for monitoring water status, sediment and biota, as do Regulation (EC) 333/2007 (on sampling and analytical methods for the control of some contaminants in foodstuffs), to SANCO/10684/2009 (on method validation and quality control procedures for pesticide residues analysis in food and feed), or to Regulation (EC) 401/2006 (on methods of sampling and analysis for the control of mycotoxins in foodstuffs). Representative examples of performance criteria for methods used to analyze patulin in foodstuffs are shown in Table 1.

This flexible approach to method performance criteria allows laboratories to quickly incorporate advances in analytical techniques and to apply new methods to address new problems when required. The crucial issues here are that the required performance criteria are met and that the method has been properly validated.

Level (μg/kg)	RSD_r % (a)	RSD_R % (b)	Recovery %
< 20	≤ 30	≤ 40	50 to 120
20 to 50	≤ 20	≤ 30	70 to 105
< 50	≤ 15	≤ 25	75 to 105

Table 1. Performance criteria for methods of analysis of patulin in foodstuffs, from Regulation 401/2006. (a: Relative standard deviation, calculated from results generated under repeatability conditions, b: Relative standard deviation, calculated from results generated under reproducibility conditions.)

Since its publication, Decision 2002/657/EC has been a key document for analytical laboratories involved in food analysis and has proven utile for laboratories in other fields, such as environmental analysis. It introduced a change of mindset, replacing reference methods with the criteria approach, and launched new definitions, such as *minimum required performance limit (MRPL)*, *decision limit (CCα)* and *detection capability (CCβ)*. Decision 2002/657/EC determines common criteria for the interpretation of test results, establishes the performance criteria requirements for screening and confirmatory methods, and presents the directives to validate the analytical methods. However, it is a complex document, and guidelines for its implementation have been published (SANCO/2004/2726-rev-4-December-2008). The most relevant aspects of Decision 2002/657/EC are further described below.

Minimum required performance limit is defined as the minimum content of an analyte in a sample that has to be detected and confirmed. It is intended to harmonize the analytical performance of methods for banned substances. The minimum required performance level for a method of a banned substance should be lower than the MRPL; however, very few MRPL values have been established to date.

The decision limit is the limit at and above which one can conclude, with an error probability of α, that a sample is non-compliant. For substances with no permitted limit α is 1%, whereas for all other substances α is 5%. Thus, the result of an analysis shall be considered non-compliant if the CCα of the confirmatory method for the analyte is exceeded.

The detection capability is the smallest content of the substance that may be detected, identified and/or quantified in a sample with an error probability of β (β is 5%). Procedures to determine the CCα and CCβ are given in Decision 2002/657/EC and its corresponding guidelines document (SANCO/2004/2726-rev-4-December-2008).
Decision 2002/657/EC also introduces the concept of *identification point (IP)*. A minimum of three IPs is required to confirm the identity of a compound that has a permitted limit, whereas at least four IPs are required for a banned compound. The number of IPs provided by the analytical method depends on the technique used. For instance, with low-resolution MS each ion earns 1 point, and with low-resolution MS^n each precursor ion earns 1 point, and each transition product, 1.5 points. More details on IPs for the different techniques can be found in Decision 2002/657/EC. This IP system has made MS an essential technique for laboratories that analyze residues and contaminants in foodstuffs.
In addition to the performance criteria requirements for screening and confirmatory methods, Decision 2002/657/EC also provides guidelines for the validation of analytical methods. Validation should demonstrate that the method complies with its performance criteria. Therefore, depending on the method category (*e.g.* qualitative or quantitative; screening or confirmatory), different performance characteristics must be determined. Table 2 shows an overview of EU legislation on analytical methods for environmental and food samples

Directive 98/83/EC	Quality of water intended for human consumption
Directive 2008/50/EC	Ambient air quality and cleaner air for Europe
Directive 2009/90/EC	Technical specifications for chemical analysis and monitoring of water status
Regulation (EC) 333/2007	Methods of sampling and analysis of lead, cadmium, mercury, inorganic tin, 3-MCPD and benzo(a)pyrene in foodstuffs.
Decision 2002/657/EC	Performance of analytical methods and interpretation of results
Regulation (EC) 401/2006	Methods of sampling and analysis of mycotoxins in foodstuffs

Table 2. Overview of EU legislation on analytical methods for environmental and food samples

5. QC management

At the LPHAB QC activities are managed by the Quality Assurance Unit (QAU), in close cooperation with the head of the Chemical Analysis Service (CAS) and the senior technicians responsible for each analytical technique or methodology.
The QAU comprises two senior technicians and one mid-level technician. Its functions include:

- Coordinating implementation and maintenance of the Quality Management System (QMS)
- Cooperating with the LPHAB's top management in the annual system review and in preparation of the annual staff training program
- Preparing and conducting an annual internal audit
- Managing any complaints received from customers or third parties

- Defining corrective and preventive actions, supervising their implementation and verifying their efficacy
- Managing documentation (Quality Manual, general procedures and SOPs, etc.), distributing and maintaining documents, and preparing lists for flexible-scope accreditation
- Approving the auxiliary equipment program control
- Advising technicians on method validation and QC activities
- Managing the LIMS

Moreover, the LPHAB's QC activities are described in several documents of its QMS. Table 3 shows these documents in a hierarchical order.

Document	Scope
Quality Manual (Section 14)	Whole laboratory
General procedure: "Assessment of the Quality of Analytical Results"	Whole laboratory
General procedure: "Management of Complaints, Non-conforming Work, and Corrective and Preventive actions"	Whole laboratory
General procedure "Management of Flexible-Scope Accreditation"	Whole laboratory
Standard Operating Procedure (SOP): "Application of the General Quality Criteria to the Chemical Analysis Service"	CAS
SOP: "Management of Standards"	CAS
Annual QC Plan	CAS
Specific SOPs (per method)	CAS
Records	CAS

Table 3. Major QC documents from the LPHAB's QMS (CAS: Chemical Analysis Service).

One of the chapters in LPHAB's Quality Manual defines the basis of QC in accordance with the requirements of the ISO/IEC 17025 standard (ISO, 2005b). General procedures, which are applicable either to the whole laboratory, or to the Microbiology Analysis Service or the CAS, outline the LPHAB's general QC activities. Standard operating procedures provide detailed specifications on the CAS's QC activities (both internal and external QC). Internal QC activities are either performed within each analytical run or are scheduled. The within-run activities are done in accordance with the specific SOP for regular samples received by the LPHAB; these encompass analysis of reagent blanks, blank samples, spiked samples and verification of instrument sensitivity. They are employed to prevent releasing of any erroneous results to customers. Scheduled activities are used to check the efficacy of within-run controls. External QC (EQC) relies on the regular and frequent participation of the LPHAB in proficiency tests organized by competent bodies (whenever possible, accredited as Proficiency Test Providers). All these activities are agreed upon by the head of the CAS, the director of the QAU and the senior technicians and are reflected in the Annual QC Plan. Table 4 shows a sample page from the CAS's Annual QC Plan.

All of the QC activities are described in the SOP entitled "Application of General Quality Criteria to the Chemical Analysis Service", as are the procedures for handling of all scheduled internal and external QC samples (in terms of ordering, analysis and evaluation). These activities are summarized in Table 8.

Technical Unit: Instrumental Organic Chemistry

Technician: XXXXXX

SOP	External Quality Control			Internal Quality Control
MA/2/19510	FAPAS	**Periodicity:**		
Determination of Chlroramphenicol	Progetto Trieste	**Blank**	v	Every run
by LC-MS/MS	CNA	**Reference Material**	v	At least once a year
		Spiked samples	v	At CCα level- every run
		Others	v	Matrix matched surrogates/ Internal standard in samples checking
MA/2/19560	FAPAS	**Periodicity:**		
Determination of Nitrofuran metabolites	Progetto Trieste	**Blank**	v	Every run/ Internal standard checking
by LC-MS/MS	Anfaco	**Reference Material**	v	At least once a year
		Spiked samples	v	At CCα level-every run
		Others	v	Calibration curve from processed samples, including MRPL (1 μg/kg). Internal Standard in samples checking
MA/2/19560	FAPAS	**Periodicity:**		
Determination of corticosteroids	Progetto Trieste	**Blank**	v	Every run/ Internal standard checking
by LC-MS/MS		**Reference Material**	v	At least once a year
		Spiked samples	v	At CCα level- every run
		Others	v	Calibration curve from processed samples, including MRL (2 μg/kg). Internal Standard in samples checking

Table 4. Sample page from the CAS's Annual QC Plan.

5.1 Management of external QC

External QC is managed through proficiency tests. Participation in each test is scheduled according to proposals by the technician responsible for each analytical procedure. Each procedure is to be tested in at least one exercise per year, if possible. In parallel, certain samples are requested in duplicate for use in scheduled internal QC.

The LPHAB tends to be extremely active in this area, since it considers external QC among the strongest point of its QC system. In the CAS, in 2010, 458 samples were analyzed in proficiency tests that encompassed 1,915 assays, 420 different analytes and 89 analytical procedures (SOPs).

Given that the market lacks universal exercises for all types of matrices and assays, the CAS aims to assess all families of analytes and all instruments. Usually, matrices included in the accreditation scope are used. Importantly, for assays included in the flexible-scope accreditation, different matrices that represent the entire assay category should be employed whenever possible. To evaluate some of the procedures for which no exercises are currently available, the CAS, together with other laboratories, has organized specific activities.

It is extremely important that any organization that aims to organize these types of evaluations be accredited according to ISO/IEC 17043 (ISO, 2010). For non-accredited entities, the quality of their exercises will be assessed.

In accordance with the aforementioned principles, CAS actively participates in the programs FAPAS® (for food) and LEAP (for water), both of which are accredited by the United Kingdom Accreditation Service (UKAS).

For each exercise, a technician is assigned to handle and follow the sample, which must be analyzed using the typical procedures and which must not be treated differently because of its interlaboratory status. Once the organizer's report has been received, an internal evaluation report is written up, which includes the results found by CAS, the mean result assigned by the organizer, and the calculated z-score for each analyte.

Upon receiving the report, each manager performs a complementary evaluation of the results obtained, considering all of the documentation referring to the analysis performed, in order to confirm that all of the QC criteria have been met. Another very important and highly utile aspect to consider is the information on the methods applied by different laboratories, which can help the CAS to improve its methods.

If the evaluation is unsatisfactory, then a report on corrective actions is written up. The results of proficiency tests are generally evaluated based on the z-scores. Nonetheless, other criteria (*e.g.* compatibility index) may also be used; these are described in the final evaluation report for the exercise.

One of the critical points for evaluating z-scores is the standard deviation used in the calculations. The standard deviation used is generally that which is documented by the organizer, which tends to the value obtained from the Horwitz equation. Nevertheless, another value can be used, as deemed necessary by the technician responsible for the evaluation, as long as it is justified in the internal evaluation report for the exercise. Fig. 5 shows a sample evaluation form for external QC samples.

The results obtained are introduced into a database, which enables tracking of any possible trends as well as confirmation of validation data over time.

The figure below illustrates moisture analysis results for various types of samples from the FAPAS® exercises in which LPHAB has participated over the past few years.

PROFICIENCY TEST REPORT

Sample No. : 11_00237	**Date of entry : 11/01/2011**
Proficiency Test : LEAP - Series: CHEM 72	**Round: C1316**
Sample : Potable water	**Laboratory id No.:**
Area : 2/0103 Chromatography	**Responsible: Dr. Laura Pineda**

ANALYTES:	**RESULTS** LIMS	Sent result	Consensus value	z-score	**SOP / Equipment used**	
Toluene	**3,9**	**3,87**	**3,2**	**1,7**	**MA/2/30470**	**2-170**
Ethylbenzene	**0,9**	**0,89**	**0,75**	**1,5**	**MA/2/30470**	**2-170**
Benzene	**1,7**	**1,65**	**1,37**	**1,6**	**MA/2/30470**	**2-170**
Xylene (mixture of isomers)	**6,1**	**6,13**	**5**	**1,8**	**MA/2/30470**	**2-170**

ASSESSMENT OF RESULTS:

Successful results: ☒

Questionable results: ______ Punctual ☐ Repetitive ☐ Preventive action ☐ PA number ☐

Unsuccessful results: ______ Non-conforming work No. ☐ Date. ☐

OTHER METHODS USED IN THE PROFICIENCY TEST:

After reviewing the methods used by the other participants, an improving opportunity has been detected:

Yes ☐ No ☒ Preventive action No ☐

GENERAL COMMENTS ON RESULTS:

Date:
Signature:

Final assessment of the head of the Chemical Analysis Section	Revision QAU	Date
Date: Signature:		

Fig. 5. Sample evaluation form for external QC samples (completed by CAS based on the organizer's report).

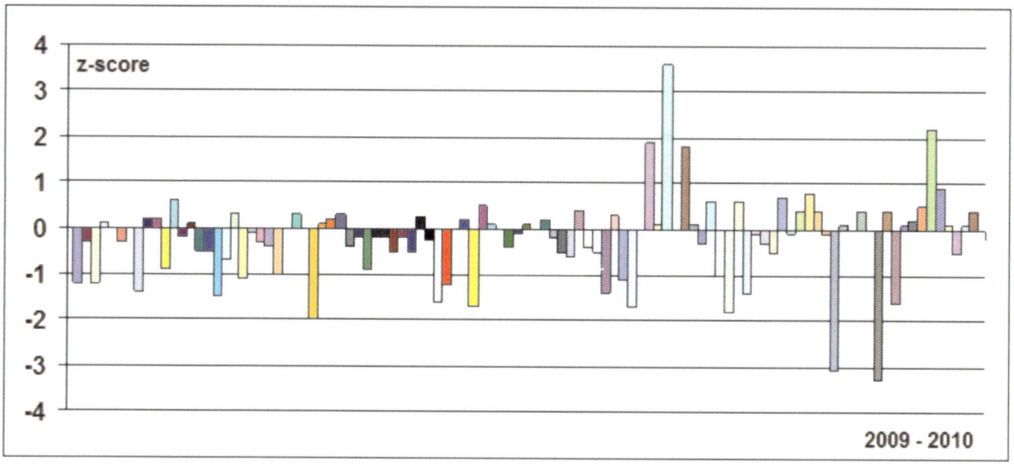

Fig. 6. Proficiency testing: results from moisture analysis of different samples performed for FAPAS® exercises.

Another important factor concerning the results obtained from proficiency tests is their utility for systematic expansion of validation data. The CAS has established a dynamic validation system in which the overall validity of the uncertainty of a procedure is checked against the different sample types and different concentration levels analyzed.

5.2 Management of internal QC

The scheduled internal QC samples generally correspond to duplicates of samples from proficiency tests (the duplicates are purchased annually at the time the test is performed).
In 2010 the CAS analyzed 99 samples for internal QC, encompassing 371 assays, 209 analytes and 77 SOPs.
Once the samples arrive at the LPHAB, their information is entered into the reference materials database, and they are carefully handled, taking their particular storage needs and expiration dates into account. The new sample is added to the sample registry in the LIMS according to the schedule. The results are analyzed using an internal evaluation form in which the z-score (accuracy) is re-calculated, and the reproducibility is calculated based on the results from the external QC and from the internal QC test. This approach enables evaluation of both accuracy and precision. Fig. 7 shows a sample evaluation form for internal QC.

5.3 Handling of any inconsistencies detected in the QC activities

Results obtained in both internal and external QC activities are suitably recorded. Out of control situations can be categorized as incidences and deviations. Incidences are sporadic events that usually do not occur in subsequent applications of the analytical method. Contrariwise, deviations are non-conforming work that must be managed through corrective actions. Detection of these events, and subsequent causal analysis, sometimes leads to proposal of preventive actions. Fig. 8 shows a general schematic of QC management.

INTERNAL QUALITY CONTROL REPORT

Sample No. : 11_00065	**Date of entry : 04/01/2011**
Sample reference : 2-R0525	**Original sample: 10_28944**
Sample : Meat	
Area : 2/0401 Food	**Responsible: José Antonio León**

ANALYTES:	**RESULTS**								**SOP**
	Result	**Certif. value**	**± U**	**Recovery**	**z-score**	**Reprod.**		**Correct**	
Nitrate (E-251, E-252)	205	226	55	90.7	-0.76	12	77	☒	**MA/2/30102**
Nitrite (E-249, E-250)	121	150	36,6	80,6	-1,58	24	51	☒	**MA/2/30101**

ASSESSMENT OF RESULTS:

PARAMETER	**ACCEPTANCE CRITERIA**	**COMMENTS**
Uncertainty		
Recovery		
Z-scores		
Reproducibility		
Cause analysis:		
Non-conforming work No.:		Date:

GENERAL COMMENTS ON RESULTS:

Date:
Signature:

Final assessment of the head of the Chemical Analysis Section	Revision QAU	Date
Date: Signature:		

Fig. 7. Sample evaluation form for internal QC samples.

Figures 9 and 10 show the number of scheduled internal QC and external QC samples in absolute values and as percentages of the total number of samples analyzed, respectively, in the CAS. These figures are testament to the LPHAB's major efforts to ensure the reliability of its results and demonstrate its commitment to quality. Moreover, this approach also implies sizeable financial investment: participation in proficiency testing costs the LPHAB roughly €60,000 per year.

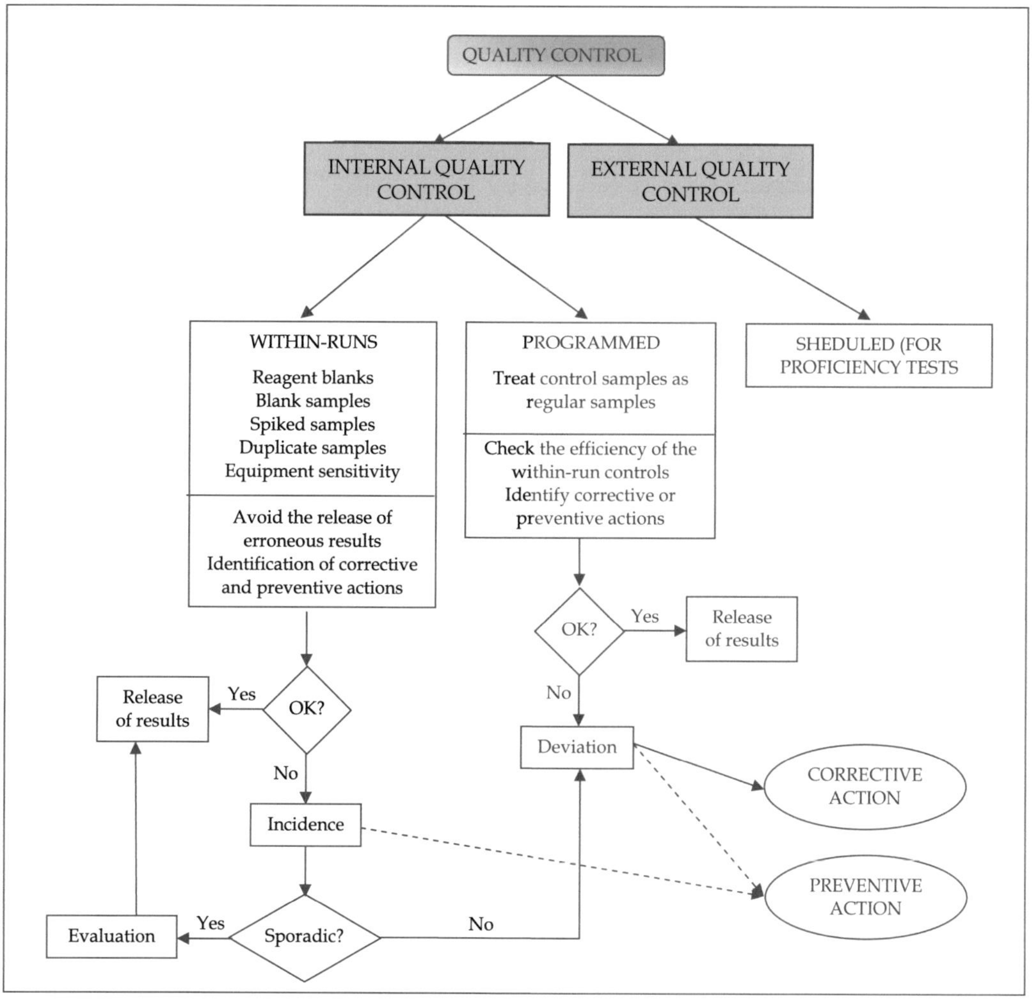

Fig. 8. General schematic of QC management.

The reliability of QC activities is greatly based on the suitability of the criteria applied. Depending on whether the limits established are too strict or too lax, α or β errors, respectively, may be committed. Over the past few recent years, the CAS has adapted the criteria applied in its internal QC to the values obtained during method validation. Improving the frequency and quality of internal QC has enabled improved detection of non-conforming results, and therefore, has enabled optimization of external QC activities.

5.4 QC in the framework of flexible scope accreditation

Accreditation of a laboratory is usually based on a concrete definition of the laboratory's scope. Thus, the technical annexes for accreditation certificates comprise detailed lists of the tests for which the laboratory has been accredited. The lists clearly specify matrices, analytes, ranges of concentration, and methods. This scheme is known as *fixed-scope accreditation*.

However, in recent years, in order to meet the needs of customers, laboratories have had to quickly expand their accreditation scope without compromising their technical competence or altering definition of the scope. Thus, highly experienced laboratories with a long history of accreditation can now adopt a new scheme, known as *flexible-scope accreditation*, whereby they perform analyses using appropriate validated methods, and then report the results as being accredited, without prior evaluation by the accreditation body. This may entail incorporation of new matrices or analytes, or inclusion of new tests within a generic method. Thus, the flexibility of the accreditation scope implies sufficient technical competence and operational capacity, which places more of the responsibility on the laboratory. This in turn means that the laboratory must endeavor to increase its QC operations in order to guarantee the quality of the results of the expanded scope. In any case, the bounds within which a scope is flexible must be precisely stated.

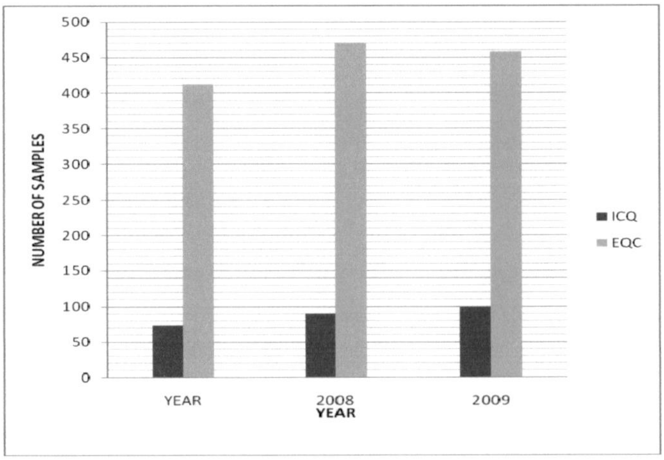

Fig. 9. Scheduled internal and external QC samples (ICQ and ECQ, respectively), expressed as number of samples.

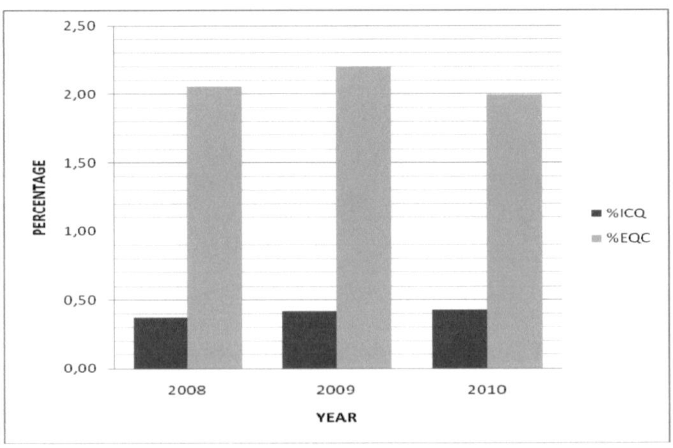

Fig. 10. Scheduled internal and external QC samples (ICQ and ECQ, respectively), expressed as percentage of total samples.

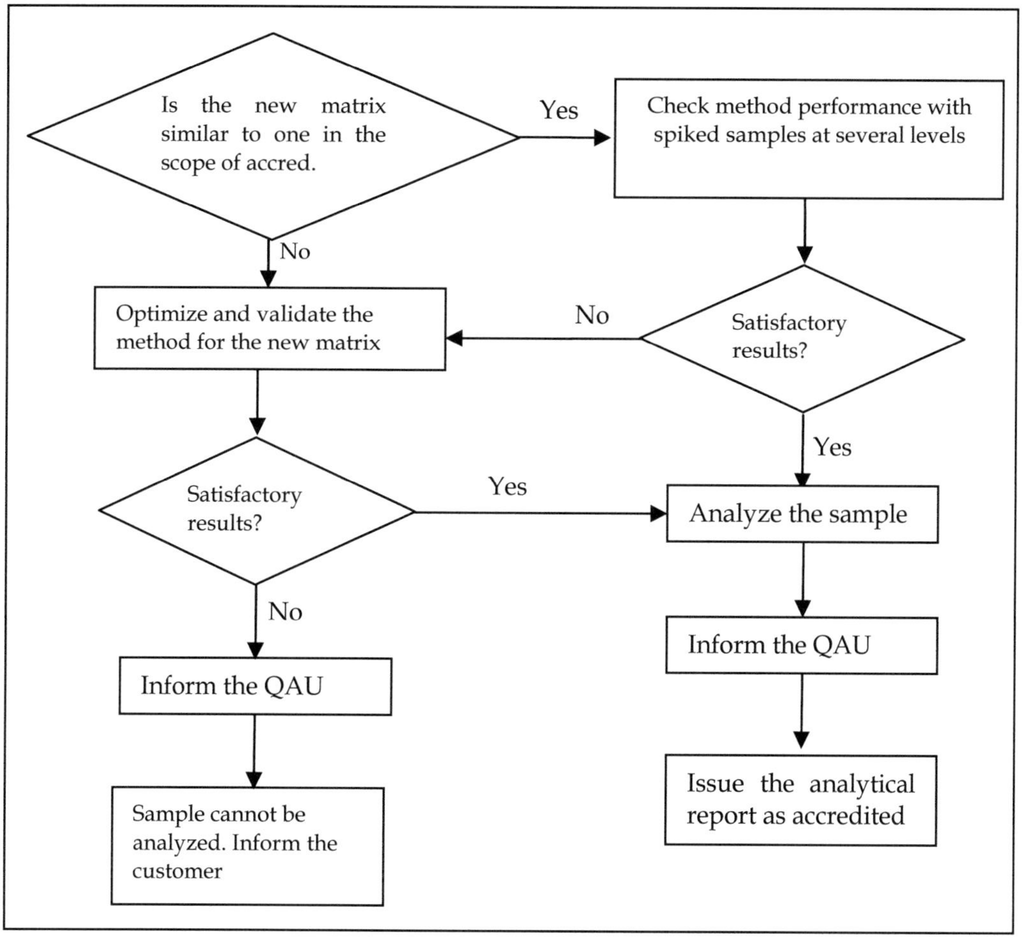

Fig. 11. Procedure for analysis of an established analyte in a new matrix.

In this context, once a laboratory receives a request for an analysis that falls within the bounds of a flexible scope, it must do the following:

- Inform the customer that the analysis will be performed in the framework of a flexible scope, and therefore, prior validation studies will be required; this will involve some delay in the delivery of results; and, if the results of the validation studies are unsatisfactory, then the report cannot be issued as being accredited.
- Perform validation studies. A scheme of this process for analysis of an established analyte in a new material is illustrated in Fig. 11. An analogous process would be employed for the opposite case (*i.e.* analysis of a new analyte in an established matrix).

Flexible-scope accreditation was initiated in 2004 for pesticide analysis and was later extended to other analyte families. The LPHAB defines these families according to the type of analyte studied and the analytical technique used. Therefore, these vary from very broad (organic compounds studied by chromatographic techniques) to rather narrow (ions studied by liquid chromatography). The CAS's current fixed-scope and flexible-scope of accreditation are summarized in Table 5.

PARAMETER MATRIX	ANALYTE OR FAMILY OF ANALYTES (NUMBER OF ANALYTES)	FIXED SCOPE	FLEXIBLE SCOPE
Additives by LC Food	Phenolic antioxidants (7), artificial dyes (14), Sudan dyes (10), preservatives (7), sweeteners (3) caffeine		X
Organic pollutants by LC and GC Food / Water Air sampling supports	Acrylamide, BFRs (8), BTEX (5), 3-MCPD, 3-MCPD ester, VOCs (15), ethyl carbamate, drugs (3), furan, PAHs (17), BADGEs (11), melamine, PBCs (7), PCNs (5) pesticides (44) , nicotine,		X
Ions by LC or segmented continuous flow analysis Food / Water	BrO_3^-, BrO^-, Cl^-, F^-, P, NO_3^-, NO_2^-, SO_4^{-2}, NH_4^+, Ca^{+2}, Mg^{+2}, K^+, Na^+,ClO^{3-}, ClO_2^-, SO_2.	X	X
Heavy metals by atomic spectroscopy Food /Water Air sampling supports	As, $As_{inorganic}$, Ba, B, Cd, Ca, Co, Cu, Cr, Sn, Fe, Mg, Mn, Hg, CH_3Hg, Ni, Ag, Pb, K, Se, Na, Ti, V, Zn		X
Mycotoxins by LC Food	aflatoxin M1, aflatoxins B i G (4), fusarium mycotoxins (6), ochratoxin A, patulin		X
Residues of veterinary drugs by LC Food of animal origin	Antibiotics: β-lactams, macrolides, quinolones, sulphonamides, tetracyclines (41), coccidiostats (10), antihelmintics (1), tireostatics (6), benzodiazepines (3), chloramphenicol, dyes (5), corticosteroids (6), synthetic hormones (9), metabolites of nitrofurans (4), nitroimidazole, β-agonists (18)		X
Residues of pesticides by GC and LC Food	Pesticides (200)		X
Physicochemical parameters Food	Mass, volume, density, acidity, pH, fat (3), alcohol degree, peroxide index, moisture, ash	X	
ELISA determinations Food	Gluten	X	
Ions by flow techniques Food / Water	P_{total}, $P_{soluble}$, N_{total}, NO_3^-, NO_2^-, Cl^-, NH_4^+, SO_2, volatile nitrogenous bases, trimethylamine	X	
Several compounds by chromatographic techniques Food	Biogenic amines (6),sugars (6), methanol, ethanol, other alcohols (5), ethyl acetate, fatty acids, steroids and estradiols	X	
Physicochemical parameters Water	Colour, turbidity, conductivity, suspended solids, dissolved matter, pH, chemical oxygen demand, alkalinity, cyanide, oxidisability	X	

Table 5. The CAS's current fixed-scope and flexible-scope of accreditation.

Managing the flexible scope implies a significant amount of extra documentation that must be completely updated. Indeed, in 2010 alone six new analytical methods were added, together with numerous matrices and analytes. The flexible scope is currently in its 22nd edition (an average of three editions are created per year).

6. QC in the analytical method SOPs: examples of general and specific QC activities

This section provides examples of the some of the QC activities summarized in Table 8, as well as the corresponding documentation for recording and evaluating the data. Generally, each analytical procedure (SOP) features a section describing internal QC activities that are performed within each run and the corresponding criteria for accepting the results, which must be evaluated by the technician responsible for the procedure before they are communicated to the customer. Several concrete examples are presented below.

6.1 Spiked samples

An example of control analysis of spiked samples is illustrated in Fig. 12, which shows a plot of arsenic analysis in food samples by inductively coupled plasma mass spectrometry (ICP-MS). The results are evaluated based on the recovery (% Rec) of samples spiked at different concentrations and with different matrices, such that the entire scope of the flexible-scope accreditation can be addressed.

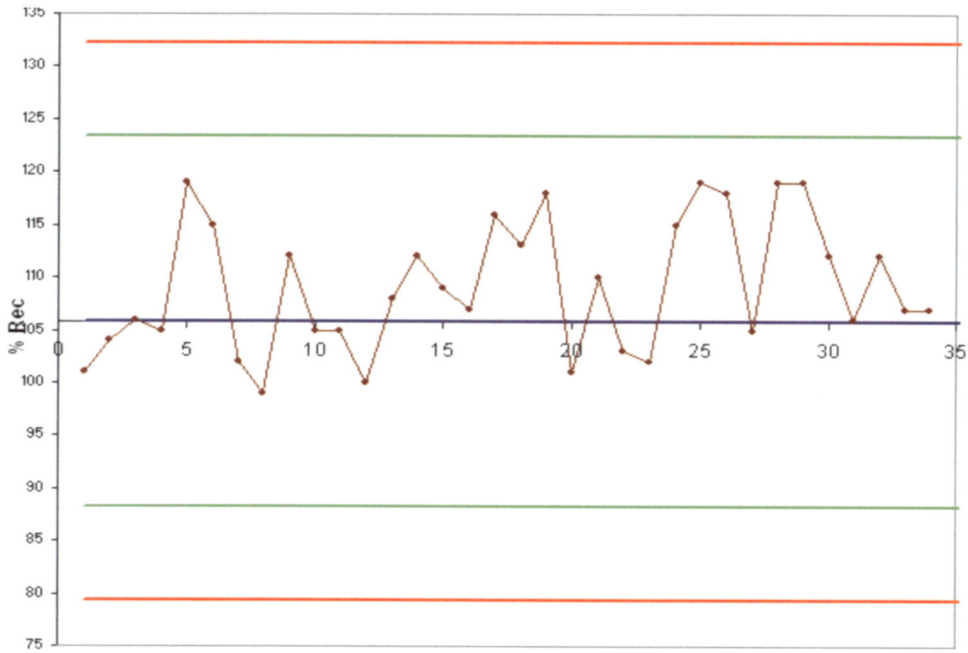

Fig. 12. Plot of arsenic recovery levels from spiked samples of different food types, as determined by ICP-MS.

6.2 Use of QC records for an LC-MS/MS procedure (detection of antibiotics)

Table 6 shows an example of a QC records for a procedure in which 44 antibiotics are analyzed in samples of products of animal origin by LC-MS/MS. The following data are recorded for representative analytes (in the case of Table 6, two antibiotics): the area of the peak corresponding to the standard used for verifying the instrument; retention time (TR) and the ratio of transitions (ion ratio [IR]) at CCα level, which are the data used for identifying and confirming the two compounds. The peak area value is checked against the minimum peak area that guarantees response at the lowest level of validation, which also verifies the confirmation. Lastly, the analytical sequence and the user's initials are also recorded.

SOP	MA/2/19210	Analysis of antibiotics by CL / MS-MS							
Instrument	2-201	CL MS-MS							
Compound	SULFAMETHAZINE				ENROFLOXACIN				Antibiotics
Average	4,50				4,30				
Maximum value	4,75				4,55				
Mminimum value	4,25	100000	0,8		4,05	1000	1,0		
DATE	TR	Area	I.R. (CC α)	OK	TR	Area	I.R.(CC α)	OK	Run/User
15/02/2011	4,35	135483	0,9		4,45	1022	1,1		110215ant/AR
22/02/2011	4,60	175563	0,8		4,20	1156	1,0		110222ant/AR
01/03/2011	4,55	115632	0,8		4,15	1325	1,2		110321ant/AR
08/03/2011	4,34	134856	0,8		4,35	1278	1,2		110308ant/AR
15/03/2011	4,57	142569	0,9		4,10	1486	1,2		110315ant/AR

Table 6. QC records from analysis of antibiotics in products of animal origin by LC-MS/MS.

In similar QC records, the responses of the internal standards (which are typically deuterated or C^{13}-labeled analogs of the test compounds) from analysis of various types of samples are recorded. This control step can also be used to broaden the validation data by incorporating new matrices (*i.e. online validation*). Based on the values of the responses of the internal standards, one can deduce the validity of the matrix-matched surrogate quantifications in the different sample types that can be incorporated into the analytical sequence.

6.3 QC records for verification of the instrument, its calibration levels, and the blank in the turbidity analysis procedure

The format of the QC records used for turbidity analysis of water samples is illustrated in Table 7 as a representative example of a physicochemical assay.

The upper and lower limits traceable to the values obtained in the validation are shown. In this case, the experimental readings obtained are recorded for each certified standard and are used to verify calibration of the instrument and to confirm the response of the blank (in this case, ASTM type I purified water).

SOP	MA/2/30504		Determination of turbidity in water				
Instrument	2-001		Turbidimeter				
STABCAL	0,5 NTU	3 NTU	6 NTU	18 NTU	30 NTU	Blank	Turbidity
Upper val.	0,48	3,11	6,43	19,1	32,5	0,094	
Lower val	0,32	2,81	5,81	17,3	29,4		
Date	Result (NTU)						User
10/03/2011	0,38	2,85	5,93	17,7	30,1	0,079	NG
11/03/2011	0,37	2,92	5,98	17,8	30,3	0,07	NG

Table 7. QC records from turbidity analysis of water samples.

QC factor	Action	Objective	Calculations and tolerance limits	Frequency at the LASPB
General internal QC				
Reagent blank	The analytical procedure is performed using only the reagents.	Enables monitoring for any contamination in materials, reagents, the environment, etc.	LOD: limit of detection Evaluation: Blank < LOD	Within-run (the type of blank depends on the method) See Table 7
Matrix blank	The analytical procedure is performed using a blank sample.	Enables monitoring for any contamination, and confirmation that the matrix is not responsible for any interference		
Duplicate samples (intermediate precision)	Full analysis of duplicate samples on different dates	Enables monitoring of the reproducibility (R) relative to the standard deviation of the validation (s_R)	$R = 2 * \sqrt{2} * s_R$ Evaluation: $x_1 - x_2 \leq R$ x_1 and x_2 are the results of duplicate samples	Scheduled: usually, once per year, using a reference material (normally, a duplicate sample from a proficiency test)
Spiked samples	The analytical procedure is performed on a sample that has been spiked with the analyte (whenever possible, previously analyzed samples containing the analyte at levels lower than the limit of detection).	Enables monitoring of the bias or the trueness based on the recovery (% Rec), and compared with the recovery (%Rec_{val}) and the standard deviation (s) obtained in the validation	$\mathrm{Re}\,c(\%) = \frac{(x_{lab})}{x_{spiked}} \times 100$ X_{lab}: obtained value X_{spiked}: spiked value Evaluation: $\mathrm{Re}\,c(\%) = \mathrm{Re}\,c_{val}(\%) \pm 2s$	Within-run See Fig. 12
Reference materials	The analytical procedure is performed on a sample which has been prepared under concrete specifications and which contains the analyte in question at a known value.	Enables monitoring of the accuracy of the results based on the compatibility index (CI), which is calculated from the reference value (x_{ref}) and the obtained value (x_{lab})	$CI = \frac{\lvert X_{ref} - X_{lab} \rvert}{\sqrt{(U_{ref}^2 + U_{lab}^2)}}$ U_{ref}: expanded uncertainty of reference material U_{lab}: expanded uncertainty laboratory Evaluation: $CI \leq 1$	Scheduled

Table 8. Part I

QC factor	Action	Objective	Calculations and tolerance limits	Frequency at the LASPB
		External QC		
Proficiency test	The analytical procedure is performed on a sample which has been part of an interlaboratory comparison scheme.	Enables monitoring of the accuracy of the results based on the z-score (z), which is calculated from the assigned value, the obtained value (x_{lab}) and the standard deviation of the participants (σ_p).	$z = \frac{\lvert x_a - x_{lab} \rvert}{\sigma_p}$ Evaluation: $z \leq 2$ satisfactory result $2 < z \leq 3$ questionable result $z > 3$ unsatisfactory result	Scheduled See Fig. 5
		Internal QC to verify equipment or reagents		
Verification of instrument at the beginning of the run and monitoring of instrument drift	A standard is injected under the instrumental conditions established in the analytical procedure.	Enables verification of proper instrument performance before the sequence is started, and at every *n* samples, based on confirmation that the response of the standard (A) falls within a pre-established range of acceptable values (x %) that guarantee the limit of quantification (LOQ)	A : response of the standard Evaluation: $A \leq \pm x\%$	Within-run See Table 7
Calibration of the instruments associated with the analytical method	The standards used to generate the calibration curve are injected.	Enables monitoring of the **quality of the fit of the calibration curve**, based on at least two different criteria: for example, the coefficient of correlation (r) and the residual error of the standard (Er %), which is the ratio of the value of the concentration of the standard in the curve (V_{curve}) to the nominal concentration value ($V_{nominal}$)	r ≥ (see specific SOP) $E_r(\%) = V_{curve}/V_{nominal} * 100$ Evaluation: Er %≤ r ≥ (see specific SOP)	Upon generation of a new calibration curve
Verification of a new lot of standards	Two different samples of the same standard are injected: one from a regularly used lot, and one from a newly prepared lot.	Enables confirmation that a standard has been correctly prepared, based on verification that the ratio of the response of the new sample (A) to the response of the sample from a previously used lot (B) falls within a pre-established range of acceptable values (x %)	A/B : response ratio Evaluation: $A/B \leq \pm x\%$	Upon preparation of new lots of standards

Table. 8. Part II

<table>
<tr><th>QC factor</th><th>Action</th><th>Objective</th><th>Calculations and tolerance limits</th><th>Frequency at the LASPB</th></tr>
<tr><td colspan="5">Specific internal QC procedures: chromatographic methods</td></tr>
<tr><td>Verification of the response of the internal standard</td><td>Addition of the internal standard to all samples, spiked samples, and other standards (matrix-matched surrogate) at the beginning of the procedure</td><td>Verification of the procedure for each sample: extraction, and performance of different matrices</td><td>Signal traceable to previous analyses Quantification based on internal standard</td><td>Within-run</td></tr>
<tr><td>Identification of the chromatographic peak</td><td>Retention time of each compound relative to that of the internal standard</td><td>Verification of the criteria described in the chromatographic method</td><td>According to chromatographic system; TR ± % tolerance limit</td><td>Performed for each chromatographic peak identified that corresponds to a standard See Table 6</td></tr>
<tr><td rowspan="3">Confirmation of the identified compounds</td><td>DAD, FLD, etc.: The compound spectra are compared to the internal standard spectra</td><td rowspan="3">Verification of the criteria described in the chromatographic method</td><td>Spectral match</td><td rowspan="3">Performed for each chromatographic peak identified See Table 6</td></tr>
<tr><td>MS (SIM): Mass spectra ion ratios</td><td>According to analysis; generally ± 20%</td></tr>
<tr><td>MS/MS: Transition ratios</td><td>According to regulations, analysis type, concentration, intensity of the transitions, etc.</td></tr>
</table>

Part III

Table 8. QC activities at the LPHAB's Chemical Analysis Service (CAS).

7. References

Shewhart (1931). *Economic Control of Quality of Manufactured Products,* Van Nostrand, New York, USA. ISBN 0-87389-076-0.

ISO (2005a). *ISO 9000:2005 standard. Quality management systems. Fundamentals and vocabulary,* ISO, Geneva, Switzerland.

ISO (2005b). *ISO 17025:2005 standard. General requirements for the competence of testing and calibration laboratories.* ISO, Geneva, Switzerland.

ISO (2010). *ISO/IEC 17043:2010 standard. Conformity assessment. General requirements for proficiency testing.* ISO, Geneva, Switzerland.

IUPAC (1998). *Compendium of analytical nomenclature: Definitive rules 1997*, 3rd ed. Blackwell Science, ISBN 978-84-7283-870-3, Oxford, United Kingdom.

Thompson, M. & Lowthian, P. J. (1993). Effectiveness of Analytical Quality Control is Related to the Subsequent Performance of Laboratories in Proficiency Tests. *Analyst*, Vol.118, No.12, (December 1993), pp. 1495-1500.

24

Laser Diffuse Lighting in a Visual Inspection System for Defect Detection in Wood Laminates

David Martin, Maria C. Garcia-Alegre and Domingo Guinea
Spanish Council for Scientific Research (CSIC), Madrid, Spain

1. Introduction

Nowadays, wood companies are ever more interested in automatic vision systems (Li & Wu, 2009), (Åstrand & Åström, 1994), for an effective surface inspection that greatly increases the quality of the end product (Smith, 2001), (Armingol et al., 2006). The inspection process, in most visual inspection systems, pursues online defects identification, to reach optimum performance (Malamas et al., 2003), (Spínola et al., 2008).

The usual wood inspection systems are visual ones, based on standard cameras and lighting (Batchelor & Whelan, 1997), (Cognex, 2011), (Parsytec, 2011), (Pham & Alcock, 2003) to operate in highly structured environments (Silvén et al., 2003). The quality control in visual surface inspection systems must be robust to cope with wood variable reflectance and high speed requirements.

The surface inspection methods proposed in the literature for visual inspection aim at ad-hoc surface inspection systems to solve each specific problem (Pham & Alcock, 1999). Usual inspection systems are based on visible lighting and few of them use diffuse components to illuminate the rough and bright surfaces. A visual wood defect detection system proposed by (Estévez et al., 2003) is composed by a colour video camera, where the standard lighting components are a mixture of two frontal halogen and ceiling fluorescent lamps. The commercial light diffusers use a light source and different components to illuminate the surface in an irregular way to eliminate shadows, but present some problems such as, short useful life, extreme sensitivity and high cost.

On the other hand, one of the major drawbacks in automated inspection systems for wood defect classification is the erroneous segmentation of defects on light wood regions, (Ruz et al., 2009). Moreover, the speed of wooden boards at the manufacturing industry is at about 1 m/s, which implies high computational costs (Hall & Aström, 1995).

Current work presents a surface inspection system that uses laser diffuse lighting to cope with different type of defects and wood laminated surfaces to improve defect detection without any previous defect information.

The work will not only highlight the specific requirements for a laser diffuse lighting in a visual inspection system but also those of unsupervised defect detection techniques to cope with the variability of wood laminated surfaces and defect types, leading to a heterogeneous and robust visual surface inspection system. The manuscript is organized as follows: section 2 displays images of different wood laminated surfaces, captured by a visual surface inspection system with standard lighting. In Section 3, an innovative surface inspection system using laser diffuse lighting is presented, as well as the acquired images. Section 4 describes an unsupervised wood defect detection and segmentation algorithm to process images acquired with the two types of lighting. Section 5 displays the results, and the conclusions are presented in Section 6.

2. Standard lighting visual systems

Most inspection systems are based on visible lighting (Guinea et al., 2000), (Montúfar-Chaveznava et al., 2001). These systems can adequately tackle defect detection, but presents some drawbacks:

1. Inspection systems composed by multiple cameras and equal standard lighting for all cameras, make difficult the inspection of the whole wood boards in the production line due to non-uniform illumination of the surface.
2. When fluorescent light is replaced by commercial light diffusers to improve lighting, the maintenance cost greatly increases.

Images acquired with a standard visual surface inspection system, composed of a CCD visual camera and fluorescent light, are displayed in Figure 1. The images show knots and splits and are manually classified by a human expert to validate the performance of the automated visual inspection system. Images show different background colour as they come from different wood laminates.

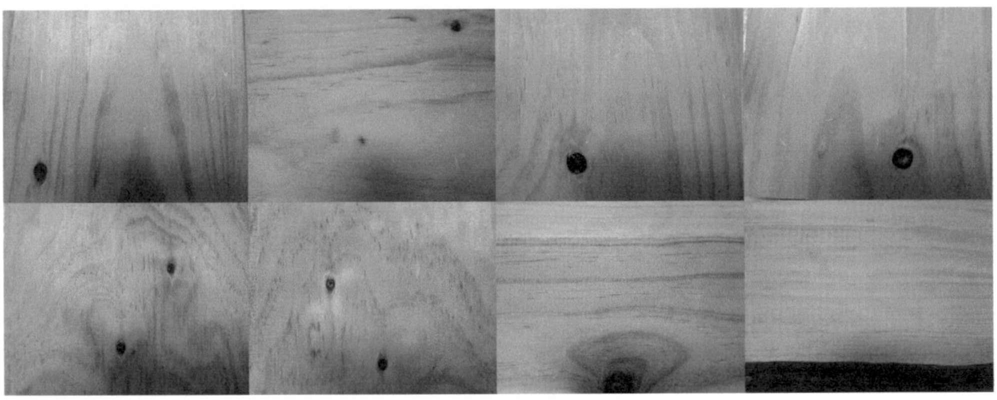

Fig. 1. Defects, knots and splits, on images acquired from a standard lighting visual surface inspection system.

3. Laser diode diffuse lighting in visual systems

Main innovation of current work is a CCD sensor-based industrial inspection system for defect detection on wood laminates, where lighting is based on a laser technique that comprises two different illumination modes: Diffuse Coaxial Lighting and Diffuse Bright-Field Lighting.

Few visual inspection systems use laser-lighting (Palviainen & Silvennoinen, 2001), (Yang et al., 2006), and even less laser diode diffuse components, to illuminate both rough and bright surfaces. The commercial light diffusers are composed of a light source and different optical components to illuminate correctly the surface in an irregular way. However, they present drawbacks such as, short useful life, extreme sensitivity and high cost, that can be overcome with the use of laser diode diffuse lighting to highlight the relevant features.

Main advantages of the proposed lighting system are:
1. Inspection of different wood surfaces without any reconfiguration of the system
2. Detection of different types of wood defects with the same diffuse lighting
3. Inspection of defects with different areas, ranging from 10 to 100 mm^2, and shapes.
4. Each CCD camera has its own laser diode diffuse lighting

The inspection vision system can be configured to work in two lighting modes (Martin et al., 2010):

- Diffuse Coaxial Lighting
- Diffuse Bright-Field Lighting

The proposed design permits to reuse the laser-optical system (laser-lighting and optical components) in both configurations to tackle high-speed and small-defect industrial surface inspections over the whole inspection region. The laser diffuse lighting provides a high intensity beam but only on a small area, removing the image shadows generated by the surface roughness, and facing the surface variable reflectance.

The wood samples are illuminated using a green diode-pumped solid-state (DPSS) laser, which provides highly uniform illumination in the region of interest (ROI). Coherent laser light is unusual for surface illumination in inspection vision systems due to the emergence of interference patterns. The two images displayed in Figure 2, are captured using only a laser-optical system without any wood surface sample, to exhibit the interference patterns caused by the optical components with coherent laser lighting.

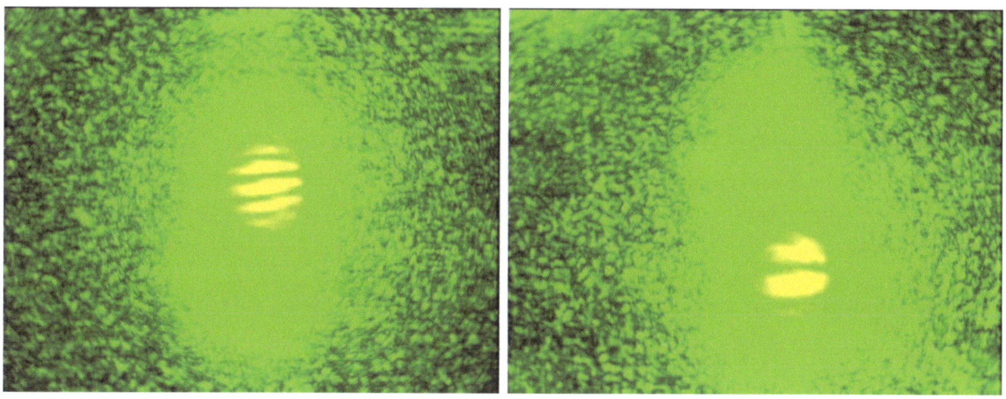

Fig. 2. Two interference patterns (yellow) caused by coherent green laser light on the surface inspection lenses.

Another interference effect appears when the laser beam is reflected on rough wood surfaces. The result of this effect in images acquired by a standard CCD camera is a bright and dark spot pattern, namely "speckle pattern", which is specific for each type of surface. The pattern is distributed randomly in the whole space and is caused by constructive and destructive interference effects.

In spite of the interference patterns, the laser lighting source is robust for integration in an industrial environment due to its low-cost and durability in comparison with commercial light diffusers. The solution to the interference problem has been achieved by means of a dispersion technique to obtain laser diffuse lighting and remove the speckle interference pattern. The components proposed for laser beam scattering are:

1. A convergent lens that increases the width of the collimated beam
2. A spinning diffuser that disperses the collimated beam

The wood defect inspection and 1 m/s wood board speed, require high intensity in the inspection region and short exposure time for suitable image acquisition. Former requirements are interrelated, as if intensity increases the exposure time, for real-time inspection, can be reduced. This high intensity allows a correct inspection as the laser diode diffuse lighting illuminates only a small inspection area, about 10-100 mm^2.

In current work, a second set of images have been acquired with diffuse bright-field lighting. The configuration is composed by the laser-optical components (laser, dispersion lens and spinning diffuser) and the imaging components (CCD sensor camera and focusing lens). The orientation of the imaging components related to the laser diffuse lighting is approximately 30º, Figure 3. A black box covers the imaging components so that the light reaching the CCD sensor only comes from the focusing lens.

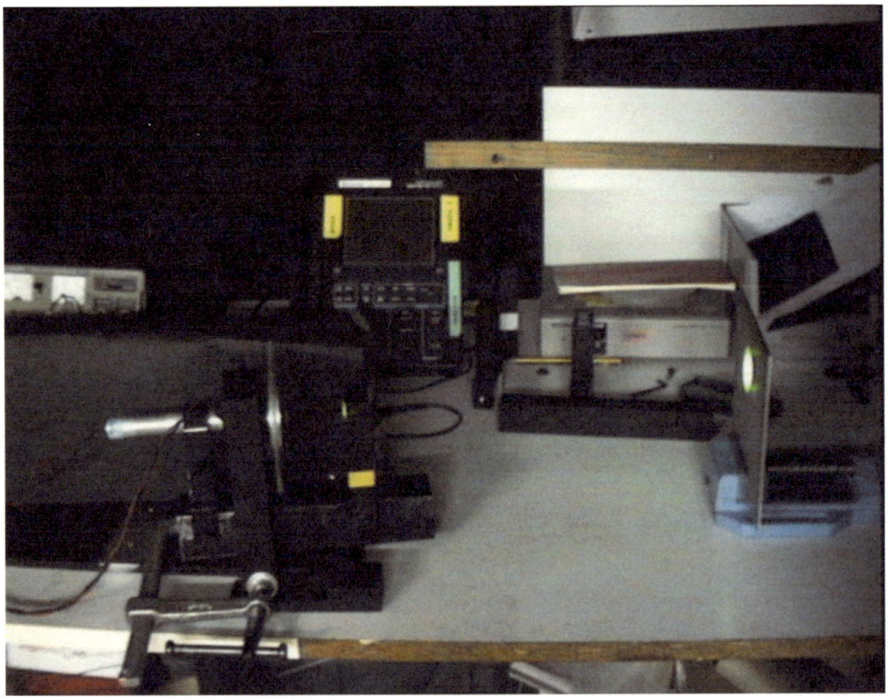

Fig. 3. Laboratory benchmark: diffuse bright-field lighting with a 160 mm focal length lens and a green DPSS laser.

The second set of images obtained with this configuration is shown in Figure 4.

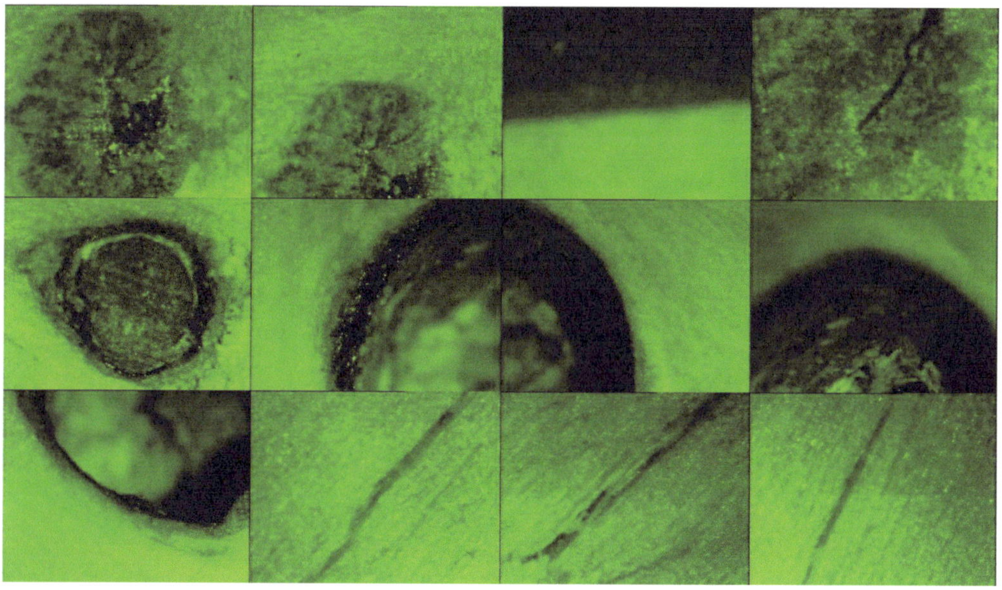

Fig. 4. Wood images acquired with a laser diffuse lighting.

The main characteristics of the proposed visual inspection system endowed of laser diffuse lighting are:

i. High intensity in the inspection region and consequently short exposure time
ii. High-speed and small-defect surface inspection on the whole inspection region
iii. Elimination of the image shadows generated by the surface roughness
iv. Deals with the variability of laminated wood such as, colour or texture
v. Algorithms required for defect detection are simpler and thus shorter the computing time, as all images present the same background colour.

4. Wood defect detection and image segmentation

Unsupervised visual processing algorithms are proposed for both wood defect detection and image segmentation. The sooner the line inspector accurately detects the appearance of defects, the shorter the problem is fixed. The defect detection process is accomplished with an algorithm that searches for the seeds of the defects, characteristic pixels belonging to the defect, which determines the location of the defects in the image. Then the image segmentation algorithm uses a region growing method to calculate the size of the defect. The region growing method is based on Active Contours (Chan & Vese, 2001). The validation of the automatic visual inspection results is performed by a human expert.
An unsupervised defect detection algorithm has been developed and tested, to cope with variations in the laminated material, such as defect type, pose, size, shape and colour, as well as with environment variations: surface illumination and speed of the laminated material in the production line. All these variables make automatic wood defect detection a challenging task. The flow chart of Figure 5 summarizes the operation of the unsupervised defect detection algorithm.

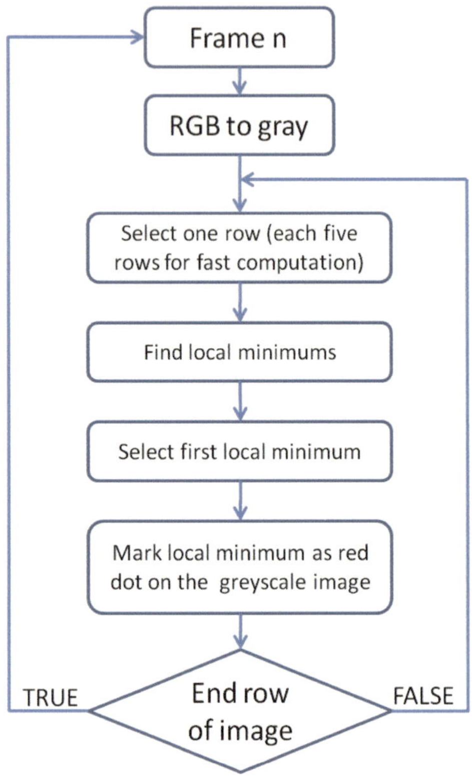

Fig. 5. Flow chart of the wood defect detection algorithm.

The wood defect detection algorithm proceeds as follows:

1. RGB colour image is converted to greyscale to reduce computational time
2. The developed algorithm searches for local minimum at each fifth row of the image and uses expert knowledge to associate defects with lower grey-level image pixels. The grey value of each pixel in an image row is compared to its neighbours and if its value is lower than the values of each neighbours, the pixel is defined as a local minimum. Two parameters are set up to configure the algorithm restrictions:
   - "Minimum valley depth", determines the minimum depth as a greyscale value. The valley depth value here selected is 190. Then, only valleys that exceed this value are returned.
   - "Minimum valley separation", specifies the minimum distance between valleys as a positive integer (pixels). That is, the algorithm ignores small valleys that occur in the neighbourhood of a deeper valley. The selected value is 100 pixels.
3. The position of the local minimum is visualised on the greyscale image with a red dot.

On the other hand, image segmentation allows for the calculation of the area of the defect. This is of great aid for quality control analyses, as the greater the size of the defects is, the lower is the quality of the wood laminates. The region growing algorithm groups pixels together into regions of similarity, beginning from an initial set of pixels (red dots). The method works iteratively for increasing the initial pixels set, comparing its grey value with that of its neighbours for increasing the area of the defect to reach the borders. These seeds, previously calculated by the unsupervised defect detection algorithm, allow applying the active contours method only in the regions where defects are detected, thus reducing computing costs.

5. Results

The wood defect detection and image segmentation algorithms, here proposed, have been tested with the two set of images acquired with both standard and laser diffuse lighting visual system.

The results of the unsupervised wood defect detection algorithm applied to the first set of images, are displayed in Figure 6. The results show that the algorithm is capable of detecting the position of the defects in the images several times (red dots). This multi-detection implies an accurate detection process caused by applying the algorithm only to each fifth row, for fast defect detection. A gap of five rows is an appropriate value for wood defect detection, as defects usually intersect in more than five rows. Then, the larger the defects are the greater the number of the detected pixels is (red dots). Knots defects have been correctly detected except in the sixth image due to the fact that the background was extremely similar to the defect grey-level. The split appearing on the eighth image has been detected. Moreover, comparing the results of the automatic defect detection algorithm with those of the expert, in the first set of images, the groundwork of the algorithm succeeded in 87.5%. Longer term processing would be required to detect a higher percentage of defects in images.

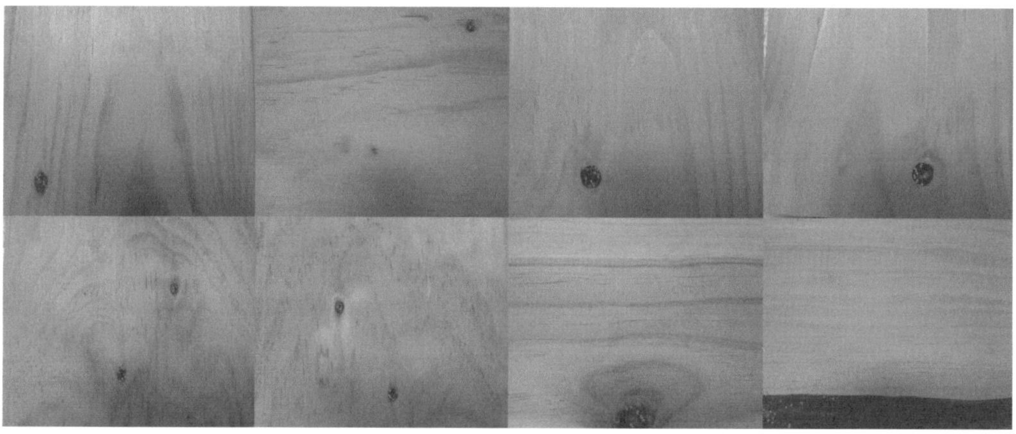

Fig. 6. Results of wood defect detection in images (first set) acquired with standard lighting visual system.

The results obtained with the unsupervised wood defect detection algorithm on the second set of images, are displayed in Figure 7. The second set of images has been captured with the laser diffuse lighting visual system, and in that case all defects were detected.

The multi-detection process enhances defect detection due to the large size of the defect in the analysed images. The different types of defects are easily perceived by the expert. In the last image, split defect was not completely detected as a consequence of the similarity between the grey-level of the defect and that of the background. The success in comparing automatic defect detection and expert visual defect detection, is 98%. This percentage is better than the one obtained in the processing of the first set of images.

Therefore, the use of a surface inspection system with laser diffuse lighting greatly improves the success of the automatic wood defect detection algorithm.

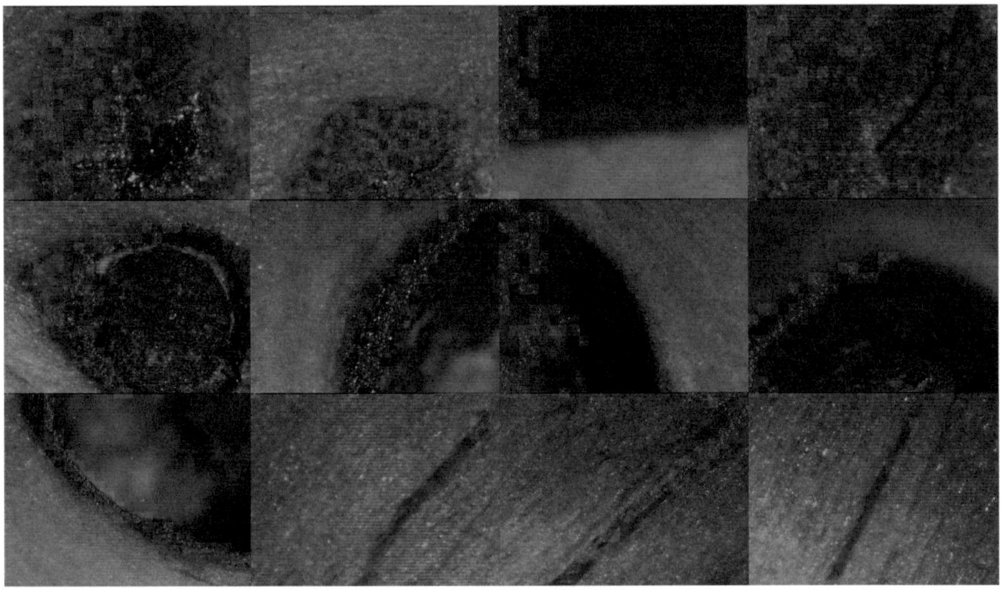

Fig. 7. Results of wood defect detection in images, second set, acquired with a laser diffuse lighting surface inspection system.

The result of the image segmentation algorithm on the first set of images is displayed in Figure 8. The image segmentation algorithm is based on a region growing method that uses Active Contours. The results are displayed in Figure 8, marking the complete border of the defect, in green colour. The defect detected pixels (red dots) grow until the border of the defect is reached, whenever the border of the defect is obtained before last iteration. Moreover, the border can be extra grown, as happens in the sixth and seventh images of Figure 8. However, the wood defects are well shaped, in spite of that present in the eighth image that would need some more iterations to complete the whole defect contour. Finally, the areas of the defects are calculated for further off-line quality control analysis.

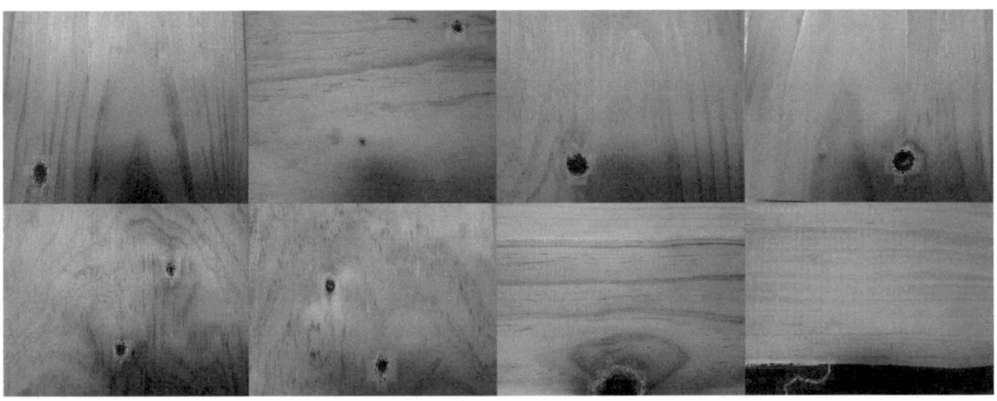

Fig. 8. Segmentation of images acquired with standard lighting visual system (first set).

The result of the image segmentation of the second set of images is displayed in Figure 9, where the partial border of the defect is marked in green colour. After 200 iterations of the algorithm, departing from the initialization pixels (red dots), the complete area of the defects present in the images is not totally segmented. Defects can be more accurately segmented by increasing the seed pixels (red dots) and the number of iterations of the region growing algorithm, but this implies a greater computing time.

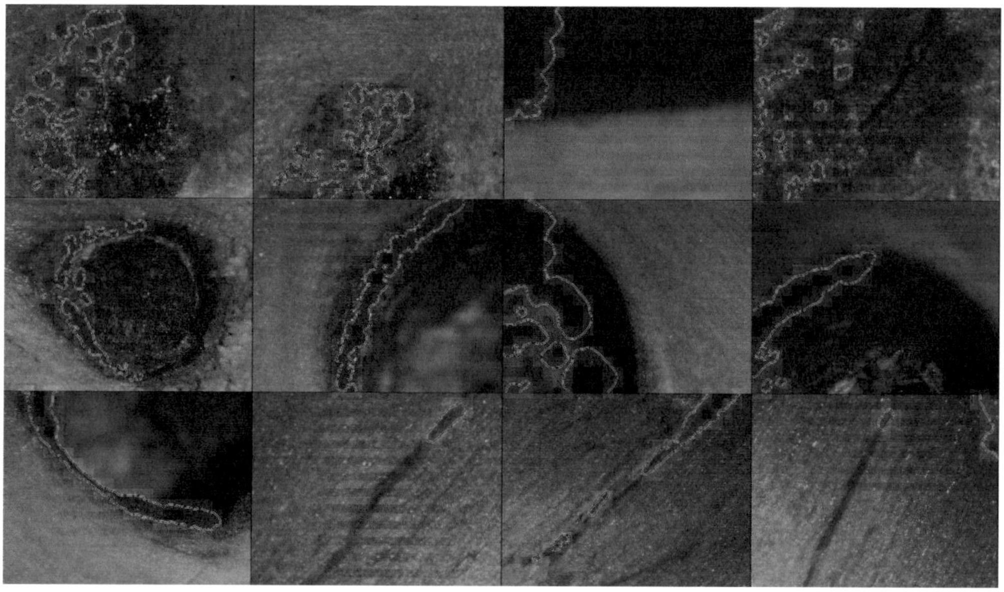

Fig. 9. Segmentation of the images acquired from the laser diffuse lighting surface inspection system (second set).

Summarizing, results displayed in Figure 9, are better than those shown in Figure 8, but total segmentation of the defects is not achieved as a greater number of seed pixels would be required.

Thus, to increase the number of seed pixels, the "Minimum valley separation" variable has been set to 10 pixels. The results obtained with the set of images acquired with laser diffuse lighting are shown in Figure 10. In first, second, fourth, tenth and twelfth images, the seeds (red dots) partially cover the wood defects, but in the third, fifth, sixth, seventh, eighth, ninth and eleventh images the defects are totally covered by the seeds. Then, the image processing would only require the wood defect detection algorithm as the calculated seeds can detect the shape of the whole defects. This point is extremely relevant as implies lower computing costs.

Next, region growing method is applied to obtain the complete segmentation of the defects (green borders), and the results are displayed in Figure 11. It can be remarked that segmentation results of the region growing method are close to those of Figure 10, obtained with the wood defect detection algorithm.

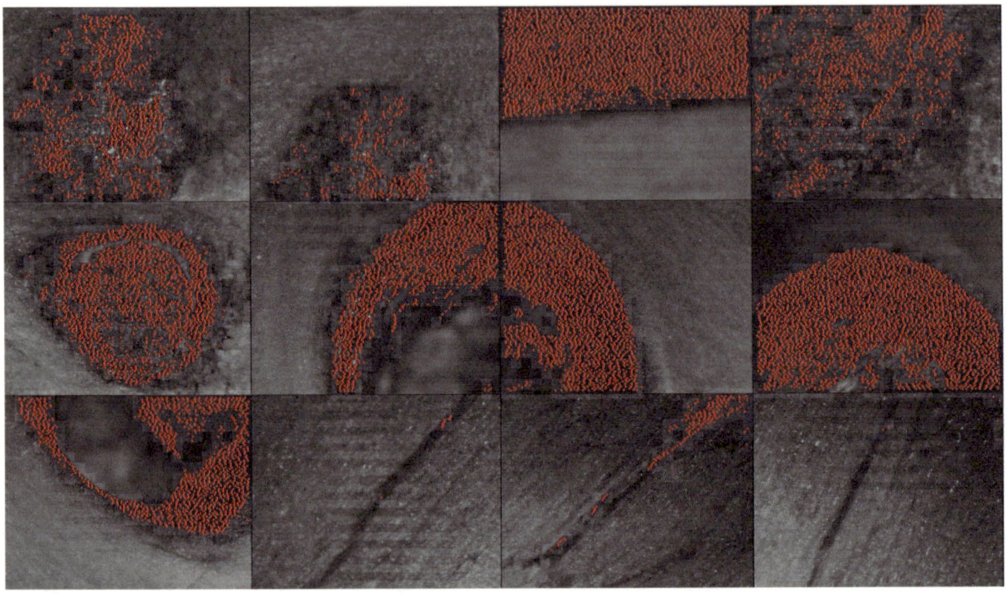

Fig. 10. Wood defect detection on images acquired with laser diffuse lighting and "Minimum valley separation" = 10 pixels.

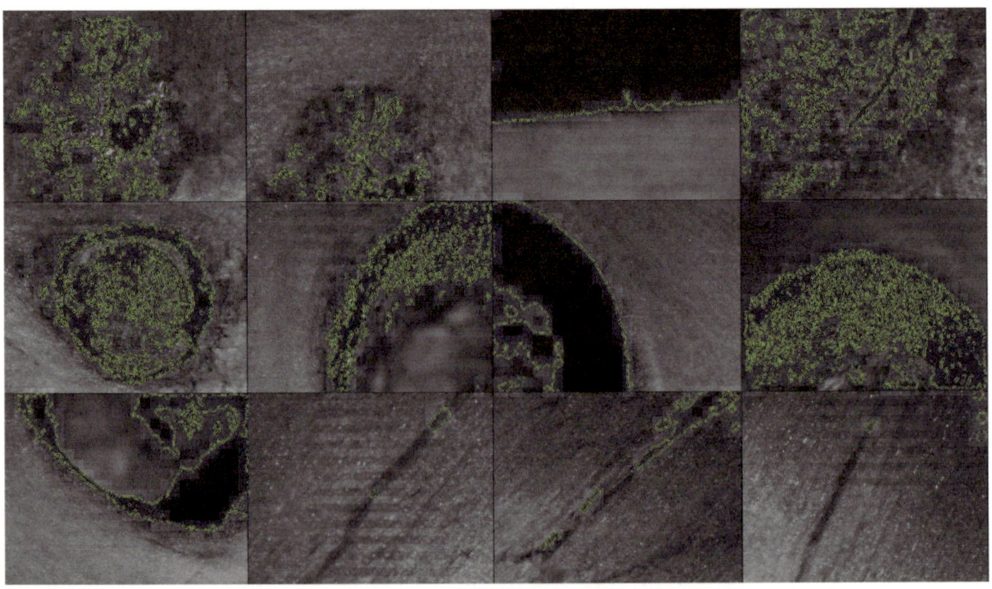

Fig. 11. Image segmentation using a large number of seeds: "Minimum valley separation"=10 pixels.

Finally, the computation time for the defect detection and segmentation algorithms is calculated for each image in Figure 10 and Figure 11, and displayed in Table 1. The second column corresponds to the computation time of the wood defect detection algorithm increasing the number of seed pixels (red dots). The third column is the computation time of the region growing algorithm for the image segmentation (green borders).

Images acquired from laser diffuse lighting surface inspection system	Defect detection algorithm (seconds)	Region growing algorithm (seconds)
Image 1	0.51	10.03
Image 2	0.23	8.65
Image 3	0.37	9.17
Image 4	0.21	9.82
Image 5	0.21	10.34
Image 6	0.21	9.90
Image 7	0.23	9.17
Image 8	0.31	12.42
Image 9	0.21	9.21
Image 10	0.23	7.96
Image 11	0.20	8.18
Image 12	0.18	8.09

Table 1. Computation time, in seconds, for each of the processing algorithm and images acquired from laser diffuse lighting.

6. Conclusions

The global goal of current work is the defect detection on wood images for quality control with short computing time. To this aim, the work compares wood images captured with both: a standard and an innovative surface inspection system that uses laser diffuse lighting.

Images from wood laminates with knots and splits defects acquired from the innovative lighting visual system are the most suitable for defect detection. The main characteristics of the proposed system to be used in the wood manufacturing industry are:

i. Detection of small defects ranging from one to few millimetres
ii. Independence from the variable reflectance of the wood laminates. Images present both a high background smoothness and high-contrast of the defects, independently of the colour of the wood laminates
iii. Defect detection time in less than 0.5 s
iv. Defect detection success is 98% on a sample of two hundred images
v. The laser diffuse light source is robust enough for integration in an industrial environment, due to its low cost and durability, in comparison with commercial light diffusers

A smart combination of algorithms for wood inspection has been proposed, based on an unsupervised method for defect detection and a region growing algorithm for defect segmentation. Results indicate that processing time for wood defect detection algorithm is better than for region growing algorithm. Moreover, first algorithm reaches both objectives: wood defect detection and segmentation.

Finally, it has been demonstrated that the use of the proposed wood defect detection algorithm with images acquired from visual surface inspection system using laser diffuse lighting, appears as the best choice for real-time wood inspection.

7. References

Armingol, J. Mª; Otamendi, J.; de la Escalera, A.; Pastor, J.M. & Rodríguez, F.J. (2006). Intelligent Visual Inspection System for Statistical Quality Control of a Production Line. *Frontiers in Robotics Research*, Nova Publishers, Chapter 1, pp. 1-33

Åstrand, E. & Åström, A. (1994). A Single Chip Multi-Function Sensor System for Wood Inspection, *Proceedings of the 12th IAPR International Conference on Pattern Recognition*, Vol. 3, pp. 300-304, ISBN: 0-8186-6275-1, Jerusalem , Israel.

Batchelor, B.G. & Whelan, P.F. (1997). *Intelligent Vision Systems for Industry*, Springer-Verlag

Chan, T.F. & Vese, L.A. (2001). Active Contours Without Edges, *IEEE Transactions on Image Processing*, Vol.10, pp. 266–277

Cognex Corporation (2011). *Machine Vision Systems*. Natick, MA 01760-2059, United States of America. Available from http://www.cognex.com/

Estévez, P.A.; Perez, C.A. & Goles, E. (2003). Genetic Input Selection to a Neural Classifier for Defect Classification of Radiata Pine Boards. *Forest Products Journal*, Vol. 53, pp. 87–94.

Guinea, D.; Preciado, V.M., Vicente, J.; Ribeiro, A. & García-Alegre, M.C. (2000). CNN based visual processing for industrial inspection, *Proceedings of SPIE Machine Vision Applications in Industrial Inspection VIII*, California, Vol. 3966, pp. 315-322

Hall, M. & Aström, A. (1995). High Speed Wood Inspection Using a Parallel VLSI Architecture, *Proceedings of the International Workshop Algorithms and Parallel VLSI Architectures III*, Leuven, Belgium, pp. 215-226

Li, M.-X. & Wu, C.-D. (2009). A VPRS and NN Method for Wood Veneer Surface Inspection, *Proceedings of the 2009 International Workshop on Intelligent Systems and Applications*, ISBN: 978-1-4244-3893-8, Wuhan, China, pp. 1-4

Malamas, E.N.; Petrakis, E.G.M.; Zervakis, M.; Petit, L. & Legat, J.D. (2003). A Survey on Industrial Vision Systems, Applications and Tools. *Image Vision Comput.*, Vol. 21, pp. 171–188

Martin, D.; Guinea, D.M.; García-Alegre, M.C.; Villanueva, E. & Guinea, D. (2010). Multi-modal Defect Detection of Residual Oxide Scale on a Cold Stainless Steel Strip. *Machine Vision and Applications*, Vol.21, pp. 653-666

Montúfar-Chaveznava, R.; Guinea, D.; Garcia-Alegre, M.C. & Preciado, V.M. (2001). CNN computer for high speed visual inspection, *Proceedings of SPIE Machine Vision Applications in Industrial Inspection IX*, California, Vol.4301, pp. 236-243

Palviainen, J. & Silvennoinen, R. (2001). Inspection of Wood Density by Spectrophotometry and a Diffractive Optical Element Based Sensor. *Measurement Science and Technology*, Vol.12, pp. 345-352

Parsytec Computer GmbH (2011). *Surface Inspection Systems*. Auf der Huels 183, 52068 Aachen, Germany. Available from http://www.parsytec.de/

Pham, D.T. & Alcock, R.J. (1999). Automated visual inspection of wood boards: selection of features for defect classification by a neural network. *Proceedings of the Institution of Mechanical Engineers -- Part E -- Journal of Process Mechanical Engineering (Professional Engineering Publishing)*, Vol. 213, pp. 231-245

Pham, D.T. & Alcock, R.J. (2003). *Smart Inspection Systems. Techniques and Applications of Intelligent Vision*, Academic Press

Ruz, G.A., Estévez, P.A. & Ramírez, P.A. (2009). Automated visual inspection system for wood defect classification using computational intelligence techniques. *International Journal of Systems Science*, Vol. 40, pp. 163–172

Silvén, O.; Niskanen, M. & Kauppinen, H. (2003). Wood Inspection with Non-supervised Clustering. *Machine Vision and Applications*, Vol.13, pp. 275-285

Smith, M.L. (2001). Surface Inspection Techniques - Using the Integration of Innovative Machine Vision and Graphical Modelling Techniques, *Engineering Research Series*. London and Bury St Edmunds, UK: Professional Engineering Publishing

Spínola, C.G.; García, F.; Martin, M.J.; Vizoso, J.; Espejo, S.; Cañero, J.M.; Morillas, S.; Guinea, D.; Villanueva, E.; Martin, D. & Bonelo, J.M. (2008). Device for Detecting and Classifying Residual Oxide in Metal Sheet Production Lines, Patent P2007_00865, Patent Cooperation Treaty (PCT): Publication No.: WO/2008/119845, International Application No.: PCT/ES2007/000768, ACERINOX S.A.

Yang, D.; Jackson, M.R. & Parkin, R.M. (2006). Inspection of Wood Surface Waviness Defects Using the Light Sectioning Method. *Proceedings of the Institution of Mechanical Engineers, Part I: Journal of Systems and Control Engineering*, Professional Engineering Publishing, Vol.220, pp. 617-626

25

Comparison Sequential Test for Mean Times Between Failures

Yefim Haim Michlin[1] and Genady Grabarnik[2]
[1]*Technion - Israel Institute of Technology*
[2]*St' Johns University*
Israel
USA

1. Introduction

The present study deals with the planning methodology of tests in which the parameters of two exponentially-distributed random variables are compared. The largest application field of such tests is reliability checking of electronics equipment. They are highly cost-intensive, and the requirements as to their resolution capability become stricter all the time. Hence the topicality and importance of an optimal plan permitting decisions at a given risk level on the basis of a minimal sample size.

Such comparison tests are required for example in assessing the desirability of replacing a "basic" object whose reliability is unknown, by a "new" one; or when the influence of test conditions on the results has to be eliminated.

This is the case when an electronics manufacturing process is transferred to another site and the product undergoes accelerated testing.

Recently, equipment and methods were developed for accelerated product testing through continuous observation of a population of copies and replacement of failed objects without interrupting the test. For such a procedure, the sequential approach is a feasible and efficacious solution with substantial shortening - on the average - of the test duration (see e.g. Chandramouli et al. 1998; Chien et al. 2007).

In these circumstances there is high uncertainty in the acceleration factor, with the same effect on the estimated reliability parameters of the product. This drawback can be remedied by recourse to comparison testing. The latter serves also for reliability matching in objects of the same design and different origins, or a redesigned product versus its earlier counterpart, or different products with the same function (see e.g. Chien & Yang, 2007; Kececioglu, 2002).

The exponential nature of the Time Between Failures (TBF) of repairable objects, or the time to failure of non-repairable ones - is noted in the extensive literature on the reliability of electronic equipment (Kececioglu, 2002; Chandramouli et al, 1998; Drenick, 1960; Sr-332, 2001; MIL-HDBK-781A, 1996). For brevity, the TBF acronym is used in the sequel for both these notations.

Mace (1974, Sec. 6.12) proposed, for this purpose, the so-called fixed sample size test with the number of failures of each object fixed in advance - which is highly inconvenient from the practical viewpoint. For example, when the "basic" object has "accumulated" the

specified number of failures, one has to wait until the "new" one has done the same, and if the latter is substantially more reliable, the waiting time may be very long.
The international standard IEC 61650 (1997) deals with two constant failure rates, which is equivalent to the problem just described. However, this standard, which forms part of an international system of techniques for reliability data analysis, does not refer to the planning aspect of the tests.
A solution to our problem was outlined in (Michlin & Grabarnik, 2007), where it was converted into binomial form, for which Wald's sequential probability ratio test (SPRT) is suitable (Wald, 1947, chap. 5). Wald and Wolfowitz (1948) also proved that this test is the most efficacious at two points of its characteristic, but it has one drawback - the sample size up to a decision can be many times larger than the average. This is usually remedied by resorting to truncation (see e.g. Wald, 1947; Siegmund, 1985).
A methodology is available for exact determination of the characteristics of such a truncated test with known decision boundaries. It was proposed by Barnard (1946) and developed by Aroian (1968). It served as basis for an algorithm and computer programmes (Michlin et al. 2007, 2009) used in examining its properties.
Hare we consider the inverse problem - determination of the test boundaries from specified characteristics.
In the absence of analytical dependences between the boundary parameters and characteristics, the search is hampered by the following circumstances:

- The number of parameter-value combinations may be very large.
- While shortening of the step makes for more combinations, it cannot be guaranteed that combinations with optimal characteristics are not missed.
- The standard optimum-search programmes are unsuitable for some of the discrete data of the type in question.

The theme of this chapter is the planning methodology for comparison truncated SPRT's. Formulae derived on its basis are presented for calculation of the test boundary parameters.
The rest of the chapter is organised as follows: In Section 2 is given a description of the test and its conversion to SPRT form. In Section 3 are described the quality indices for a truncated test and criteria for the optimal test search. In Section 4 are discussed the discrete nature of the test boundaries and its characteristics; a search algorithm is presented for the oblique boundaries. Section 5 describes the planning methodology, and approximative dependences are presented for calculation of the boundary parameters. Section 6 deals with planning of group tests. Section 7 presents a planning example and applications. Section 8 - the conclusion.

2. Description of test and its SPRT presentation

2.1 Description of test procedure in time domain. Checked hypothesis

In the proposed test two objects are compared - one "basic" (subscript "*b*") and the other "new" (subscript "*n*"). In the course of such tests, the "null" hypothesis is checked, that the ratio of the mean TBF (MTBF) of these objects exceeds or equals a prescribed value Φ_0, versus the alternative of it being smaller than the latter. The compared objects work concurrently (Figure 1). When one of them fails, it is immediately repaired or replaced. The unfailed object is not replaced but allowed to continue working until it fails in turn (in which case it is neither replaced nor repaired), or until the test terminates. A situation may occur in which there has been no failure in one object and it kept working throughout the whole test, as against several failures in the other object. The total work times T are equal for both objects.

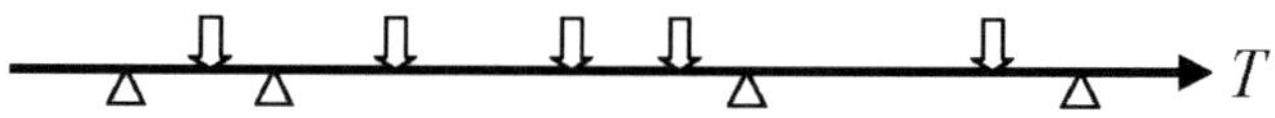

Fig. 1. Scheme of test course (Upward marks - failures of basic item; downward marks - those of new item; T - time, common to both systems) (Michlin et al., 2011).

The probability density of the TBF for each of the compared objects has the form:

$$f_{TBF}(t)=(1/\theta)*\exp(-t/\theta)$$

where θ is the MTBF for the "new" (θ_n) and "basic" (θ_b) objects respectively. At each failure, a decision is taken - continuing the test versus stopping and accepting the null hypothesis, or rejecting it in favour of the alternative (Michlin & Migdali, 2002; Michlin & Grabarnik, 2007):

$$\begin{aligned} &H_0: \ \Phi \geq \Phi_0 \quad \left(P_a(\Phi_0)=1-\alpha\right) \\ &H_1: \ \Phi < \Phi_0 \quad \left(P_a(\Phi_1)=\beta\right) \end{aligned} \tag{1}$$

where

$$\Phi = \theta_n / \theta_b \tag{2}$$

α and β are the probabilities of I- and II-type errors; in the sequel, their target values will be denoted by the subscript "*tg*", and their actual values - by the subscript "*real*".
$P_a(\Phi)$ is the probability of acceptance of H_0, which is the Operating Characteristic (OC) of the test;

$$\Phi_1 = \Phi_0 / d \tag{3}$$

d>1 being the discrimination ratio.
Mace (1974 , Sec. 6.12) presents the following estimate $\hat{\Phi}$ for Φ, obtained with the aid of the maximum likelihood function (for the proof, see Kapur & Lamberson 1977, Sec. 10.C):

$$\hat{\Phi} = (T_n / r_n)/(T_b / r_b)$$

where r_n and r_b - the accumulated number of failures over times T_n and T_b.
As in this test T_n=T_b=T, we have:

$$\hat{\Phi} = r_b / r_n \tag{4}$$

Figure 2 shows an example of the test field. In the course of the test, it can reside at a point of this field characterised by an integer number of failures of each of the objects. When one of them fails, the test "jumps" to a neighbouring point located above (failure of "n") or to the right (failure of "b"). With the test course thus described, shifts from point to point occur only on failures in one of the objects, i.e. the time factor is eliminated from the analysis. When the Accept boundary is crossed, the test stops at the Accept Decision Point (ADT), and when its Reject counterpart is crossed - at the RDP. The boundaries consist of two parallel oblique straight lines (accept line (AL) and reject line (RL)) and the truncation lines parallel to the coordinate axes and intersecting at the Truncation Apex (TA).

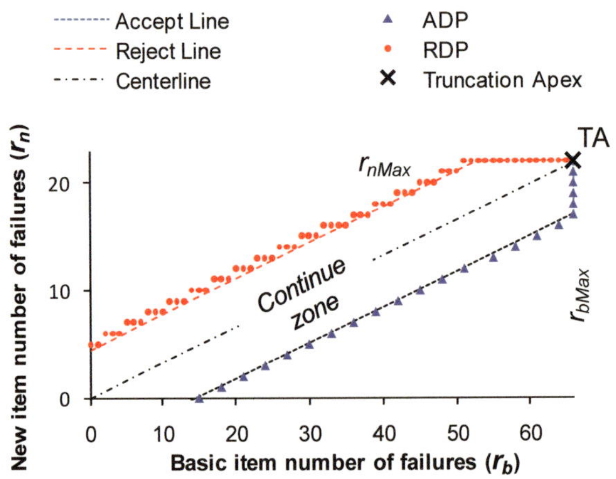

Fig. 2. Truncated test field for Φ_0=4.3, d=2, α_{real}=0.098, β_{real}=0.099, R_{ASN} =9.2%, (Michlin et al., 2011).

2.2 Binomial presentation of test and SPRT solution

For all points of the test field, the probability of the next failure occurring in the new object (i.e. of a step upwards) is constant and given by the following expression (for proof see Michlin & Grabarnik, 2007):

$$P_R(\Phi) = 1/(1+\Phi) \tag{5}$$

A binomial SPRT is available for such a test (Wald, 1947, chap. 5), whose oblique boundaries are:

$$\text{Accept line (AL):} \qquad r_b = r_n/s + h'_b \tag{6}$$

$$\text{Reject line (RL):} \qquad r_n = r_b \cdot s + h_n \tag{7}$$

where s is their slope, uniquely determined by the SPRT theory depending on α, β, Φ_0, d (Wald,1947; Michlin & Grabarnik, 2007), and given by:

$$s = -\ln q/(\ln q + \ln d) \tag{8}$$

where

$$q = (1+\Phi_0)/(d+\Phi_0) \tag{9}$$

The absolute terms of (6) and (7) are given by:

$$h_a = \ln\left(\beta^*/(1-\alpha^*)\right)/(\ln q + \ln d) \tag{10}$$

$$h_n = \ln\left(\left(1-\beta^*\right)/\alpha^*\right)/\left(\ln q + \ln d\right) \qquad (11)$$

$$h_b' = -h_a / s \qquad (12)$$

The expressions (10)-(12) have one drawback: the parameters α^* and β^* are unknown. Their dependence on α_0, β, Φ_0, d, and on the TA coordinates is available only in the form of the limits between which the parameters lie (Michlin et al., 2009). Still, these limits suffice for determining - from the above expressions - corresponding search limits for h'_b and h_n. A search methodology, within these limits, for exact values ensuring the target characteristics - is, basically, the goal of this work.

2.3 Calculation of test characteristics acc. to given boundaries

The probability of hitting a given point of the test is given by (Barnard, 1946; Michlin & Grabarnik, 2007):

$$P_{r_b,r_n}(\Phi) = P_{r_b,r_n-1}(\Phi)\cdot P_R(\Phi) + P_{r_b-1,r_n}(\Phi)\cdot\left[1-P_R(\Phi)\right] \qquad (13)$$

while that of hitting the given ADP is:

$$P_{ADP}(r_n,\Phi) = P_{r_b-1,r_n}(\Phi)\cdot\left[1-P_R(\Phi)\right] \qquad (14)$$

and that for the given RDP is:

$$P_{RDP}(r_b,\Phi) = P_{r_b,r_n-1}(\Phi)\cdot P_R(\Phi) \qquad (15)$$

$P_a(\Phi)$ is the sum of all the probabilities $P_{ADP}(r_n,\Phi)$ of hitting all ADP, hence the actual values of α and β, namely α_{real} and β_{real}, are given by:

$$\alpha_{real} = 1 - P_a(\Phi_0);\quad \beta_{real} = P_a(\Phi_1) \qquad (16)$$

The Average Sample Number (ASN) of a truncated test is calculated as:

$$ASN(\Phi) = \sum_{r_b=0}^{TA_b}\left[r_b + r_{nRDP}(r_b)\right]P_{RDP}(r_b,\Phi) + \sum_{r_n=0}^{TA_n}\left[r_n + r_{bADP}(r_n)\right]P_{ADP}(r_n,\Phi) \qquad (17)$$

where $r_{nRDP}(r_b)$ is the r_n-coordinate of the RDP with given r_b.
The Average Test Duration (ATD) for each object is:

$$ATD(\Phi) = \theta_b \cdot ASN(\Phi)/(1+1/\Phi) \qquad (18)$$

3. Comparative characteristics and optimality of test

In this Section the optimality criteria for the test, on which the comparison- and selection algorithm is based, are substantiated, and the problems of the study are clarified.
In (Michlin & Grabarnik, 2007) were presented three optimality criteria which can be calculated for the specified boundaries:

- Closeness of the test OC to the prescribed one. For given *d*, the measure of this closeness is R_D:

$$R_D = \sqrt{\left[\left(\alpha_{real} - \alpha_{tg}\right)/\alpha_{tg}\right]^2 + \left[\left(\beta_{real} - \beta_{tg}\right)/\beta_{tg}\right]^2} \tag{19}$$

with a_{real} and β_{real} as per (16).
- The degree of truncation, which characterises the maximum test duration whose measure can be, for example, the sum of the TA coordinates.
- The efficacy of the test according to Wald (1947) and to Eisenberg & Ghosh (1991), as the measure of which R_{ASN} was adopted (Michlin et al., 2009) – the relative excess of the function $ASN(\Phi)$ of the truncated test over $ASN_{nTR}(\Phi)$, its non-truncated counterpart which can be taken as ideal:

$$R_{ASN} = \left\{\sum_{i=1}^{5}\left[ASN(\Phi_i) - ASN_{nTr}(\Phi_i)\right]\right\} \Big/ \sum_{i=1}^{5} ASN_{nTr}(\Phi_i) \tag{20}$$

where Φ_i - values of Φ in geometric progression:

$$\Phi_0 \cdot \left(\sqrt{d}\right)^{i-4} \quad \text{for } i = 1 \ldots 5 \tag{21}$$

$ASN(\Phi)$ – calculated as per the recursive formulae (17), (13…15) $ASN_{nTr}(\Phi)$ – calculated by Wald's formulae (1947, chap. 3) obtained for a non-truncated test of the type in question:

$$ASN_{nTr}(\Phi) = \frac{(1+\Phi(\eta))\left[P_a(\eta)\ln B + (1 - P_a(\eta))\ln A\right]}{(1+\Phi(\eta))\ln\left[(1+\Phi_0)/(d+\Phi_0)\right] + \ln d} \tag{22}$$

where

$$\Phi(\eta) = \frac{d^{\eta}(1+\Phi_0)^{\eta} - (d+\Phi_0)^{\eta}}{(d+\Phi_0)^{\eta} - (1+\Phi_0)^{\eta}} \tag{23}$$

$$P_a(\eta) = \left(A^{\eta} - 1\right)/\left(A^{\eta} - B^{\eta}\right) \tag{24}$$

$$A = (1 - \beta_{real})/\alpha_{real} \tag{25}$$

$$B = \beta_{real}/(1 - \alpha_{real}) \tag{26}$$

η– an auxiliary parameter calculated by (23) for Φ values as per the progression (21).

The choice criterion for the optimal test is:

$$\min(TA_n + TA_b) \tag{27}$$

subject to:

$$(\min R_d \text{ at given TA}) \& (R_d < R_{d0}) \& (R_{ASN} < R_{ASN0}) \tag{28}$$

where R_{d0} and R_{ASN0} – threshold values of R_d and R_{ASN}.
The TA of such a test called Optimal TA (OTA). Section 5 presents approximative formulae for determination of those OTA coordinates which permit reduction of the search field to 2 to 6 points. A particular problem in this context is: for a given TA, find h'_b and h_n (eqs. (6), (7)) ensuring min R_d.

4. Discreteness of test boundaries and their search at given TA

This Section deals with the interrelationships between the boundary parameters of the test on the one hand, and the characteristics of the test itself (namely, a_{real} and β_{real}) and those of its quality (introduced in the preceding Section, R_d and R_{ASN}) – on the other. These interrelationships lack analytical expression and are further complicated by the discreteness of the test. Thus one had to make do with typical examples of their behaviour in the vicinity of the optimum. With this behaviour clarified, an efficacious search algorithm could be developed for the optimum in the discrete space in question. Clarity of the picture is essential both for the developer of the planning methodology and for the practitioner planning the binomial test in the field.

4.1 Discreteness of test boundaries

As the slope s of the oblique test boundaries, described by eqs. (6) and (7), is unrelated to a and β (see eq. (8)), the search for them under the min R_d stipulation reduced to finding the absolute terms in the describing equation, namely the intercepts h'_b and h_n on the coordinate axes (Figure 3).

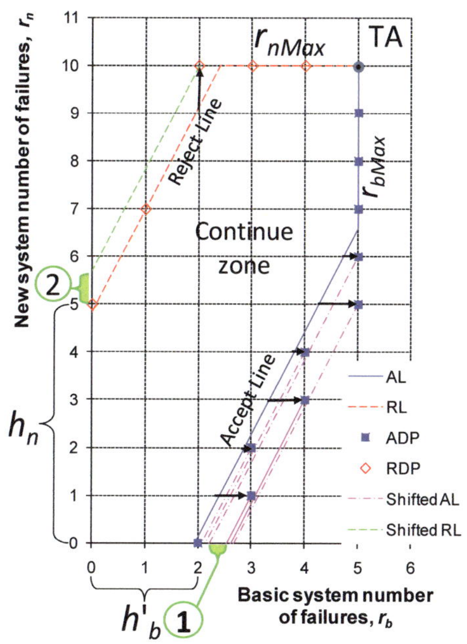

Fig. 3. Test Plane (Michlin & Grabarnik, 2010). 1 – Example of interval of h'_b values over which the test ADP's do not change. 2 – Ditto for h_n and RDP.

Stopping of the test occurs not on the decision lines, but at points with integer coordinates, with ADP to the right of the AL and RDP above the RL. If the AL is shifted from an initial position (solid line in Figure 3) to the right, the test characteristics remain unchanged until it crosses an ADP, which in turn is then shifted in the same direction by one failure. The AL positions at these crossings are shown as dot-dashed lines, and its shifts are marked with arrows. Projecting the termini of these arrows, parallel to the AL, on the r_b axis, we obtain the values of h'_b at which the changes occur. An analogous process takes place when the RL is shifted upwards.

The intervals of h'_b and h_n over which the test characteristics remain unchanged are marked in Figure 3 by the circled numbers 1 and 2 respectively.

When the AL is shifted to the right (h'_b increased) $P_a(\Phi)$ is reduced, i.e. a_{real} increases and β_{real} decreases. When the RL is shifted upwards, the effects are interchanged. These relationships are monotonic and stepwise, and differ in that change of h'_b is reflected more strongly in β_{real} and more weakly in a_{real}. With h_n the pattern is reversed.

4.2 Basic dependences between oblique boundaries and test characteristics

In (Michlin et al., 2009, 2011; Michlin & Kaplunov, 2007) were found the limits within which a^* and β^* of the optimal tests should be sought. These limits can also serve for determining the search limits of h'_b and h_n, as per (10) - (12).

Figure 4 shows an example of the above, with the limits for h'_b and h_n calculated, according to the data of (Michlin et al., 2009), for d=2, Φ_0=1, $a_{tg}=\beta_{tg}$=0.1, TA_b=27, TA_n=38, R_{ASN}≤12%. In the figure, the points mark the centres of rectangles within which the characteristics remain unchanged. The resulting picture is fairly regular, even though the spacings of the columns and rows are variable. In space, the R_d points form a cone-shaped surface.

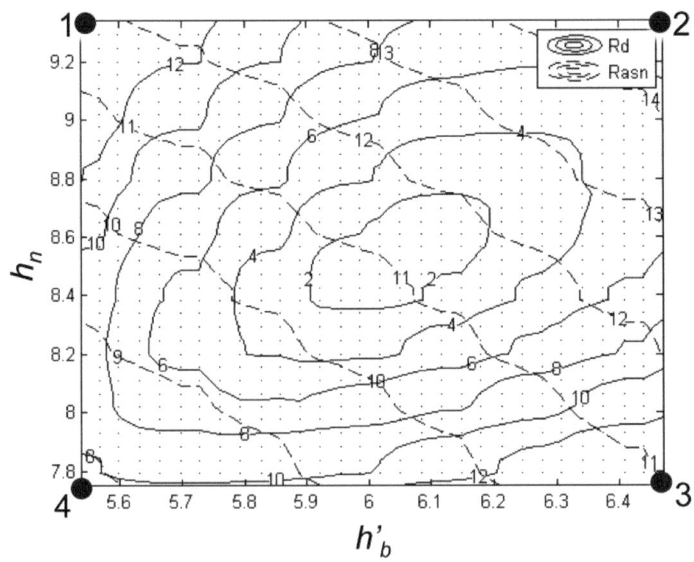

Fig. 4. Contours of R_{ASN} (dashed lines) and R_D (solid lines) vs. h'_b and h_n. (Michlin & Grabarnik, 2010). The dots mark the centres of rectangles within which the test characteristics do not change. 1 – 4 are the corner points at which the test characteristics are calculated in the search for the optimum (Subsection 4.3, stage ‹1.1›).

The figure also contains the contours (isopleths) of R_d (solid lines) and R_{ASN} (dashed lines), given as percentages. In macro the R_d contours can be described as oval-shaped, whereas in micro they are quite uneven, so that derivatives calculated from a small set of points would show large jumps, which would hamper the search for the minimum R_d . It is seen that in the vicinity of that minimum, R_{ASN}≈11%.

Figure 5 shows two projections representing a_{real} and β_{real}, calculated according to the coordinates of Figure 4, so that to each point of the latter corresponds one of a_{real} and β_{real}. These points form intersecting almost-plane surfaces. In the upper figure the coordinate axes are oriented so that the intersection zone (a_{real} - β_{real}) is perpendicular to the page; in the lower figure. the orientation is such that the rows of β_{real} points reduce in projection to a single point – in other words, they form parallel or almost-parallel lines.

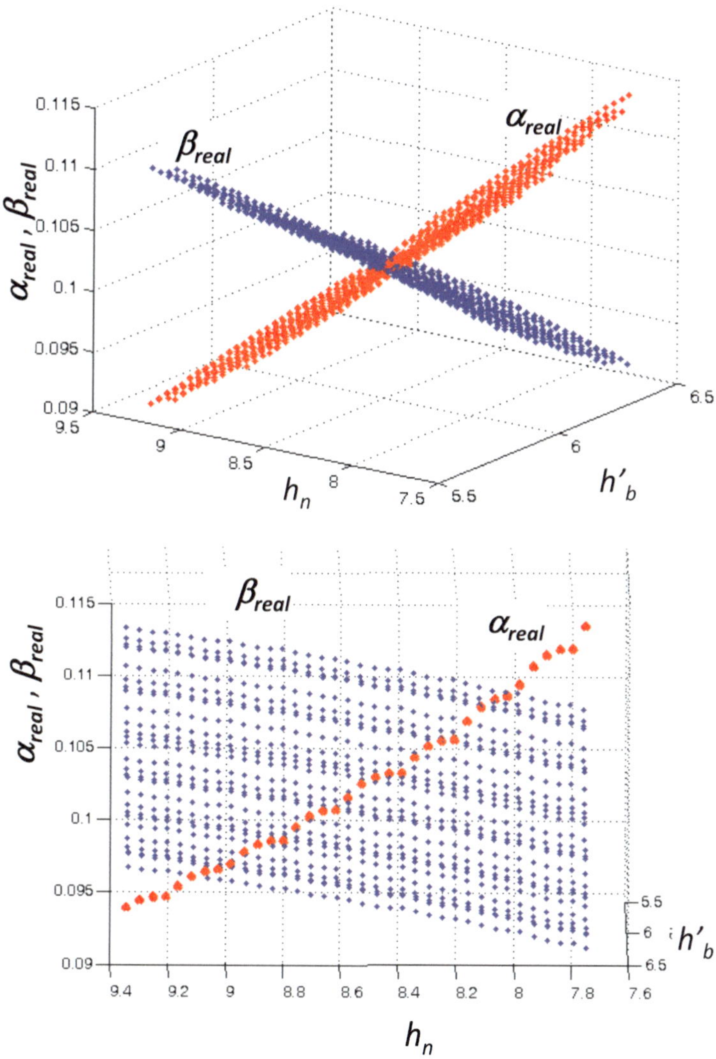

Fig. 5. Two projections of α_{real} and β_{real} "planes". (Michlin & Grabarnik, 2010).

Figure 6 shows analogous projections for R_{ASN}, and we again have an almost-plane surface, monotonic and uneven in micro.
The provided examples show that the described patterns characterise the dependences of α_{real}, β_{real} and R_{ASN} on h'_b and h_n within the limits determined in Subsection 5.3 (Michlin et al., 2009, 2011; Michlin & Kaplunov, 2007). Over small intervals these dependences are stepwise, and the lines through the step midpoints are uneven as well.

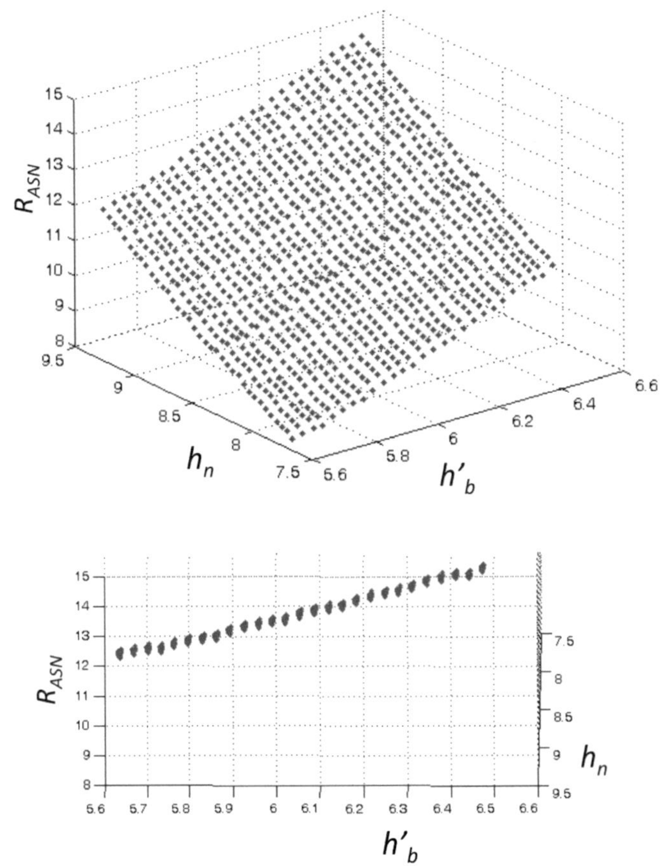

Fig. 6. Two projections of R_{ASN} "plane". (Michlin & Grabarnik, 2010).

4.3 Search algorithm for oblique test boundaries

Standard search programmes for minima (such as those in Matlab) operate poorly, or not at all, with discrete data of the type in question. Availability of known regularities in the behaviour of the functions α_{real}, β_{real}, R_{ASN}, R_D makes it possible to construct a fast and efficacious algorithm.
These known regularities are:

- The values of h'_b and h_n at which the test characteristics change.
- The limits of h'_b and h_n, yielding tests with the specified characteristics.
- Almost-plane monotonic dependences of α_{real}, β_{real} and R_{ASN} within the above limits, stepwise and unstable but also monotonic in narrower intervals.

- Stronger dependence of a_{real} on h_n than on h'_b; the reverse – for β_{real}.

In expanded form, the search algorithm for min R_d consists in the following:

1st stage.

‹1.1› Calculation of the test characteristics at the four vertices of a rectangle (Figure 4) whose coordinates are obtained from the relationships presented in Subsection 5.3.
‹1.2› Approximation of $a_{real}(h'_b, h_n)$ and $\beta_{real}(h'_b, h_n)$ as planes, and determination of the first estimate h'_{b1}, h_{n1} yielding min R_D (point 5, Figure 7). Checking for $R_D \le R_{D0}$. If satisfied, stopping of search.

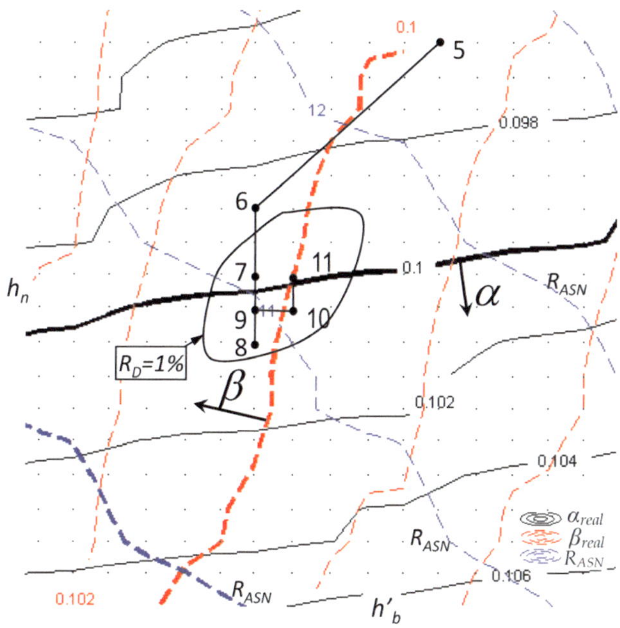

Fig. 7. Example of search scheme for min(R_D). (Michlin & Grabarnik, 2010). 5 – 11 are points of test characteristics calculation.

2nd stage.

Determination of point 6 – from a_{real5}, β_{real5} and the slopes of the a-, β-planes as per ‹1.2›. Re-checking for $R_D \le R_{D0}$.

3rd stage.

Alternating advance parallel to the h'_b- and h_n-axes. In view of the discreteness and complexity of the R_D function, the search for its minimum was reduced to one for the points h'_b and h_n where Δa and $\Delta\beta$ change sign:

$$\Delta a = a_{real} - a_{tg}; \quad \Delta\beta = \beta_{real} - \beta_{tg}$$

This problem is easier to solve, as both Δa and $\Delta\beta$ are monotonic functions of h'_b and h_n.
The search can be stopped at every step, subject to $R_D \le R_{D0}$.
‹3.1› If at point 6 (‹2› above) $\Delta a_6 > \Delta\beta_6$, a path parallel to the h_n-axis is taken in uniform steps Δh_n, until Δa changes its sign (points 6,7,8 on Figure 7), $\Delta h_n = \Delta a_6 / a_3$, where a_3 is the

coefficient in the equation of the a-plane as per ‹1.2›. Beyond that point, the root $\Delta a(h_n)$ is searched for by the modified Regula Falsi method (point 9). (The term "root" refers here to one of a pair of adjoining points at which the function changes its sign and has the smaller absolute value). The special feature of this procedure is accounting for the discreteness of the solution.

‹3.2› At the point of the root Δa, a right-angled turn is executed and a path parallel to the h'_b-axis is taken, searching for the $\Delta\beta$ root (point 10).

‹3.3› The alternating procedure is continued until a situation is reached where two consecutive turns involve only movement to an adjoining point. This point 10 corresponds to min(R_D). If in ‹3.1› $\Delta a_6 < \Delta\beta_6$, we begin from ‹3.2›.

4.4 Efficacy of algorithm

With a view to assessing the efficacy of the proposed algorithm, a search for the h'_b and h_n values yielding min R_d was conducted with the aid of a Matlab programme which realised this algorithm, and alternatively with the Matlab fminsearch command, with the same function WAS (Michlin & Grabarnik, 2007) referred to in both cases. This function determines the test characteristics according to its specified boundaries. The run covered different tests with R_{ASN0}=5 and 10%.

The calculation results are shown in Figure 8.

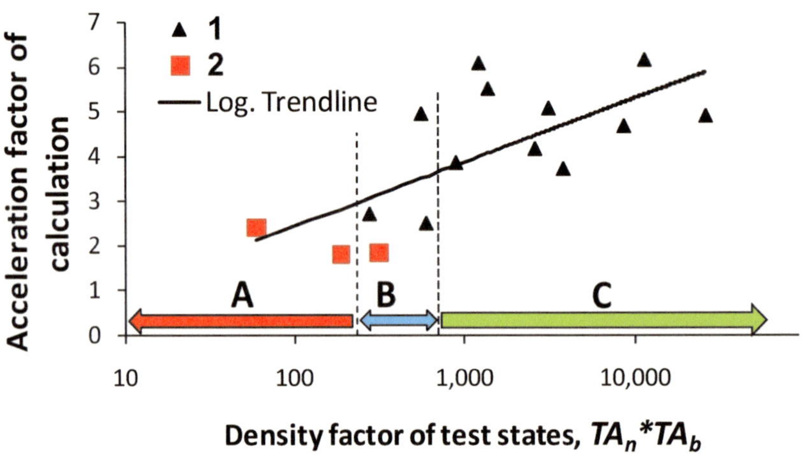

Fig. 8. Comparative efficacy of proposed algorithm. (Michlin & Grabarnik, 2010).
1 - fminsearch (Matlab) found min R_D or stopped close to it;
2 - fminsearch failed to find R_D;
A, B, C - short, medium and long tests, respectively.

In it, the abscissa axis represents the product TA_b*TA_n, which we term "density factor of test states". The higher the latter, the denser the disposition of the test points in the search zone (see Figure 4), the smaller the changes in the test characteristics from point to point, and the closer the search to one over a continuous smooth surface. A small value of the product is associated with a short test, due to be completed at small sample size and moderate computation times for the characteristics; a large value - with long tests, completed on the average at large sample sizes and long computation times.

The ordinate axis represents the "acceleration factor of calculation", which is the ratio of references to the WAS-function by fminsearch and the proposed algorithm respectively. The larger the ratio, the faster the algorithm compared with the standard Matlab function.
The diagram shows that at low densities (short tests, zone A) fminsearch fails to find min R_d. In zone C (long tests) the command finds it or stops close to it, but with 3 to 6 times more references to WAS. In zone B (medium tests) the minimum is either not found, or found with 2.5 to 5 times more references to WAS. By contrast, the programme based on the proposed algorithm found the minimum in all cases.
Accordingly, for the present task - searching for the optimum in a discrete space - the proposed algorithm accomplishes it much faster than the Matlab standard fminsearch command, thus saving computation time in long tests. Moreover, it guarantees a solution - a critical aspect in short tests, where fminsearch usually fails to find one.

5. Estimates for boundary parameters

5.1 Search methodology for optimal test boundaries

In (Michlin & Grabarnik, 2007) it was established that for Φ_0=1 and $\alpha=\beta$, the OTA lie on the centreline (which runs through the origin parallel to the AL/RL), so that

$$r_n = s \cdot r_b \tag{29}$$

This was checked for different Φ_0. With given $\alpha_{tg}=\beta_{tg}$, d, and $R_D \le 1\%$, a search was conducted for three location zones of the TA - namely, with $R_{ASN} \le 5\%$, $5\% < R_{ASN} \le 10\%$, and $R_{ASN} > 10\%$, the last-named being restricted by the above requirement on R_D, i.e. achievability of α_{tg} and β_{tg}.
A typical example of such zones for $\alpha_{tg}=\beta_{tg}=0.05$, d=2, and Φ_0=1, 2, 3 is shown in Figure 9. The fan-shaped zones have their apices on the corresponding centrelines. These apices are the OTA locations, as with the imposed limits satisfied they are closest to the origin (heaviest truncation). In these circumstances the search zone is narrowed, the location problem being converted from two- to one-dimensional.
To study the relationships between the sought boundary parameters (TA, α^*, β^*) and the specified test characteristics (Φ_0, d, $\alpha_{tg}=\beta_{tg}$, $R_{ASN\,max}$), a search was run over a large population of optimal tests with the characteristics given in the Table below.

	Lower limit	Upper limit	Number of levels
Φ_0	0.3	5	9
d	1.5	5	12
$\alpha_{tg}=\beta_{tg}$	0.05	0.25	5
$R_{ASN\,max}$	5%	10%	2

Table 1. Regions of characteristics covered by search

5.2 Search results for OTA and their curve fitting

The dots in Figure 10a mark the OTA for $\alpha_{tg}=\beta_{tg}=0.05$, $R_{ASN} \approx 10\%$, and wide intervals of d and Φ_0. Figure 10b is a zoom on the domain in 3a representing the "short" tests, namely those with small ASN and - correspondingly - low TA coordinates. It is seen that all curves smooth out as the distance from the origin increases (the tests become longer), the reason

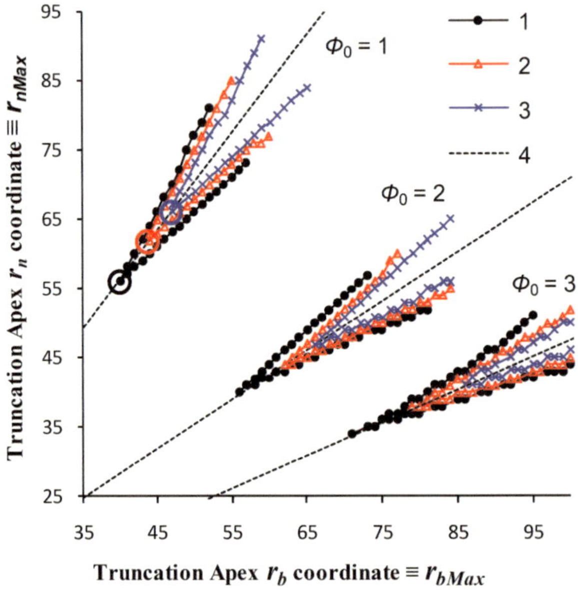

Fig. 9. TA zone boundaries for three Φ_0 values and three R_{ASN} zones (Michlin et al., 2011): 1 = Boundary beyond which $R_D \leq 1\%$ is unachievable at any R_{ASN}; 2 = Boundary for $R_{ASN} \leq 10\%$; 3 = Boundary for $R_{ASN} \leq 5\%$; 4 = Centreline. Remark 1. $\Phi_0=1$ subgraph: OTA for each R_{ASN} zone circled. Remark 2. For this figure: $d=2$, $a_{tg}=\beta_{tg}=0.05$, $R_D=1\%$.

being the weakening influence of discreteness of the test characteristics (Michlin et al., 2009).

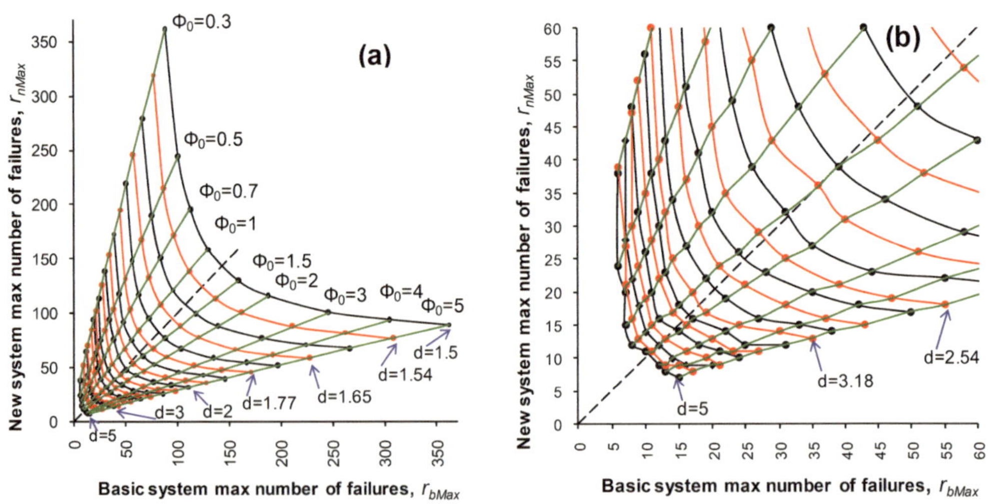

Fig. 10. (a) OTA locations for different d and Φ_0, and for $a_{tg}=\beta_{tg}=0.05$, $R_{ASN}\approx 10\%$. (b) Zoom on short test zone. (Michlin et al., 2011).

The Φ_0-isopleths in the figures are broken radial lines, whereas their d-counterparts are symmetrical about the $r_n= r_b$ line and approximate neatly to a hyperbola:

$$r_n(r_b) = \left\{\left[k(d)\cdot q(x, R_{ASN})\right]^{-1} - r_b^{-1}\right\}^{-1} \tag{30}$$

where

$$k(d) = \exp\left[5.58(d-1)^{-1/4}\right] - 1\,; \tag{31}$$

$$q(x, R_{ASN}) = -\frac{1}{20}(1 + 1.10\ln x + 0.41R_{ASN} - 1.03R_{ASN}\ln x) \tag{32}$$

x - common target value for α and β, $x= \alpha_{tg}=\beta_{tg}$;
R_{ASN} - in relative units rather than in percent.
The formulae indicate that the approximate curves differ only in the scale factor $k(d)$, common to both axes - it remains the same for any pair (x, R_{ASN}).
As the formulae do not contain Φ_0, the OTA is searched for through its required adherence to the centreline, whose expression (29) is uniquely determined by d and Φ_0. Accordingly, the sought OTA is the integer point closest to the intersection of the curve (30) and the centreline (29) (Figure 11).

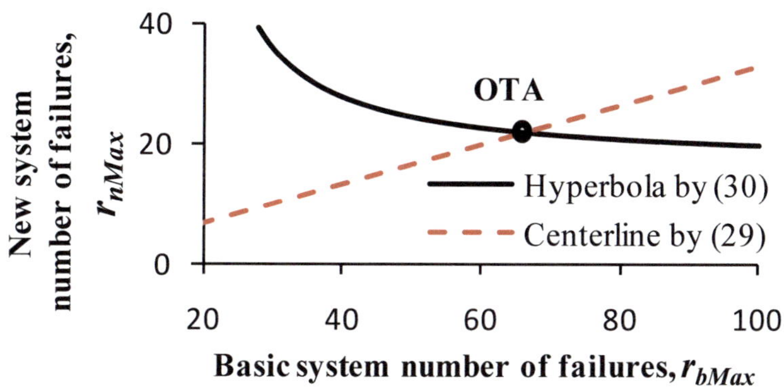

Fig. 11. Determination of OTA. (Michlin et al., 2011).
The coefficients in (31) and (32) were found through the requirement of minimal root mean square error (*RMSE*) - the difference between the OTA's found as per eqs. (29) and (30). For the data in the Table, *RMSE*=0.88, indicating high estimation accuracy for such a broad domain.

5.3 Estimates for α^* and β^*

As already mentioned, the problem of finding the oblique boundaries reduces to that of finding α^* and β^*. This Subsection presents regressional dependences of the latter on the test characteristics Φ_0, d, $x=\alpha_{tg}=\beta_{tg}$, and $R_{ASN\ max}$, as well as their counterparts for the upper and lower limits (α_U^* and α_L^*, β_U^* and β_L^*) of these parameters. These dependences, determined on the basis of the total data on optimal tests with the characteristics in the Table, were sought in the form:

$$\alpha_M^* = c_\alpha \cdot x; \qquad (33)$$
$$\beta_M^* = c_\beta \cdot x.$$

The Matlab tool for stepwise regression yielded the coefficients for the above:

$$c_\alpha = 1.10 - 0.021(\ln x)^2 - 0.0081\Phi_0^2 + 0.036\Phi_0 d + 1.07R_{ASN}\ln x \qquad (34)$$

where *RMSE*=0.061 and R^2=0.83, the latter being the coefficient of determination, and

$$c_\beta = 1.09 + 0.096\ln x + 0.14d - 0.018\Phi_0 d + 1.11R_{ASN}\ln x \qquad (35)$$

with *RMSE*=0.069 and R^2=0.80.
The limit formulae read

$$\left.\begin{matrix}\alpha_U^* \\ \alpha_L^*\end{matrix}\right\} = (1 \pm c_{\alpha B})\alpha_M^* \qquad (36)$$

$$\left.\begin{matrix}\beta_U^* \\ \beta_L^*\end{matrix}\right\} = (1 \pm c_{\beta B})\beta_M^* \qquad (37)$$

where

$$c_{\alpha B} = -0.045 + 0.14\ln d - 0.031\ln x \qquad (38)$$

$$c_{\beta B} = -0.059 + 0.16\ln d - 0.048\ln x \qquad (39)$$

and such that all a^* and β^* obtained for the Table are included.
Figure 12 shows example dependences for the regressional value α_M^* and the upper and lower limits, versus $x=a_{tg}$ for Φ_0=3 and d=1.5, 3. Also included are the actual values of a^*. (The graphs for β^* are analogues). The bounded zone becomes narrower as d and a_{tg} decrease. It is seen that at low d, α_M^* and β_M^* can serve as the calculation values without undue deviation of a_{real} and β_{real} from their targets.
The search methodology for a^* and β^* of the optimal test, described in detail in Section 4, is based on knowledge of the limits (36), (37), which is one of the reasons for its high efficacy.

5.4 Accuracy assessment of proposed planning

The accuracy of the proposed planning, using eqs. (30) – (32) and (33) – (35) – was assessed by applying them in calculating the test boundaries for all characteristic values in the Table. This was followed by calculation of a_{real}, β_{real} and R_{ASN} for these tests and their deviation from the targets. The *RMSE*'s of a_{real} and β_{real} decrease with decreasing d and R_{ASN}. For $d\leq2$ they do not exceed 3 to 4% of the target value and for large d they reach 8 and 10% at R_{ASN}=5 and 10% respectively. In the former case this is very satisfactory accuracy, while in the latter case it may become necessary to find more accurate values of the boundary parameters – for which the methodology outlined in Section 4 is recommended, using eqs. (30) – (32) and (34) – (39) for the search limits.

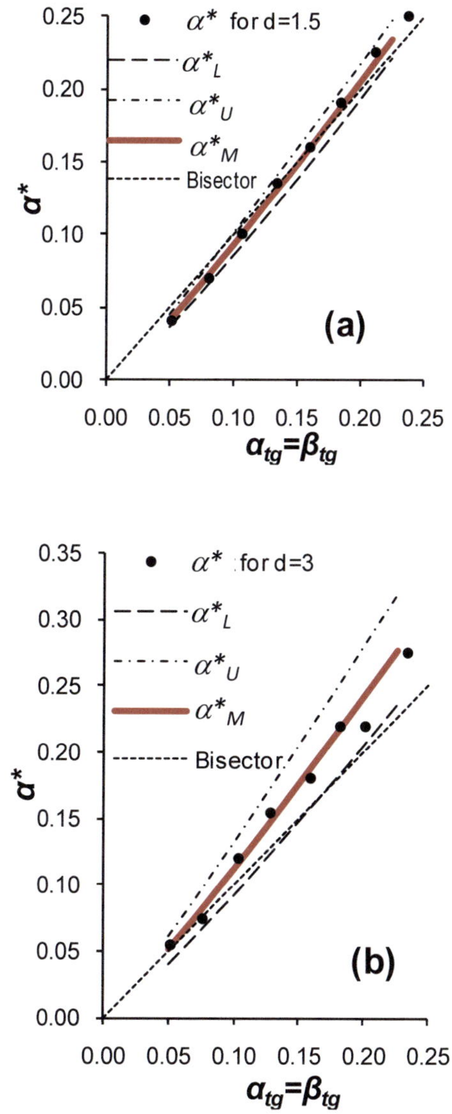

Fig. 12. Actual α^*, regressional dependence, and upper and lower search limits. R_{ASN}=5%. (a) d=1.5. (b) d=3. (Michlin et al., 2011).

6. Group tests

In this case the items are compared groupwise, which makes for economy in the time to a decision. The items of the respective subgroups, N_b and N_n in number, are drawn at random from their respective populations with exponential TBF's, and tested simultaneously. On failing, they are immediately replaced or repaired – just as in the two-item tests. The subgroup can be treated as a single item with an N-times shorter MTBF (Epstein & Sobel, 1955). The planning procedure remains the same, except that Φ in the calculations is replaced by Φ_g:

$$\Phi_g = \Phi \cdot N_b / N_n \tag{40}$$

Thus when N_b=N_n= N, the test boundaries remain as in the two-item case, except that the test duration is also N times shorter (see (18)). When $N_b \neq N_n$, it is recommended to check the efficacy of larger groups, e.g. in terms of a shorter average test duration $ATD_g(\Phi)$. By (18) and (40) we obtain:

$$ATD_g\left(\Phi_g\right) = \left(\theta_b / N_b\right) \cdot ASN_g\left(\Phi_g\right) / \left(1 + 1/\Phi_g\right) \tag{41}$$

where $ASN_g(\Phi_g)$ is the ASN of the group test as per (17) or (22), except for Φ_g replacing Φ of (40).

The planning example covers also the problem of choice of N_b and N_n, while ensuring min ATD_g and satisfying additional essential test-planning conditions.

7. Example of test planning

A large organisation operates a correspondingly large body of mobile electronic apparatus whose MTBF is substantially shortened under the stressful exploitation conditions. The manufacturer offers to modify this equipment, thereby significantly improving its resistance to external impacts, albeit at increased weight and cost.

In a fast laboratory test the modified (hereinafter "new") apparatus exhibits high reliability, but so does the original ("basic") one. Accordingly, it is decided to check the MTBF increase under field conditions on an experimental batch.

The requirements regarding the test OC are established as follows. If the MTBF of the new product is 5 times that of the basic (Φ=5), replacement is beneficial; at Φ=2.5 it does no harm; but at Φ=1.5 it is unacceptable. These findings follow from the OC_{nTr} of a non-truncated SPRT with a=β=0.1, d=2, Φ_0=5 (Figure 13), constructed as per (23) - (24).

The apparatus are operated in sets of 28 items, so that conditions within a set are practically uniform. Each set comprises both new and basic items, so as to offset the influence of fluctuating conditions.

A "failure" in this context is defined as any event that necessitates repair or re-tuning of the item, with enforced idleness for more than 20 seconds. The failed item is either treated in situ - or replaced by a spare, repaired and stored with the spares. Thus the size of the operative set remains 28.

The assignment is - planning a truncated test with the proportions of new and basic items in the test group chosen so as to ensure a minimal ATD. Below is the planning procedure:

a. As the OC's are practically the same for truncated and non-truncated tests when their Φ_0, d, a_{real} and β_{real} coincide (Michlin & Grabarnik, 2007) - we chose the initial parameters given above:

$$\Phi_0=5,\ d=2,\ a_{tg}=\beta_{tg}=0.1 \tag{42}$$

and specified

$$R_{ASN\ max} = 10\% \tag{43}$$

Thus the test has an ASN and ATD close to that of the non-truncated SPRT, and at the same time its maximal duration is heavily restricted, a fact of practical importance for the organisation.

b. Eqs. (41) and (18) yielded the approximate dependences of $ATD_g(\Phi)/\theta_b$ on N_n for different Φ, given $N_n+N_b=28$. A minimum was found at $N_n\approx18$. Figure 14 shows examples of these dependences at $\Phi=\Phi_0$ and $\Phi=\Phi_1$, which are seem to be almost flat over a wide interval around the minimum, and $N_n=15$ was chosen accordingly. With this choice, $ATD_g(\Phi)$ only slightly exceeds the minimum, while the number of new items is lower, with the attendant saving in preparing the experimental batch. By (40) we have

$$\Phi_{0g}=\Phi_0\cdot 13/15=4\tfrac{1}{3} \tag{44}$$

The values of $ASN_g(\Phi)$ and $ATD(\Phi)$, obtained by (41) and (18) with allowance for (44) – confirmed the practicability of the test.

c. Eq. (8) yielded s=0.330. Simultaneous solution of (29) and (30) yielded, after rounding-off, the TA coordinates: r_{bMax}=66, r_{nMax}=22.
Eqs. (33) through (35) yielded a^*=0.0909, β^*=0.074, which in turn, by (11) and (12), yielded h'_b=4.804, h_n=4.453.

The decision boundaries for a test planed on the basis of these parameters are shown in Figure 2.

Figure 13 shows the exact values of the functions $OC(\Phi)$ and $ASN(\Phi)$ as per eqs. (13) - (18), which in turn yield the test's real characteristics: Φ_0=5, d=2, a_{real} =0.098, β_{real} =0.099, R_{ASN}=9.2%, in very close agreement with the given (42) and (43) - evidence of the high accuracy of eqs. (30) - (35).

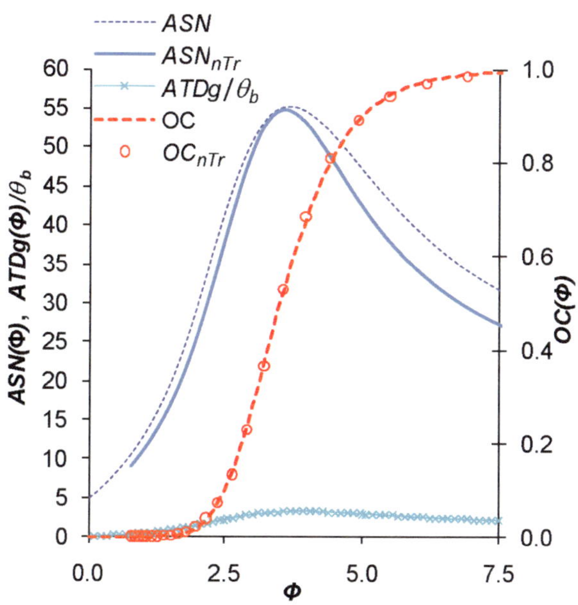

Fig. 13. OC and ASN of truncated group test and of non-truncated theoretical (subscript nTr) test; normalised expected duration of group test $ATDg(\Phi)/\theta_b$ for Φ_0=5, Φ_{0g}=4$^1/_3$, d=2, α_{real}=0.098, β_{real}=0.099, r_{bMax}=66, r_{nMax}=22. (Michlin et al., 2011).

The $OC(\Phi)$ of the planned test (Figure 13) practically coincides with that of the non-truncated test $OC_{nTr}(\Phi)$ with the same α_{real} and β_{real}. The ASN of the former is higher than that of the latter, in accordance with R_{ASN}=9.2%. The diagram also shows the estimate for the normalised ATD, i.e. the ratio $ATD_g(\Phi)/\theta_b$. Assuming $\hat{\theta}_b = 10$ hr, the time requirement of the test should be reasonable. In practice, it ended with acceptance of the null hypothesis in 16 hr, following the twenty-first failure in the basic subgroup, by which time a total of 2 failures in the new subgroup had been observed.

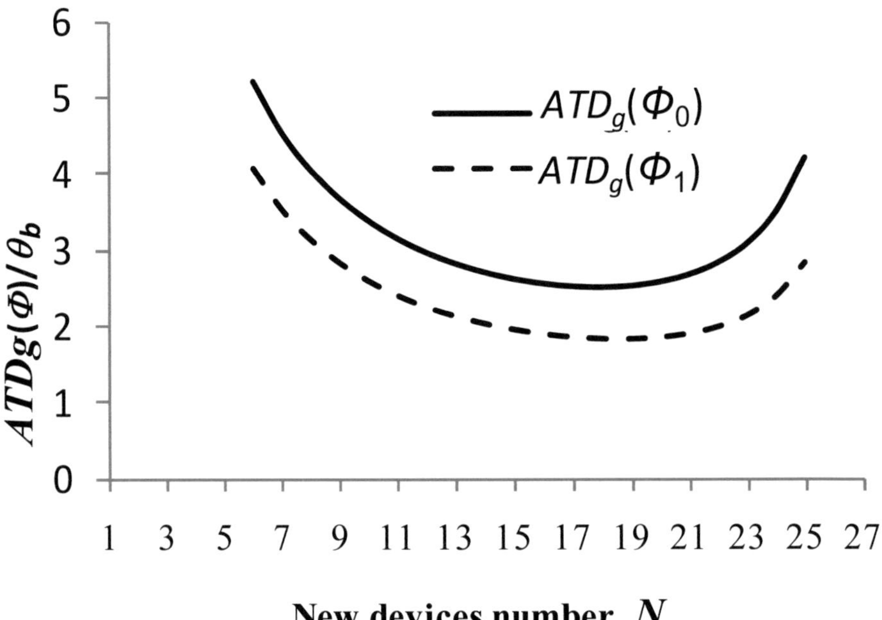

Fig. 14. Normalised expected group test duration vs. number of new devices, for $\Phi=\Phi_0$, and $\Phi=\Phi_1$. (Michlin et al., 2011).

8. Conclusion

The example in Section 7 demonstrated the potential of the proposed planning methodology for a truncated discrete SPRT. An innovative feature in it are the test-quality characteristics R_{ASN} and R_D - which represent, respectively, increase of the ASN on truncation and closeness of the test OC to the non-truncated one. This innovation permitted comparison of different SPRT and automatisation of the optimum-choice process. It was found that over a large domain about the solution, the R_{ASN} and boundary parameters are linked monotonically and almost linearly. This implies sound choice of this characteristic and simplifies the planning. An efficacious search algorithm was developed for the optimal test boundaries, incorporating the obtained interrelationships.

The findings can be summed up as follows:

- A truncated SPRT was studied with a view to checking the hypothesis on the ratio of the MTBF of two objects with exponential distribution of TBF.

- It was established that the basic test characteristics α_{real}, β_{real}, R_{ASN} depend monotonically on the absolute terms in the equations of the oblique test boundaries.
- At the search limits for these absolute terms, determined in Section 5, these dependences are almost plane.
- α_{real} and β_{real} change stepwise with the smooth changes in the absolute terms of the oblique boundaries; expressions are derived for the minimal intervals of these terms, over which α_{real} and β_{real} remain unchanged.
- These and other established regularities yielded an efficacious algorithm and programme for determining the optimal location of the test boundaries.
- The found links between the input and output characteristics of the test, and the fast-working algorithm for its planning, permit improvement of the planning methodology and its extension to all binomial truncated SPRT.
- On the basis of the above body of information, regressional relationships were derived for determining the TA coordinates and oblique-boundary parameters of the optimal tests. Also derived were formulae for the limits of the latter parameters. These are very close at low d and R_{ASN} and draw apart as the characteristics increase; the reason being increasing influence of the test's discreteness. The regressional relationships and boundary-parameter limits permit quick determination of these boundaries for the optimal test with specified characteristics.
- The methodology is also applicable in group tests, with the attendant time economy; moreover, it permits optimisation of the respective group sizes.
- A planning and implementation example of this test is presented.

9. Acknowledgements

The authors are indebted to Mr. E. Goldberg for editorial assistance, and to MSc students of the "Quality Assurance and Reliability" Division of the Technion: Mrs. E. Leshchenko, and Messrs. Y. Dayan, D. Grinberg, Y. Shai and V. Kaplunov, who participated in different stages of this project.

The project was supported by the Israel Ministry of Absorption and the Planning and Budgeting Committee of the Israel Council for Higher Education.

10. Acronyms

ADP	accept decision point
AL	accept line
ASN	average sample number
ATD	average test duration
MTBF	mean TBF
OC≡ $P_a(\Phi)$	operating characteristic
OTA	truncation apex of the optimal test
RDP	reject decision point
RL	reject line
RMSE	root mean square error
SPRT	sequential probability ratio test

TA	truncation apex
TBF	time between failures or time to failure
WAS	program name

11. Notations

$ASN(\Phi)$	exact value of ASN for a truncated test, obtained recursively (17)
$ASN_{nTr}(\Phi)$	ASN calculated via an analytical formula (22) for a non-truncated test
$ATD(\Phi)$	ATD function for given Φ
c	with the appropriate subscripts, coefficients in the approximative equations
$d = \Phi_0 / \Phi_1$	discrimination ratio
h'_b, h_n	absolute terms of Accept, and Reject oblique boundaries, respectively
N_b, N_n	item numbers of "basic" and "new" subgroups in group test
$P_a(\Phi) \equiv OC$	acceptance probability of H_0 at given Φ
$P_{ADP}(r_n,\Phi)$, $P_{RDP}(r_b,\Phi)$	probabilities of reaching the given points ADP, RDP
$P_R(\Phi)$	probability of new system failing next during test
r_b, r_n	system number of failures observed up to time T
$r_{bADP}(r_n)$	r_b-coordinates of ADP for given r_n
$r_{nRDP}(r_b)$	r_n-coordinates of RDP for given r_b
R^2	coefficient of determination
R_{ASN}	relative excess of the ASN of the truncated test over its non-truncated counterpart
R_D	relative deviation a_{real} and β_{real} from their targets
R_{d0} and R_{ASN0}	threshold values of R_d and R_{ASN}
s	slope of oblique boundaries
T	current test time
TA_b, TA_n	r_b- and r_n-coordinates of TA, respectively
x	common target value for a and β, $x = a_{tg} = \beta_{tg}$
a, β	probabilities of I- and II-type errors in test
a_{real}, β_{real}	exact real values of α and β computed for prescribed stopping boundaries
a_{tg}, β_{tg}	target values of a, β
a^*, β^*	parameters determining the constant terms of initial boundary lines
$\alpha_M^*, \alpha_U^*, \alpha_L^*, \beta_M^*, \beta_U^*, \beta_L^*$	regressional value, upper and lower search limits of a^* and β^*
$\theta, \theta_b, \theta_n$	MTBF, same for the basic system θ_b, and for the new system θ_n respectively
$\Phi = \theta_n / \theta_b$	true MTBF ratio
Φ_0	Φ value for which the null hypothesis is rejected with probability a
Φ_1	Φ value for which the null hypothesis is rejected with probability $1-\beta$
Φ_g	Φ for group test

12. References

Aroian L. A. (1968). Sequential analysis-direct method. *Technometrics*. Vol. 10, pp. 125-132.

Barnard, G. A. (1946). Sequential test in industrial statistics, *Journal of the Royal Statistical Society*. Suppl., Vol. 8, pp. 1-21.

Chandramouli, R.; Vijaykrishnan N. & Ranganathan, N. (1998). Sequential Tests for Integrated-Circuit Failures, *IEEE Transactions on Reliability*. Vol. 47, No. 4, pp. 463–471.

Chien, W. T. K. & Yang, S. F. (2007). A New Method to Determine the Reliability Comparability for Products, Components, and Systems in Reliability Testing. *IEEE Transactions on Reliability*, Vol. 56, No. 1, pp. 69–76.

Drenick, R. F. (1960). The failure law of complex equipment. *The Journal of the Society for Industrial Applications of Mathematics*, Vol. 8, No. 4, pp. 680-689.

Eisenberg, B., & Ghosh, B. K. (1991). The sequential probability ratio test. In: *Handbook of Sequential Analysis*, B. K. Ghosh, Sen P.K (Ed.), pp. 47-66, Marcel Dekker, NY.

Epstein, B. & Sobel, M. (1955). Sequential life test in the exponential case. *The Annals of Mathematical Statistics*, Vol. 26, pp. 82-93.

IEC 61650 (1997) *Reliability Data Analysis Techniques - Procedures for Comparison of Two Constant Failure Rates and Two Constant Failure (Event) Intensities.*

Kapur, K. C. & Lamberson, L. R. (1977). *Reliability in Engineering Design*. Wiley, NY, pp. 342-363.

Kececioglu, D. (1993). *Reliability & Life Testing: Handbook*. Vol. 1, Prentice Hall, NJ, pp. 133-156.

Mace, A. E. (1974). *Sample Size Determination*. Robert E. Krieger Pub. Co., NY, pp. 110-114.

Michlin, Y. H. & Grabarnik, G. (2007). Sequential testing for comparison of the mean time between failures for two systems. *IEEE Transactions on Reliability*, Vol. 56, No. 2, pp. 321-331.

Michlin, Y. H.; Grabarnik, G., & Leshchenko, L. (2009). Comparison of the mean time between failures for two systems under short tests. *IEEE Transactions on Reliability*, Vol. 58, No. 4, pp. 589-596.

Michlin, Y. H. & Grabarnik, G. (2010). Search boundaries of truncated discrete sequential test. *Journal of Applied Statistics*. Vol. 37, No. 05, pp. 707-724.

Michlin, Y. H.; Ingman, D. & Dayan, Y. (2011). Sequential test for arbitrary ratio of mean times between failures. *Int. J. of Operations Research and Information Systems*, Vol. 2, No. 1, pp. 66-81.

Michlin, Y. H. & Kaplunov, V. (2007). Optimal truncation of comparison reliability tests under unequal types I and II error probabilities, *Proceedings of the 9th Conference of Israel Society for Quality*, Tel-Aviv, Nov. 2007, 6 pp.

Michlin, Y. H. & Migdali, R. (2004). Test duration in choice of helicopter maintenance policy. *Reliability Engineering & System Safety*, Vol. 86, No. 3, pp. 317-321.

MIL-HDBK-781A (1996). *Reliability test methods, plans, and environments for engineering, development, qualification, and production*. US DOD, pp. 32-42.

Siegmund, D. (1985). *Sequential Analysis: Tests and Confidence Intervals*, Springer, NY, pp. 34-63.

Sr-332. (2001). *Reliability prediction procedure for electronic equipment.*, Telcordia Technologies Inc.,. Red Bank, NJ, Section 2.4.

Wald, A. (1947). *Sequential Analysis*, John Wiley & Sons, NY.

Wald, A. & Wolfowitz, J. (1948). Optimum character of the sequential probability ratio test. *The Annals of Mathematical Statistics*, Vol. 19, No. 3, pp. 326-339.

26

Dependence of Determination Quality on Performance Capacity of Researching Technique, Exemplified by the Electron Probe X-Ray Microanalysis

Liudmila Pavlova
Vinogradov Institute of Geochemistry, Siberian Branch of Russian Academy of Sciences, Irkutsk Russia

1. Introduction

The quality of results obtained by any analytical method depends on every stage of data acquisition: representativeness of study object; appropriate sample preparation; optimum conditions for analytical signal excitation and registration; availability of reference materials for comparison; procedure to process acquired values referred to the content to be determined. These characteristics vary in different analytical methods. In some cases, the sample preparation represents the major source for analytical errors, in some others, the complexities arise from the incorrect selection of calibrating plot, and thus availability of reference materials for comparison is essential. The technique specifications pose the requirements to every stage of analysis.

The problem of quality has ever been critical in analytical work, and every time it depends on the level of progress in the theory and application of selected technique.

Quality has emerged and remained the dominant theme in management thinking since the mid-twentieth century (Beckford, 2010).

2. Quality of the electron probe X-ray microanalysis at each stage

The electron probe X-ray microanalysis is a fairly young technique. The first papers describing the basics of electron probe microanalysis (EPMA) (Castaing, 1951; Borovskii, 1953) and original designs and constructions of microanalyzers (Castaing & Guinier, 1953; Borovskii & Il'in, 1956) were published in 1951-1956 in France and the USSR. The technique was rapidly progressing. In 1973 Borovskii (Borovskii, 1973), the founder of EPMA in Russia, admitted that one could hardly identify the fields of science and engineering, where the EPMA had not been successfully used. The instruments and theory were developing simultaneously.

At present it is difficult to overestimate the significance and application of the method. It is one of the leading methods in mineralogy, and similarly in metallurgy and biology. The investigations of the micro-level are required at approbation of technological processes in all fields of science and engineering.

The quality of results is of prime importance for researchers in all fields of science. The specifics of the method related to obtaining information suggest it to be the research technique, thus intensifying the problem of determination quality, rather than the analytical method. The studies on developing the theoretical foundation of the method and upgrading the instruments, reported in numerous publications, and partly mentioned in the articles reviewed by authors of articles (Szaloki et al., 2004; Pavlova et al., 2000), allow the way of improving quality of the electron probe X-ray microanalysis to be observed. The topical studies undertaken by the author are included into this chapter.

2.1 Representativeness of materials

The representativeness of the sample substance is initially defined when posing problem, and it depends on the requirements of particular analytical technique. These requirements imply the subsample weight, homogeneity of components, and particle size in the subsample, solubility and miscibility of subsample substance with a binding substance or a solvent, and others.

In case of EPMA the representativeness of determinant, namely the quantity of samples and size of the surface, prepared for examination, depend on the frequency of component occurrence in material, probability of their occurrence on the prepared surface and their phase distribution over the investigated surface. In electron probe microanalysis, when we study the inclusions rarely occurring in the groundmass, it is essential to have a sufficient amount of the geological substance to be examined.

In EPMA a correct solution of the problem posed is dependent on the frequency of determinant occurrence within the observation zone. With EPMA, an absence on the studied surface of the element to be defined is not suggestive of its complete lack in the sample. It might be assumed, that the sought element has skipped from the study zone.

It can be exemplified by searching for the invisible gold in lithochemical stream sediments of the Dukat gold-silver deposit in northeastern Russia. Initially, the studies of rock in thin and polished sections did not provide wanted results - fine gold inclusions have not been detected. The probability of gold inclusion occurrence on the studied surface was negligibly small. Only having extracted the heavy fraction and prepared the briquette thin sections and after locating grains on the surface and polishing thin sections and thoroughly studied numerous grains in the thin section we managed to obtain positive results.

If the sought inclusion had not been found on the surface, a thin layer of substance was removed, and researched through entire sample. The process was being repeated until the sought inclusion was found (Fig. 1). If the grains of heavy fraction were completely polished down a new briquette thin section of the same sample was prepared and searching was continued. Figure 1 presents the gold inclusion found in the mineral only after the third try of surface polishing. The size of the inclusions is 10 μm, while the area of thin section is about $25 mm^2$. In such cases the quality of investigations and conclusion correctness result from both correctly selected and prepared material and thorough search.

2.2 Sample preparation

In any technique sample preparation is truly important, that is decisive for analytical procedures. In different analytical methods the laboriousness of sample preparation varies. Because the EPMA, in effect, is the analysis of a surface, the sample surface is required to be flat, well polished, smooth and clean and to have a good conductivity. Sample preparation

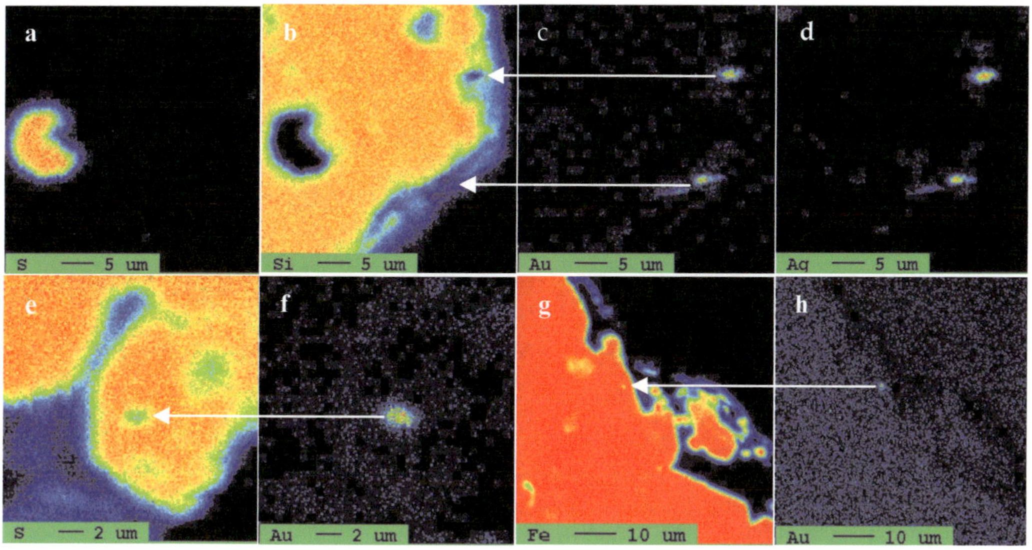

Fig. 1. The gold inclusion found in the mineral only after the third try of surface polishing.

for the electron probe microanalysis is in many ways still as much of an art as a science (Goldstein et al., 1992). Different procedures are applied in EPMA for sample preparation. The solid samples for EPMA may be thin sections, thick micro sections, briquette sections and isolated particles. The preparation of polished thin sections and thick micro sections consists in selecting sized solid material, its cutting, polishing selected surface with diamond pastes.

2.2.1 Particle preparation

When examining the grinded substance the particles should be fixed so that they are not scattered in air and retain the representativity of material when investigated in the vacuum microanalyzer. Various preparation procedures are used to analyse different solid particles (Pavlova et al., 2001): (1) fixing particles on substrate with colloid; (2) pasting particles on the carbon double-faced adhesive tape; (3) preparing briquette sections. In the first case particles are fixed on the polished surface of the substrate with collodion. The particles are distributed as a thin layer on the surface of the substrate, which is preliminarily covered by a thin layer of the liquid collodion. While drying the collodion fixes the particles on the surface. Gluing particles on the carbonic adhesive tape is commonly applied when studying conductive materials. In this case the glued particles cannot be covered by the conductive layer.

Two techniques to prepare briquette thin sections: 1) Particles of any solid material are mixed with epoxy resin, and after the surface is hardened it is polished. 2) Grains of any solid material are glued on the adhesive tape and coated with epoxy resin. After the resin is hardened, the sample is removed from the adhesive tape and the surface with particles is polished to make it mirror-smooth. Due to this procedure plenty of grains are included in the same puck. The briquette thin sections with particles are often used for quantitative determinations of particle composition using the wave spectrometers applied for studying horizontal well polished surfaces. Table 1 presents the data on the comparison of two garnet particles composition processed by different methods of sample preparation.

Preparation			Non-polished particles				Polished particles			
Analytical method			EDS		WDS		EDS		WDS	
Sample	Oxide	Certified concent-ration, wt. %	Concentration, wt. %	Relative error, %	Concentration, wt. %	Relative error, %	Concentration, wt. %	Relative error, %	Concentration, wt. %	Relative error, %
Garnet O-145	*MgO*	20	20,59	*2,95*	21,325	6,63	20,91	*4,55*	19,93	*-0,35*
	Al_2O_3	23,4	23,62	*0,94*	21,286	-9,03	23,19	*-0,90*	23,36	*-0,17*
	SiO_2	42,3	42,55	*0,59*	43,527	2,9	42,07	*-0,54*	42,56	*0,61*
	CaO	4,03	4,14	*2,73*	3,06	-24,02	4,34	*7,69*	4,03	*0,00*
	MnO	0,17	0,07	*-58,82*	0,12	-29,41	0,16	*-5,88*	0,16	*-5,88*
	FeO	10,1	9,3	*-7,92*	9,63	-4,65	9,36	*-7,33*	10,04	*-0,59*
Garnet C-153	*MgO*	21,09	22,61	*7,21*	23,46	11,24	21,25	*0,76*	20,93	*-0,76*
	Al_2O_3	18,09	17,73	*-1,99*	19,97	10,39	18,86	*4,26*	17,84	*-1,38*
	SiO_2	41,52	42,24	*1,73*	42,98	3,52	42,27	*1,81*	41,94	*1,01*
	CaO	3,4	3,7	*8,82*	3,94	15,88	3,62	*6,47*	3,38	*-0,59*
	Cr_2O_3	7,41	6,59	*-11,07*	6,89	-7,02	7,09	*-4,32*	7,3	*-1,48*
	MnO	0,32	0,21	*-34,38*	0,27	-15,63	0,534	*66,88*	0,34	*6,25*
	FeO	7,59	6,85	*-9,75*	6,57	-13,44	6,163	*-18,8*	7,41	*-2,37*

Table 1. Compared compositions of two garnet particles processed by different methods of sample preparation. Relative error=100*(C-Ccer)/Ccer; C- is concentration; Ccer – is certified concentration.

The composition of non-polished particles is not determined with wave spectrometers splitting the X-ray spectrum by the wave length. The quality of determinations is low and the error can reach as high as tens percent. When the particles are glued on the substratum and adhesive tape it is feasible to study the shape and size of particles; when using the energy-dispersive spectrometer it is possible to identify (i) what elements compose grains, (ii) how elements are distributed over the surface, (iii) element contents.

2.2.2 Preparation of biological samples

The surface suitable for the analysis is hard to receive in examining porous samples, which are often biological samples. Preparation of biological materials is differently approached. For example, sponges are first rinsed in distilled water, then dehydrated in alcohol, freed from alcohol with acetone and impregnated with epoxy resin (Pavlova et al., 2004). The samples obtained are fit in one or some briquette sections. The briquette sections with specimens are polished with diamond paste to get the surface flat and mirror-smooth. The Figure 1 displays the sponge image. The solid part of the sponge, its skeleton consists of spicules. This kind of preparation of fragile biological specimens ensures intact solid part of sponge. It avoids distortions due to destructions when polishing. With this procedure we determine silicon concentration in the center of spicule cross-section and on its margins: they are higher in the center than in the margins.

2.2.3 The influence of surface on the quality of results

The distortion of horizontal position of surface (effect of absence of flat horizontal surface) is the cause of false conclusions and considerable deterioration of analytical results. If the surface is either not flat or not horizontal the analytical signal is distorted.

Figure 3 presents the pattern of x-ray radiation distribution of manganese in Mn-pure (a) and gold in Au-pure (b). Samples of pure Mn and Au have flat polished, even if not horizontal surfaces, displaying the x-ray intensity distortion. Manganese shows decrease in intensity to the right and to the left of center (Fig. 3a), because the sample surface is tilted relative to the center: the right part is higher, and the left one is lower. In Fig. 3b the gold particle bottom is in focus, but the top part is elevated toward horizon, therefore AuL_{α} – intensity is deformed in the top markedly lower, though the sample is homogeneous, but some inclusions. The gold image in backscattered electrons does not exhibit the surface inclination (Fig. 3c). Both gold images in Fig. 3b and Fig. 3c were simultaneously produced.

Because the grains of majority of natural samples are dielectrics, a layer of carbon (20-30 nm thick) is vacuum-sprayed onto the polished surface of all tablets to make it conductive and to remove the accumulative charge.

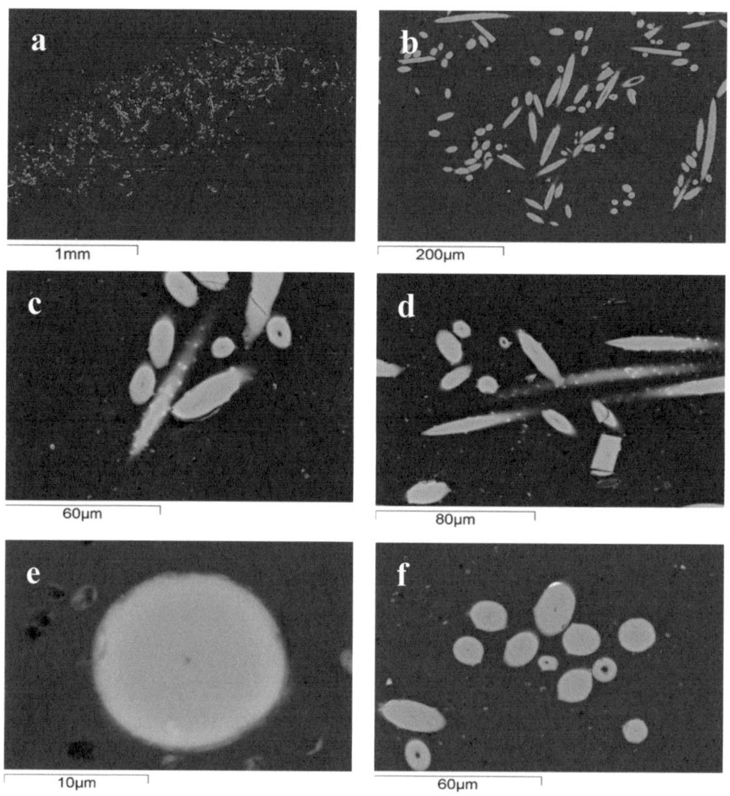

Fig. 2. The sponge surface prepared for EPMA studies. The image is given in back- scattered electrons. Section of whole sponge in epoxy resin (a). Section of sponge part (b). Sections several separate sponge spicules (c, d). Cross-sections of sponge spicules (e, f).

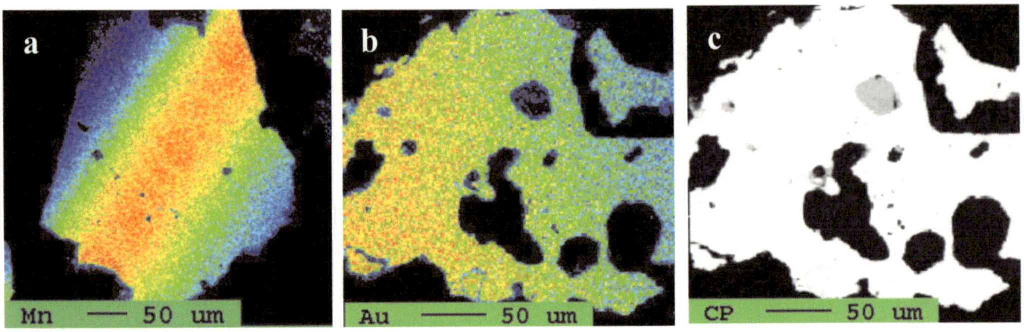

Fig. 3. Patterns of x-ray radiation distribution of manganese (a) and gold (b) in Mn- and Au-pure samples. The samples have flat polished but not horizontal surfaces.

The mass absorption coefficients are basic quantities used in calculations of the penetration and the energy deposition by photons of x-ray in biological, shielding and other materials. The different authors offer special absorption coefficients of x-ray radiation in the same element, specifically in carbon. Varying mass absorption coefficients defined in about twenty articles are discussed in the work (Hubbell et al., 2011). In Table 2 the x-ray absorption cross sections, determined by different authors for energy 0,277 keV in carbon have been compared with the experimental data contained in the National Bureau of Standards collection of measured x-ray attenuation data (Saloman et al., 1988)
The x-ray attenuation coefficients have been approximated by Finkelshtein and Farkov (Finkelshtein & Farkov, 2002). The article by Farkov with co-autors (Farkov et al., 2005) reports the data on compared absorption coefficients of nitrogen and oxygen radiation in carbon from different literature sources (Table 2). Table 3 presents the absorption coefficients of carbon, nitrogen and oxygen radiation in carbon, as well as relative deviations in the data of different authors.

Authors (Reference from the article of Saloman et al., 1988)	Fomichev V.A. & Zhukova I.I. (68FO1)	Henke et al. (67HE1)	Denne D.R. (70DE2)	Denne D.R. (70DE1)	Weiswe-iler W. (68WE1)	Messner R.H. (33ME1)	Duncumb P. & Melford D.A. (65DU1)	Kurtz H. (28KU1)	Dershem, E. & Schein, M. (31DE1)
XACS, barns/atom $*10^4$	5,600	4,569	3,850	4,09	5,056	3,43	4,587	4,200	4,400
Relative error, %	3,7	-17	-40	-31,8	-6,6	-57,2	-17,5	-28,4	-22,5

Table 2. X-ray absorption cross sections, determinated by different authors (Saloman et al., 1988). XACS - is X-ray absorption cross sections of authors; Relative error= -EXPS - XACS)/ EXPS)*100; EXPS - is experimental data contained in the National Bureau of Standards collection of measured x-ray attenuation data.

Emitter		CK_α			NK_α			OK_α		
Authors		(Heinrich, 1986)	Henke*, 1967	(Henke, 1993)	(Heinrich, 1986)	Henke*, 1967	(Henke, 1993)	(Heinrich, 1986)	Henke*, 1967	(Henke, 1993)
Carbon absorber		2147	2292	1960	23585	21784	25000	11575	9079	12100
Relative deviation, %	Athor-Heinrich, 1986	-	6,3	-9,5	-	-8,3	5,7	-	-27,5	4,3
	Athor - Henke*, 1967	-6,7	-	-16,9	7,6	-	12,9	21,6	-	25,0
	Athor - Henke, 1993	8,7	14,5	-	-6,0	-14,8	-	-4,5	-33,3	-

Table 3. X-ray absorption coefficients (centimeter^2/g) in the carbon and relative deviations in the data of different authors. *Henke's data from work of Saloman (Saloman et al., 1988). Relative deviation=(Author-Comp)/Author*100%. Author - is data of author; Comp – is comparison data.

It might be assumed how the quality light elements (carbon, nitrogen and oxigen) determination depends on choice of absorption coefficients.

The thickness of carbonic film on the studied surface and on the sample for comparison should be the same not to deteriorate the quality of results.

2.3 Optimum conditions to excite and register analytical signals

Ensuring a high quality of results depends on uniformity of analytical signal through the time of observation. The conditions of exciting and registering analytical signal are to provide such uniformity. Selection of optimum conditions to excite and register analytical signals is the major condition to obtain correct information on a sample. It is one of the main conditions for any method of analytical chemistry.

With EPMA applied the surface of different samples: glassy materials, sponges, bones, argentiferous samples and others can be destroyed during excitation with x-radiation, when electron probe falls on the sample surface, and as this takes place, the analytical signals become heavily contorted. Selection of optimum conditions of measurement can be based on the criteria of (a) minimum detection limit of sought elements, (b) uniformity of analytical signal during measurement, (c) variations of element intensity depending on the electron beam density or (d) sample stability while performing measurements (Buseck @ Goldstein, 1969).

2.3.1 The criterion of minimum detection limit

The detection limit variations depending on measurement conditions were identified for scandium, strontium and barium in basalt glasses. Figure 4 illustrates the plotted variations of detection limits for scandium, strontium and barium in basalt glasses depending on measurement conditions (Paradina @ Pavlova, 1999).

Radiation of SrL_α -, BaL_β - and ScK_α - lines is measured with the PET crystal for different accelerating voltages, counting times and probe currents. The curves show that the optimum conditions for electron probe microanalysis with wavelength spectrometers (WD EPMA) of

Ba, *Sc* and *Sr* are accelerating voltage 25 kV, sample current 80-100 nA and counting time 20 s. With the other measurement conditions the detection limits of above-indicated elements are higher, thus the determination quality of these elements in glasses is worse.

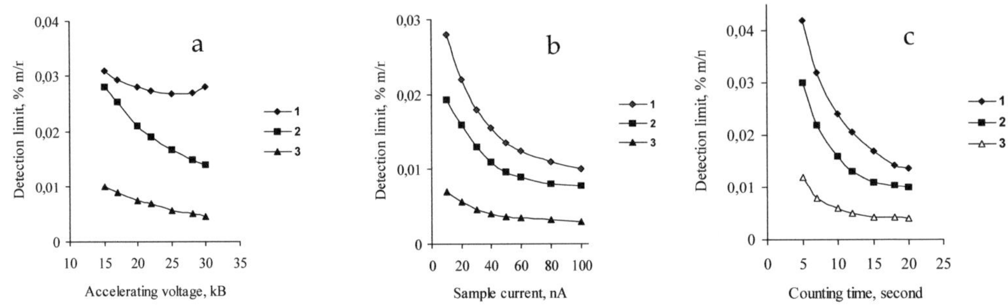

Fig. 4. Detection limits for SrL_{α} - (1), BaL_{β} - (2), and ScK_{α} - (3) as the functions of: (a) accelerating voltage (sample current 15 nA, counting time 10 s); (b) sample current (accelerating voltage 25 kV, counting time 10 s); (c) counting time (probe current 15 nA, accelerating voltage 25 kV). Measurements were made by Camebax-micro and Camebax SX-50 microprobes (Paradina @ Pavlova, 1999).

2.3.2 Variations of x-ray intensity versus counting time

Intensity variations of the elements present in the bones versus counting time are examined by microprobe Superprobe-733 (Pavlova et al., 2001). Intensity variations of elements present in bone tissue are shown in the fig. 5.

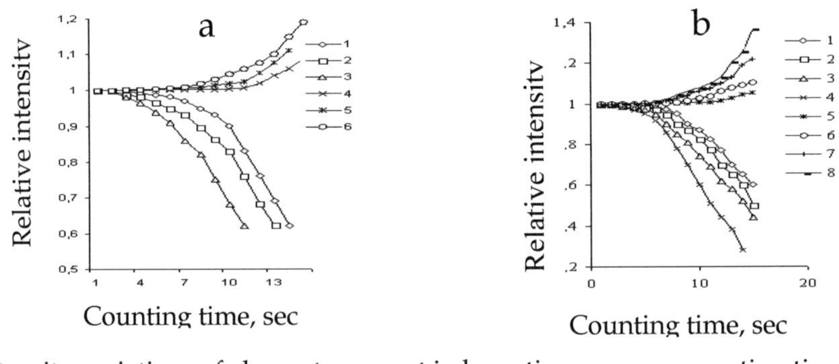

Fig. 5. Intensity variations of elements present in bone tissue versus counting time: (a) NaK_{α} (curves 1-3 correspond to accelerating voltage 10, 15 and 20 kV, respectively, and probe current 15 nA) and MgK_{α} (curves 4–6 correspond to accelerating voltage 10, 15 and 20 kV, respectively, and probe current 15 nA); (b) NaK_{α} (curves 1–4 correspond to sample current 10, 15, 20 and 25 nA, respectively, and accelerating voltage 15 kV) and MgK_{α} (curves 5–8 correspond to sample current 10, 15, 20 and 25 nA, respectively, and accelerating voltage 15 kV). Measurements were made by Superprobe-733 microprobe (Pavlova et al., 2001).

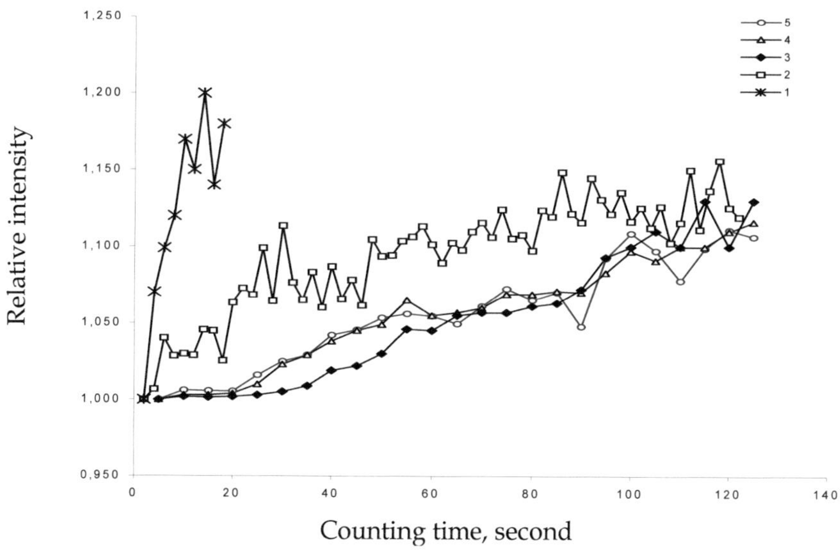

Fig. 6. The change of CaK_α - line intensity in accordance with the electron beam power densities ($\mu W / \mu m^2$): 1 - 191,00; 2 - 3,82; 3 - 2,54; 4 - 2,86; 5 - 3,06.

The X-ray intensity of calcium versus counting time (Fig. 6) with varying electron beam power density assessed for omul fish otoliths (Pavlova et al., 2003). This change of CaK_α - line intensity versus the electron beam power densities has been chosen as the criterion for selecting optimum conditions for analyzing the elements in omul fish otoliths. As seen on the plot the CaK_α intensity of otoliths remains constant during twenty seconds, if the beam power density is below 2.8 $\mu W / \mu m^2$. The increase of measuring time causes signal distortion and deterioration of result quality.

2.3.3 Criterion of sample stability while performing measurements

The stability of argentiferous samples can be studied quantitatively by Buseck's technique (Buseck @ Goldstein, 1969), and it can also be used as the criterion for selecting optimum conditions for EPMA and thereby refinement of substance test (Pavlova @ Kravtsova, 2006). According to the recommendations (Buseck and Goldstein, 1969) the sample is stable (analytical signal is constant for counting time), if it obeys the relationship

$$S_c < \sigma_c \tag{1}$$

where the repeatability of standard deviation of the measured x-ray counts for the major elements Sc is given by:

$$S_c = \sqrt{\sum_{i+1}^{n} (N_i - \overline{N})^2 / (n-1)} \tag{2}$$

where N_i is the number of x-ray counts in each measurement i and n is the number of individual measurements, and average number of x-ray counts is

$$\overline{N} = \sum_{i+1}^{n} (N_i / n) \quad (3)$$

the population standard deviation stipulated by the Poisson counting statistics σ_c is:

$$\sigma_c = \sqrt{\overline{N}} \ . \quad (4)$$

The stability characteristics of argentiferous sample were determined according to the recommendations (Buseck and Goldstein, 1969). Table 4 provides stability characteristics of argentiferous samples.

It was found that the response from argentiferous sample was stable through period from 1 to 1.5 min depending on the beam power densities.

The beam power density <2.55 $\mu W / \mu m^2$ is admissible for the analysis of argentiferous samples. The best compromise conditions for exciting the x-ray radiation and recording analytical signal for WD EPMA of argentiferous samples are: accelerating voltage of 15-20 kV, beam current of 10 nA, probe diameter of 10 μm and counting time of 10 s. Hence, so as to obtain the correct results of analysis it is necessary to select optimum conditions for exciting and registering analytical signals.

Counting time, second				10		20		30		40		50		60	
Conditions of measurements															
E_0 kV	i nA	d_b µm	P µW /	S_c	σ_c	S_c	σ_c	S_c	σ_c	S_c	σ_c	S_c	σ_c	S_c	σ_c
2	1	2	0,64	15	33,	16,	33,	15,	34,	16,	33,	16,	34,	15,	33,
2	1	2	0,96	16	35,	15,	35,	16,	35,	16,	35,	16,	35,	16,	34,
2	2	2	1,27	15	37,	16,	37,	16,	37,	15,	37,	15,	37,	42,	38,
2	5	1	1,27	16	37,	16,	37,	16,	37,	16,	37,	16,	37,	42,	38,
2	8	1	2,04	15	39,	16,	39,	16,	38,	18,	39,	50,	39,	49,	39,
2	1	1	2,55	16	40,	16,	40,	16,	39,	49,	41,	52,	41,	62,	42,
2	1	1	3,06	16	41,	48,	41,	51,	42,	53,	42,	59,	43,	63,	43,
2	1	1	3,82	47	45,	69,	44,	69,	43,	59,	44,	63,	44,	65,	44,
2	5	1	127,	49	48,	96,	48,	98,	49,	97,	50,	98,	56,	99,	49,

Table 4. The stability characteristics of argentiferous sample. d_b - beam diameter; E_0 - accelerating voltage; i - probe current; P - beam power densities; S_c is repeatability standard deviation of the measured X-ray counts; σ_c is population standard deviation, stipulated by Poisson counting statistics.

2.4 Reference samples and standard materials for electron probe microanalysis

The accuracy of results of any analytical technique depends on the application of a sufficient number of required reference and control samples of top quality. An important condition to acquire appropriate analytical results of required quality is the application of a sufficient number of required reference samples meeting all requirements of analytical method employed. It is desirable to use certified standards as control and reference samples.

A specific feature of electron probe X-ray microanalysis is its locality, which is 10^{-13} g of the substance, causing toughening the requirements for standard and reference materials claimed for EPMA. In the case of EPMA, reference and connrol samples should correspond with the following requirements simultaneously: (1) have known chemical composition; (2) be uniform at macro and micro levels of spatial resolution; (3) remain stable under action of electron probe; (4) do not decay in vacuum (up to 10^{-16} mm Hg); (5) must be well polished. Evidently, only the samples being standard samples of structure and properties at the same time may become standard and reference samples for EPMA. The standard samples widely utilized in the other analytical techniques are often inapplicable in the case of EPMA, because they are not homogeneous both at macro and micro levels.

2.4.1 Homogeneity and stability of samples

The homogeneity (at macro- and micro-level) of space resolution and substance stability resulting from the effect of electron beam is of particular interest for the EPMA. There are different approaches to assess the substance homogeneity (Borkhodoev, 2010; Buseck @ Goldstein, 1969; MI, 1988). Two methods to test the homogeneity of elements distribution at macro and micro levels (Buseck @ Goldstein, 1969; MI, 1988) are described for the copper-rich alloys and basaltic glass.

According to the recommendations of Buseck and Goldstein (Buseck @ Goldstein, 1969) the sample is homogeneous on micrometer- to millimeter scale if

$$S_c / 2\sigma_c < 1 \tag{5}$$

the homogeneity is doubtful if

$$1 < S_c / 2\sigma_c < 2\,, \tag{6}$$

the sample is inhomogeneous if

$$2 < S_c / 2\sigma_c\,, \tag{7}$$

S_c and σ_c values are determined from (2)-(4) formulae.

The macro homogeneity at the micrometer to millimeter scale was evaluated by recording EPMA profiles across a few different profile lines of sample.

The beam diameter ranged from 1 to 10 μm, the profile length was about 1000 μm, the point spacing was 10 μm and the line spacing was 20 μm. $S_c / 2\sigma_c$ values in Table 5 demonstrate that at micron-mm level in alloys CA-2 and CA-6 all elements are distributed irregularly and the tin homogeneity in CA-4 alloy is doubtful.

According to the National Standard for homogeneity assessment (MI, 1988) the material is uniform on a micrometer scale, if

$$\sigma_i = \sqrt{\sigma_b^2 + \sigma_l^2} < RSD\,, \tag{8}$$

Sample		Element	C_{cer}, wt. %	Characteristics					
				Stability (for 2 min)	homogeneity				
				σ_c / S_c	$S_c / 2d$	σ_b	σ_l	σ_i	$\sigma / 8$
Coppery alloys	CA-1	*Sn*	1,62	1,38	0,77	0,03	0,71	0,71	0,71
		Ni	0,10	n.d	0,76	0,08	1,94	1,94	2,10
		Fe	0,05	n.d	0,88	0,09	2,05	2,05	2,50
	CA -2	*Cu*	68,74	1,48	2,96	n.d	n.d	n.d	n.d
		Sn	0,97	1,75	2,95	n.d	n.d	n.d	n.d
		Zn	30,00	1,13	2,27	n.d	n.d	n.d	n.d
	CA -3	*Zn*	0,71	n.d	0,80	0,12	1,12	1,12	0,26
		Cu	98,53	1,49	0,98	0,14	1,12	1,13	0,31
		Sn	0,19	n.d	0,79	0,00	1,55	1,55	1,56
		Ni	0,26	n.d	0,79	0,07	1,13	1,13	1,20
		Fe	0,30	n.d	0,76	0,06	1,16	1,16	0,81
	CA -4	*Zn*	1,56	1,36	0,73	0,04	0,68	0,68	0,85
		Sn	0,09	n.d	1,48	n.d	n.d	n.d	n.d
	CA -5	*Zn*	1,00	1,46	0,92	0,11	1,12	1,12	1,13
		Sn	0,12	n.d	0,83	0,00	1,55	1,55	1,56
		Ni	0,51	n.d	0,89	0,07	1,13	1,13	1,20
		Fe	3,63	1,45	0,87	0,14	0,79	0,80	0,81
	CA -6	*Cu*	63,00	1,23	2,46	n.d	n.d	n.d	0,31
		Zn	33,80	1,43	2,87	n.d	n.d	n.d	n.d
Basalt glass		*Ba*	2,30	1.94	0,47	0,00	1,73	1,73	3,80
		Sr	2,01	1.75	0,375	0,00	1,56	1,56	2,80
		Si	21,84	1.84	0,42	0,02	0,13	0,13	0,13
		Al	9,50	1.72	0,36	0,20	0,34	0,40	0,44
		Ca	7,34	1.68	0,34	0,13	0,38	0,40	0,40
		Mg	3,49	1.88	0,44	0,17	0,49	0,52	0,58
		Fe	8,94	1.66	0,33	0,12	0,32	0,35	0,35
		Na	2,02	1.58	0,29	0,02	0,96	0,96	1,00
		Ti	0,87	n.d	0,39	0,17	0,81	0,83	0,88

Table 5. Characteristics of stability and uniformity of copper alloys and basalt glass. n.d. means that the value was not determined. CA stands for coppery alloys; S_c is repeatability standard deviation of the measured X-ray counts; σ_c is population standard deviation, stipulated by Poisson counting statistics; σ_i is total root mean square deviation; σ_b is the root mean square deviation of random components of error for some parts of the specimen; σ_l is the root mean square deviation of random components of error for the analytical areas; *RSD* is the relative standard deviation (Thompson et al., 1997).

where σ_b is the root mean square deviation of random components of error for some parts of the specimen; σ_l is the root mean square deviation of random components of error for the analytical areas; RSD is the relative standard deviation used for certification (Thompson et al., 1997).

The total X-ray counts were recorded at each point of the sample by the static beam to detect possible micro heterogeneity at 1–10 μm scale. The X-ray intensities of analytical lines were measured from 20 to 30 points of a sample. The pattern of these points was chosen from the table of random numbers. At each point two measurements were made using the optimum conditions. The homogeneity of the material in the surface layers was estimated by repeating measurements after the repeated polishing. The results for each element were processed by a dispersion analysis with a three-step grouping of material. Evidently the samples are homogeneous on micrometer scale excepting alloys CA-2, CA-6 and CA-4.

At EPMA the stability of reference samples and standards under the microprobe is significant both during measuring the analytical signal and using the sample for studies. The data on stability of alloys and glasses, given in Table 5, are obtained by the following method (Pavlova et al., 2001). The X-ray radiation of the most intensive lines was excited by the electron beam at 5 arbitrary points of the selected sample. This was the x-ray radiation of copper Kα-line for alloys and calcium Kα-line for basalt glass recorded by the WD EPMA technique. The measurements were made 30 times and lasted for 5 minutes per point. Relationship (1) was applicable for some elements of all copper alloys and basalt glass. Table 5 indicates that all copper alloys and basalt glass are resistant to electron action during 2 minutes.

2.4.2 Valuation of laboratory reference samples

The necessity to have certified standard samples is obvious; they help to achieve the requisite quality of results. But in some cases the certified standard samples are absent and researchers are forced to apply laboratory reference materials after having the control study fulfilled and these materials assessed (Wilson & Taggart, 2000).

One of the techniques to assess quality of the EPMA results obtained using laboratory reference samples instead of standards is described below for the silicate mineral analysis (Pavlova et al., 2003). Measurements were performed on the Jeol Superprobe-733 electron microprobe using accelerating voltage of 15 kV and probe current of 20 nA with an electron beam diameter of 1 and 10 μm. Assessment was done by analyzing 26 control samples (glass K-412, garnet IGEM, Ti - glass, Mn-glass, Cr –glass, basalt glasses BHVO-2G, BCR-2G, BIR-1G, diopside, ilmenite, garnet O-145, olivine, spinel, garnet C-153, albite, garnet UD-92, orthoclase, chromite UV-126, oxides MgO, MnO, Fe_2O_3, Cr_2O_3, Al_2O_3, TiO_2, SiO_2 and $CaSiO_3$). The glass K-412 was supplied by the National Institute of Standards and Technology of the USA (NIST, 1990). Garnet IGEM, Ti - glass, Mn-glass and Cr -glass were prepared as laboratory reference samples at the Institute of Geology and Mineralogy, Moscow, Russia. Three basalt glasses BHVO-2G, BCR-2G andBIR-1G were produced by the US Geological Survey (Denver Federal Center) (Wilson & Taggart, 2000). The diopside, ilmenite, olivine, spinel, garnets C-153 and UD-92, albite and orthoclase were certified as the laboratory reference samples at the Institute of Geology and Geophysics, Novosibirsk,

Russia. The oxides MgO, MnO, Fe_2O_3, Cr_2O_3, Al_2O_3, TiO_2, SiO_2 and $CaSiO_3$ were supplied by JEOL Ltd., Tokyo, Japan. Each reference sample producer has reported that all samples meet the requirements needed for the samples for comparison at EPMA.

These control samples comprised the following. (I) Seven glasses, in which the contents of the elements analyzed varied over the following ranges (wt.%): SiO_2, 45.35–54.11; MgO, 3.48–19.33; Al_2O_3, 1.40–16.68; FeO, 8.46–12.04; CaO, 6.87–23.38; TiO_2, 0.81–9.11; MnO, 0.17–8.48; and Cr_2O_3, 10.20. (II) Eight oxides: MgO, SiO_2, Al_2O_3, TiO_2, Fe_2O_3, MnO, Cr_2O_3 and $CaSiO_3$. (III) Twelve minerals with concentrations varying as follows (wt.%): MgO, 1.02–49.20; Al_2O_3, 0.50–26.10; SiO_2, 0.30–55.50; CaO, 2.24–53.80; TiO_2, 0.33–50.00; Cr_2O_3, 0.08–44.80; MnO, 0.17–30.76; and FeO, 0.05–62.40.

All samples were at first used as control samples before being accepted as reference samples. The available reference samples were divided into three groups (Table 6). Group I comprised the basic components (Si, Al, Fe, Mg and Ca) which were defined using glass K-412, while Ti, Mn and Cr were defined from Ti-glass, Mn-glass and Cr-glass. The second complete set (II) consisted of simple minerals: diopside (for Si and Ca); ilmenite GF-55 (for Ti); olivine (for Mg); spinel $MnFe_2O_4$ (for Fe); garnets C-153 (for Al), UD-92 (for Cr) and IGEM (for Mn). Simple oxides MgO, MnO, Fe_2O_3, Cr_2O_3, Al_2O_3, TiO_2, SiO_2 and $CaSiO_3$ (Ca) represent the third complete set of reference samples (III). In all three cases the sodium content was determined from albite and that of potassium from the orthoclase standard sample. The analyzed values were corrected for matrix effects using the PAP method (Pouchou @ Pichoir, 1984) through the MARCHELL program (Kanakin @ Karmanov, 2006) adapted for the Superprobe-733 operating system. When applying three complete sets of calibration samples the three series of concentration data (C_I, C_{II} and C_{III}) have been received. Table 6 shows the analytical results for USGS TB-1 glass measured by EPMA. The same analytical results have been acquired for every control sample.

The results obtained in this study (Table 6) showed that deviations from the recommended/certified value varied in value and sign, however, they did not depend on the group of the reference samples selected for calibration. The largest deviations were observed in the elements with concentrations close to the detection limit. The relative standard deviations do not exceed the target precision (σ_r) in all cases. The values of z-scores for all elements determined lie within permissible limits ($-2<z<2$) for the elements ranging in concentration from 0.1 to 100%. The relative standard deviation for each element assessed in all control samples depends on the concentration and varies as (%): Na_2O, 0.30–2.89; MgO, 0.42–1.76; Al_2O_3, 0.29–2.4; SiO_2, 0.11–2.32; K_2O, 0.43–2.00; CaO, 0.37–1.91; TiO_2, 0.84–2.16; Cr_2O_3, 0.71–2.25; MnO, 0.72–2.59; and FeO, 0.45–2.80. The relative standard deviations for each element in all control samples were not higher than the admissible relative standard deviations (σ_r), defined for 'applied geochemistry' category of analysis (category 2) in the GeoPT proficiency testing program (Thompson et al., 1997). Fig. 7 shows the correlation of concentrations for MgO and Cr_2O_3 analyzed using the different sets of reference samples vs. their recommended or certified values.

Each trend of data plotted in Figure 7 is well described as a straight line. In all cases the correlation coefficients (R^2), describing the reliability of the linear dependence, is close or equal to 1 (Table 7). This confirms the absence of systematic differences and confirms the reliability of each set of reference samples in the calibration of the EPMA instrument.

Group of reference samples	Metrological performance	Na_2O	MgO	Al_2O_3	SiO_2	P_2O_5	K_2O	CaO	TiO_2	MnO	FeO
I	C_{cer}, wt. %	3,3	3,5	13,6	54,1	0,36	1,7	7	2,2	0,18	12
	n	17									
	C_{av}, wt.%	3,3	3,5	13,6	54	0,35	1,7	7	2,2	0,18	12
	s	0,1	0,1	0,17	0,6	0,01	0	0,1	0	0,01	0,2
	s_r, %	1,4	1,4	1,25	1,1	2,31	1,8	1,5	1,7	2,53	1,3
	σ_r, %	1,7	1,7	1,35	1,1	2,33	1,9	1,5	1,8	2,59	1,4
	ΔC , wt. %	0	0	0,08	0,3	0	0	0,1	0	0	0,1
	z	-0	0,4	-0,5	-0,2	-1,2	0,3	0,4	-1	-0,4	-0,1
II	n	8									
	C_{av}, wt.%	3,3	3,5	13,7	54	0,36	1,7	6,9	2,3	0,18	12
	S	0,1	0,1	0,16	0,6	0,01	0	0,1	0	0,01	0,2
	s_r, %	1,7	1,6	1,17	1	2,31	1,8	1,5	1,7	2,54	1,3
	σ_r, %	1,7	1,7	1,35	1,1	2,33	1,9	1,5	1,8	2,59	1,4
	ΔC, wt. %	0	0	0,11	0,4	0,01	0	0,1	0	0	0,1
	z	0,7	-0	0,38	0,2	-0,1	-1	-1	0,5	-0,7	-0,7
III	n	7									
	C_{av}, wt.%	3,3	3,6	13,7	54	0,37	1,7	7,1	2,2	0,18	12
	s	0,1	0	0,09	0,4	0,01	0	0,1	0	0,01	0,1
	s_r, %	1,5	1,1	0,64	0,7	2,27	1,8	1,4	1,7	2,5	1,2
	σ_r, %	1,7	1,7	1,35	1,1	2,33	1,9	1,5	1,8	2,59	1,4
	ΔC , wt. %	0,1	0,1	0,1	0,4	0,01	0	0,1	0	0,01	0,2
	z	1	1,8	0,46	0,4	1,19	-1	1,7	-1	0	1,2

Table 6. Analytical results for USGS TB-1 glass measured by EPMA with the calibrations obtained from three separate groups of reference materials for each of samples: C_{cer} is the certified concentration; n is number of measurements; $\Delta C = t * s / \sqrt{n}$ is confidence intervals; $s_r = s / C_{av}$ is standard deviations, where $s = \sqrt{\{[n\sum_{i=1}^{n}(C_i)^2 - (\sum_{i=1}^{n}C_i^2)] / [n(n-1)]\}}$; $C_{av} = \frac{1}{n}\sum_{i=1}^{n} C_i$ is average concentrations; $z = (C_{av} - C_{cer}) / \sigma_r$ is z-scores (Thompson et al., 1997); $\sigma_r = 100 * 0.02 * C^{0.8495} / C$ is acceptable standard deviation (Thompson et al., 1997), where C is the concentration expressed as fraction.

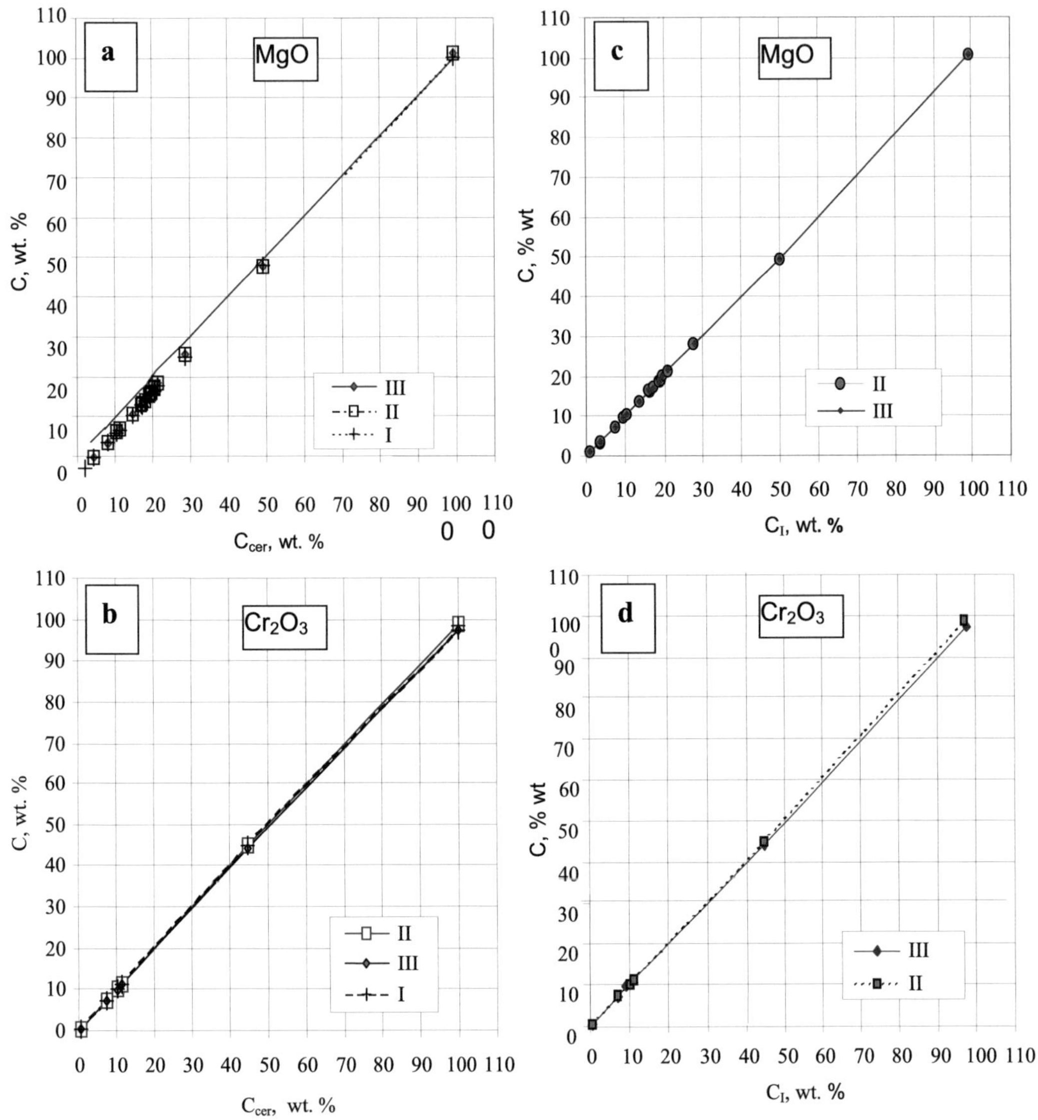

Fig. 7. The graphic correlation of concentrations for MgO and Cr_2O_3. a, b – the graphic correlation of concentrations (C) received according to three different reference sample sets (I, II, III), to their certified values (C_{cer}). Graphic representation of the ratio between the concentrations determined using laboratory reference samples II (C_{II}) and III (C_{III}) and C_I determined using standard sample glasses I. Graphs correspond to the concentrations, determined using: fist set of reference samples (I) – glass K-412, Ti-glass, Mn-glass, Cr-glass, albite and orthoclase; the second set (II) – diopside, ilmenite GF-55, olivine, spinel, garnets C-153, UD-92, IGEM albite and orthoclase; for the third set (III) – MgO, MnO, Fe_2O_3, Cr_2O_3, Al_2O_3, TiO_2, SiO_2, $CaSiO_3$, albite and orthoclase.

Oxides	C_{III} (C_{cer})	C_{II} (C_{cer})	C_I (C_{cer})	C_{III} (C_I)	C_{II} (C_I)
Al_2O_3	$C_{III} = 1.003\,C_{cer}$ - 0.0985; R^2 =0.9999	$C_{II} = 1.0020\,C_{cer}$ - 0.0782; R^2 = 0.9999	$C_I = 1.0001\,C_{cer}$ + 0.0674; R^2 = 0.9999	$C_{III} = 1{,}0396\,C_I$ - 0,0105; R^2 = 0,9998	$C_{II} = 1{,}0414\ C_I$ - 0,0104; R^2 = 0,9997
MgO	$C_{III} = 1.0081\,C_{cer}$ - 0.1342; R^2 =0.9999	$C_{II} = 1.0068\,C_{cer}$ - 0.0978; R^2 = 0.9999	$C_I = 0.9977\,C_{cer}$ - 0.1246; R^2 = 0.9998	$C_{III} = 1{,}0101\,C_I$ - 0,0048; R^2 = 0,9998	$C_{II} = 1{,}0089\ C_I$ + 0,0280; R^2 = 0,9998
SiO_2	$C_{III} = 0.9943\,C_{cer}$ + 0.234; R^2 =0.9997	$C_{II} = 0.9966\,C_{cer}$ + 0.1482; R^2 = 0.9997	$C_I = 0.9964\,C_{cer}$ + 0.0369; R^2 = 0.9991	$C_{III} = 1{,}0044\,C_I$ - 0,4210; R^2 = 0,9991	$C_{II} = 1{,}0034\ C_I$ - 0,3896; R^2 =0,9991;
Na_2O	$C_{III} = 1.005\,C_{cer}$ - 0.0361; R^2 =0.9987	$C_{II} = 1.0001\,C_{cer}$ + 0.0193; R^2 = 0.9997	$C_I = 1.0042\,C_{cer}$ - 0.0415; R^2 = 0.9996	$C_{III} = 1{,}0016\,C_I$ + 0,0838; R^2 = 0,9999	$C_{II} = 0{,}9991\,C_I$ + 0,2071; R^2 =0,9989;
K_2O	C_{III} =0.9974 C_{cer} + 0.0532; R^2 =0.9998	$C_{II} = 0.9992\,C_{cer}$ + 0.0359; м = 0.9990	$C_I = 0.9821\,C_{cer}$ + 0.0581; R^2 = 0.9999	$C_{III} = 1{,}0148\,C_I$ - 0,0107; R^2 = 0,9998	$C_{II} = 1{,}0157\,C_I$ - 0,0126; R^2 = 0,9990
CaO	C_{III} =0.9925 C_{cer} + 0.0026; R^2 = 0.9999	$C_{II} = 0.9946\,C_{cer}$ + 0.0351; R^2 = 0.9999	$C_I = 1.0057\,C_{cer}$ - 0.0203; R^2 = 0.9999	$C_{III} = 0{,}9855\,C_I$ + 0,0584; R^2 = 0,9998	$C_{II} = 0{,}9889\,C_I$ + 0,0720; R^2 = 0,9999
TiO_2	$C_{III} = 1.0140\,C_{cer}$ - 0.0105; R^2 = 1.0000	$C_{II} = 1.0058\,C_{cer}$ + 0.0202; R^2 = 0.9999	$C_I = 1.0051\,C_{cer}$ + 0.0339; R^2 = 1.0000	$C_{III} = 1{,}0093\,C_I$ - 0,0389; R^2 = 1,0000	$C_{II} = 1{,}0005\,C_I$ + 0,0036; R^2 = 1,0000
Cr_2O_3	$C_{III} = 0.9747\,C_{cer}$ + 0.0672; R^2 = 0.9999	$C_{II} = 0.9872\,C_{cer}$ + 0.1281; R^2 = 0.9999	$C_I = 0.9780\,C_{cer}$ + 0.2218; R^2 = 0.9998	$C_{III} = 0{,}9950\,C_I$ + 0,1126; R^2 = 0,9999	$C_{II} = 1{,}0144\,C_I$ - 0,1526; R^2 = 0,9999
MnO	$C_{III} = 1.0130\,C_{cer}$ - 0.0512; R^2 = 0.9999	$C_{II} = 1.0075\,C_{cer}$ + 0.0027; R^2 = 1.0000	$C_I = 1.0140\,C_{cer}$ - 0.0066; R^2 = 1.0000	$C_{III} = 1{,}0013\,C_I$ - 0,0551; R^2 = 0,9990	$C_{II} = 0{,}9935\,C_I$ + 0,0099; R^2 = 0,9999
FeO	$C_{III} = 1.0031\,C_{cer}$ - 0.0332; R^2 = 0.9999	$C_{II} = 1.0013\,C_{cer}$ - 0.0112; R^2 = 0.9999	$C_I = 0.9880\,C_{cer}$ + 0.1251; R^2 = 0.9999	$C_{III} = 1{,}0107\,C_I$ - 0,1443; R^2 = 0,9999	$C_{II} = 1{,}0090\,C_I$ - 0,1295; R^2 = 0,9999

Table 7. Evaluation of the linear function and correlation coefficients (R^2) between determinations made against the three sets of calibration samples (C_I , C_{II} , C_{III}) and the certified concentrations of control samples (C_{cer}) as well as between C_{II} , C_{III} and C_I . Concentrations C_I , C_{II} , C_{III} are determined using the respective sets of reference samples: (I) – glass K-412, *Ti* -glass, Mn-glass, Cr-glass, albite and orthoclase; (II) – diopside, ilmenite GF-55, olivine, spinel, garnets C-153, UD-92, IGEM, albite and orthoclase; (III) – MgO , MnO , Fe_2O_3 , Cr_2O_3 , Al_2O_3 , TiO_2 , SiO_2 , $CaSiO_3$, albite and orthoclase.

Figure 7 includes the data for elements in which the maximum differences were observed between the results. It is evident that in all cases there are no essential systematic differences between the data sets. This confirms the absence of systematic differences and confirms the reliability of each set of reference samples in the calibration of the EPMA instrument.

Table 8 gives the results of calculations designed to check the hypothesis that there are no differences between sets of results obtained from any of these three sets of calibration samples.

The series of concentrations determined using calibrations established using three sets of reference samples have been compared with certified values using a two-tailed selective Student's *t*-test. Numerical values of probabilities for each pair of the series are significant, showing that the populations of results from all three series are statistically indistinguishable certified values.

The closeness in value of the significance data listed in Table 8 demonstrates the absence of systematic differences between the certified concentrations and the analyzed data for every series. Thus, the sets of reference samples tested in such a way can be successfully applied in EPMA for obtaining high-quality results.

Oxides	$C_{cer} - C_I$	$C_{cer} - C_{II}$	$C_{cer} - C_{III}$	$C_I - C_{II}$	$C_I - C_{III}$
MgO	0.312	0.355	0.299	0.359	0.268
AL_2O_3	0.244	0.237	0.215	0.251	0.267
SiO_2	0.310	0.363	0.328	0.265	0.368
CaO	0.455	0.444	-	0.292	-
MnO	-	0.503	0.490	-	-
FeO	0.359	0.390	0.346	0.374	0.328

Table 8. Results from checking the hypothesis that there is no significant difference between determinations made by any of the three sets of calibration and certified/recommended values using a 2-pair selective Students t-test. C_{cer} – certified concentrations of control samples; C_I, C_{II}, C_{III} – concentrations, determined using the following sets of reference samples: (I) – glass K-412, Ti-glass, Mn-glass, Cr-glass, albite and orthoclase; (II) – diopside, ilmenite GF-55, olivine, spinel, garnets C-153, UD-92, IGEM, albite and orthclase; (III) – MgO, MnO, Fe_2O_3, Cr_2O_3, Al_2O_3, TiO_2, SiO_2, $CaSiO_3$, albite and orthoclase.

2.5 The dependence of EPMA quality on homogeneity of reference samples

The influence of inhomogeneity of reference samples on the EPMA quality has been exemplified by copper-containing alloys (Pavlova, 2009). Ten copper-rich alloys, the standards for chemical, optical and x-ray fluorescence analysis have been quantitatively evaluated as the reference materials to be employed in the electron probe microanalysis.

The optimum conditions for measurements were selected considering the dependence of intensity and detection limit on conditions of the x-ray radiation excitation and analytical signal recording.

The reference samples were divided into the three groups: the control group 1 consisted of the certified glass (Fe), simple minerals (Zn, Sn), metals (Cu, Ni); two groups of samples for comparison included the assessed alloys: group 2 comprised the alloys MC76 (Fe), MC44 (Sn, Cu), MC104 (Ni), MC153 (Zn); and group 3 included MC71 (Zn), MC74 (Fe, Ni), MC42 (Sn) and metallic copper (Cu).

Three data sets (1, 2 and 3 - in agreement with the groups of reference samples) comprising the average concentrations, standard deviations, relative standard deviations, confidence interval and the z-score of data quality were calculated for 10 copper-rich alloys.

The average concentrations for all elements of every control sample were being defined from 8 to 18 times. The measured values were corrected for matrix effects using the PAP method (Pouchou and Pichoir, 1984) and applying the MARCHELL program (Kanakin & Karmanov, 2006) adapted for the microanalyzer Superprobe-733 operating system. Table 9 presents the data for one alloy.

The relative standard deviations obtained for each element were lower than the target values for all determinations in all cases except for set 2, where group 2 of reference samples was used. In two sets of data the z-score values for all elements determined lie within acceptable limits ($-2<z<2$) for concentrations ranging from 0.1 to 100%.

Figure 8 shows the graphic dependence of certified Ni, Fe, Sn, Zn concentrations and values obtained from different (1, 2 and 3) groups of reference samples. Obtained sets of concentrations were compared between each other and with the certificated/recommended values using 2-pair selective Student's t-test. The table 9 gives the results of calculations to

Group of reference samples	Metrological performance	*Zn*	*Cu*	*Sn*	*Ni*	*Fe*
	C_{cer}, wt. %	1,96	88,8	4,89	1,09	2,46
1	n	14				
	C_{av}, wt.%	1,94	88,5	4,85	1,11	2,48
	s	0,07	1,78	0,14	0,04	0,08
	s_r, %	3,56	2,01	2,98	3,89	3,41
	σ_r, %	3,61	2,04	3,15	3,95	3,49
	ΔC, wt. %	0,04	1,05	0,09	0,03	0,05
	z	-0,3	-0,2	-0,3	0,46	0,24
2	n	17				
	C_{av}, wt.%	1,75	85,7	4,56	1,34	2,95
	s	0,07	2,25	0,16	0,06	0,11
	s_r, %	3,78	2,63	3,45	4,22	3,76
	σ_r, %	3,61	2,04	3,15	3,95	3,49
	ΔC, wt. %	0,05	1,67	0,12	0,04	0,08
	z	-3,2	-1,4	-2,1	4,42	4,34
3	n	12				
	C_{av}, wt.%	1,98	89,3	4,92	1,08	2,45
	s	0,07	1,81	0,15	0,04	0,08
	s_r, %	3,54	2,03	3,12	3,92	3,46
	σ_r, %	3,61	2,04	3,15	3,95	3,49
	ΔC, wt. %	0,04	1,03	0,09	0,02	0,05
	z	0,29	0,26	0,2	-0,2	-0,1

Table 9. Analytical results for copper alloys measured by EPMA using calibrations obtained from three separate groups (1, 2 and 3) of reference materials for each of samples: C_{cer} is the certified concentration; n is number of measurements; $s_r = s / C_{av}$ is standard deviations, where $s = \sqrt{\{[n\sum_{i=1}^{n}(C_i)^2 - (\sum_{i=1}^{n} C_i^2)] / [n(n-1)]\}}$; $C_{av} = \frac{1}{n}\sum_{i=1}^{n} C_i$ is average concentrations; $\Delta C = t * s / \sqrt{n}$ is confidence intervals; $z = (C_{av} - C_{cer}) / \sigma_r$ is z-scores (Thompson et al., 1997); $\sigma_r = 100 * 0.02 * C^{0.8495} / C$ is acceptable standard deviation (Thompson et al., 1997), where C is the concentration expressed as fraction.

check the hypothesis that there are no differences between the series of data obtained from any of the three groups of reference samples. We can see, the numerical values of probabilities in columns II, III and VI are not significant, thus indicating that the population of results from series C_2 is statistically different from the recommended values (column II),

and the values C_1 and C_3 obtained with groups 1 and 3 of reference samples (columns III and VI).
No systematic divergence was the case between the concentrations obtained from set 1 and set 2, when analyzed results were compared with the certified compositions. The set (2) of reference samples for copper-rich alloys yields erratic data because this set contained the inhomogeneous reference samples.
Lack of close values of probabilities (columns II, III and VI) demonstrates the presence of systematic differences between the concentration series C_2 and recommended concentrations (column II) as well as between the results of the second series C_2 and the data on the series C_1 and C_3 (columns III and VI). This confirms incorrectness of reference samples in group 2. It was previously shown that alloys CA-6 and CA-2 are not homogeneous (table 5).
The numerical values of probabilities are significant in columns I, IV, V and exhibit that the concentrations from selected sets are statistically indistinguishable. Similar values of probabilities listed in these columns for alloys demonstrate the absence of systematic differences of the first and third concentration series both between each other and the recommended concentrations.

Samples	Elements	Numerical values of probabilities for each pair of the series					
		I	II	III	IV	V	VI
		C_{cer} & C_1	C_{cer} & C_2	C_1 & C_2	C_{cer} & C_3	C_1 & C_3	C_2 & C_3
Alloys	*Fe*	0.341	0.031	0.013	0.421	0.296	0.021
	Ni	0.324	0.012	0.009	0.367	0.311	0.008
	Cu	0.336	0.003	0.026	0.383	0.382	0.017
	Sn	0.445	0.0154	0.0142	0.435	0.278	0.007
	Zn	0.425	0.009	0.001	0.398	0.318	0.011

Table 10. Comparison of concentrations using a coupled selective Student's t-test; C_{cer} is the certified concentration; C_1, C_2, C_3 are three series of concentration data obtained for each of copper-rich alloys using three groups of reference samples.

As a result, the systematic error is very small in the case when the compositions of control samples are calculated using groups 1 and 3 of reference samples. These experiments show that the quality of data obtained from alloy reference samples of group 3 is not inferior to that from the certified reference samples. The quality of results obtained from group 3 of reference samples corresponds to the 'applied geochemistry' type of analysis (category 2) as defined in the GeoPT proficiency testing program (Thompson et al., 1997)

2.6 Method to recalculate experimental values into concentrations

The choice of method of processing experimental values, when the concentration to be determined is concerned, influences determination of a true composition and thus the quality of analysis. Almost every analytical method of recalculating of experimental values into concentrations is developed for certain samples and conditions. Different methods of analytical chemistry have several ways of recalculating experimentally measured values in the concentration. Every way has its own advantages and disadvantages.

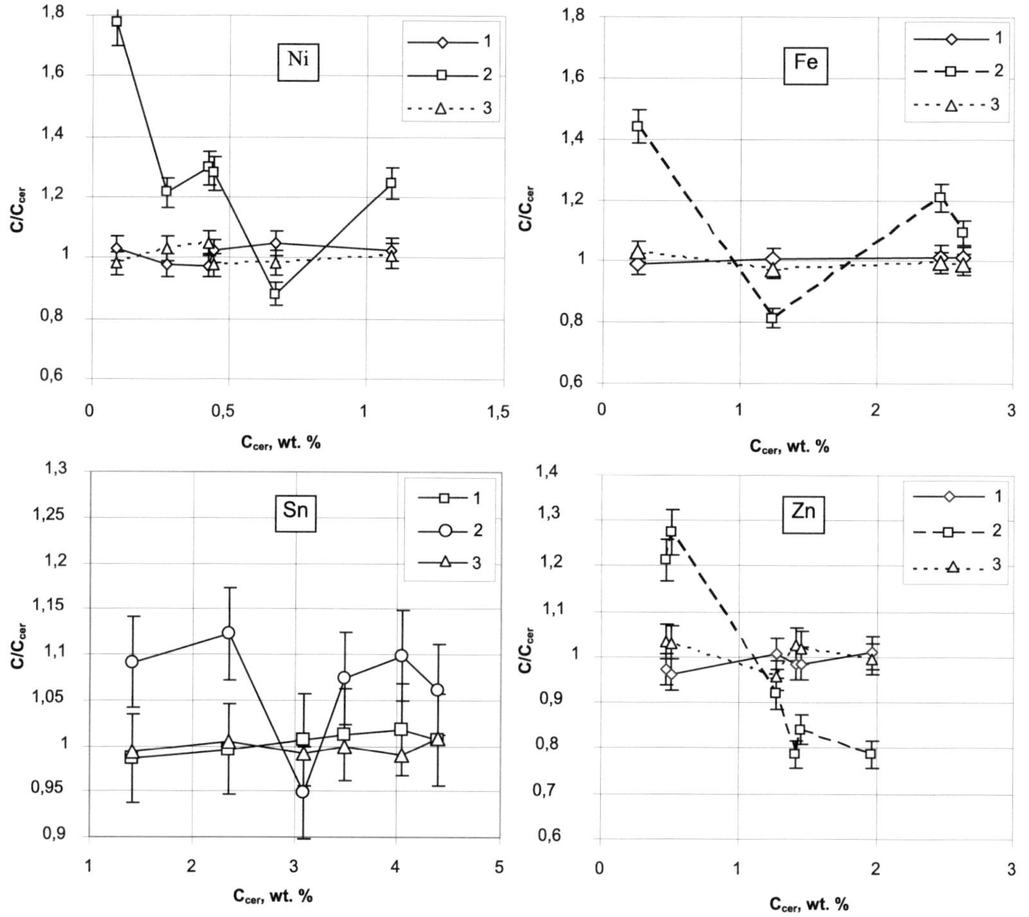

Fig. 8. Dependence of certified Ni , Fe , Sn , Zn concentrations and values obtained from different (1, 2 and 3) groups of reference samples.

When applying EPMA, the recalculation of x-ray radiation intensity in the concentration of silver group elements depends on the correct choice of x-ray radiation absorption coefficients and ability to consider the matrix effects. The determination of the composition of silver-containing compounds can exemplify how the counting procedure influences the quality of composition determination. The results on the composition of silver minerals, differently obtained by calculating absorption coefficients, are compared in Table 10 (Pavlova @ Kravtsova, 2006). The measured intensities, acquired by microprobes JXA8200 and JCXA-733 (Japan), were corrected for matrix effects using programs of JXA8200 Unix platform and the PAP method (Pouchou & Pichoir, 1984) applying the original controlling computer program MARshell (Kanakin & Karmanov, 2006) adapted for JCXA-733 microprobe.

One can observe how available results differ in various methods of calculating matrix effects and how the quality of results depends on the correct calculation of absorption coefficients. It is evident, that the best results are gained by the PAP method (Pouchou & Pichoir, 1984), when the Marenkov's absorption coefficients are applied (Marenkov, 1982), as well as program of JXA8200 Unix platform by PPX method (Pouchou & Pichoir, 1991) from software of microprobe JXA8200 (Japan).

The change of the subject under study often requires the change in the method of recalculating of experimental intensities into concentrations. Thus, a correct determination of the composition of particles comparable in size with the area of generation of x-ray radiation is dependent on the particle size. The quality of obtained results for such particles depends on a correct consideration of size factor (Table 12). The dependence of quality of particle composition versus the method of calculation of their size is exemplified in the article (Belozerova, 2003). Here it is seen that the results obtained for particles with size 3–5 μm are closer to the stoichiometrical composition than those obtained for particles sized 1–2 μm size. One of the best commonly used methods of calculating element content in bulk samples (PAP-method) provides a relative error of determining the composition of one micron particles ranging from 0.5 to 45 %. Using the exponent model of particle composition calculation lowers this inaccuracy.

Method to calculate matrix effects	Authors of absorption coefficients	Sample 1		Sample 2		Sample 3	
		Concent-ration, wt. %	Relative error, %	Concent-ration, wt. %	Relative error, %	Concent-ration, wt. %	Relative error, %
(Pouchou & Pichoir, 1984)	(Marenkov, 1982)	65.56	-0.23	62.69	0.11	77.52	-0.51
	(Pouchou @ Pichoir, 1984)	67.16	-2.67	62.91	-0.24	79.51	-3.08
	(Heinrich, 1986)	66.03	-0.95	61.70	1.69	77.72	-0.77
(Lavrentiev et al., 1980)	(Marenkov, 1982)	73.20	-11.91	68.54	-9.20	86.65	-12.35
	(Pouchou @ Pichoir, 1984)	73.78	-12.79	68.95	-9.87	87.32	-13.21
	(Heinrich, 1986)	68.45	-4.64	62.49	0.42	78.82	-2.19
(Brizuela & Riveros,1990)	(Marenkov, 1982)	67.33	-2.94	63.36	-0.96	79.75	-3.40
	(Pouchou @ Pichoir, 1984)	67.93	-3.86	63.79	-1.63	80.45	-4.30
	(Heinrich, 1986)	66.39	-1.49	62.56	0.32	78.65	-1.97
(Sewell et al., 1985)	(Marenkov, 1982)	68.78	-5.15	64.26	-2.39	81.40	-5.53
	(Pouchou @ Pichoir, 1984)	69.35	-6.03	64.67	-3.04	82.06	-6.39
	(Heinrich, 1986)	67.85	-3.73	63.48	-1.14	80.32	-4.13
programs of JXA8200: (Philibert, 1963); (Reed, 1965); (Duncumb & Reed, 1968); (Berger & Zeltzer, 1964)	(Henke, 1966) (Henke, et al., 1982) (Heinrich, 1966) (Pouchou &Pichoir, 1991).	66,02	-0,93	63,18	-0,67	78,03	-1,16
programs of JXA8200: (Pouchou & Pichoir, 1991)		65,94	-0,81	62,57	0,3	77,67	-0,71
Certified concentration, wt. %		65.41		62.76		77.13	

Table 11. Comparison of methods to compute matrix effects and coefficients of absorption to measure silver by EPMA.

Particle size. μm	Calculation technique	Sample	Albite			Quart	Calcit	Corun	Rutile	Pyrite	
		Element	*Na*	*Al*	*Si*	*Si*	*Ca*	*Al*	*Ti*	*Fe*	*S*
1	(Pouchou, & Pichoir, 1984)	C_i, wt. %	4.78	6.12	19.40	30.13	25.43	42.95	50.68	44.09	49.2
		R. er., %	45.50	40.50	39.62	35.54	36.49	18.89	15.46	5.28	7.82
	(Belozerova et al., 1998)	C_i, wt. %	5.54	8.06	25.04	35.58	36.87	47.23	57.48	46.56	52.8
		R. er., %	36.83	21.67	22.07	23.88	7.92	10.77	4.12	-0.02	1.08
	(Belozerova et al., 2003)	C_i, wt. %	6.82	9.93	30.86	44.15	42.28	51.40	58.42	46.67	52.9
		R. er., %	22.23	3.50	3.95	5.54	-5.59	2.89	2.55	-0.24	0.86
2	(Pouchou, & Pichoir, 1984)	C_i, wt. %	7.06	8.50	28.32	40.83	34.03	50.71	57.44	45.66	53.1
		R. er., %	19.50	17.39	11.86	12.64	15.00	4.19	4.19	1.91	0.56
	(Belozerova et al., 1998)	C_i, wt. %	7.12	8.87	29.59	43.62	36.86	50.85	58.05	45.67	53.3
		R. er., %	18.81	13.80	7.91	6.68	7.94	3.93	3.17	1.89	0.22
	(Belozerova et al., 2003)	C_i, wt. %	7.38	9.20	30.68	45.33	37.47	51.19	58.07	45.68	53.3
		R. er., %	15.85	10.59	4.51	3.02	6.42	3.28	3.14	1.86	0.22
3	(Pouchou, & Pichoir, 1984)	C_i, wt. %	8.28	9.93	31.54	45.50	38.80	52.26	59.15	46.09	52.9
		R. er., %	5.59	3.50	1.83	2.65	3.10	1.26	1.33	0.94	0.91
	(Belozerova et al., 1998)	C_i, wt. %	8.28	10.01	31.82	46.17	39.54	52.15	59.16	46.02	52.9
		R. er., %	5.59	2.72	0.96	1.22	1.25	1.47	1.32	1.14	1.01
	(Belozerova et al., 2003)	C_i, wt. %	8.33	10.08	32.04	46.52	39.62	52.18	59.16	46.02	52.9
		R. er., %	5.02	2.04	0.28	0.47	1.05	1.42	1.32	1.14	0.99
4	(Pouchou, & Pichoir, 1984)	C_i, wt. %	8.57	10.26	31.71	46.00	40.02	52.37	59.25	46.39	53.1
		R. er., %	2.28	0.29	1.31	1.58	0.05	1.06	1.17	0.34	0.60
	(Belozerova et al., 1998)	C_i, wt. %	8.56	10.27	31.72	46.19	40.19	52.25	59.19	46.34	53.0
		R. er., %	2.39	0.19	1.28	1.18	-0.37	1.28	1.27	0.45	0.71
	(Belozerova et al., 2003)	C_i, wt. %	8.57	10.28	31.76	46.25	40.20	52.25	59.19	46.34	53.0
		R.er., %	2.28	0.10	1.15	1.05	-0.40	1.28	1.27	0.45	0.71
Massive grain	(Pouchou, & Pichoir, 1984)	C_i, wt. %	8.71	10.27	32.06	46.69	40.57	53.00	59.60	46.59	53.4
		R. er., %	0.68	0.19	0.22	0.11	-1.32	-0.13	0.58	-0.09	0.07
	(Belozerova et al., 1998)	C_i, wt. %	8.71	10.27	32.07	46.76	40.62	52.97	59.63	46.55	53.3
		R. er., %	0.68	0.19	0.19	0.04	1.45	0.07	0.53	0.03	0.16
	(Belozerova et al., 2003)	C_i, wt..	8.71	10.27	32.07	46.76	40.62	52.97	59.63	46.55	53.3
		R. er., %	0.68	0.19	0.19	0.04	1.45	0.07	0.53	0.02	0.16
Stoichiometric composition		C_s, wt..	8.77	10.29	32.13	46.74	40.04	52.93	59.95	46.55	53.4

Table 12. Particle composition versus the method of calculation of their size. C_i is obtained concentration; $\text{R.er.} = 100(C_s - C_i)/C_s\ \%$.

The EPMA technique (Belozerova, 2003) for calculating composition developed for approximately spherical particles, comparable in size to the X-ray generation volume, takes into account the particle-size factor. The size factor correction significantly improves the results in spite of a simple analytical function of average atomic number and particle size. Taking into account the particle-size factor reduces the error of composition determination from 0.5–45 to 0.2–22% relative percent, for particles sized as 1–3 μm.
The relative error increases with decreasing the element concentration from 0.02% for bulk sample to 22.2% for 1 μm particle. The size-factor introduction markedly improves the quality of determinations of particles comparable in size with the area of x-ray radiation excitation.

3. Conclusion

This chapter shows the quality dependence of EPMA on every analysis stage, beginning from the representativeness of the material, sample preparation and conditions for analytical signal excitation and registration to the availability of reference samples and the calculation methods.
We have shown how important it is to correctly select the study area and to have properly prepared samples.
It has been found that the quality of study performance is dependent on the optimum conditions of measuring and processing analytical signal.
The comparison of different methods of taking into account the processes occurring in substance in the electron-excited x-ray radiation proves the necessity to correctly select the methods of their consideration in every study.
The method of assessing the selected set of reference samples and defining their appropriateness for EPMA is described.
The urgency to develop and certify new control samples is critical for the methods of analytical chemistry and especially EPMA.
We have shown the influence of inhomogeneity of samples for comparison on the quality of EPMA results.
Thus, the example of EPMA for the case study of determination content quality suggests, that every aspect of analytical technique is responsible for the quality of element tests.

4. Acknowledgment

The author is grateful to Mrs. T. Bunaeva and Mrs. M. Khomutova for editing the English version of this.

5. References

Beckford, J. L. W. (2010). Quality: a Critical Introduction, Routledge, ISBN 978-0415996358, New York.

Belozerova, O.Y., Afonin, V.P. & Finkelshtein, A.L. (1998). Modified biexponential model and its application to the X-ray microanalysis of gold-containing alloys. *Russian Journal Analytical Chemistry*, Vol. 53, No. 10, pp.1060–1065, ISSN: 0044-4502.

Belozerova, O.Yu., Finkelshtein, A.L. & Pavlova, L.A. (2003). Electron-Probe X-Ray Microanalysis of Individual Particles of Solid Snow Sediment with Size Factor Correction. *Micron*, Vol. 34, No. 1, pp. 49-55, ISSN: 0968-4328.

Berger M.J. & Zeltzer S.M. (1964). *Tables of energy losses and ranges of electrons and positrons.* National Academy of Sciences - National Reseach Council. Publ. 1133, DC, pp. 205-267, Washington.

Borkhodoev, V. Ya. (2010). Assessment of reference samples homogeneity in electron microprobe analysis. *X-ray Spectrometry,* Vol. 39, No. 1, pp. 28–31, ISSN: 1097-4539.

Borovskii, I.B. (1953). To the 70th Anniversary of I.P. Bardin, In: *Problems of Metallurgy,* Borovskii, I.B. pp. 135-153, AN SSSR, Moscow.

Borovskii, I.B. & Il'in N.P., (1956). New method of chemical content investigation in micro volume. *Dokl. Akad. Nauk SSSR,* Vol. 106, No. 4, pp. 654-657, ISSN: 0869-5652.

Borovskii, I.B., Vodovatov, F. F., Zhukov, A.A. & Cherepin, V. T. (1973). *Lokal'ny Metody Analiza Mineralov*, Metallurgiya, Moscow.

Brizuela, H. & Riveros, J.A. (1990). Study of mean excitation energy and K-shell effect for electron probe microanalysis. *X-Ray Spectrometry*, Vol. 19, № 4, p. 173–176, ISSN: 1097-4539.

Buseck, P.R. & Goldstein, J.I. (1969). Olivine Compositions and Cooling Rates of Pallasitic Meteorites. *Geological Society of America Bulletin*, Vol. 80, No. 11, pp. 2141-2158, ISSN: 1943-2674.

Castaing, R. (1951). Application of electron probes to local chemical and crystallographic analysis (Thesis, Univ. of Paris), Translated by P. Duwez and D.B, wittry, CalTech, 1955 (*Special Technical Report* under U.S. Army).

Castaing, R. & Guinier, A. (1953). Point-by-point chemical analysis by X-ray spectroscopy: Electronic Microanalyzer, In: *Analytical Chemistry special issue "Symposium on X-rays as analytical chemical tool"*, Vol. 25, pp. 724-726.

Duncumb, P. & Reed S.J.B.(1968). The calculation of stopping power and backscatter effects in electron probe microanalysis. *Quantitative electron probe microanalysis : proceedings of a seminar held at the National Bureau of Standards*, pp. 133 -154, Gaithersburg, Maryland, June 12-13, 1967.

Farkov, P.M., Il'icheva, L.N., & Finkelshtein, A.L. (2005). X-ray Fluorescence Determination of Carbon, Nitrogen, and Oxigen in Fish Plant Samples. *Russian Journal Analytical Chemistry*, Vol. 60 , No. 5, pp. 485-489, ISSN: 0044-4502.

Finkelshtein, A.L. & Farkov, P.M., (2002). Approximation of X-ray Attenuation Coefficients in Energy Range 0,1 to 100 keV. *Analytic and control*, Vol. 6, No. 4, pp. 377-382, ISSN: 2073-1442 (Print), ISSN: 2073-1450 (Online).

Hubbell, J. H. & Seltzer, S. M. (April 2011). Tables of X-Ray Mass Attenuation Coefficients and Mass Energy-Absorption Coefficientsfrom 1 keV to 20 MeV for Elements Z=1 to 92 and 48 Additional Substances of Dosimetric Interest, In: *NIST X-ray Attenuation Datadases*, 05.04.2011, Available from: http:// NIST Home/PML/Physical Reference Data/X-Ray Mass Attenuation Coefficients (http://www.nist.gov/pml/data/xraycoef/index.cfm).

Heinrich, K.F.J. (1966). X-ray absorption uncertainty. *The electron microprobe; proceedings of the symposium sponsored by the Electrothermics and Metallurgy Division, the Electrochemical Society*, pp. 296-377, Washington, D.C., October, 1964.

Heinrich, K.F.J. (1986). Mass absorption coefficients for electron probe microanalysis. *11th International congress on X-ray optics and microanalysis. Proceedings of a conference held in London, Ontario, Canada*, pp. 67-119, 4–8 August 1986.

Henke, B.L. (1966). Application of multilayer analyzers to 15-150 angstrom fluorescence spectroscopy for chemical and valence band analysis. *Advance in X-ray Analysis*, Vol. 9, n.d., pp. 430-440, ISSN: 0376-0308

Henke, B.L., Lee, P., Tanaka, T.J., Shimabukuro, R.L. and Fujikawa, B.K. (1982). Low Energy X-ray Interaction Coefficients: Photoabsorption, Scattering and Reflection.E=10-2000 eV. Z=1-94. *Atomic Data and Nuclear Data Tables*, Vol. 27, pp. 1-144, ISSN: 0092-640X.

Henke, B.L., Gullikson, E.M., & Davis, J.C. (1993). X-ray interactions: photoabsorption, scattering, transmission, and reflection at E=50-30000 eV, Z=1-92. *Atomic Data and Nuclear Data Tables, Vol.* 54, No. 2, pp. 181-342.

Goldstein, J.I., Newbury, D.E., Echlin, P., Joy, D.C., Lyman, C.E., Fiori, C., Romig A.D. Jr. & Lifshin, E. (1992). *Scanning Electron Microscopy and X-Ray Microanalysis*, Plenum Press, NewYork.

Kanakin S.V. & Karmanov, N.S. (2006). ICA and main scope of the software package MARshell32, *Proceedings of the V All-Russian X-Ray analysis Conference*, Irkutsk, May–June, 2006.

Lavrentiev, Yu.G., Berdichevskiy, G.V., Chernyavsriy, L.I. & Kuznetsova, A.I. (1980). KARAT – is program for quantitative electron probe X-ray microanalysis, *Apparatus and Methods for X-Ray Analysis,* Mashinostroenie, Leningrad, Vol. 23, n.d., pp. 217–224.

Marenkov O.S. (1982). *Tables and Formulas X-ray analysis. Methodical recommendations.* Mashinostroenie, Leningrad.

MI. (1988). Standard sample homogeneity of monolithic materials for the spectral analysis. Technique of measurement. Standards Publish, Moscow.

NIST (1990) Standard Reference Materials Catalog 1990-1991, Ed: R.S. McKenzie, NIST Special Publication 260, U.S. Department of Commerce, National Institute of Standards and Technology.

Pavlova, L. A. Belozerova, O. Yu., Paradina, L. F. & Suvorova, L. F. (2000). *X-Ray Electron Probe Analysis of Environmental and Ecological Objects*, Nauka, ISBN 5-02-031533-8, Novosibirsk, Russia.

Pavlova, L. A., Paradina, L. F. & Belozerova, O. Yu. (2001) Electron Probe Microanalysis of Environmental Samples: Preparation of Reference Materials, Correction for Particle Size Effects and Representative Results on Sediments Recovered from Snow, Power Station Fly Ash and Bone Samples. *Geostandard Newsletter: the Journal of Geostandard and Geoanalysis*, Vol. 25, No 2-3, pp. 333-344, ISSN: 0150-5505.

Pavlova, L.A., Pavlov, S. M, Anoshko, P.N. & Tyagun, M.L. (2003). Determination of calcium and sodium in Baykal omul otoliths by Electron probe microanalysis, *Analytic and control*, Vol. 7, No. 3, pp. 242-247, ISSN: 2073-1442 (Print), ISSN: 2073-1450 (Online).

Pavlova, L. A., Suvorova, L. F, Belozerova, O. Yu. & Pavlov, S.M. (2003). Quality of determinations obtained from laboratory reference samples used in the calibration of X-ray electron probe microanalysis of silicate minerals. *Spectrochimica Acta Part B*, Vol. 58, No. 2, pp. 289-296, ISSN: 0584-8547.

Pavlova, L.A., Pavlov, S. M, Paradina, L. Ph., Karmanov, N.S., Kanakin, S.V., Anoshko, P.N. & Levina, O.V. (2004). Investigation of Baykal flora and fauna representative using the techniques of electron probe microanalysis and electron microscopy. *Ecological chemistry*, Vol. 13, No. 4, pp. 249-256, ISSN: 0869-3498.

Pavlova, L.A. & Kravtsova, R.G. (2006). Studing modes of silver occurrence, discovered in the lithochemical stream sediments of Dukat gold-silver deposit by electron probe x-ray microanalysis. *Methods and objects of chemical analysis*, Vol. 1, No. 2, pp. 132-140, ISSN: 1991-0290.

Pavlova, L. A. (2009). Quality of electron probe X-ray microanalysis determinations obtained from laboratory reference materials of the coppery alloys and basaltic glasses. *Spectrochimica Acta Part B: Atomic Spectroscopy*, Vol. 64, No 8, pp. 782-787, ISSN: 0584-8547.

Paradina, L.F. & Pavlova, L.A. (1999). Definition of vanadium, rubidium and strontium impurities in a vitreous substance of Siberia. *Russian Journal Analytical Chemistry*, Vol. 54 , No. 1, pp. 78–82, ISSN: 0044-4502.

Philibert J.A. (1963). A method for calculation of the absorption correction in electron probe microanalysis. In: *X-Ray Optics and X-Ray Microanalysis*, V.E. Cosslet & A. Engstrom, pp. 379-392, Academic Press, ISSN: 0028-0836, N.Y.

Pouchou J.L. & Pichoir F. A. (1984). New model for quantitative X-ray microanalysis, Part 1. Applications to the analysis of homogeneous samples. *La Recherche Aerospatiale*, n.d., No. 3, pp. 13–38, ISSN: 0034-1223.

Pouchou J.L. & Pichoir F. A. (1991). Quantitative analysis of homogenous or stratified microvolumes applying the model "PAP". In: *Electron Probe Quantitation*, K.F.J. Heinrich and D.E. Newbury (Eds.), 31-75, Springer, ISBN: 0306438240, Plenum Press.

Reed S.J.B. (1965). Characteristic fluorescence correction in electron probe microanalysis. *British Journal of Applied Physics*, Vol. 16, No 7, pp. 913–926, ISSN: 0508-3443.

Saloman, E.B., Hubbell, J. H. & Scofield J.H. (1988) X-ray attenuation cross srctions for energies 100 eV to 100 keV and elements Z=1 to Z=92. *Atomic Data and Nuclear Data Tables*, Vol. 38, pp. 1-197, ISSN: 0092-640X/88.

Sewell, D.A., Love, G. & Scott, V.D. (1985). Universal correction procedure for electronprobe microanalysis. I. Measurement of X-ray depth distribution in solids. *Journal of Physics D: Applied Physics*, Vol. 18, No. 7, pp. 1233–1243, ISSN: 0022-3727.

Szaloki, I., Osan, J. & Van Grieken, R. E. (2004). X-ray Spectrometry. *Analytical Chemistry*, Vol. 76, No. 12, pp. 3445-3470, ISSN: 1097-4539.

Thompson, M., Potts, P.J. & Webb, P.C. (1997). GeoPT1. International proficiency test for analytical geochemistry laboratories – report on round 1 (July 1996). *Geostandard Newsletter: the Journal of Geostandard and Geoanalysis*, Vol. 21, No. 1, pp. 51–58, ISSN: 0150-5505.

Wilson, S.A. & Taggart, J.E. (2000). Development of USGS microbeam reference materials for geochemical analysis, *Proceedings of the 4th International Conference on the Analysis of Geological and Environmental Materials*, Pont a Mousson, France, 29th Aug.-1st Sep., 2000.

Quality Control Through Electronic Nose System

Juan C. Rodríguez-Gamboa, E. Susana Albarracín-Estrada and Edilson Delgado-Trejos
MIRP, Research Center, Instituto Tecnológico Metropolitano (ITM), Medellín Colombia

1. Introduction

Quality control is defined as: "a process selected to guarantee a certain level of quality in a product, service or process. It may include whatever actions a business considers as essential to provide for the control and verification of certain characteristics of its activity. The basic objective of quality control is to ensure that the products, services or processes provided meet particular requirements and are secure, sufficient, and fiscally sound"1 In order to apply Quality Control through the Electronic Nose System, all the stages involved in the process must be taken into account, this case refers to the use of electronic nose systems as a tool for quality control tasks. Therefore best practices must be implemented that will lead to obtaining good quality measures, which will later become good results (Badrick, 2008; Duran, 2005)

Section 2 of this chapter presents an overview of the parts or subsystems involved in an electronic nose system and the operating principle.

Section 3 deals with the issue of food quality control using electronic nose systems. This section discusses how to use the electronic nose system for these types of applications, and also presents some issues for consideration when analyzing products such as coffee, fruits and alcoholic beverages.

Section 4 covers other applications of electronic nose systems, especially applications in the medical field for detection and diagnosis of diseases. This section focuses more on viable alternatives for the detection of diseases, rather than on quality control.

It is important to note that quality control is mainly used to find errors in processes, so the deductions presented here have gone through a series of tests and experiments to obtain the desired results and thus facilitate further research and shed light on the question of how these types of applications should be addressed.

2. A look at the electronic nose systems

Existing systems for electronic olfaction (EOS), also commonly known as electronic noses, are basically arrays of chemical sensors, connected to a computer or processing systems

[1] Applications and experiences of Quality Control. Preface. www.intechweb.org Copyright 2011 Intech.

which apply advanced techniques of digital signal processing and statistical pattern recognition. Their main objective is to enable the qualification of odours through classification tasks, discrimination, prediction, and even quantification of products, elements or components according to their organoleptic characteristics (Duran & Baldovino, 2009; Wilson & Baietto, 2009; Zhou et al., 2006).

2.1 Elements of an Electronic Nose System (EOS)

An electronic nose system can be seen as an instrument or measuring equipment of artificial olfaction, consisting of a series of modules that work together, which analyzes gas samples, vapours and odours. An instrument or equipment of this type has at least 4 parts, each with specific functions which are detailed below (Duran & Baldovino, 2009; Tian et al., 2005).

2.1.1 Matrix or array of gas sensors

In general, the gas sensors are devices that consist of two main parts, the first is an active element which changes its physical or chemical properties in the presence of that which it detects and the second part is a transducer, which converts the changes in the properties of the active element into an electrical signal. These sensors typically have a selective membrane, preventing passage of particles or unwanted material, acting as a first noise filter. In Figure 1 shows a simplified diagram of a device of this type, in which the main parts of a gas sensor and the nature of the inputs and outputs can be seen. (Grupo E-Nose, 2011, Tian et al., 2005).

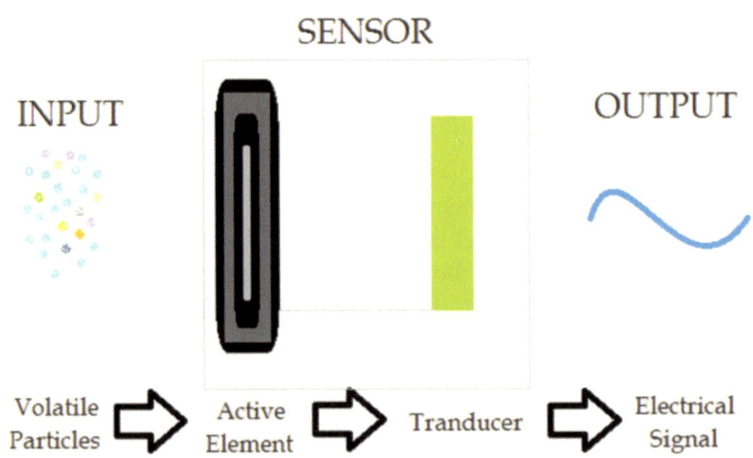

Fig. 1. Simplified schematic diagram of a gas sensor.

There are different types of gas sensors for use in EOS, the most common are: MOX (Metal Oxide Semiconductor), QCM (Quartz Crystal microbalance), SAW (Surface Acoustic Waves), MOSFET (Metal Oxide Semiconductor Field Effect Transistor), CP (Conducting Polymers), and FO (Fiber Optics). This chapter deals specifically with MOX sensors built with semiconductor materials such as Tin oxide (SnO2), Zinc oxide (ZnO), Titanium oxide (TiO2), among others. Their operating principle is based on the change of conductivity of a sensitive material when it absorbs or reacts with the gases in the environment, Figure 2 shows several commercial sensors of this type (Berna, 2010).

Fig. 2. Commercial gas sensors manufactured by Figaro and FIS, with different sizes and pin configuration.[2]

The majority of gas sensors are general purpose and usually have high sensitivity, detecting very low concentrations of volatile, but have disadvantages when trying to determine concentrations of a single component, because the output signal cannot be unambiguously assigned to the component by its generality (Duran, 2005, 2009).

Due to the fact that all EOS have a gas sensor array, it is desirable that the array be located in a special chamber or compartment in which the right conditions can be ensured for the proper operation. Mainly adequate insulation must be ensured to prevent pollutants from entering and the appropriate temperature and pressure must be maintained, these parameters are important or critical depending on the type of sensor used (Duran, 2005). Another advantage of using a chamber of sensors is that it facilitates the measurement process, because the volatiles will be in a higher concentration and they will have more contact with the active element of the sensor, which enables better and faster response from the sensors. It has also been experimentally determined that if the chamber of sensors is more hermetic, it can further exploit these advantages. Figure 3 shows a photograph of a chamber of sensors used in one of our projects with EOS (Velásquez et al., 2009).

2.1.2 Volatile delivery system

Basically it is a system that is responsible for transporting volatiles emitted by the samples or elements to be scanned into the chamber of sensor. Sometimes the sample is manually injected into the chamber of sensors, which results in error and delays; other times an automated system is responsible for transporting the volatile odorous molecules to the chamber of sensors, with the injection of a gas or air (Duran 2005, 2009).

---

[2] Images of sensors were taken from different Internet pages.

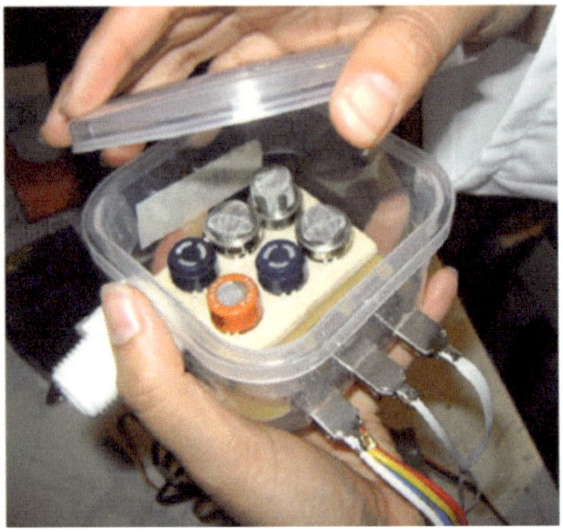

Fig. 3. The chamber of sensors provides hermetic isolation and guarantees reliable measurements.

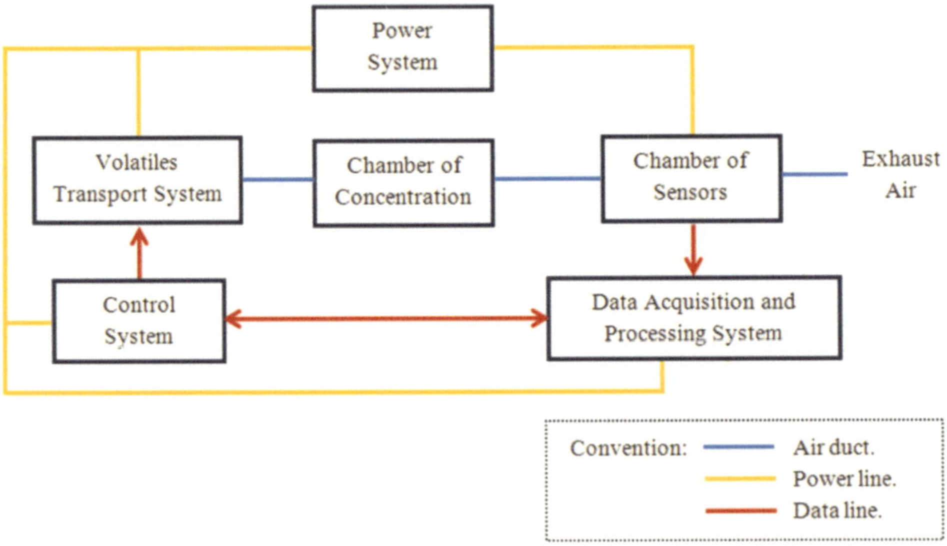

Fig. 4. Block diagram representing the electronic nose system.

Additionally most EOS have some kind of cleaning mechanism for the chamber of sensors, so that subsequent measures are made based on the same initial conditions and thus reproducibility of results is ensured. We recommend a different camera or hermetic compartment be used, called the "Chamber of Concentration", for containing the sample to be analyzed provided the environmental and physical conditions of the system allow it. Figure 4 shows the representation of an electronic nose system; note that the volatiles transport system is fundamental because it affects the operation of the EOS in the 3 different processes that can be carried out: concentration of volatiles, measurement and cleanup. (Rodriguez & Duran, 2008).

2.1.3 Control system and data acquisition

The control system takes care of proper handling of the Volatiles Transport System, for example: valves, air pump and other devices that are part of this system. It is also in charge of controlling additional subsystems or variables that the electronic nose system may have, such as temperature and humidity control, among others (Duran, 2005).

The data acquisition system is responsible for capturing the signals provided by the gas sensors and then delivering them to the process processing or computing system that has the appropriate software for processing such information (Rodriguez & Duran, 2008).

The control and data acquisition systems can be integrated into a single device, which can be a data acquisition card, a microcontroller, a DSP (Digital Signal Processor) or a computer; it must also have adequate power stage to handle the elements that consume more power and must have the proper memory settings to store large amounts of data obtained from the sensors.

We recommend working with a data acquisition card connected to a computer, to achieve good storage capacity, correct handling of information processing and graphical representation. Although in some cases when portability is required, a DSP or microcontroller can be used.

A significant part of the control system is the power source, which must be of a few amps, depending on the number of gas sensors and additional elements used; a source of 3 Amps is enough when working with an EOS that contains an array of 8 gas sensors.

2.1.4 Processing system

The processing system in most cases consists of a computer with an appropriate software for manipulating the data obtained by the sensors. Pre-processing techniques are applied to the data in order to extract the static parameters of the measures and reduce the amount of information to be analyzed. Subsequently multivariate analysis techniques and pattern recognition can be applied, such as PCA (Principal Component Analysis) and ANN (Artificial Neural Networks) to perform tasks such as: classification, discrimination, prediction, quantification of samples according to their organoleptic characteristics (Wilson & Baietto, 2009; Berna, 2010).

2.2 Operation of an electronic nose system

The operation of an electronic nose system depends on the component parts and the features of the equipment. In order to obtain measurements with a EOS the first step is to adjust the adequacy of the sample to be examined, this depends on the type of element to be analyzed, which sometimes must be heated, cut, mixed with other elements and simply placed near the sensors array or in the chamber of concentration ready to be analyzed. (Duran & Baldovino, 2009).

The concentration process begins when the sample is placed in the chamber of concentration. After this a few minutes should be given to allow the sample to release enough volatile particles, only then can the measurement process begin, for which the volatiles must be deposited or transported from then chamber of concentration to the chamber of sensors. During the measurement process, the data acquisition system records all the changes in the output signal of each of the gas sensors. When the measurement process is finished the cleaning process of chamber of sensors begins and the stored data can be processed and analyzed immediately (off-line processing), using the pre-processing software and signal processing, in order to obtain an olfactory footprint that represents the sample, to perform the tasks of classification, discrimination and other (Berna, 2010; Wilson & Baietto, 2009).

3. Quality control of food using electronic nose systems

A great part of electronic nose system applications are used in the food industry, where studies can be found with meat, milk and dairy products, eggs, different grains, fruits, oils, alcoholic and non alcoholic beverages, among others (Berna, 2010).

a)

b)

Fig. 5. a) Growing Coffee.[3]. b) Image of green coffee beans.[4]

Food quality control is one of the many applications that can benefit from the use of electronic nose (EOS). E.g., it can determine the type of product that is being analyzed, it can be classified by region, quality, time of ripening or storage, the food life span can be determined or predicted, as well as the level of deterioration or decomposition, the food life span can be determined or predicted and can determine flavors (Berna, 2010; El Barbri et al., 2008).

This chapter discusses the use of EOS in the quality control of foodstuffs such as coffee, fruits and alcoholic beverages.

3.1 Quality control of coffee with an electronic nose system

In the quality control of coffee, the organoleptic characteristics are a determinant of its quality and therefore, they are significant to locating the predominant defects of coffee beans, as they negatively affect its flavor and odor (Rodriguez et al., 2010; Pardo et al., 2000).

It is important to keep in mind that coffee production (Fig. 5. a.) is such an artisan process, that its control is somewhat complex and highly dependent on the historical traditions and cultural knowledge of those involved in the process, the lack of modernization of coffee farms, the incidence of fungal and other diseases in the crops and the need for chemicals sometimes influences the product quality (Rodriguez et al., 2010).

Another important factor to take into consideration in the quality of coffee is the climatic and edaphological conditions or nature of the soil. The Colombian coffee zone is located on

[3] Photo owned by FNC, by Patricia Rincon Mautner. http://www.colombia.travel/es/turista-internacional/actividad/590-clima-y-ubicacion-geografica-del-cafe

[4] Taken from the website of Herbolario Esencia. http://herboesencia.es/a-e/cafe-verde-coffea-arabica/

hillsides between 1000 m and 2000 m above sea level, with temperatures between 17ºC (290ºK) and 23ºC (296ºK) and relative humidities between 70% and 85%. The Table 1, show other data associated with optimal climatic conditions for growing coffee ([CENICAFE], 2011).

Average Solar Radiation	Solar Brightness	Temperature	Rainfall	Relative Humidity	Daily evaporation	Winds
Between 300 and 450 cal/cm ^{2} per day.	Between 4 to 5 hours daily.	Between 17 and 23 ° C or 290ºK and 296ºK	Between 1800 and 2800 mm annually.	Between 70 and 85%	Between 3 to 4 mm.	Below 5 km/h.

Table 1. Average Climatic conditions in coffee growing regions ([CENICAFE], 2011).

The CENICAFE web page (2011) states that: "The soils of the Colombian coffee region are relatively young, e.g. they are still under development and the nature of the material which is derived from petrographic material is grouped into the following classes: Metamorphic, igneous and sedimentary, which occur on different levels and patterns of coverage with volcanic ash. These soils are highly variable in their characteristics and their distribution throughout the coffee zone, its location in reliefs from flat or gently undulating to steep with 75% slope, and the variety of their physical (from rocky and sandy to loam and clay) and chemical conditions (low to high content of organic matter and minerals)".

3.1.1 What should be taken into account when considering a product such as coffee?

Coffee is a product that is collected manually and subjected at certain processes before obtaining the green coffee beans (Fig. 5. b.), in this condition it is more difficult to make an organoleptic analysis of coffee; therefore the best way to analyze coffee is the same way as tasters do, who perform tests on toasted and ground coffee, therefore the coffee must be subjected to a process of roasting and grinding in order to obtain a powder which is mixed with water at an average temperature of 60ºC (333ºK), which enables the emission of volatile particles. This mixture is introduced into the chamber of concentration (Fig. 6) in order to cluster the volatile particles which are then carried to the chamber of sensors for the measurement process. Figure 7 shows in detail the procedure used for the preparation of the mixture before the measurements (Falasconi et al., 2005).

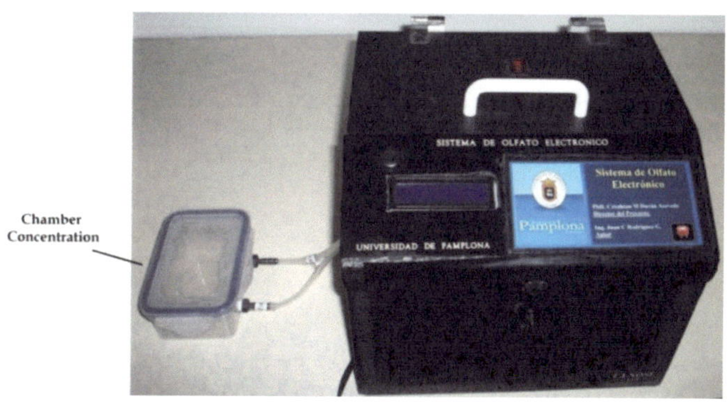

Fig. 6. Chamber of concentration, container for the different samples to be analyzed.

3.1.2 Some results obtained with the coffee

There have been several tests of different varieties of export quality "Excelso" coffee, with "regular" coffee and coffee with marked defects in the grains. Tests have been accompanied by personnel trained in coffee tasting, who issued their concept based on their personal perception of each coffee sample, helping in the designation of various patterns for facilitating subsequent classification tasks with different measures (Rodriguez et al., 2010).
In one of the tests measures were taken from samples of export quality coffee of two different varieties, Excelso Europe and Excelso UGQ-10%, which are classified as good quality cafes. These measurements were compared to those of regular coffee (for domestic consumption, commonly known as "Pasillas"), in which the experts detected some defects such as traces of fermentation, chemical contamination and signs of "Repose" (caused by prolonged storage or storage in unfavorable conditions). It is noteworthy that the defects mentioned are those most commonly found in coffee, influenced by poor product handling techniques (Rodriguez et al., 2010).

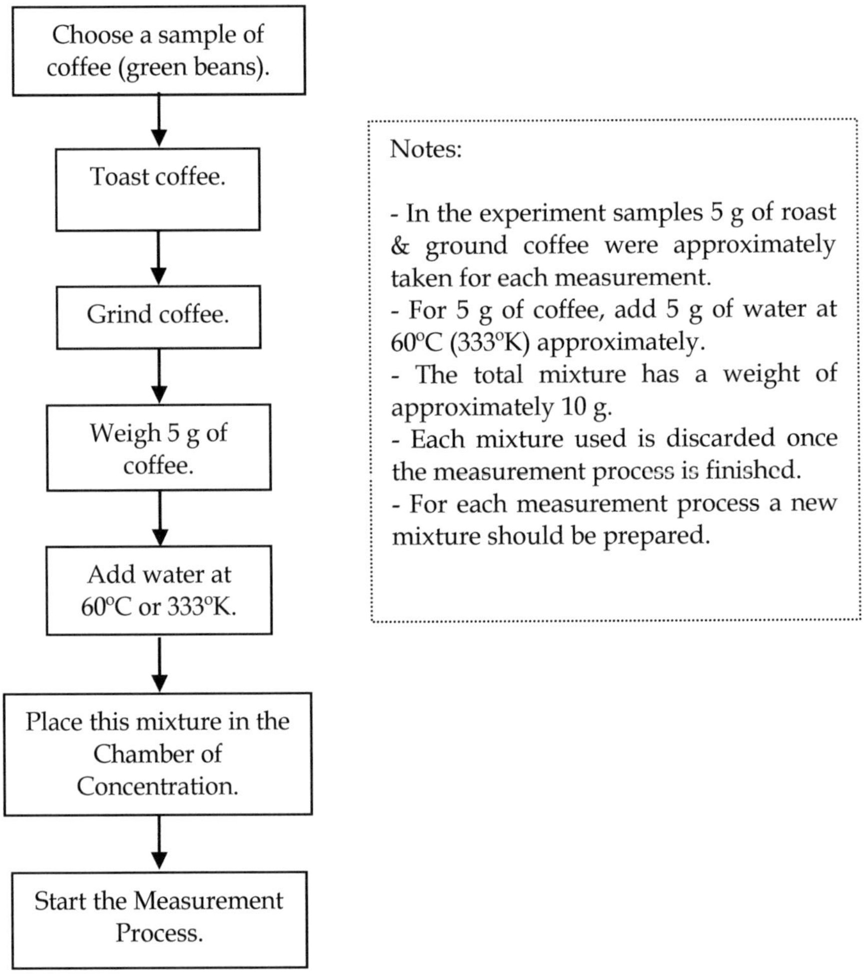

Fig. 7. Procedure used for preparing samples of coffee before starting the measurement process.

Figure 8 shows the analysis of these measurements, using the technique PCA (Principal Component Analysis). The different measurement groups can be seen, clearly differentiated in samples of regular and export type coffee. The measurements taken from export quality coffee are highlighted in green circles, while the measurements taken from regular coffee are in red circles.

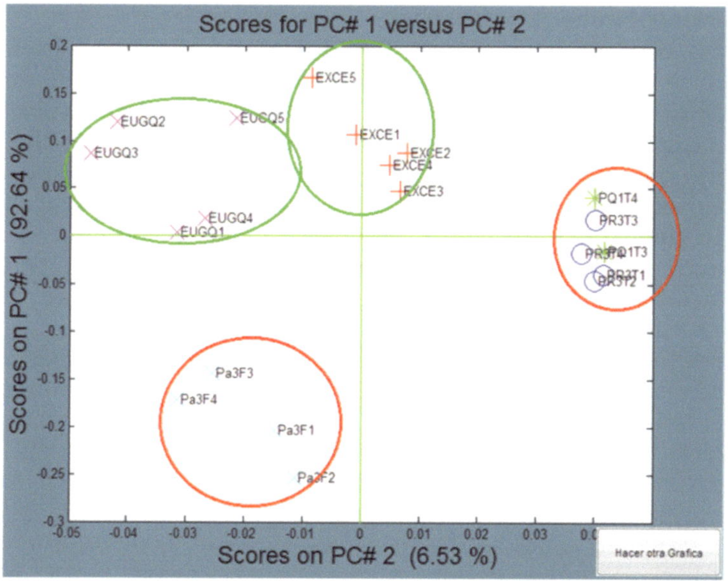

Fig. 8. Results of PCA analysis between measures of good quality coffee (green circles) and coffee with defects (red circles).

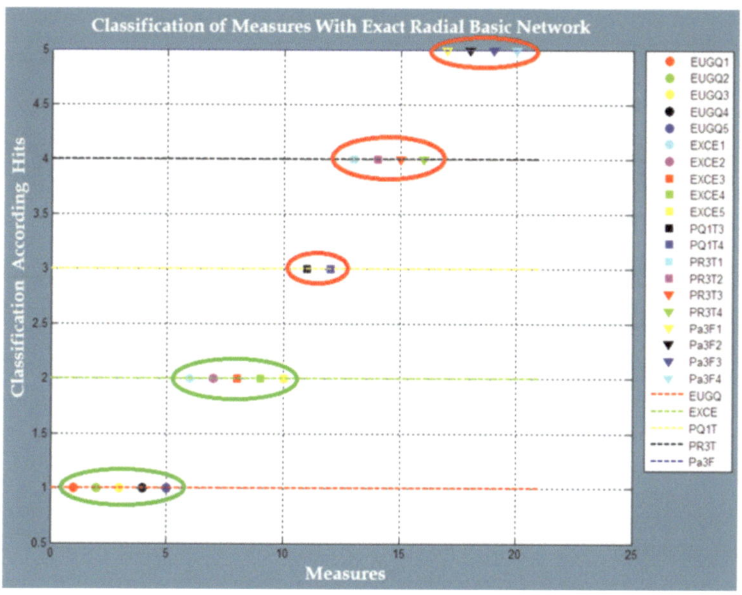

Fig. 9. Classification results of the measurements with a radial basis neural network, between good quality coffee (green circles) and coffee with defects (red circles).

This group of measurements was classified using a radial basis neural network (Figure 9). It can be seen how the various measurements of the same type are located within a horizontal axis, forming 5 different subgroups, which belong to two major groups of export type coffee (green circles) and regular coffee (red circles).

3.2 Quality control of fruits

For the analysis of fruits invasive and noninvasive techniques can be used. Invasive techniques involve damaging the fruit to take a sample, in order to perform various tests with the same fruit at the same moment and also facilitate extraction of volatile particles, as manipulation helps to release more volatile particles, which facilitates the measurement process. A drawback of this technique is that once the product is handled it can only serve in the measurement process, because handling accelerates the decomposition process. Meanwhile, the noninvasive techniques, just take the fruit for testing without inflicting damage therefore the same fruit can be used for further testing in order to analyze maturity stages and study the processes of decomposition. (Rodriguez & Duran, 2008; Duran & Baldovino, 2009; Berna, 2010).

Figure 10 shows the results of the analysis of some measurements made on samples of passion fruit, peaches and apples, using the PCA technique. The 2 measurement groups can be seen, which can be clearly differentiated in samples of passion fruit and peaches, in addition 2 measurements of apple were introduced as a test (Creole apple and Chilean apple), in order to test the classification accuracy of the system and the similarity that may exist between different varieties of a fruit. Also Figure 11 shows the validation of the measurements using an Artificial Neural Network Feed Forward Back Propagation, applying the technique "Leave one out", it can be seen how the system responds to the eventual absence of a measure in the training of neural network, the most significant result occurs with measurements of apples which are classified successfully despite having so few measures.

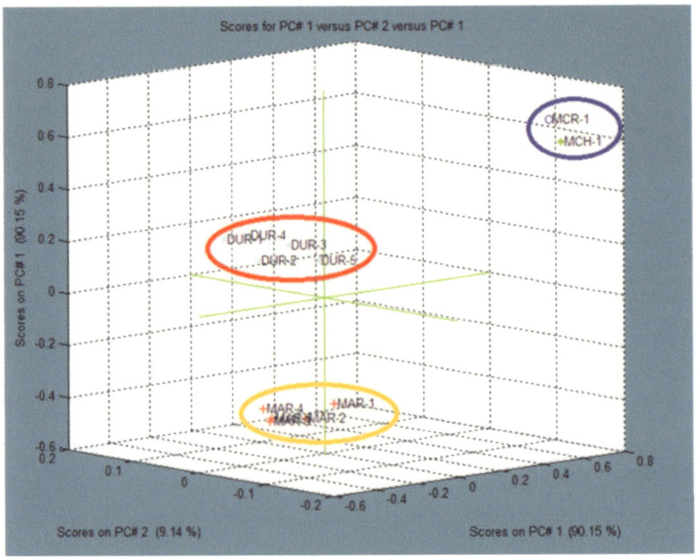

Fig. 10. Results of PCA analysis between passion fruit (yellow circle), peaches (red circle) and apples (blue circle).

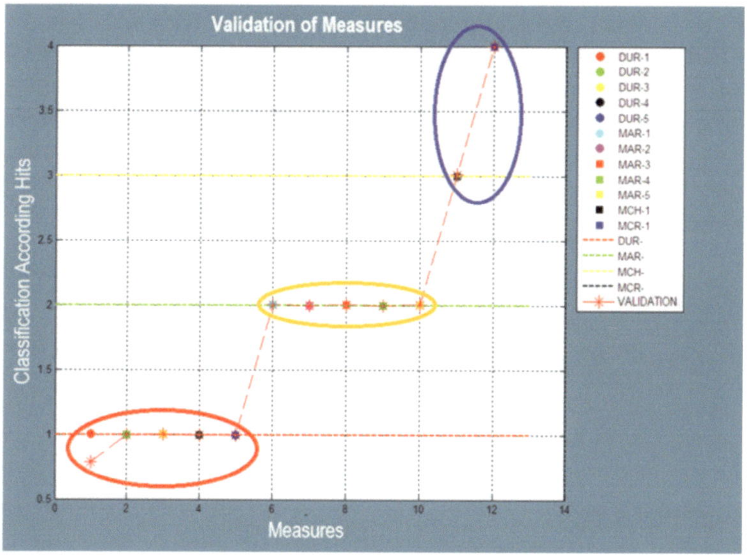

Fig. 11. Validation Results of measurements with passion fruit (yellow circle), peaches (red circle) and apples (blue circle) using a neural network Feed Forward Back Propagation.

3.3 Quality control of alcoholic beverages

Electronic nose systems have been widely used for classification, discrimination of characteristics and detection of different elements or compounds considering the organoleptic characteristics, but its application in quantification tasks has not been widely explored. In some of these studies the least square regression method is used to consider the gas concentration (Khalaf et al., 2009) and for the quantification of mixed contaminants in the air (Zhou et al., 2006), also the new technologies have been used as systems based on micro-electromechanical sensors for the quantification of components in vapor mixtures (Zhao et al, 2007).

Below is a study with an electronic nose system, where a digital signal processor DSP was adapted and artificial neural network "Feed-forward back propagation" was implemented, which was trained with the aim of identifying and quantifying levels of Ethanol and Methanol in different samples. As result the percentage of Ethanol and Methanol of the samples were obtained, and the electronic nose system was improved, called "A-NOSE" (Rodriguez et al., 2010), when the processing software was implemented in a different device from the personal computer.

The artificial neural network that was used to perform the identification and quantification of Ethanol and Methanol was Multi Layer Perceptron (MLP) Feed-forward back propagation network, which was trained and tried in R2006a Matlab software; as a result of training of the artificial neuronal network the weight matrices and bias vectors were obtained, that were used to codify the artificial neural network program in C++ language with software CodeWarrior and subsequently downloaded this program in the digital signal processor DSP56F801 of Motorola.

The initial samples were 95% Ethanol and 95% Methanol, which were diluted with distilled water to obtain 50%, 25% and 10% concentrations. Different measurements with the Ethanol and Methanol were realized in their different concentrations to realize the training of the artificial neuronal network and additionally other measures with wines (red, white, fruity, orange wine) and Aguardiente (national drink) were performed. The percentages of wine were close to 10% Ethanol and 0% Methanol and Aguardiente was close to 30% Ethanol and 0% Methanol, values in accordance to the values specified on the product labels.

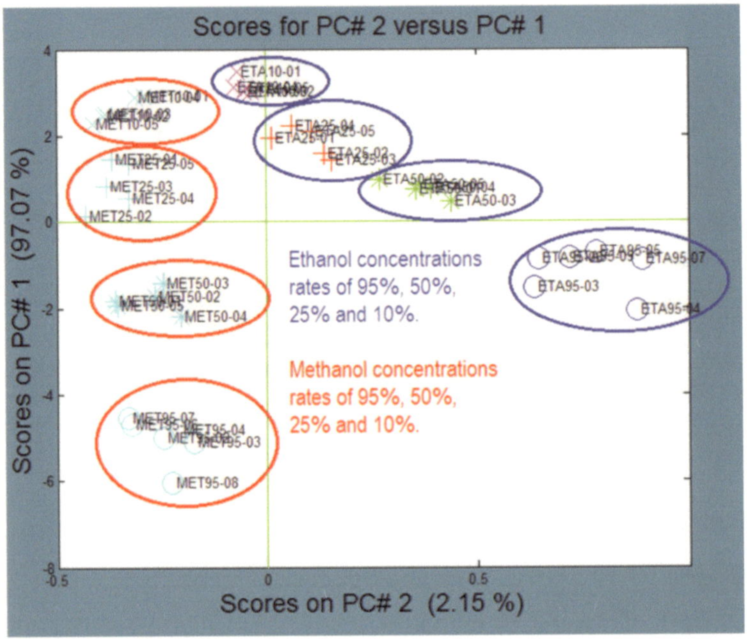

Fig. 12. Results of PCA analysis between Ethanol measurements (blue circles) and Methanol measurements (red circles).

Figure 12 the results of PCA analysis can be seen applied to measurements of different ethanol and methanol concentrations. It can be inferred that the measurements follow a trend, which yield a characteristic equation that models the behavior for different concentrations of Ethanol and Methanol. It can also be seen that as the concentration of Ethanol and Methanol is lower measurements tend to find a common point, this may be because they have are both alcohol.

Another test analyzed samples of different kinds of wines (e.g.: red wine, white wine, orange wine) and Aguardiente (national drink), the results are shown in Figures 13 and 14. It can be seen that wine measurements are close to 10% Ethanol, the results obtained by the neural network (feed-forward back propagation) that was trained for this purpose showed results very close to 10% and 0% Ethanol Methanol. It should be clarified that the neural network was trained with data from measurements of different ethanol and methanol concentrations and were then tested with data from measurements of different drinks.

4. Other applications of electronic nose systems

The applications of electronic nose systems are very diverse. The previous sections have covered some of the possible applications in the agro-food industry, but there are many other still to be mentioned, for example: The identification and diagnosis of respiratory

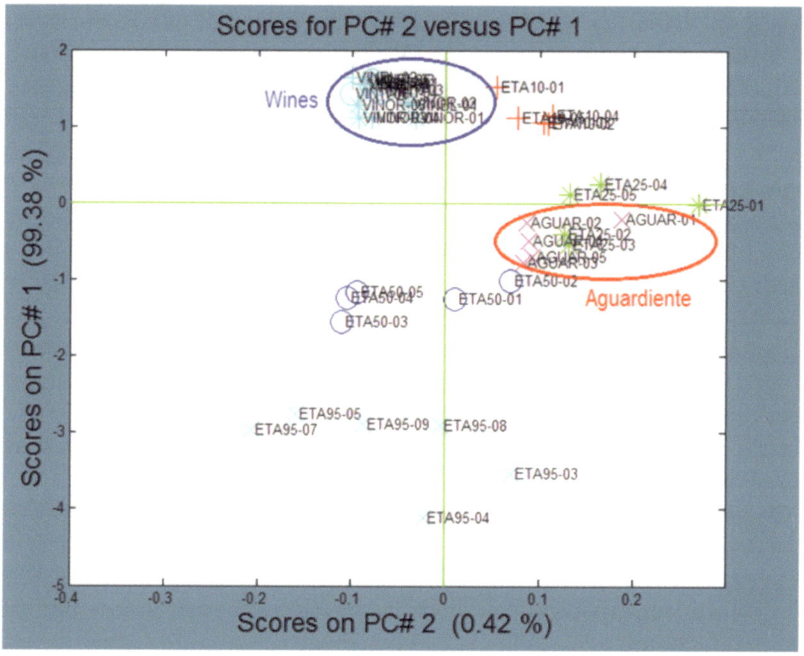

Fig. 13. Results of PCA analysis for the classification of wines and Aguardiente, according to the concentration of Ethanol.

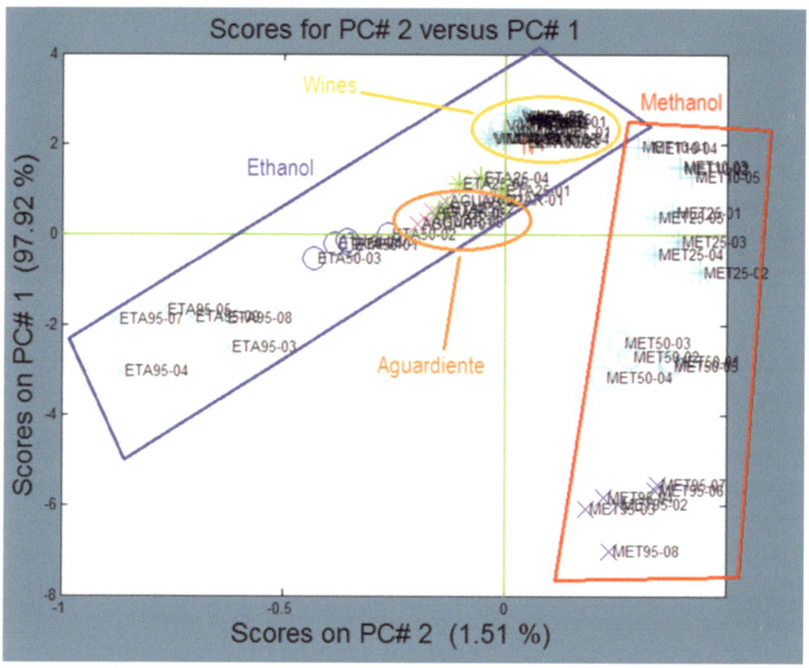

Fig. 14. Results of PCA analysis for the classification of wines and Aguardiente, according to the concentration of Ethanol and Methanol.

diseases (Xu et al., 2008, Velasquez et al ., 2009), the detection of narcotics and explosive substances (Oakes, L.; Dobrokhotov, V., 2010), determination of air quality and the environment (Zhou et al., 2006), among others, although there are still many possible applications to be explored.

4.1 Detection of diseases using electronic nose systems

There are a variety of respiratory diseases, which in some cases are caused by smoking and exposure to contaminated environments. This is the case of Chronic Obstructive Pulmonary Disease COPD, which has a mortality rate exceeding 15.9% (Velásquez et al., 2009; Velásquez, 2008).

COPD is a chronic lung disease characterized by airflow limitation that is not fully reversible, with progressive deterioration and is associated with abnormal lung inflammatory response to noxious particles or gases (Velásquez et al., 2009; Velásquez, 2008). The main cause of COPD is prolonged consumption of cigarettes, it is said that up to 20% of smokers have COPD.

This disease is more common in:

- White people.
- People over 60 years of age.
- People who work in environments polluted by chemical vapors and harmful dust that can damage the lungs.
- People who suffer from chronic asthma.
- People with a family history of emphysema.

Other COPD risk factors include:

- Passive smoking.
- Air pollution.
- Low birth weight and other lung infections (Velásquez, 2008).

4.1.1 Analysis of measurements taken from people with COPD and healthy controls

Below are some results of the analysis of measurements taken from healthy controls, nonsmokers and patients diagnosed with COPD. All patients diagnosed with COPD were long time smokers from 16 years up to even 50 years and most of them have already quit smoking, due to the fact that many receive medical treatment. (Velásquez et al., 2009; Velásquez, 2008).

Figure 15 shows the results of PCA analysis of samples of healthy controls and patients with COPD. The low dispersion of measurements of healthy controls and high dispersion of measurements of patients with COPD can be seen; due the fact that to not all patients have the disease at the same level.

Figure 16 has separated the samples from different patients with COPD, from these results it can be inferred that according to the health of the person the different patients can be classified. Future research should conduct more measurements on patients at different stages of the disease and could also be extended to other respiratory diseases and even gastric related diseases.

A study that deserves attention is (Xu et al., 2008) who developed a solid trap/thermal desorption-based odorant gas condensation system designed and implemented for measuring low concentration odorant gas. The results showed that the technique was successfully applied to a medical electronic nose system. The developed system consists of a flow control unit, a temperature control unit and a sorbent tube. The theoretical analysis and experimental results indicate that gas condensation, together with the medical electronic nose system can significantly reduce the detection limit of the nose system and increase the system's ability to distinguish low concentration gas samples.

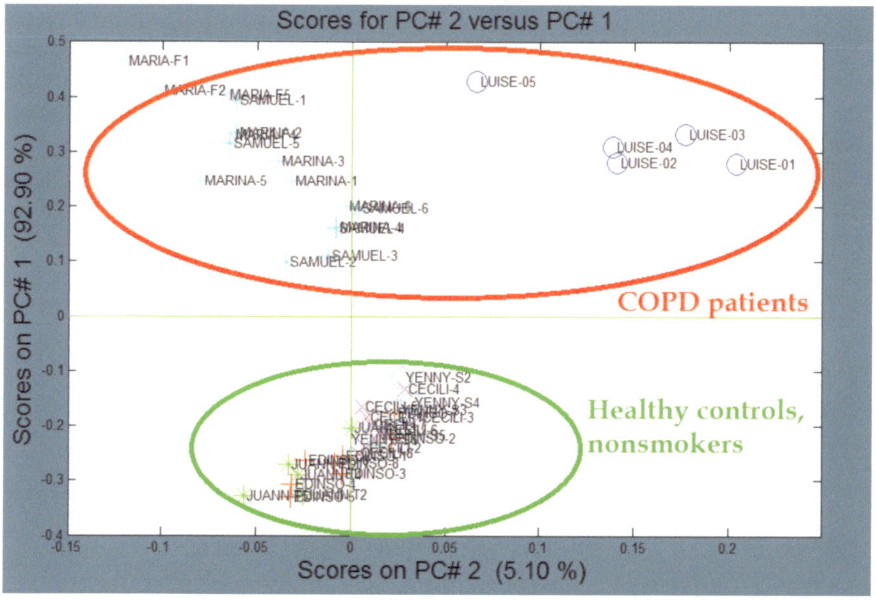

Fig. 15. Results of PCA analysis for the classification of COPD patients and non-smokers.

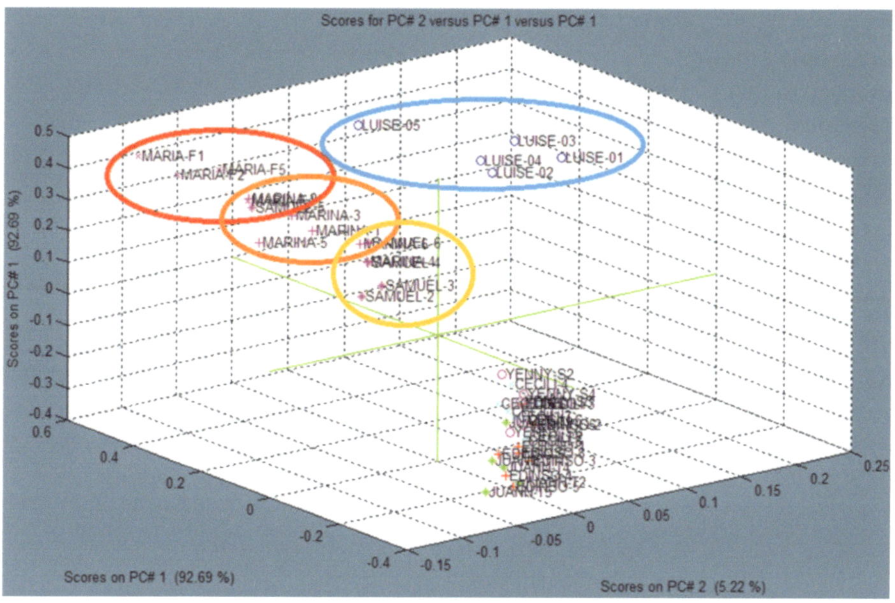

Fig. 16. Results of PCA analysis with emphasis on measurements of patients with COPD. Using DGN (Center), displaying the first 2 components.

4.2 Determination of air quality using electronic nose systems

Such applications have a bright future in the industry, because the environment is very susceptible to leakage and contamination by gases, which in many cases can be harmful and even lethal to humans.

NASA has done some work on this issue, for example (Ryan et al., 2009), Whom Developed an Electronic Nose to be used in Environmental Monitoring in the International Space Station, the Electronic Nose (Enose) is an array of 32 polymer sensors, the pattern of response may identify contaminants in the environment. An engineering test model of the ENose was used to monitor the air of the Early Human Test experiment at Johnson Space Center for 49 days. Examination of the data recorded by the ENose shows that major excursions in the resistance recorded in the sensor array may be correlated with events recorded in the Test Logs of the Test Chamber. The ability to monitor the constituents of breathing air in a closed chamber in which air is recycled is important to NASA for use in closed environments such as the space shuttle and the space station.

In the same way an electronic nose system could be developed for places such as airports or customs, in order to detect narcotics or prohibited hallucinogenic substances and in hostile or war environment to detect explosives or mines planted in the soil.

5. Conclusions

The operation of the electronic nose system depends on the component parts and the features of the equipment. Inside we find the gas sensor array, the volatile particle delivery system, control system, data acquisition and data processing system.

We recommend a different chamber or hermetic compartment be used for containing the sample to be analyzed, called "Chamber of Concentration", provided the environmental and physical conditions of the system allow it.

The volatile particle transport system is fundamental because it affects the operation of the electronic nose system in the 3 different processes: concentration of volatile particles, measurement and cleanup.

Measurements with electronic nose systems begin by ensuring the adequacy of the sample to be examined, this depends on the type of element to be analyzed, which sometimes must be heated, cut, mixed with other elements or simply placed near the sensor array or in the chamber of concentration.

During the measurement process, the data acquisition system records all the changes in the output signal of each of the gas sensors. When the measurement process is finished the cleaning of the chamber of sensors begins, which is very important to restore the initial conditions of the system and to ensure the reproducibility of the measurements.

Once the measurement process is finished the stored data is processed and analyzed using the pre-processing software which allows to extraction of static parameters from the measurements and reduces the amount of information to be analyzed. Subsequently the processing software is applied, in order to obtain an olfactory footprint that represents the sample, to perform classification, discrimination and other tasks.

Coffee is preferably analyzed in the same way as by tasters, who perform tests on toasted and ground coffee, therefore the coffee must be roasted and ground in order to obtain a powder which is mixed with hot water, to facilitate the emission of volatile particles and this mixture is introduced into the chamber of concentration for the measurement process. This procedure for the preparation of the mixture can be applied similarly to other elements before the start of the measurements.

To identify measurement patterns or to carry out the training applications using computational intelligence it is very important to have expert staff on hand, as in the case of coffee quality control which had the support of trained coffee tasters, who issued their concept based on personal perception of each coffee sample, helping in the designation of the various patterns to facilitate subsequent classification tasks with different measurements.

Electronic nose systems have been widely used for classification, discrimination of characteristics and detection of different elements or compounds considering the organoleptic characteristics, and even for quantification tasks. This can be carried out with tools like multivariate analysis techniques and pattern recognition, such as PCA (Principal Component Analysis) and ANN (Artificial Neural Networks).

The processing software of electronic nose system can be implemented on a digital signal processor DSP using an artificial neural network like the alcohol research case presented in section 3 which used a Feed-forward back propagation network, which was trained with the aim of identifying and quantifying Ethanol and Methanol of different samples. As result the percentage of Ethanol and Methanol of the samples were obtained.

The artificial neural network that was used for the identification and quantification of Ethanol and Methanol was trained and tried using R2006a Matlab software; the training results were used to codify the the artificial neural network program in C++ language and subsequently downloaded this program in the digital signal processor DSP56F801 of Motorola.

The results of PCA analysis for samples of healthy controls and patients with COPD showed differences in the low dispersion of the measurements taken of healthy controls and high dispersion of the measurements taken of patients with COPD; due to the fact that not all patients have the disease at the same level.

6. Acknowledgements

This study was supported by the P09225 grant and carried out within the MIRP Research Group, Research Center, Instituto Tecnológico Metropolitano ITM, Medellín-Colombia.

7. References

Badrick, T. (2008). The Quality Control System. The Clinical Biochemist - Reviews, (August,2008), p.p. 67-70, ISSN 0159 - 8090.

Berna, A. (2010). Metal Oxide Sensors for Electronic Noses and Their Application to Food Analysis. Sensors, Vol.10, (April 2010), pp. 3882-3910, ISSN 1424-8220.

[CENICAFE] Centro Nacional de Investigaciones del Café Colombia (March 2011). Sistemas de Producción, 25.03.2011, Available from http://cenicafe.org/modules.php?name=Sistemas_Produccion&file=condclim

Duran, C. & Baldovino, D. (2009). Monitoring System to Detect the Maturity of Agro-industrial Products Through of an Electronic Nose. Revista Colombiana de Tecnologías de Avanzada, Vol.1, No.13, (December 2009), pp. 1-8, ISSN 1692-7257.

Duran, C. (2005). Diseño y optimización de los subsistemas de un sistema de olfato electrónico para aplicaciones agroalimentarias e industriales. Universitat Rovira i Virgili. Tarragona, España.

El Barbri, N.; Llobet, E.; El Bari, N.; Correig, X. & Bouchikhi, B. (2008). Electronic Nose Based on Metal Oxide Semiconductor Sensors as an Alternative Technique for the

Spoilage Classification of Red Meat. Sensors, Vol.8, (January 2008), pp. 142-156, ISSN 1424-8220.

Falasconi, M.; Pardo, M.; Sberveglieri, G.; Ricco, I. & Bresciani, A. (2005). The novel EOS[835] electronic nose and data analysis for evaluating coffee ripening, Sensors and Actuators B: Chemical, Volume 110, Issue 1, (September 2005), p.p 73-80, ISSN 0925-4005.

Grupo E-Nose, (March 2011) ¿Qué es una Nariz Electrónica?, 15.04.2011, Available from http://www.e-nose.com.ar/paginas/funcionamiento.htm

Khalaf, W.; Pace, C. & Gaudioso, M. (2009). Least Square Regression Method for estimating gas concentration in an Electronic Nose System. Sensors, Vol. 9, pp. 1678-1691, ISSN 1424-8220.

Oakes, L.; Dobrokhotov, V. (2010). Electronic Nose for Detection of Explosives. American Physical Society. (March 2010).

Pardo, M.; Niederjaufner, G.; Benussi, G.; Comini, E.; Faglia, G.; Sberveglieri, G.; Holmberg, M. & Lundstrom, I. (2000). Data preprocessing enhances the classification of different brands of Espresso coffee with an electronic nose, Sensors and Actuators B: Chemical, Volume 69, Issue 3, (October 2000), p.p 397-403, ISSN 0925-4005.

Ryan, M.; Homer, M.; Buehler, M.; Manatt, K. & Zee, F. (2009). Monitoring the Air Quality in a Closed Chamber Using an Electronic Nose. Jet Propulsion Laboratory, California Institute of Technology. Johnson Space Center, NASA, Houston TX 77058.

Rodríguez, J.; Durán, C.; Reyes, A. (2010). Electronic Nose for Quality Control of Colombian Coffee through the Detection of Defects in "Cup Tests". Sensors, Vol.10, (December 2009), pp. 36-46, ISSN 1424-8220.

Rodriguez, J. & Duran, C. (2008). Electronic odor system to detect volatile compounds. Revista Colombiana de Tecnologías de Avanzada, Vol.2, No.12, pp. 20-26, ISSN 1692-7257.

Tian, F.; Yang, S. & Dong, K. (2005). Circuit and Noise Analysis of Odorant Gas Sensors in an E-Nose. Sensors, Vol.5, (February 2005), pp. 85-96, ISSN 1424-8220.

Velásquez, A.; Durán, C.; Gualdron, O.; Rodríguez, J. & Manjarres, L. (2009). Electronic Nose to Detect Patients with COPD From Exhaled Breath. Proceedings of the 13th International Symposium on Olfaction and Electronic Nose. AIP Conference Proceedings, Volume 1137, pp. 452-454, ISBN: 978-0-7354-0674-2, Brescia, Italy, April 15-17, 2009.

Velásquez, A. (2008). Sistema Multisensorial Electrónico No Invasivo Para La Detección De La Patología Respiratoria Epoc. Universidad de Pamplona. Pamplona, Norte de Santander, Colombia.

Wilson, A. & Baietto, M. (2009). Applications and Advances in Electronic-Nose Technologies. Sensors, Vol.9, (June 2009), pp. 5099-5148, ISSN 1424-8220.

Xu, X.; Tian, F.; Yang, S.; Jia, Q. & Ma, J. (2008). A Solid Trap and Thermal Desorption System with Application to a Medical Electronic Nose. Sensors, Vol.8, (November 2008), pp. 6885-6898, ISSN 1424-8220.

Zhao, W.; Pinnaduwagel, L.; Leis, J.W.; Gehl, A.C.; Allman, S.L.; Shepp,A. & Mahmud, K.K. (2007) Quantitative analysis of ternary vapor mixtures using a microcantilever-based electronic nose. Applied Physics Letters, 91 (4). ISSN 0003-6951.

Zhou, H.; Homer, M.; Shevade, A. & Ryan, M. (2006). Nonlinear Least-Squares Based Method for Identifying and Quantifying Single and Mixed Contaminants in Air with an Electronic Nose. Sensors, Vol.6, (December 2005), pp. 1-18, ISSN 1424-8220.

28

Mammographic Quality Control Using Digital Image Processing Techniques

Mouloud Adel and Monique Rasigni
Université Paul Cézanne, Institut Fresnel, UMR-CNRS 6133
Domaine Universitaire de Saint Jérôme,
France

1. Introduction

Breast cancer is the leading cause of cancer mortality among middle aged women. Survival and recovery depend on early diagnosis. At present mammography is one of the most reliable methods for early breast cancer detection. However relevance of diagnosis is highly correlated to image quality of the mammographic system. Hence periodic controls in mammographic facilities are necessary in order to make sure they work properly. In particular global image quality is evaluated from a mammographic phantom film. A phantom is an object with the same anatomic shape and radiological response as an average dense fleshed breast and in which are embedded structures that mimic clinically relevant features such as microcalcifications, nodules and fibrils. For each category of features, the targets have progressively smaller sizes and contrast so that the largest one is the most readily visible and the next is less visible and so on. Using a phantom makes it possible to free from the variable of differences in breast tissue positioning and level of compression from patient to patient. The process is as follows: the mammographic phantom film is analysed independently by several readers and a score is obtained by each of them depending on the number of objects they see. The independent object visibility scores are then averaged and the resulting score is assigned to the phantom film.
Automating this score by using computer image processing of digitized phantom films should make the evaluation of mammographic facilities easier and less subjective. In addition image processing should enable us to take into account other parameters such as, for instance, noise, texture and shape of the targets that a reader eye cannot estimate quantitatively, and so to perform a more elaborate analysis. In collaboration with ARCADES (Association pour la Recherche et le Dépistage des Cancers du Sein et du col de l'utérus) which set, since 1989, a breast cancer screening program in South of France, a project aimed at automating phantom film evaluation is in progress. Such a project consists first in digitizing phantom films with the adequate spatial resolution and then in processing the obtained images in order to detect, segment and characterize the objects contained in the phantom.
Little work has been done to automate quality control in mammographic facilities. Fast Fourier transform is used (Brooks et al., 1997) to establish some visibility criteria for the phantom test object. (Chakraborty et al., 1997) compares phantom images with a pattern image to obtain relations between some of the image parameters and the physical conditions

in which the images have been obtained. His work concerned only microcalcifications. (Dougherty, 1998) studies the most prominent microcalcification group and the most prominent mass using a manual threshold and some mathematical morphology operators. (Castellano et al., 1998) used binary masks to locate microcalcifications and they studied image resolution scales contained in the phantom. (Blot et al., 2003) used grey level co-occurrence matrices to score structures embedded in the phantom. (Mayo et al., 2004) used region growing and morphological operators to segment and characterize microcalcifications. They also studied horizontal resolution areas using morphological operators. This chapter presents a feasibility study aiming at automating phantom scoring using image processing techniques on digitized phantom films.

In the following sections, we describe the phantom used, the mammographic phantom image acquisition and digitization, the noise reduction and the contrast enhancement schemes used for processing phantom images, the segmentation step for each object (microcalcifications, masses and fibres) and the results obtained on nine phantoms films.

2. Description, acquisition and digitization of phantom films

2.1 Phantom description

The phantom used in this study is the MTM 100/R (Computerized Imaging Reference Systems, Inc., 2428 Almeda Avenue Suite 212, Norfolk, VA 23513, U.S.A., Phantom Serial Number: 2788). The MTM 100/R is used in France for Mammography Accreditation. It is made of tissue equivalent material in which are embedded objects simulating 7 pentagonal-shaped groups of microcalcifications (M1 to M7), 7 masses (N1 to N7) and 7 fibres (F1 to F7). For each category of features, the targets have progressively smaller size and contrast so that the largest one is the most readily visible, the next is less visible and so on. For convenience M1 stands for the microcalcification group containing the largest specks and M7, the smallest ones, with a similar convention for the other target structures. Inside are also present vertical (V) and horizontal (H) spatial resolution scales (line pair target : 20 lp/mm), a delimited zone (Z) for measuring a reference optical density, two different optical density contiguous areas (C1 and C2) for defining contrast, three cavities (D) for x-ray dose measurement and at last small balls (B) for x-ray alignment control. Figure 1 shows a schematic diagram of the MTM 100/R phantom.

2.2 Acquisition and digitization of phantom films

Phantom films are digitized with an ultra high resolution drum scanner (Scanmate 11000-ScanView A/S.Meterbuen 6. DK-2740 Skovlunde. Denmark) which may digitize from 50 to 11000 dpi (dots per inch) and code images on 256 (8bits /pixel) or 16384 (14bits /pixel) grey levels. Spatial scanning resolution was chosen so that it approximately corresponds to the resolving power of a viewer (with a standard resolving power ~ 4.10^{-3}rd) using a twice magnifying lens (such lenses are often used by radiologists for reading clinical mammograms). So phantom films were digitized with a resolution of 50 µm per pixel (or 508 dpi) and were coded on 256 grey levels.

For each category of structures (microcalcifications, masses and fibers) a sub-image was extracted from the digitized phantom image so that each sub-image contained one target and was roughly centered on it. Subimages sizes were 256×256-pixels for microcalcification groups and masses and 412×412-pixels for fibres. Fig. 2 shows an example of subimages extracted from a digitized phantom film.

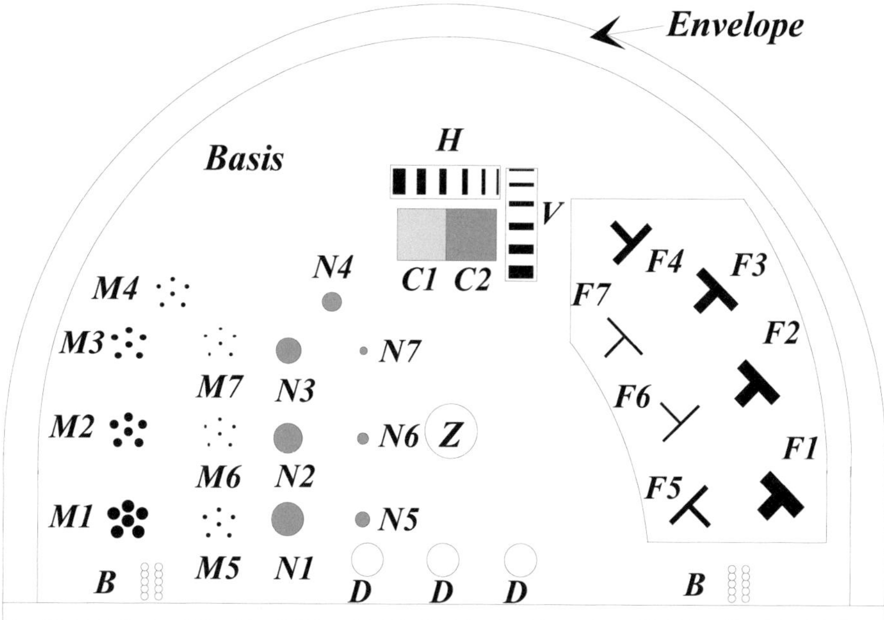

Fig. 1. Schematic diagram of the phantom MTM 100/R showing the locations and the relative size of features.

Fig. 2. Subimages extracted from a digitized phantom film. (a) : group of microcalcifications (b) : mass; (c) : fibre

3. Image processing of digitized phantom films

3.1 General description

Two pre-processing steps are applied to each extracted subimage before the segmentation step as shown in Fig. 3. Because of the noisy nature of these subimages, a noise reduction method is used as a first processing step. A contrast enhancement step is then applied. At last image segmentation is done.

3.2 Local contrast modification method description

In classical image processing techniques, a fixed shape and a fixed size window around each pixel is used in order to convolve it with a defined filter. In order to take the local features

around each pixel into account, a variable shape and size neighbourhood is defined (Dhawan et al., 1986; Dhawan & Le Royer, 1988) using local statistics.
Noise reduction step consist in filtering two kinds of noise. The non uniform background considered as a ''low frequency noise'' and the radiographic noise (high frequency noise): film granularity and quantum mottle. Shadow correction of the background is adapted to each object (microcalcifications, masse and fibre), whereas the radiographic noise filter and the contrast enhancement steps are based on the same method, described in the next section. This method consists in computing a local contrast around each pixel using a variable neighbourhood whose size and shape depend on the statistical properties around the given pixel. The obtained image is then transformed into a new contrast image using various contrast modification functions. At last an inverse contrast transform is applied on the new contrast image to yield an enhanced version of the original image. Contrast enhancement step consists in enhancing image features while preserving details for the segmentation step. Image Segmentation is adapted to the objects to be detected and is presented in the following sections.

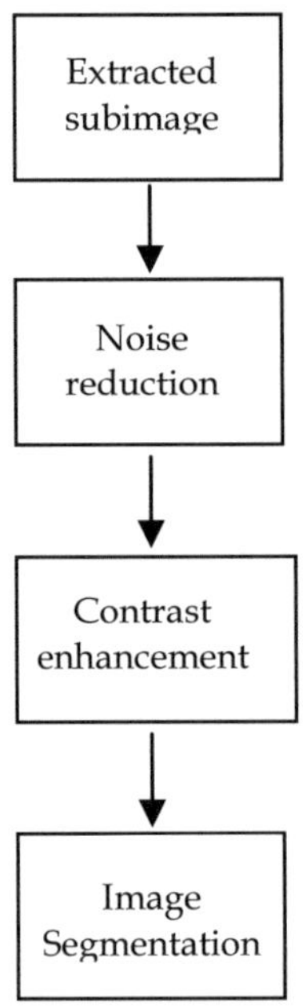

Fig. 3. Flowchart of the image processing steps applied to phantom images.

Each pixel (i,j) is assigned an upper window W_{max} centered on it, whose size is $(2N_{max}+1)\times(2N_{max}+1)$. We also define an inner area around the pixel (i,j) whose size is (c×c) and an external area whose size is (c+2) × (c+2), where c is an odd number. Let I(i,j) be the grey level of pixel (i,j) in image I, and T a given threshold. Pixel (k,l) within W_{max} is assigned a binary mask value "0" if $|I(k,l) - I(i,j)| > T$, else it is assigned a binary mask value "1". Then the percentage P_0 of zeros is computed over the region between the external (c+2)×(c+2) and the inner (c×c) areas, for each c in the range $[1, 3, 5, \dots, 2N_{max} - 1]$. The process stops if this percentage is greater than 60% or if the upper window W_{max} is reached. The value of 60% has been chosen because beyond this limit, we may consider too many pixels "0" are surrounding the inner area and so the notion of neighbourhood with the central pixel (i,j) in

terms of grey levels is no longer satisfactory. Let c_0 be the upper c value beyond which the percentage P_0 is greater than 60%. The pixel (i,j) is assigned the window $W=(c_0+2)\times(c_0+2)$.
In the window W such as W≤Wmax we finally define the "center" as the set of pixels having the mask value "1", and the "background" as the set of pixels having both the mask value "0" and which are 8-neighbourhood connected at least to a pixel "1". Pixels "0" which do not verify the previous constraint belong neither to the "center" nor to the "background" and are not taken into account later on. Fig. 4 gives an example the way the "center" and the "background" areas are determined.

13	14	10	9	8	10	11	13	10
13	10	15	10	9	8	13	13	8
12	15	14	14	9	14	15	7	8
8	5	15	14	16	17	10	7	8
7	6	20	15	**1**	16	10	16	8
13	21	22	21	15	16	10	9	8
13	20	19	14	16	15	10	8	9
11	15	14	15	17	16	7	6	8
10	9	15	14	15	16	4	5	6

(a)

1	1	0	0	0	0	0	1	0
1	0	1	0	0	0	1	1	0
1	1	1	1	0	1	1	0	0
0	0	1	1	1	1	0	0	0
0	0	0	1	1	1	0	1	0
1	0	0	0	1	1	0	0	0
1	0	0	1	1	1	0	0	0
0	1	1	1	1	0	0	0	0
0	0	1	1	1	0	0	0	0

(b)

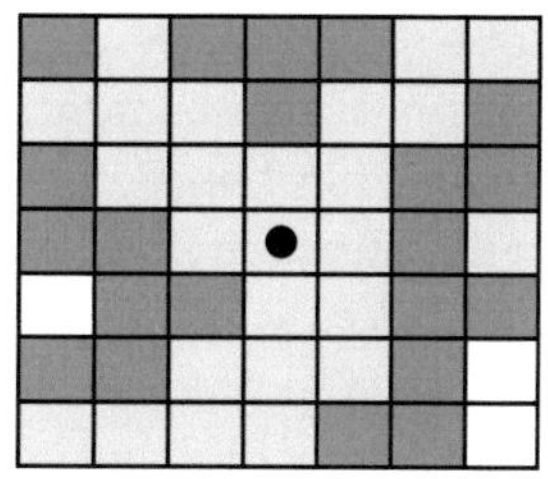

(c)

Fig. 4. Adaptive neighbourhood selection with a threshold value T=5; (a) W_{max} window around bold-faced pixel value 15, N_{max}=4; (b) Mask values associated to test pattern in (a); (c) The "center" (light grey)and the "background" (dark grey) areas around the bold-faced pixel. White cells correspond to pixels belonging neither to the "center" nor to the "background".

After determining the "center" and the "background" regions around each pixel (i,j), a local contrast image C is computed from :

$$C(i,j)=\frac{|M_c(i,j)-M_b(i,j)|}{\max(M_c(i,j),M_b(i,j))} \tag{1}$$

where $M_c(i,j)$ and $M_b(i,j)$ are the mean values, in image I, of pixels labelled as the "center" and as the "background" regions around pixel (i,j) respectively. Note that C(i,j) is within the range [0,1].
The local contrast image C obtained in the previous step is transformed into a new image C′ such as $C'(i,j)=\psi(C(i,j))$, where ψ is a contrast modification function depending on features to be detected. This function meets some requirements in the interval [0,1] :
- $\psi(0) = 0$ and $\psi(1) = 1$.
- $\psi(x) \geq 0$ for $x \in [0,1]$.
- ψ is an increasing function in the range [0,1].

In image C′ each pixel value is a contrast value. In order to obtain the corresponding image in the grey level domain, an inverse contrast transform of the process used to obtain image C from I (Eq. 1) is used as follows:

$$E(i,j)=M_b(i,j)(1-C'(i,j)) \text{ if } M_b(i,j) \geq M_c(i,j)$$

$$E(i,j) = \frac{M_b(i,j)}{(1-C'(i,j))} \quad \text{if } M_b(i,j) < M_c(i,j)$$

This transform gives a new image E which is an enhanced version of image I. It is then possible to evaluate the efficiency of the method from comparison between images E and I.

3.2.1 Performance evaluation on simulated images

Several functions including square root, exponential, polynomial and trigonometric were tested (Guis et al., 2003). Actually functions which are over the line y=x increase the contrast but enhance the noise too. In the other hand, functions which are under the line y=x yield noise reduction. Because of the noisy nature of real images of phantom, the second kind of functions was chosen for enhancing the objects contained in these images. To choose suitable function ψ, computer simulated images containing objects similar to those observed in the phantom film were generated with various contrast and noise levels. The aim of this simulation was to perform a quantitative evaluation of the noise reduction method described in previous sections. For each target, 6 noise-free images were generated each of them with a different contrast level. Three noise levels were then assigned to each contrast level image. Contrast level is defined as the difference between the mean grey level of the object and the mean grey level of the background divided by the mean grey level of the background. According to studies on radiographic noise, two types of noise sources, namely film granularity and quantum mottle, are present in an X-ray image. Spatially correlated Poisson noise model has to be considered in the case of mammographic films. In our simulations, a signal-dependent spatially uncorrelated Gaussian noise is used as a first-order approximation of the Poisson noise model (Quian et al., 1994; Aghadasi et al., 1992; Kuan et al., 1985) namely: $n(i,j) = \sqrt{f(i,j)}u(i,j)$, where f is the noise free image and where u is a zero-mean Gaussian noise with standard deviation σ. The computer simulated image or noisy image g is then given by $g(i,j) = f(i,j) + n(i,j)$.
Contrast levels of noise free images were in the range [10%; 60%] with a step size of 10%. Background grey-level was set to 128. Concerning noisy images, standard deviation σ of the zero-mean Gaussian noise u was adjusted so that the signal to noise ratio (SNR) takes the values 21dB, 15dB and 9dB which simulate a low, an intermediate and a high noise level respectively. Computer simulated images consist of 256×256-pixels for microcalcification groups and nodules, and 336×336-pixels for fibres. The whole computer simulated images were coded on 256 grey levels.
Two criteria are used to test the effectiveness of the algorithm on computer simulated images. The first one, namely output to input Signal to Noise Ratio (SNR) ρ, quantifies noise suppression, and the second one, namely the mean-squared error (MSE), in addition to quantify noise removal gives also an information on structure distortion and therefore better interprets the first criterion. One can notice that parameter ρ is all the more higher as the method removes much more noise, whereas MSE parameter is all the smaller as the method denoises and preserves structures in the image.

Results obtained on simulated images show that the trigonometric function $\psi(x) = \tan\left(\frac{\pi}{4}x\right)$ gives the best balance between noise reduction and edge sharpness preservation and that $\psi(x) = \sqrt{x}$ is suitable for contrast enhancement.

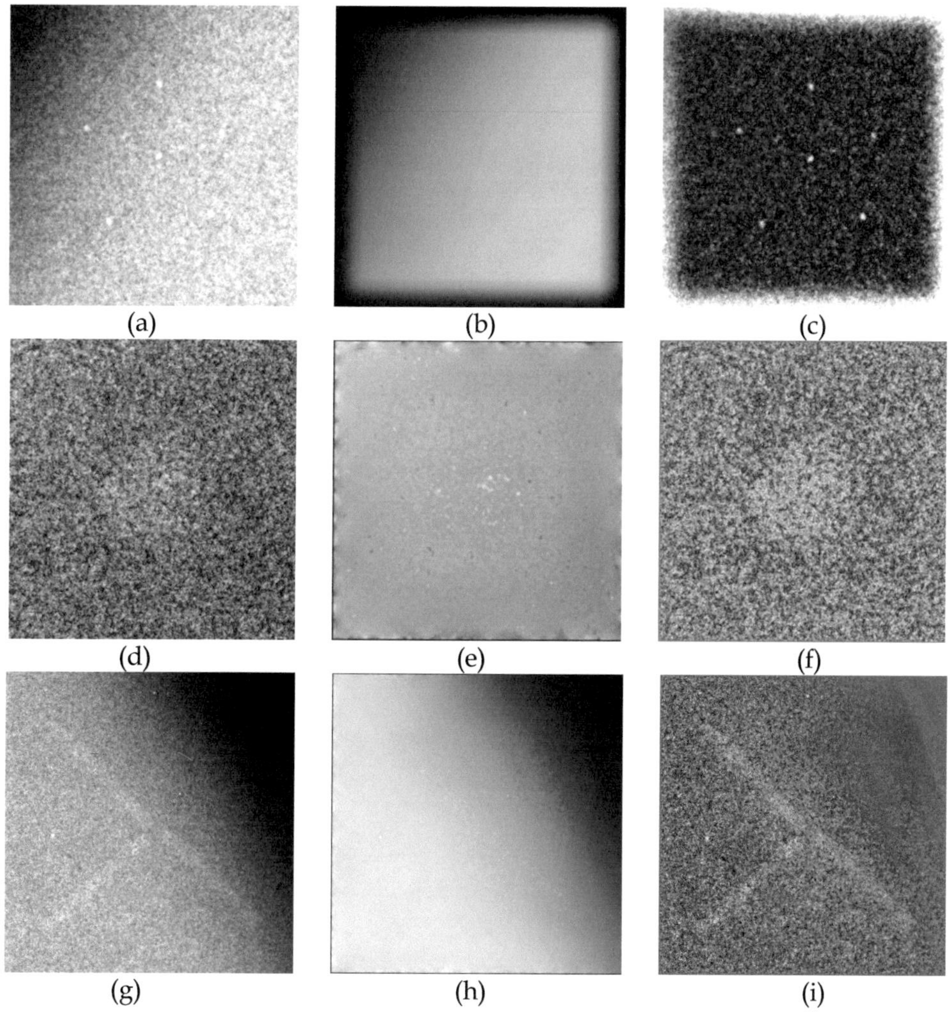

Fig. 5. Correction of Background inhomogeneity on extracted subimages. (a), (d) and (g) : original extracted subimages; (b), (e) and (h) background images; (c), (f) and (i) corrected images.

3.2.2 Preprocessing of real phantom images

Before applying noise reduction and contrast enhancement steps, inhomogeneous background of subimages containing microcalcification is extracted using the classical multiresolution Burt decomposition into level 3 (level 0 is the original image). A linear interpolation is then applied to obtain background image. The same method was not suitable to correct background for subimages containing masses and fibres, due to the object to image size ratio. Using the local contrast modification method described above with a big window size for W_{max}, enabled us to obtain background image. At last corrected image for the whole objects is obtained by subtracting the background image from the original image as shown in Fig. 5.

Applying the noise reduction and contrast enhancement steps described in the previous section, yield the images shown in Fig. 6.

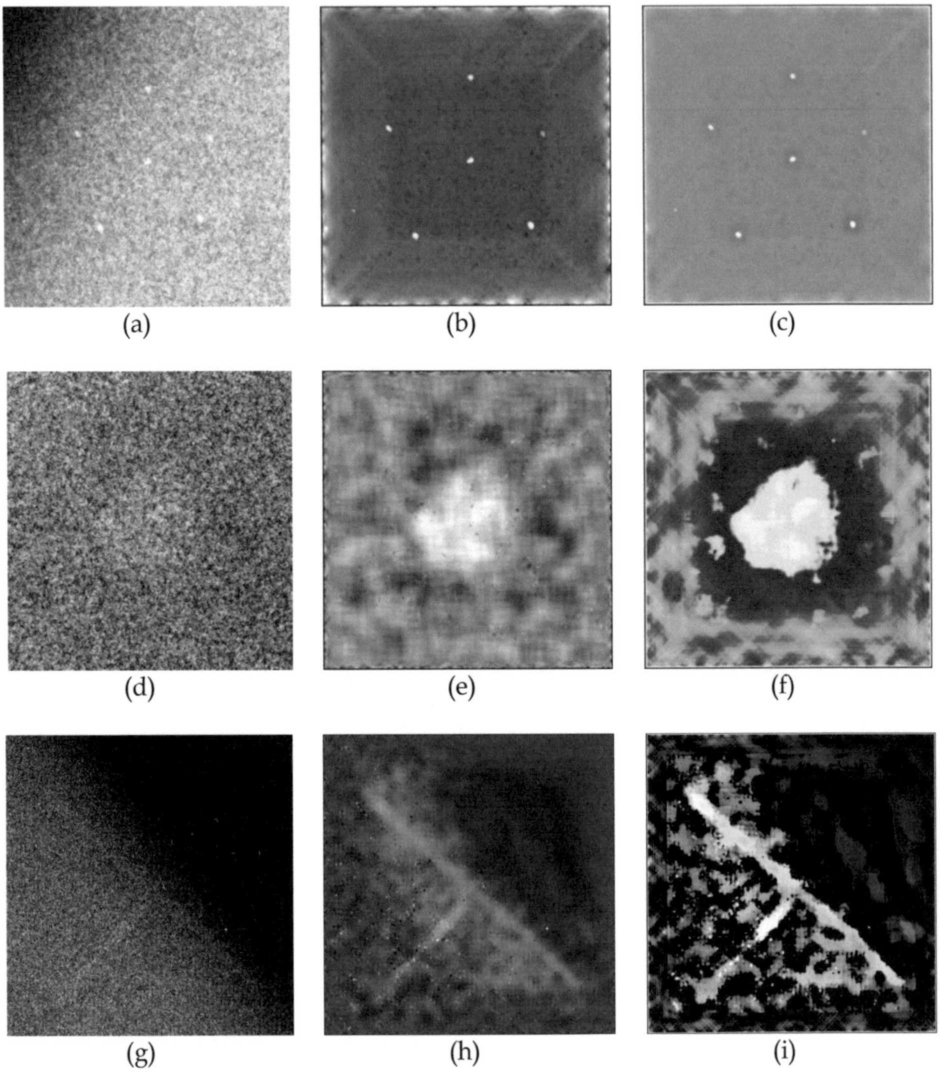

Fig. 6. Noise reduction and contrast enhancement steps on extracted subimages (a), (d) and (g): original extracted subimages; (b), (e) and (h) denoised images; (c), (f) and (i) contrast enhanced images.

4. Segmentation of extracted subimages

4.1 Microcalcification segmentation case

Microcalcifications segmentation was based on the computation of a cross-correlation between a template image M_{mic} and the preprocessed resulting image after noise reduction and contrast enhancement I_{net}. The different steps of microcalcification groups are summerized in figure 7. The template image was built after a supervised learning on real phantom images. A global thresholding was then applied on the thresholded image for extracting microcacifications. The connected component labelling step is done to determine the number of detected objects in I_{seuil} image. The microcalcification extraction step consisted in defining around each detected object, a window centered on it and using a threshold based on the computation of the mean and standard deviation of pixels within this window. Fig 8 shows an example of microcalcifications group segmentation.

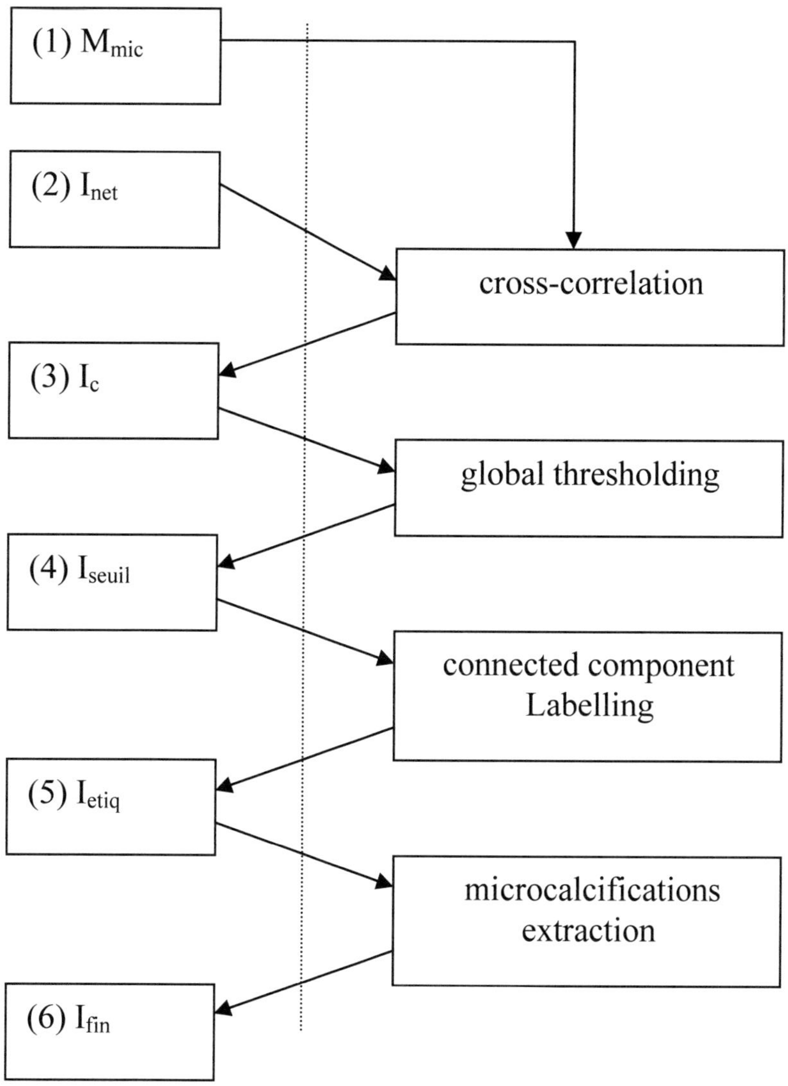

Fig. 7. General scheme of microcalcification group segmentation. (1) M_{mic} : template image, (2) I_{net} : resulting image after noise reduction and contrast enhancement, (3) I_c : resulting image after cross-correlation,(4) I_{seuil} : thresholded cross-correlated image, (5) I_{etiq} : connected component labelled image, (6) I_{fin} : resulting image with detected microcalcifications.

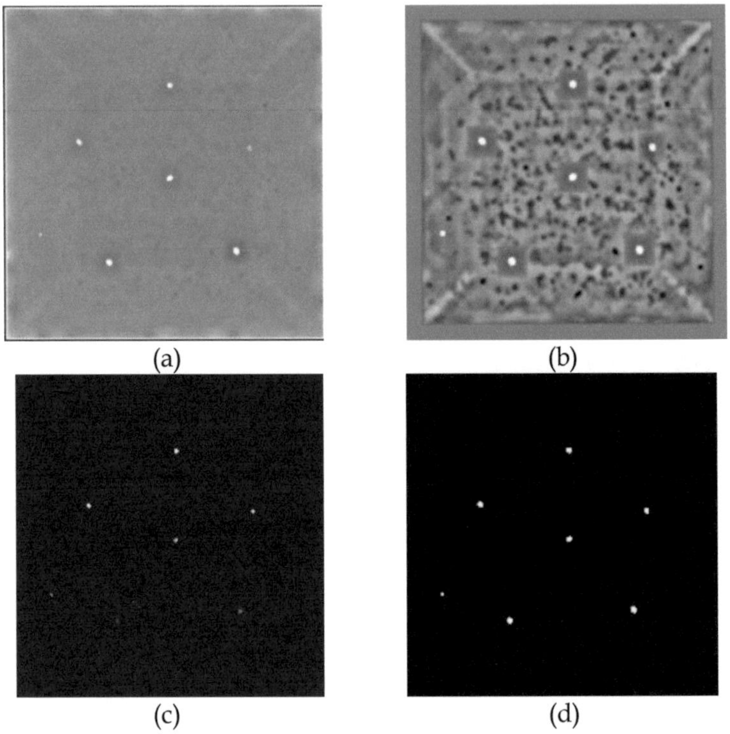

Fig. 8. Segmentation of microcalcifications. (a): extracted subimage after noise reduction and contrast enhancement I_{net} ; (b): result of template matching I_c; (c): connected component labelled image I_{etiq} ; (d): resulting image I_{fin}.

4.2 Mass segmentation case

Mass segmentation was done by using an active contour. First a square was set as an initial contour and the energy used depended only on image gradient. The algorithm used for that purpose is described as follows:

Step 1: Each point i of the active contour evolved along the normal of segment (i-1,i+1) until it met a mass edge.

Step 2: When the four initial points reached the mass edges, other points were added between each couple of points (i and i+1).

Step 3: Each added point in the previous step evolved as initials points in step 1.

The algorithm stopped when a fixed but great number of iterations was reached. Fig. 9 shows an example of a mass segmentation.

4.3 Fibre segmentation case

As for microcalcifications, fibre segmentation used a template matching between two template images (see Fig. 10) and the preprocessed resulting fibre image after applying noise reduction and contrast enhancement steps. An automatic global thresholding is then used, followed by a logical filter OR and a connected component labelling step. Figs. 11 and 12 show the general scheme of a fibre segmentation and an example of fibre segmentation respectively.

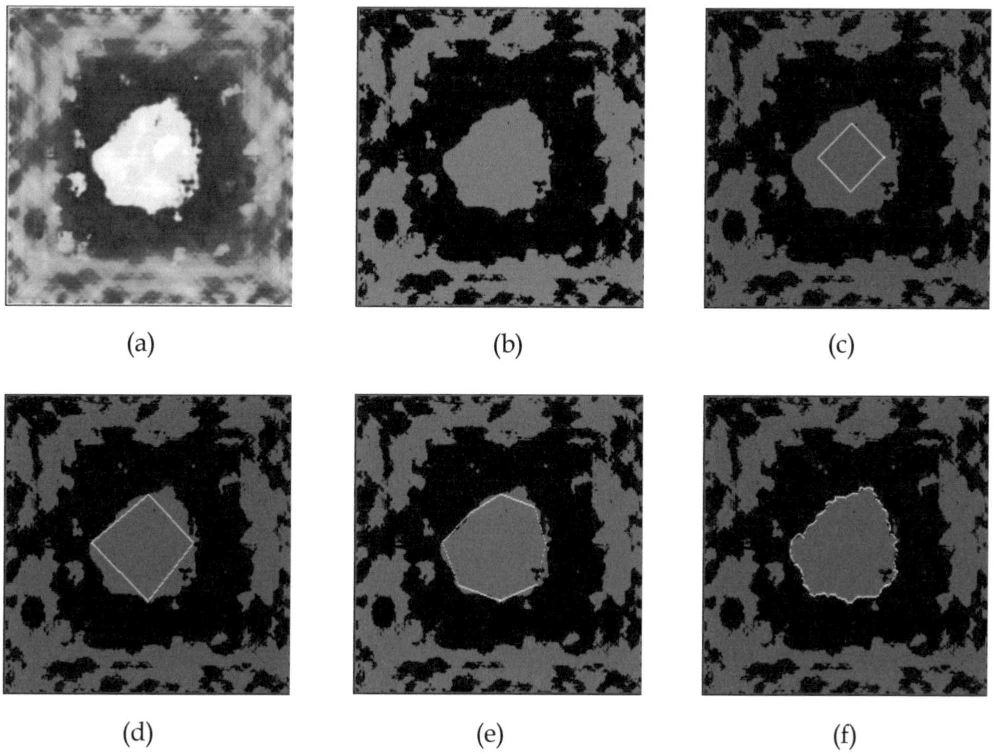

Fig. 9. Segmentation of a mass subimage (a): extracted subimage after noise reduction and contrast enhancement; (b): Thresholded image. (c): Initialization of active contour. (d) First iteration of the active contour. (e): Second iteration of active contour. (f): final segmentation.

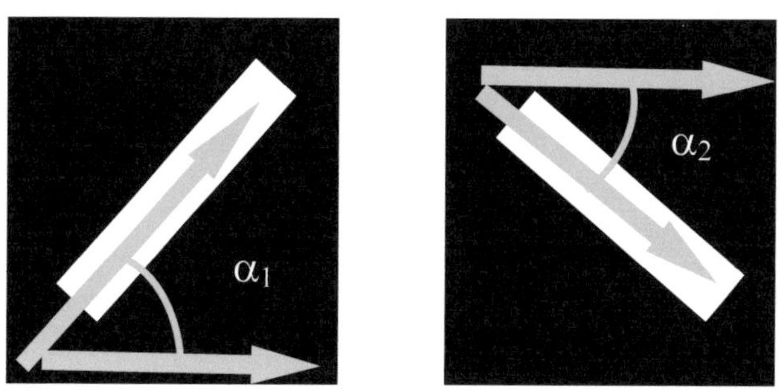

(a) (b)

Fig. 10. Template images for fibre segmentation.

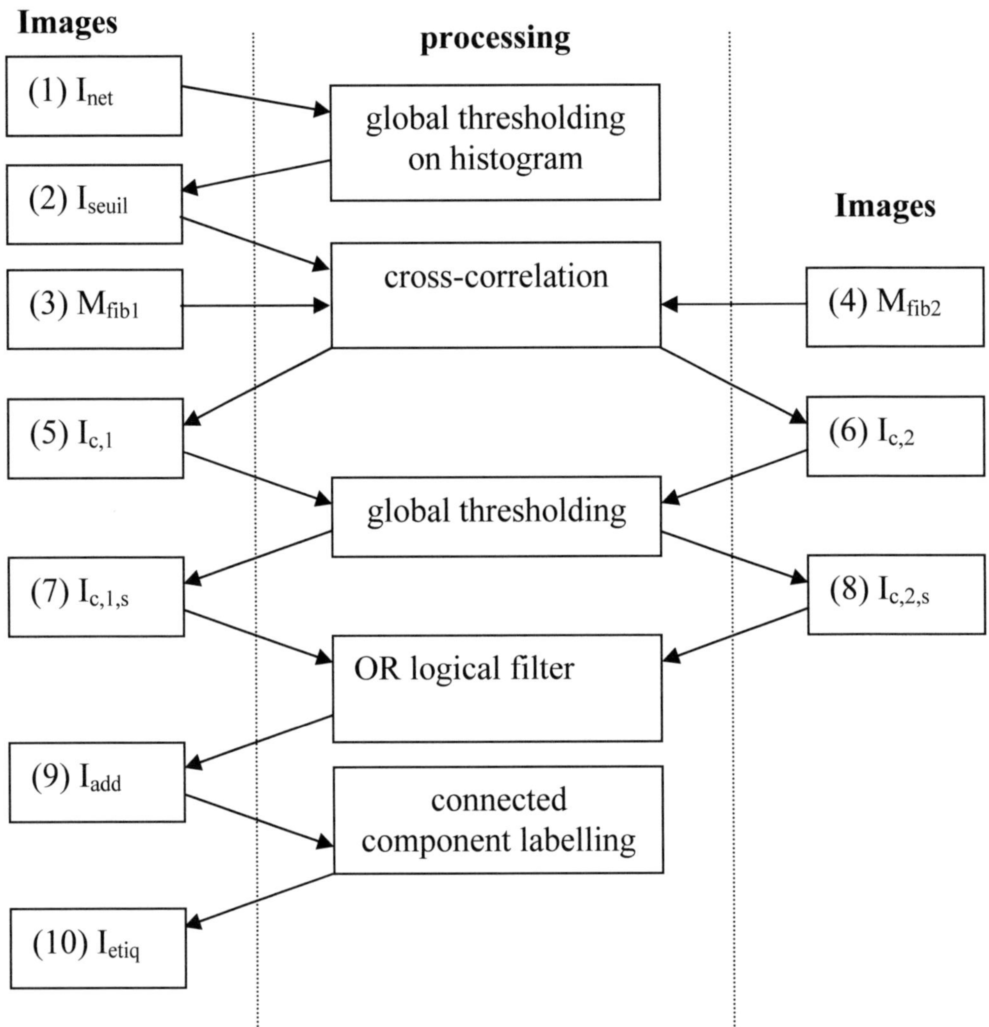

Fig. 11. General scheme of fibre segmentation. (1) I_{net} : resulting image after noise reduction and contrast enhancement , (2) I_{seuil} : image I_{net} thresholded , (3) M_{fib1} : template image 1, (4) M_{fib2} : template image 2, (5) $I_{c,1}$: obtained image after cross correlation between M_{fib1} and I_{seuil} , (6) $I_{c,2}$: obtained image after cross correlation between M_{fib2} and I_{seuil} (7) $I_{c,1,s}$: image $I_{c,1}$ thresholded , (8) $I_{c,2,s}$: image $I_{c,2}$ thresholded , (9) I_{add} : resulting image after logical filter OR between $I_{c,1,s}$ and $I_{c,2,s}$ (10) I_{etiq} : connected component labelled image.

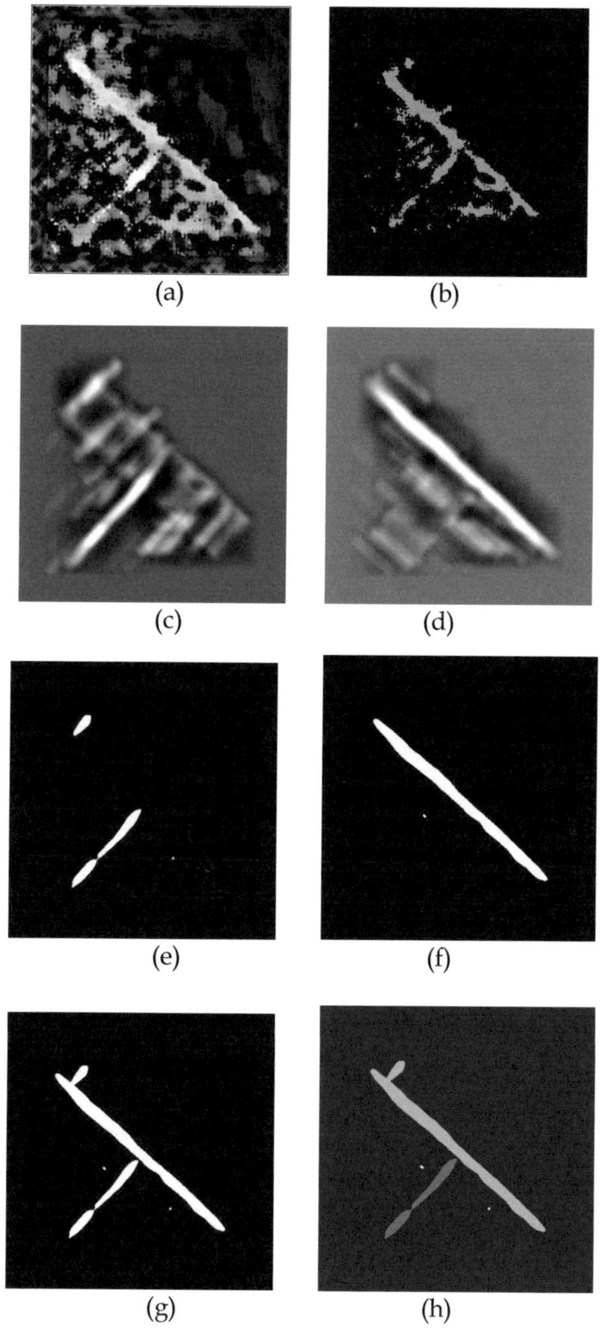

Fig. 12. Segmentation of a fibre subimage. (a): extracted subimage after noise reduction and contrast enhancement; (b): Thresholded image (c): Result of template matching with template image in Fig. 10(a); (d): Result of template matching with template image in Fig. 10(b); (e): Resulting image after thresholding image (c); (f): Resulting image after thresholding image (d); (g): Resulting image after applying OR logic filter on images (e) and (f); (h): Connect component labelled image.

5. Results and discussion

Nine phantom images from different mammographic facilities were tested. For each phantom image, only 4 subimages of each target were extracted. Two main reasons leaded us to do this choice: first, readers could not detect more than 4 objects on the phantoms used in our study, and second, a mammographic facility is considered to have good quality phantom films if at least 4 objects are detected from each embedded target.
216 microcalcifications (36 microcalcification groups) were studied. The most prominent microcalcification group M1 and M2 were almost all detected. Microcalcifications that were not detected were those with poor contrast. False detections were due to film emulsion tearing and appeared on M3 and M4 groups. Table 1 summarizes results obtained on these nine phantom films.
Thirty six masses and 36 fibres were studied. Three masses among 36 were not detected because of the non convergence of the iterative active contour algorithm. This appeared on masses containing holes when being preprocessed. When better initialising the active contour, it was possible to detect the whole masses.
The whole fibres were detected but as seen in Fig 12 some other objects appeared on the final segmentation image. These small objects will be removed in a further processing.

		1	*2*	*3*	*4*	*5*	*6*	*7*	*8*	*9*
M1 group	*Detected microcalcifications number*	**6**	**6**	**6**	**6**	**6**	**6**	**6**	**6**	**6**
	False detection									
M2 group	*Detected microcalcifications number*	**6**	**6**	**6**	**6**	**6**	**6**	**6**	**6**	**6**
	False detection	**1**					**1**		**3**	
M3 group	*Detected microcalcifications number*	**6**	**6**	**6**	**6**	**5**	**6**	**5**	**6**	**5**
	False detection	**5**	**4**	**1**	**3**	**2**				
M4 group	*Detected microcalcifications number*	**6**	**6**	**6**	**6**	**3**	**6**	**5**	**4**	**6**
	False detection	**3**	**3**	**2**			**1**			**1**

Table 1. Detection results on microcalcification groups.

6. Conclusion

This chapter presents a feasibility study of automating breast phantom scoring using image processing techniques. The main contribution in this project is noise reduction and contrast enhancement of noisy images extracted from digitized phantom films. The segmentation step which uses known methods shows that quality control in mammographic facilities could be done using image processing techniques. Next step in this project is to adapt image processing techniques used for digitized film to digital phantom images acquired directly from Full-Filed Digital Mammograms. In this case it will be possible to control the quality of digital mammographic systems using software similar to the one described in this study.

7. References

K. W. Brooks, J. H. Trueblood, K. J. Kearfott, D. T. Lawton "Automated analysis of the American College of Radiology mammographic accreditation phantom images". Medical Physics 24 : 907-923, 1997.

D. P. Chakraborty "Computer analysis of mammography phantom images (CAMPI). Application to the measurement of microcalcification image quality of directly acquired digital images". Medical Physics 24 : 1269-1277, 1997.

G. Dougherty "Computerised evaluation of mammographic image quality using phantom images". Computerized Medical Imaging and Graphics 22: 365-373, 1998.

A. D. Castellano Smith, I. A. Castellano Smith, D. R. Dance. "Objective assessment of phantom image quality in mammography: a feasibility study". The British Journal of Radiology 71 : 48-58, 1998

L. Blot, A. Davis, M. Holubinka ,R. Marti, R. Zwiggelaar "Automated quality assurance applied to mammographic imaging". EURASIP Journal on Applied Signal Processing 7 : 736-745, 2003

P. Mayo, F. Rodenas, G. Verdu, J. I. Villaescusa, J. M. Campayo. "Automatic evaluation of the image quality of a mammographic phantom". Computers Methods and Programs in Biomedicine 73, 115-128, 2004

A. P. Dhawan, G. Buelloni R. Gordon, "Enhancement of mammographic features by optimal adaptative neighborhood image processing," IEEE Trans. Medical Imaging 5(1), 8-15,1986.

A. P. Dhawan E. Le Royer "Mammographic feature enhancement by computarized image processing". Computer Methods and Programs in Biomedecine 27, 23-25, 1988.

V. Guis, M. Adel, M. Rasigni, G. Rasigni, B. Seradour, P. Heid "Adaptive neighborhood contrast enhancement in mammographic phantom images". Optical Engineering 42: 357-366, 2003.

W. Qian, L.P. Clarke, M. Kallergi, R. A. Clark, "Tree-Structured Nonlinear Filters in Digital Mammography". IEEE Trans. Medical Imaging 13(1), 25-36, 1994.

F. Aghadasi, R. K. Ward, B. Palcic, "Noise filtering for mammographic images", *14th Annual International Conference of the I.E.E.E. Engineering in Medecine and Biology Society* 1877-1878, 1992.

D. T. Kuan, A. A. Shawchuk, T. C. Strand, P. Chavel, "Adaptive noise smoothing filter for images with signal-dependent noise". IEEE Trans. Patt. Anal. Mach. Intell. 7(2), 165-177, 1985.